Laser Electronics

SECOND EDITION

JOSEPH T. VERDEYEN

Department of Electrical and Computer Engineering
University of Illinois at Urbana-Champaign
Urbana, Illinois

Prentice Hall, Englewood Cliffs, New Jersey 07632

LIBRARY OF CONGRESS
Library of Congress Cataloging-in-Publication Data

Verdeyen, Joseph Thomas.
  Laser electronics/Joseph T. Verdeyen.—2nd ed.
    p.   cm.--(Solid state physical electronics series)
  Bibliography: p.
  Includes index.
  ISBN 0-13-523630-4
  1. Lasers. 2. Semiconductor lasers. I. Title. II. Series.
TA1675.V47 1989   88-17592
621.36'61--dc19   CIP

SOLID STATE PHYSICAL ELECTRONICS SERIES
Nick Holonyak, Jr., *Editor*

Editorial/production supervision and
  interior design: CRACOM Corporation
Cover design: Wanda Lubelska Design
Manufacturing buyer: Mary Noonan

© 1989, 1981 by Prentice-Hall, Inc.
A Division of Simon & Schuster
Englewood Cliffs, New Jersey 07632

All rights reserved. No part of this book may be reproduced, in any form
or by any means, without permission in writing from the publisher.

Printed in the United States of America
10 9 8 7 6 5 4 3 2

ISBN 0-13-523630-4

Prentice-Hall International (UK) Limited, *London*
Prentice-Hall of Australia Pty. Limited, *Sydney*
Prentice-Hall Canada Inc., *Toronto*
Prentice-Hall Hispanoamericana, S.A., *Mexico*
Prentice-Hall of India Private Limited, *New Delhi*
Prentice-Hall of Japan, Inc., *Tokyo*
Simon & Schuster Asia Pte. Ltd., *Singapore*
Editora Prentice-Hall do Brasil, Ltda., *Rio de Janeiro*

Contents

PREFACE, xi

0 PRELIMINARY COMMENTS, 1

1

REVIEW OF ELECTROMAGNETIC THEORY, 6
1.1 Introduction, 6
1.2 Maxwell's Equations, 7
1.3 Wave Equation for Free Space, 8
1.4 Algebraic Form of Maxwell's Equations, 9
1.5 Waves in Dielectrics, 10
1.6 The Uncertainty Relationships, 11
1.7 Spreading of An Electromagnetic Beam, 13
1.8 Wave Propagation in Anisotropic Media, 15
1.9 Elementary Boundary Value Problems in Optics, 18
    1.9.1 Snell's Law, 19
    1.9.2 Brewster's Angle, 20

1.10 Coherent Electromagnetic Radiation, 21
1.11 Example of Coherence Effects, 27

2
RAY TRACING IN AN OPTICAL SYSTEM, 34

2.1 Introduction, 34
2.2 Ray Matrix, 34
2.3 Some Common Ray Matrices, 36
2.4 Applications of Ray Tracing—Optical Cavities, 38
2.5 Stability—Stability Diagram, 42
2.6 The Unstable Region, 43
2.7 Example of Ray Tracing in a Stable Cavity, 44
2.8 Repetitive Ray Paths, 47
2.9 Initial Conditions—Stable Cavities, 47
2.10 Initial Conditions—Unstable Cavities, 48
2.11 Astigmatism, 49
2.12 Continuous Lens-Like Media, 51
    2.12.1 Propagation of a Ray in an Inhomogeneous Medium, 53
    2.12.2 Ray Matrix for a Continuous Lens, 54
2.13 Wave Transformation by a Lens, 56

3
GAUSSIAN BEAMS, 62

3.1 Introduction, 62
3.2 Preliminary Ideas—TEM Waves, 62
3.3 Lowest-Order $TEM_{0,0}$ Mode, 65
3.4 Physical Description of $TEM_{0,0}$ Mode, 69
    3.4.1 Amplitude of the Field, 69
    3.4.2 Longitudinal Phase Factor, 70
    3.4.3 Radial Phase Factor, 71
3.5 Higher-Order Modes, 72
3.6 Divergence of the Higher-Order Modes—Spatial Coherence, 76

4
GAUSSIAN BEAMS IN CONTINUOUS MEDIA, 81

4.1 Introduction, 81
4.2 The Dielectric Slab Waveguide, 82
4.3 Numerical Aperture, 84
4.4 Modes in a Step-Index Fiber (or a Heterojunction Laser), 85
    4.4.1 TE Mode ($E_z = 0$), 87
    4.4.2 TM Modes ($H_z = 0$), 89

Contents   v

      4.4.3   Graphic Solution for the Propagation Constant, 90
- 4.5   The *ABCD* Law for Gaussian Beams, 91
- 4.6   Fiber-Optic Waveguide—An Approximate Analysis, 95
- 4.7   Fiber-Optic Waveguide—An Exact Analysis, 97
- 4.8   Higher-Order TEM Modes, 100
- 4.9   Dispersion In Fibers, 101

5

OPTICAL RESONATORS, 109

- 5.1   Introduction, 109
- 5.2   Gaussian Beams In Simple Stable Resonators, 109
- 5.3   Application of the *ABCD* Law to Cavities, 112
- 5.4   Mode Volume In Stable Resonators, 116

6

RESONANT OPTICAL CAVITIES, 123

- 6.1   General Cavity Concepts, 123
- 6.2   Resonance, 123
- 6.3   Sharpness of the Resonance: Q, Finesse, Photon Lifetime, 127
- 6.4   Photon Lifetime, 132
- 6.5   Resonance of the Hermite-Gaussian Modes, 134
- 6.6   Diffraction Losses, 136
- 6.7   Cavity with Gain—An Example, 137

7

ATOMIC RADIATION, 149

- 7.1   Introduction and Preliminary Ideas, 149
- 7.2   Blackbody Radiation Theory, 150
- 7.3   Einstein's Approach—The A and B Coefficients, 156
  - 7.3.1   Definition of Radiative Processes, 156
  - 7.3.2   Relationship Between the Coefficients, 158
- 7.4   Introduction to the Rate Equations—Lifetime Broadening, 160
- 7.5   The Line Shape, 164
  - 7.5.1   Homogeneous Broadening, 164
  - 7.5.2   Inhomogeneous Broadening, 169
  - 7.5.3   General Comments on the Line Shape, 171
- 7.6   Transition Rates For Monochromatic Waves, 172
- 7.7   Amplification by an Atomic System, 173
- 7.8   Review, 176

8 LASER OSCILLATION AND AMPLIFICATION, 183

8.1 Introduction—Threshold Condition for Oscillation, 183
8.2 Laser Oscillation and Amplification in a Homogeneous Broadened Transition, 185
8.3 Gain Saturation in a Homogeneous Broadened Transition, 188
8.4 Laser Oscillation in an Inhomogeneous System—An Example, 199
8.5 Multimode Oscillation, 205
8.6 Gain Saturation in Doppler-Broadened Transition—Mathematical Treatment, 206
8.7 Amplified Spontaneous Emission (ASE), 210
8.8 Laser Oscillation—A Different Viewpoint, 214

9 GENERAL CHARACTERISTICS OF LASERS, 231

9.1 Introduction—Quantum Efficiency, 231
9.2 Output Power (Energy)—The Coupling Problem, 233
    9.2.1 Traveling Wave Ring Laser, 233
    9.2.2 Optimum Coupling, 236
    9.2.3 Standing Wave Lasers, 239
        9.2.3(a) The High Q Approximate Solution, 240
        9.2.3(b) The "Exact" Analysis (after Rigrod), 241
    9.2.4 Extraction Efficiency of a Pulse Excited Amplifier, 244
9.3 Transient Effects, 247
    9.3.1 Q Switching, Q Spoiling, or Giant Pulse Behavior, 248
    9.3.2 Gain Switching/Relaxation Oscillations, 260
    9.3.3 Pulse Propagation in Amplifiers, 264
9.4 Multimode Effects—Mode Locking, 270
    9.4.1 Simplified Mathematical Analysis, 273
    9.4.2 Gaussian Amplitude Distribution, 275
    9.4.3 Spatial Variation of the Field, 278
    9.4.4 Active Mode Locking in a Homogeneously Broadened Laser, 280
9.5 Dispersion Effects, 286
    9.5.1 Material Dispersion, 286
    9.5.2 Anomalous Dispersion, 288

10 LASER EXCITATION, 305

10.1 Introduction, 305

Contents vii

 10.2 Three- and Four-Level Lasers, 306
 10.3 Optical Pumping, 308
  10.3.1 Overview, 308
  10.3.2 Ruby Lasers, 308
  10.3.3 Neodymium Lasers, 315
  10.3.4 Neodymium-YAG Lasers, 316
  10.3.5 Neodymium-Glass Lasers, 319
 10.4 Dye Lasers, 321
 10.5 Gaseous-Discharge Lasers, 325
  10.5.1 Overview, 325
  10.5.2 Helium-Neon Laser, 326
  10.5.3 Ion Lasers, 333
  10.5.4 CO_2 Lasers, 334
 10.6 Chemical Lasers, 341
  10.6.1 Background, 341
  10.6.2 HF Chemical Lasers, 342
 10.7 Excimer Lasers: General Considerations, 343
  10.7.1 Formation of the Excimer State, 343
  10.7.2 Excitation of the Rare Gas–Halogen Excimer Lasers, 347
   10.7.2.1 E-Beam (or Nuclear) Excitation, 347
   10.7.2.2 Discharge Pumping, 348
 10.8 The Free Electron Laser, 349

11

SEMICONDUCTOR LASERS, 361

 11.1 Introduction, 361
  11.1.1 Overview, 361
  11.1.2 Populations in Semiconductor Laser, 363
 11.2 Review of Elementary Semiconductor Theory, 365
  11.2.1 Density of States, 365
 11.3 Occupation Probability: Quasi-Fermi Levels, 370
 11.4 Optical Absorption and Gain in a Semiconductor, 372
  11.4.1 Gain Coefficient in a Semiconductor, 375
  11.4.2 Spontaneous Emission Profile, 380
  11.4.3 The Inverted Semiconductor: An Example, 382
 11.5 The Diode Laser, 386
  11.5.1 The Homojunction Laser, 386
  11.5.2 Heterojunction Lasers, 389
 11.6 Quantum Size Effects, 391
 11.7 Modulation of Semiconductor Lasers, 399
  11.7.1 Static Characteristics, 402
  11.7.2 Frequency Response of Diode Lasers, 403

12 GAS-DISCHARGE PHENOMENA, 413

- 12.1 Introduction, 413
- 12.2 Terminal Characteristics, 415
- 12.3 Spatial Characteristics, 416
- 12.4 Electron Gas, 418
    - 12.4.1 Background, 418
    - 12.4.2 "Average" or "Typical" Electron, 418
    - 12.4.3 Electron Distribution Function, 426
    - 12.4.4 Computation of Rates, 428
    - 12.4.5 Computation of a Flux, 430
- 12.5 Ionization Balance, 432
- 12.6 Example of Gas-Discharge Excitation of a CO_2 Laser, 434
    - 12.6.1 Preliminary Information, 434
    - 12.6.2 Experimental Detail and Results, 434
    - 12.6.3 Theoretical Calculations, 436
    - 12.6.4 Correlation Between Experiment and Theory, 439
    - 12.6.5 Laser-Level Excitation, 443
- 12.7 Electron Beam Sustained Operation, 445

13 ADVANCED TOPICS IN ELECTROMAGNETICS OF LASERS, 453

- 13.1 Introduction, 453
- 13.2 Unstable Resonators, 454
    - 13.2.1 General Considerations, 454
    - 13.2.2 The Unstable Confocal Resonator, 460
- 13.3 Integral Equation Approach to Cavities, 463
    - 13.3.1 Mathematical Formulation, 463
    - 13.3.2 The Fox and Li Results, 467
    - 13.3.3 The Confocal Resonator, 468
- 13.4 Semiconductor Cavities, 474
    - 13.4.1 TE Modes, 476
    - 13.4.2 TM Modes, 478
    - 13.4.3 Physical Comparison of TE and TM Modes, 479
- 13.5 Gain Guiding: An Example, 481
- 13.6 Distributed Feedback, 487
- 13.7 Laser Arrays, 497
    - 13.7.1 System Considerations, 497
    - 13.7.2 The Semiconductor Laser Array—Physical Picture, 497
    - 13.7.3 The Supermodes of the Array, 499
    - 13.7.4 The Radiation Pattern, 502

Contents ix

14
QUANTUM THEORY OF THE LASER: AN INTRODUCTION, 511

14.1 Introduction, 511
14.2 The Classical Model of an Atom, 512
    14.2.1 The Antenna Problem, 512
    14.2.2 The "Bound" Electron, 513
    14.2.3 The "Driven" Oscillator, 514
    14.2.4 Dispersion: Mode Pulling, 517
14.3 Quantum Viewpoint of the Interaction of the Atom with a Classical Field, 520
    14.3.1 General Formulation, 520
    14.3.2 Perturbation Solution of Schrödinger's Equation, 521
14.4 Derivation of Einstein Coefficients, 524
14.5 Time-Dependent Populations: The Rabi Approach, 527
14.6 The Density Matrix, 530
    14.6.1 Definition, 530
    14.6.2 Equation of Motion for the Density Matrix, 533
    14.6.3 The Density Matrix Equations for a Two-Level System, 535
    14.6.4 Relaxation Terms in the Density Matrix, 536
    14.6.5 The Polarization Current, 537

15
SPECTROSCOPY OF COMMON LASERS, 548

15.1 Introduction, 548
15.2 Atomic Notation, 549
    15.2.1 Energy Levels, 549
    15.2.2 Transitions—Selection Rules, 550
15.3 Molecular Structure—Diatomic Molecules, 551
    15.3.1 Preliminary Comments, 551
    15.3.2 Rotational Structure and Transitions, 553
    15.3.3 Thermal Distribution of the Population in Rotational States, 553
    15.3.4 Vibrational Structure, 555
    15.3.5 Vibration-Rotational Transitions, 555
    15.3.6 Relative Gain on P and R Branches—Partial and Total Inversions, 557
15.4 Electronic States in Molecules, 559
    15.4.1 Notation, 559
    15.4.2 The Franck-Condon Principle, 560
    15.4.3 Molecular Nitrogen Lasers, 560

16 DETECTION OF OPTICAL RADIATION, 565

16.1 Introduction, 565
16.2 Quantum Detectors, 566
    16.2.1 Vacuum Photodiode, 566
    16.2.2 The Photomultiplier, 568
16.3 Solid-State Quantum Detectors, 570
    16.3.1 The Photoconductor, 570
    16.3.2 The Junction Photodiode, 572
    16.3.3 The p-i-n Diode, 575
    16.3.4 The Avalanche Photodiode, 576
16.4 Noise Considerations, 576
16.5 The Mathematics of Noise, 578
16.6 Sources of Noise, 582
    16.6.1 Shot Noise, 583
    16.6.2 Thermal Noise, 584
    16.6.3 Noise Figure of Video Amplifiers, 586
    16.6.4 Background Radiation, 587
16.7 Limits of Detection Systems, 588
    16.7.1 Video Detection of Photons, 588
    16.7.2 The Heterodyne System, 593

APPENDIX I "DETAILED BALANCING" OR "MICROSCOPIC REVERSIBILITY," 600

APPENDIX II THE KRAMERS-KRONIG RELATIONS, 604

INDEX, 609

Preface

The underlying philosophy of *Laser Electronics* in this second edition remains as in the first edition: lasers are simple devices—much simpler than common electronic devices. They were developed as a natural extension of lower frequency technology, and the student should not hesitate to apply his background to the optical domain. It is just a matter of the decimal point.

However, the second edition of *Laser Electronics* has expanded on three topics that were given only a brief treatment in the first edition: semiconductor lasers (Chapter 11), advanced electromagnetics of lasers (Chapter 13), and a more detailed introduction to a formal quantum description of a laser using the density matrix (Chapter 14).

The advances in theory and performance of semiconductor heterostructure and quantum well lasers has dictated that this topic become one of the central themes of optical electronics. Indeed, such lasers have become so convenient and applicable to such a wide variety of cases—even high power ones—that one could justify devoting the entire book to it. However, it is my opinion that it is easier for the student to grasp the essential ideas of lasers using the atomic structure of gases and solids first before graduating to the bands of a semiconductor (Chapter 11).

Most students have a fair grasp of the beauty and elegance of electromagnetic theory but have the mistaken view that the word *photon* somehow weakens its applicability. That is unfortunate. After all, even the lowest power laser generates literally billions of photons per second, and the classical field description is quite adequate. Many of the advances in semiconductor lasers, in particular, can be traced to classical hard-core electromagnetic theory. Chapter 13 is included to introduce the student to some of the more advanced topics—possibly to be studied in a second course.

Chapter 14 is an attempt to provide a bridge between the simple rate-equation description of a laser and the more formal quantum theory using the density matrix. The two approaches agree, precisely, for the case of a two-level system, but the former is much easier and more akin to the student background. The latter will handle the advanced cases, Raman, two-photon phenomena, etc., at the expense of considerably more mathematics. The serious student should become aware of the transition between the two approaches, have confidence in both, and be aware of the pitfalls and limitations, again in a second course.

Many more problems are included in this second edition, with the primary goal of driving home the transparent simplicity of the rate-equation description. Rate equations are no more difficult than simple coupled circuit equations; both provide a tremendous insight into the operation of the devices.

ACKNOWLEDGMENTS

It is a pleasure to publicly thank my colleagues in physical electronics at University of Illinois for their association and help on the many phases of this work. I am particularly grateful to Prof. J. J. Coleman for his encouragement some years ago to write the book, and for his recent discussions and data on quantum well lasers. I am also grateful to Prof. Roland Sauerbrey of Rice University for many valuable suggestions for this edition.

The task of revising the book has been helped significantly by the close association with Mary Espenschied of CRACOM Corporation. She has been most helpful and attentive to detail that would have escaped me.

Joseph T. Verdeyen

ROMAN SYMBOLS

A^*	Complex conjugate of A
$\mathbf{a}_n$	Unit vector in direction of n
a	Attachment rate per unit of drift
$[A]$	Density of A
A_{21}	Einstein coefficient for spontaneous emission
A, B, C, D	Components of ray matrix
$\mathbf{b}, \mathbf{B}$	Magnetic induction vector
B_e	Rotational constant
B_{21}	Einstein coefficient for stimulated emission
B_{12}	Einstein coefficient for absorption $B_{12} = g_2 B_{21}/g_1$
c	Velocity of light in vacuum
c'	c/n, velocity of light in a material with index n
$\mathbf{d}, \mathbf{D}$	Displacement vector
D_T	Transverse diffusion coefficient
$D_{n(p)}$	Diffusion coefficients for electrons (holes)
E	Energy
e	Electronic charge
$\mathbf{e}, \mathbf{E}$	Electric field intensity
$E_{c(v)}$	Energy of the conduction (valence band)
E_F	Fermi energy
$\bar{e}^2, \bar{i}^2$	Equivalent noise generators
F	Finesse = $\text{FSR}/(\Delta \nu_{1/2})$
f	Focal lengths
$F(\epsilon)$	Electron distribution function per unit of energy
$f(E)$	Fermi function
$F(J)$	Rotational energy
$F_{n(p)}$	Quasi-Fermi level for electrons (holes)
FSR	Free spectral range = $c/2d$
$f(z)$	Forward wave
$F(v)$	Electron distribution function per unit of velocity
FWHM	Full width at half maximum
f_{12}	Absorption oscillation strength
$g(\nu)$	Lineshape
$\bar{g}(\nu)$	Lineshape normalized to unity at line center, i.e., $\bar{g}(\nu_0) = 1$
$G(v)$	Vibrational energy
G	Power gain
$1/q(z)$	Complex beam parameter = $1/R(z) - j\lambda/[\pi w^2(z)]$
$g_{1,2}$	$(1 - d/R_{1,2})$, the g parameter of a cavity
g	$2J + 1$, the degeneracy of a quantum state
h	Planck's constant
$\hbar$	Planck's constant divided by 2π
$\mathbf{h}, \mathbf{H}$	Magnetic field intensity

$H_n(u)$	Hermite polynomial of order n argument u
H_{op}	Operator corresponding to the Hamiltonian
$Im(\ )$	Imaginary part of the quantity ( )
I_ν	Intensity at a frequency ν
$I(\nu)$	Intensity per unit of frequency at ν
J	Angular momentum quantum number
j	The imaginary number $(-1)^{1/2}$
j, J	Conduction current
k	$2\pi n/\lambda_0$
k	Wave vector $= k\mathbf{a}_n$
K.E.	Kinetic energy
l_g	Length of gain medium
ln	Natural log of ( ) (set in italic for ease in reading)
$m^*_{c(v)}$	Effective mass in the conduction (valence) band
N	Density (number/volume)
n	Population difference $(N_2 - N_1)V$
n	Index of refraction $(\epsilon_r)^{1/2}$
$n(\nu)$	Frequency dependent refractive index
$n_c(E)$	Density of electrons in conduction band per unit of energy
n_e	Electron density
n_{th}	Threshold value of population difference
N_p	Number of photons in the laser cavity
N_ν	Number of modes in a volume V between 0 and ν
$N_2(v)$	Nitrogen in a vibrational state v
$N_{2(1)}$	Density of atoms in state 2(1)
$\langle P \rangle$	Average power (averaged over many cycles)
$\langle p \rangle$	Instantaneous power (averaged over a few cycles)
p	Pressure
p	Hole density
p	g_2/g_1; ratio of degeneracies
p	Mode index
p, P	Polarization vector
p_{el}	Power in electron gas
P_f	Fluorescence power
$p(\nu)$	Mode density
P_s	Probability of belonging to a class s
$p_v(E)$	Density of holes in the valence band per unit of energy
Q	Quality factor $= \omega W/(-dW/dt)$
q	Axial mode number of a resonant mode in a cavity, also the number of half-wavelengths between the mirrors
R	Resistance
R	If the numerical value >1, radius of curvature; if <1, power reflectivity

List of Symbols

r	Position vector $= x\mathbf{a}_x + y\mathbf{a}_y + z\mathbf{a}_x$
Re ( )	Real part of the quantity ( )
$r(z)$	Reverse wave
$R(z)$	Radius of curvature of phase front
r_{min}	Equilibrium spacing in a stable molecule
S	Poynting vector
S/N	Signal to noise ratio
$S_T(\nu)$	Power per unit of frequency
T	Ray matrix
T	Temperature (°K)
T_1	Lifetime of the inversion ($\rho_{22} - \rho_{11}$)
T_2	Mean time between dephasing collisions
T_e	Electronic term energy
TE	Transverse electric
TEM	Transverse electric and magnetic mode
TM	Transverse magnetic
V	Voltage or volume
$V_{m,n}$	Volume of the TEM$_{m,n}$ mode
v	Speed, velocity or vibrational quantum number
v_g	Group velocity
v_z	Velocity in z direction
VSWR	Voltage standing wave ratio
W	Energy
w	Drift velocity
W_e	Energy of the electron gas
w_0	Minimum spot-size
w_s	Saturation energy
$w(z)$	Spot-size as a function of z
z_0	Characteristic length parameter of a Gaussian beam
Z	Impedance

Greek Symbols

α	Absorption coefficient (loss per length)
α	Townsend ionization coefficient (ionization rate per unit of drift) (Chapter 12)
α_e	Correction to the rotational constant due to vibration
β	Phase constant (rad/length) of a guided wave
β_m	A phase constant satisfying the Bragg condition (section 13.5)
Γ_p	Photon flux ($I/h\nu$)

$\gamma_0(\nu)$	Small signal gain coefficient
$\gamma(\nu)$	Intensity dependent gain coefficient
∂	Partial derivative
δ	Secondary emission ratio
δ	Fraction of the electron's excess energy lost in an elastic collision
$\Delta\nu_D$	Doppler line width
$\Delta\nu_h$	Homogeneous line width
$\Delta\nu_n$	Natural line width
$\Delta\nu_H$	Hole line width
$\Delta\omega$	Line width in radian frequency units
$\Delta t_{1/2}$	Pulse wide (FWHM)
ϵ_k	Characteristic energy of electrons $= D_T/u$
ϵ_A	Characteristic energy of atoms or molecules
ϵ'	Real part of the relative dielectric constant
ϵ''	Imaginary part of the relative dielectric constant
ϵ_0	Permittivity of free-space
ϵ_r	Relative dielectric constant
ϵ	Electron energy
η_0	Wave impedance of free-space $(\mu_0/\epsilon_0)^{1/2}$
η_{qe}	Quantum efficiency
η_{xtn}, η_x	Extraction efficiency
θ_p	Photon fluence
Λ	Characteristic length
λ_q	Wavelength of the TEM$_{m,m,q}$ mode
λ_0	Free-space wavelength
λ	Wavelength λ_0/n
μ	Mobility
μ_{21}	Electric dipole moment
μ_0	Permeability of free-space
ν, f	Frequency (hertz)
ν_i	Ionization rate (per electron)
ν_0	Line center
$\bar{\nu}$	Wave number (# of wavelengths per centimeter)
ν_c	Collision of frequency
$\rho(\nu)$	Energy per unit of volume per unit of frequency at ν
ρ_ν	Energy per unit of volume at a frequency ν
ρ	Reflection coefficient for the electric field
ρ_{11}	Component of the density matrix corresponding to the fraction of excited atoms in state 1
ρ_{22}	Component of the density matrix corresponding to the fraction of excited atoms in state 2
ρ_{22}	Diagonal element of density matrix corresponding to the polarization

List of Symbols

$\sigma(\nu)$	Stimulated emission cross-section
$\sigma_c(\epsilon)$	Collision cross-section
σ_i	Ionization cross-section
τ_1	Lifetime of state 1
τ_2	Lifetime of state 2
$1/\tau_{1,2}$	Decay rate of state 1 (or 2)
τ_{21}^{-1}	Decay rate of state 2 into state 1
τ_p	Photon lifetime
τ_r	Radiative lifetime
τ_{RT}	Time for a round trip
ϕ, θ	Phase shift or geometric angles
ϕ_{12}	Branching ratio
χ	Electric susceptibility
χ'	Real part of susceptibility
χ''	Imaginary part
$\nabla, (\nabla \cdot), (\nabla \times)$	Vector differential operation
$\mathcal{L}(\ )$	Laplace transform of ( )
ψ	Wave function
ω	Radian frequency $= 2\pi\nu$
Ω	Rabi frequency ($\mu_{21} E/2\hbar$)

To my wife, Katie
who, in addition to creating and maintaining a
wonderful home life for our family, has contributed an
enormous amount of work at all phases of this effort—
typing manuscript, proofreading the galleys and page proofs,
and helping with the index—making this book a joint effort
in every respect

0

Preliminary Comments

Before the 1960s, optics formed the basis for a relatively small industry involving rather sedate and mature topics such as optical instruments, cameras, microscopes, and scientific applications. Then the laser came on the scene, first the solid-state (ruby) laser, then the gas laser, then the semiconductor injection laser—and now optics forms the basis for many more functions, products, and services.

At first, the standard joke was: "The laser is a solution in search of a problem." More seriously, almost everybody recognized the potential of the laser in communications, in data processing, storage, and retrieval, and even in eye surgery. The question to be answered was, Could the laser do things that had not been done before, or could it do things better and more economically than had previous devices and technologies? It is interesting to observe that the initial applications of the laser have not been in the rather obvious fields listed above but in new applications by ingenious people who understood the principles of the laser and who understood the problems to be solved. Hence, we have laser transits, laser pattern cutting, laser cutting of steel, and laser fusion, and we are starting to make inroads in optical communications. The history of the laser in the field of communications illustrates the point about "obvious" applications.

The frequency of the first laser, ruby, at $\lambda = 694.3$ nm is 4.32×10^{14} Hz, a quantity of interest to any communication engineer. If only 1% of this carrier frequency is used for the information bandwidth, we have a communication channel that has two to three orders of magnitude (10^2 to 10^3) more capacity than the widest band channel in existence. Some of the microwave radio-relay links used by the telephone company have channel widths as large as 10% of the carrier frequency. Consequently, one laser beam should be able to carry a huge number of telephone conversations (bandwidth required per telephone conversation, 4 kHz) and many television programs (bandwidth, ~5 MHz) simultaneously.

There are, however, a few problems. We do not know how to modulate this carrier at a 4×10^{12} Hz rate; nor does the technology exist to demodulate at this rate; nor could our terminal equipment handle information at this rate. If that is not enough, we are not overly confident of being able to transmit the information from point A to a distant point B with the reliability afforded by microwave links. Finally, there was some doubt as to the reliability of the laser. Consequently, communications by lasers with the same degree of sophistication as is done at microwave frequencies lies in the future, but some inroads are being made.

The invention of glass fibers exhibiting very low loss has made laser communications a viable alternative to wired links. After all, the world has practically exhausted high-grade copper ore supplies, but we have not really touched the primary ingredient of glass—SiO_2 (i.e., sand). Thus the first "obvious" application of the laser, communications, had to wait until 1977 for trial runs over short-haul links.

Some 10 years later, in 1987, fiber optics is no longer a laboratory curiosity but rather appears to be headed toward a dominant position in the communication and control aspects of electrical technology.

The point to be made is that obvious applications are not so straightforward and simple. Most often, it is the materials that are the major impediment, but this should not stop one from looking for other uses. For instance, a first major use of the ruby laser was in the "trimming" of solid-state circuit components. In that case, one uses the ability to focus the energy of the laser onto a very small spot so as to vaporize the excess material.

Recently, there has been the marriage of xerography, word processing with computers, and the ability to modulate, deflect, and focus a laser beam to produce manuscript with nearly the quality of offset printing, but at a fraction of the cost in capital equipment.

In view of the above, then, we will not look at a laser from an applications standpoint. Rather, we will try to introduce the elementary and *simple* principles of the laser itself, the propagation of its radiation, and the elements of the detection problem. The word "simple" was italicized to emphasize the goal of this textbook: to make the reader feel comfortable with the following issues:*

1. What physical principles are involved in *generation* of laser radiation? (2)
2. What peculiarities can be anticipated in the *transmission* of laser beams? (1)

*The numbers in parentheses represent the order of the topics covered in this book.

3. What are the characteristics and limitations of common detectors? (3)

Once the reader feels comfortable with an initial understanding, he or she can then read the more advanced texts and current literature to obtain the finer points of the field of quantum electronics.

One final comment should be made about the material. Lasers are quantum devices—there is no avoiding that fact. However, it is not necessary to be familiar with every rule, regulation, philosophy, and theorem of quantum mechanics to develop a good understanding of lasers. Lasers could have been invented in 1908 before the vacuum triode tube.

All the necessary quantum theory was available by 1930—indeed the formula for the amplification or gain coefficient can be found in reference 1 of Chapter 7 (first published in 1934)—and thus the laser could have been invented then. Why did scientists take so long (about 30 years) to obtain an oscillator at optical frequencies? The answer is, of course, speculation, but the question truly does generate wonderful philosophy. According to Townes,[1] one of the Nobel Prize winners for the laser, the time lag can be attributed to physicists knowing the atom-field interaction but being less than dexterous with the feedback requirement for an oscillator, whereas the electrical engineers had the opposite strength and weakness. (References 1 and 4 are papers published for the twenty-fifth anniversary of the laser and are highly recommended.) Thus it was not until between 1950 and 1960 that the field of quantum electronics boomed.

Much of the credit for this boom must be given to the microwave industry, a technology that was developed during and after World War II (and is still going strong). Many of the concepts can be taken over into the laser field en masse, and after complicated microwave tubes, lasers are, in some respects, quite simple.

Indeed, lasers are much easier to understand than is the low-frequency oscillator. Any oscillator is just an amplifier with feedback. If we break this feedback system at an arbitrary point and start a wave around the loop, then a *necessary* condition for oscillation is

Round-trip gain must exceed unity.

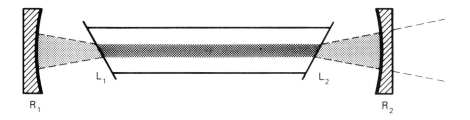

Figure 0.1 Schematic of a simple laser.

At low frequencies, it is sometimes hard to distinguish what is gain and what is feedback. For a simple laser, it is obvious. For instance, consider the simple He/Ne gas laser operating at 632.8 nm shown in Fig. 0.1. If R represents the power reflectivity of the mirrors, L the loss per pass through the windows, and G the power gain through the tube per pass, the laser will oscillate provided that

$$G(1 - L_2)R_2(1 - L_2)G(1 - L_1)R_1(1 - L_1) \geq 1 \tag{0.1}$$

In writing this equation, we have broken the "loop" at the right of window 1 and followed a wave around the path. The equation is trivial and transparent!

Some of the interesting problems are (1) How do we excite the system to get the gain G? (2) Are there special techniques to construct the mirrors? (3) Why use curved mirrors? (4) Why orient the windows as shown? (5) What is the beam spread? (6) How much power do we obtain? (Obviously, it must be less than we put into the system.)

Interestingly enough, quantum theory enters only in the choice of the gases involved, helium and neon, and then, only to provide two energy states separated by

$$E = h\nu \qquad \nu = \frac{c}{\lambda}, \qquad \lambda = 632.8 \text{ nm} \tag{0.2}$$

Then a few relatively simple equations relate the gain to the number of atoms in each of these two states. All the other problems listed can be discussed to an unusual degree of precision without once invoking the quantum nature of the device.

Most readers will be familiar with the theory of the simple pn junction for rectification of AC signals and as an integral part of transistors and other solid-state devices. These are the "complicated" applications of semiconductor electronics which depend, to a major degree, on the differences between "forward" and "reverse" bias. The semiconductor injection laser uses this same pn junction in the forward direction to promote the stimulated recombination of the electrons and holes.

$$e + h \rightarrow h\nu \tag{0.3}$$

The basic physics is quite simple; the technology has benefitted from some rather ingenious thinking, so that now the semiconductor laser is the overwhelming choice for low-power communication and control applications.

However, the point remains: lasers are quantum devices—a fact that we accept, live with, enjoy, and frequently ignore.

REFERENCES

See the Centennial Issue of IEEE J. Quant. Electron. *QE-20*, 1984.

1. C. H. Townes, "Ideas and Stumbling Blocks in Quantum Electronics," IEEE J. Quant. Electron. *QE-20*, 547, No. 6., 1984.

2. W. E. Lamb, Jr., "Laser Theory and Doppler Effect," IEEE J. Quant. Electron. *QE-20*, 551, 1984.
3. N. Bloembergen, "Non-Linear Optics," IEEE J. Quant. Electron. *QE-20*, 556, 1984.
4. A. L. Schawlow, "Lasers in Historical Perspective," IEEE J. Quant. Electron. *QE-20*, 558, 1984.
5. See also the historical section of IEEE J. Quant. Electron. *QE-20*, 1987—25th Anniversary of Semiconductory Laser.
6. The five-volume set entitled *The Laser Handbook* (New York: North-Holland Publishing Company).

    Volume 1, Eds. F. T. Arecchi and E. O. Schulz-Dubois, 1972.
    Volume 2, Eds. F. T. Arecchi and E. O. Schulz-Dubois, 1972.
    Volume 3, Ed. M. L. Stitch, 1974.
    Volume 4, Eds. M. L. Stitch and M. Bass, 1979.
    Volume 5, Eds. M. Bass and M. L. Stitch, 1985.

7. R. J. Pressley, Editor in-Chief, *Handbook of Lasers* (Cleveland, Ohio: Chemical Rubber Co.).
8. There are "thousands" of known lasers spaning the wavelength range of far infrared to the UV and x-ray portion of the spectrum. A reasonably complete listing for gases is given by R. Beck, W. Englisch, and K. Gürs, *Tables of Laser Line in Gases and Vapors*, Springer Series in Optical Sciences, 3rd ed. (New York: Springer-Verlag, 1980).

1

Review of Electromagnetic Theory

1.1 INTRODUCTION

We will be dealing with electromagnetic waves in that part of the spectrum where optical techniques have played a historical role. One uses lenses to focus the radiation, mirrors to direct it, and free space to transmit it. Yet it is still electromagnetic radiation, it obeys Maxwell's equations, and all the laws studied at low frequencies apply at the "optical" portion of the spectrum.

The major difference lies in the size of the components used. For instance, a 1-cm-diameter capacitor used at 1 MHz is less than 10^{-4} of a free-space wavelength ($\lambda = 300$ m), whereas a 1-cm-diameter "contact lens" for your eyes is greater than 10^4 wavelengths for visible radiation. The small size of the capacitor compared to a wavelength is a requirement for the validity of circuit theory; however, the large size of the lens makes life easy for the more exact field theory.

Before we go into field theory at optical frequencies, let us mention the question of units. The *rationalized* SI (or MKSA) system will be used throughout in analytical developments. However, numerical answers will almost always be expressed in cm, cm^{-3}, or cm/sec. This does not mean we are using a CGS system of units but merely that we are expressing an answer in a more convenient and intuitively comfortable form, as well as conforming to most modern and traditional literature. Only if ϵ_0, the permittivity of free space, or μ_0, the permeability of free space, appears in the equation must one go through the exercise

1.2 MAXWELL'S EQUATIONS

of converting centimeters to meters. Most of the time, the product appears (i.e., $\mu_0\epsilon_0$), which is, of course, equal to $1/c^2$. In that case, one can keep c as $\sim 3 \times 10^{10}$ cm/sec in all the equations, provided that the other quantities are also measured in centimeters.

In order to describe an electromagnetic wave, we need two field-intensity vectors, **e** and **h**, which are related to each other by

$$\nabla \times \mathbf{h} = \mathbf{j} + \epsilon_0 \frac{\partial}{\partial t} \mathbf{e} + \frac{\partial}{\partial t} \mathbf{p} \quad (1.2.1a)$$

$$\nabla \times \mathbf{e} = -\mu_0 \frac{\partial \mathbf{h}}{\partial t} \quad (1.2.1b)$$

where **p** is the polarization current induced by the electric field. (A term of the form $\partial \mathbf{m}/\partial t$ can be added to (1.2.1b) but will be ignored for now.) We use lowercase letters to represent vectors that are explicit functions of time t and the three spatial coordinates x, y, and z. Most of the time we will be talking about sinusoidal variations of the field and use the phasor representation:

$$\begin{aligned} \mathbf{e}(\mathbf{r}, t) &= Re[\mathbf{E}(\mathbf{r})e^{j\omega t}] & \mathbf{h}(\mathbf{r}, t) &= Re[\mathbf{H}(\mathbf{r})e^{j\omega t}] \\ \mathbf{j}(\mathbf{r}, t) &= Re[\mathbf{J}(\mathbf{r})e^{j\omega t}] & \mathbf{p}(\mathbf{r}, t) &= Re[\mathbf{P}(\mathbf{r})e^{j\omega t}] \end{aligned} \quad (1.2.2)$$

where

$$Re = \text{real part}$$
$$\mathbf{r} = x\mathbf{a}_x + y\mathbf{a}_y + z\mathbf{a}_z$$
$$\mathbf{a}_i = \text{unit vector in the } i^{\text{th}} \text{ direction}$$

and the capital letters **E** and **H** are complex vector quantities depending on space coordinates but not on time. We recognize that if we want the complete field, we must take the real part of the product $\mathbf{E} \exp(j\omega t)$.

If we substitute (1.2.2) into (1.2.1) we obtain the time-independent form of Maxwell's equations:

$$\nabla \times \mathbf{H} = \mathbf{J} + j\omega\epsilon_0 \mathbf{E} + j\omega \mathbf{P} = \mathbf{J} + j\omega \mathbf{D}$$
$$\nabla \times \mathbf{E} = -j\omega\mu_0 \mathbf{H} \quad (1.2.3)$$

with $\quad \mathbf{D} = \epsilon_0 \mathbf{E} + \mathbf{P}$

where the common factor of $\exp(j\omega t)$ has been canceled from each side of the equation. The polarization term is related to the electric field by a constitutive relation:

$$\mathbf{P} = \epsilon_0 \chi \mathbf{E} \quad (1.2.4)$$

where the term χ is the complex susceptibility of the medium through which the wave is propagating.

After we have become familiar with the simple approach to lasers, we will find that the atoms enter Maxwell's equations via an "equation of motion" for $\mathbf{P}$; but for now we assume that the coefficient χ is a given parameter of the medium. For instance, the form of the polarization given by (1.2.4) suggests that it and the vacuum displacement term can be combined into a single term:

$$\begin{aligned}\mathbf{D} &= \epsilon_0 \mathbf{E} + \mathbf{P} \\ &= \epsilon_0 (1 + \chi) \mathbf{E} \\ &= \epsilon_0 \epsilon_r \mathbf{E} = \epsilon_0 n^2 \mathbf{E}\end{aligned} \quad (1.2.5)$$

Thus the relative dielectric constant ϵ_r is related to the susceptibility by $1 + \chi$, and it in turn is equal to the square of the index of refraction. In the interest of simplicity, $\mathbf{P}$ was assumed to be in the same direction as $\mathbf{E}$, but this is not true for many of the interesting electro-optic materials.

Actually, we have done something very important in going from (1.1.1) to (1.2.3, 1.2.4, and 1.2.5). We have gone from the time domain to the angular-frequency domain, ω, by the application of the Fourier transform, defined by

$$F(\omega) = \int_{-\infty}^{+\infty} f(t) e^{-j\omega t} \, dt \quad (1.2.6)$$

If we follow the prescription of (1.2.6) as applied to Maxwell's equations, we obtain (1.2.3) directly. Consequently, $\mathbf{E}$, $\mathbf{H}$, $\mathbf{P}$, $\mathbf{D}$, and $\mathbf{B}$ are also spectral representations of the respective quantities and should be written as $\mathbf{E}(r, \omega)$ $\mathbf{H}(r, \omega)$, etc. However, since we are usually dealing with a single frequency, we are often lazy and do not bother to show that dependence. Unfortunately, this laziness will return to haunt us unless we are forewarned.

1.3 WAVE EQUATION FOR FREE SPACE

Let us consider free space, so that the conduction current $\mathbf{J}$ is zero. If we take the curl of (1.2.1b) and eliminate $\mathbf{h}$ by the use of (1.2.1a) we obtain

$$\left.\begin{aligned}\nabla \times (\nabla \times \mathbf{e}) &= -\mu_0 \frac{\partial}{\partial t} (\nabla \times \mathbf{h}) = -\mu_0 \epsilon_0 \frac{\partial^2 \mathbf{e}}{\partial t^2} \\ \text{or} \quad & \\ \nabla^2 \mathbf{e} - \frac{1}{c^2} \frac{\partial^2 \mathbf{e}}{\partial t^2} &= 0\end{aligned}\right\} \quad (1.3.1a)$$

Sec. 1.4 Algebraic Form of Maxwell's Equations

where $c^2 = 1/\mu_0 \epsilon_0$ is the square of the velocity of light. If the procedure is reversed to eliminate **e**, we obtain the same equation with **h** substituted for **e** in (1.3.1a):

$$\nabla^2 \mathbf{h} - \frac{1}{c^2} \frac{\partial^2 \mathbf{h}}{\partial t^2} = 0 \qquad (1.3.1b)$$

It is most important to realize that *any function of the form* $f(t - \mathbf{a}_n \cdot \mathbf{r}/c)$ *is a solution*, where $\mathbf{a}_n$ is a unit vector. It is easy to show this in one dimension, and only slightly more complicated to do so for the general case. Physically, it merely means that the wave propagates in the direction $\mathbf{a}_n$ with a velocity of c.

For sinusoidal representation, we say that there is a phase change as the wave propagates along the direction described by $\mathbf{a}_n$.

$$\mathbf{e}(\mathbf{r}, t) = Re\left\{[\mathbf{E}(\omega, \mathbf{k}_0)] \exp\left[j\omega\left(t - \frac{\mathbf{a}_n \cdot \mathbf{r}}{c}\right)\right]\right\} \qquad (1.3.2a)$$

$$\mathbf{e}(\mathbf{r}, t) = Re\{[\mathbf{E}(\omega, \mathbf{k}_0)] \exp(j\omega t)(-j\mathbf{k}_0 \cdot \mathbf{r})\} \qquad (1.3.2b)$$

$$|\mathbf{k}_0| = \frac{\omega}{c} = \frac{2\pi}{\lambda_0} \qquad (1.3.3)$$

where λ_0 is the wavelength in free space.

In writing (1.3.2) we took the functional form of (1.2.2) and, in every place that t appeared, we replaced it by $t - \mathbf{a}_n \cdot \mathbf{r}/c$, just as the solution to the wave equation demanded. We also combined ω, c, and the unit vector $\mathbf{a}_n$ into a new vector $\mathbf{k}_0$. Obviously, the equation for **h** is modified in the same manner. We will use $\mathbf{k}_0$ to denote the wave vector in free space and **k** (without the subscript) to indicate it in a dielectric medium.

We could have been much more formal in our approach and started with $\mathbf{k}_0$ as a three-dimensional Fourier transform variable with respect to the three spatial coordinates. Again, we tend to be somewhat lazy and not bother to state that E is now a function of $\mathbf{k}_0$ in addition to being a function of ω. Thus, most of the time, we say that we are representing a wave of constant amplitude E propagating along the direction $\mathbf{k}_0$.

1.4 ALGEBRAIC FORM OF MAXWELL'S EQUATIONS

If we take (1.3.2) and the corresponding one for **h** and insert them into Maxwell's equation for free space, we obtain the algebraic form of Maxwell's equations:

$$\left\{\begin{matrix}\mathbf{e}\\ \mathbf{h}\end{matrix}\right\} = \left\{\begin{matrix}\mathbf{E}\\ \mathbf{H}\end{matrix}\right\} \exp(j\omega t) \exp(-j\mathbf{k}_0 \cdot \mathbf{r}) \qquad (1.4.1)$$

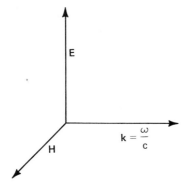

Figure 1.1 Geometric orientation of the vectors **E**, **H**, and **k** according to (1.4.2).

where

$$\mathbf{k}_0 \cdot \mathbf{r} = (k_x \mathbf{a}_x + k_y \mathbf{a}_y + k_z \mathbf{a}_z) \cdot (x \mathbf{a}_x + y \mathbf{a}_y + z \mathbf{a}_z)$$
$$= k_x x + k_y y + k_z z$$

In (1.4.1), **E** and **H** are not functions of x, y, or z. Some fortitude and patience with the rules for the curl operation yield

$$\mathbf{k}_0 \times \mathbf{E} = +\omega \mu_0 \mathbf{H}$$
$$\mathbf{k}_0 \times \mathbf{H} = -\omega \epsilon_0 \mathbf{E} \quad (1.4.2)$$

The main utility of (1.4.2) lies in the geometric interpretation. For instance, **H** is obviously perpendicular to both $\mathbf{k}_0$ and **E** from the first one, and **E** is also perpendicular to **H** and $\mathbf{k}_0$ by the second. This geometric arrangement is shown in Fig. 1.1. The vectors **E** and **H** are related to each other (and are obviously in phase, since j is absent from the equations).

$$\frac{|\mathbf{E}|}{|\mathbf{H}|} = \frac{\omega \mu_0}{|\mathbf{k}_0|} = \frac{|\mathbf{k}_0|}{\omega \epsilon_0} = \left(\frac{\mu_0}{\epsilon_0}\right)^{1/2} \cong 377 \, \Omega \quad (1.4.3)$$

$$\eta_0 = \left(\frac{\mu_0}{\epsilon_0}\right)^{1/2} = \text{wave impedance of free space}$$

For free space, the Poynting vector $S = \frac{1}{2} \mathbf{E} \times \mathbf{H}^*$ points in the same direction as $\mathbf{k}_0$.

$$S = \frac{1}{2} \mathbf{E} \times \mathbf{H}^* = \frac{1}{2} \mathbf{E} \times \frac{(\mathbf{k}_0 \times \mathbf{E})^*}{\omega \mu_0} = \frac{1}{2} \mathbf{E} \cdot \mathbf{E}^* \frac{\mathbf{k}_0}{\omega \mu_0} \quad (1.4.4)$$

since $\mathbf{k}_0 \cdot \mathbf{E} = 0$.

1.5 WAVES IN DIELECTRICS

Let us examine (1.2.3) in more detail for cases that are commonly encountered in solid-state lasers or electro-optic materials. For instance, the active atoms in a ruby laser are chromium which is added (doped) to a level of 5% into the Al_2O_3 host crystal, the details of which are covered in Chapter 10. The point to be made here is that both Al_2O_3 lattice and the active atoms contribute to the polarization, and that it is useful to separate their effects on the propagation of waves. Accordingly, we rewrite Maxwell's equations (with $\mathbf{j} = 0$):

$$\nabla \times \mathbf{h} = \epsilon_0 \frac{\partial \mathbf{e}}{\partial t} + \frac{\partial \mathbf{p}_l}{\partial t} + \frac{\partial \mathbf{p}_a}{\partial t} = \epsilon_0 n^2 \frac{\partial \mathbf{e}}{\partial t} + \frac{\partial \mathbf{p}_a}{\partial t} \qquad (1.5.1a)$$

$$\nabla \times \mathbf{e} = -\mu_0 \frac{\partial \mathbf{h}}{\partial t} \qquad (1.5.1b)$$

where we have combined the lattice polarization term $\mathbf{p}_l$ with the vacuum displacement $\epsilon_0 \mathbf{e}$ with the aid of (1.2.5) to obtain the term involving the square of the index of refraction n^2. Now we repeat the mathematics used to derive the homogeneous wave equation of Sec. 1.3: take the curl of (1.5.1b)

$$\nabla \times \nabla \times \mathbf{e} = -\mu_0 \frac{\partial [\nabla \times \mathbf{h}]}{\partial t}$$

$$\nabla(\nabla \cdot \mathbf{e}) - \nabla^2 \mathbf{e} = -\mu_0 \epsilon_0 n^2 \frac{\partial^2 \mathbf{e}}{\partial t^2} - \mu_0 \frac{\partial^2 \mathbf{p}_a}{\partial t^2} \qquad (1.5.2)$$

$$\nabla^2 \mathbf{e} - \left(\frac{n}{c}\right)^2 \frac{\partial^2 \mathbf{e}}{\partial t^2} = \mu_0 \frac{\partial^2 \mathbf{p}_a}{\partial t^2}$$

This differs from 1.3.1 in two ways: (1) the velocity of popagation is c/n, a result which could be anticipated from elementary electromagnetic theory; and (2) the right-hand side is no longer zero—the equation is now an inhomogeneous one, with the source, i.e., the right-hand side, being the time derivative of the polarization contributed by the active atoms. In other words, the active atoms are the *source* for the optical *fields*. This is the proper approach, but it is also somewhat tedious and thus will be postponed until Chapter 14.

1.6 THE UNCERTAINTY RELATIONSHIPS

It was mentioned in Sec. 1.2 that the representation of a sinusoidal function by the real part of a complex phasor is a shorthand way of taking the Fourier transform of the time function. This is very important from many standpoints. First, it is the formal way of handling nonsinusoidal functions and paves the way for a general transient analysis. But most of all, it leads most naturally into the concept of minimum beam spread from a given aperture.

To appreciate this, let us recall the "uncertainty" relation as it pertains to communication:

$$\Delta \omega \Delta t \geq \tfrac{1}{2} \qquad (1.6.1)$$

In communications, this theorem says that a minimum bandwidth $\Delta\omega$ is required to pass a pulse with a rise time Δt. If we multiply both sides of the equation by $\hbar = h/2\pi$, we obtain formally a relation equivalent to the Heisenberg* uncertainty principle:

$$\Delta E\,\Delta t \geq \frac{h}{4\pi} \quad (1.6.2)$$

It is not a very interesting exercise in transform theory to prove that any two conjugate variables (such as ω and t), which are related by the Fourier transform, obey (1.6.1). The genius of Heisenberg was in relating a physical problem to a mathematical abstraction.

Let us now turn to other conjugate variables. For instance, k_x is the Fourier transform variable with its conjugate x, k_y with y, and k_z with z. Once (1.6.1) is accepted, the same theory of Fourier transforms yields

$$\Delta k_x \Delta x \geq \tfrac{1}{2}$$
$$\Delta k_y \Delta y \geq \tfrac{1}{2} \quad (1.6.3)$$
$$\Delta k_z \Delta z \geq \tfrac{1}{2}$$

If we again multiply by $\hbar = h/2\pi$ and identify $\hbar\mathbf{k}$ as the momentum, we obtain the conventional form of Heisenberg's uncertainty relations. These relationships are summarized in Table 1.1. Note that the uncertainty principle says nothing whatsoever about the relation between nonconjugate variables.

Before we leave this topic, it is worthwhile to have a more precise definition of the term "uncertainty": it is the rms value of the deviation of the parameter from its average value. For instance, if the transverse variation of the electric field of an optical beam were given by

$$E(y) = E_0 \exp\left[-\left(\frac{y}{w_0}\right)^2\right] \quad (1.6.4)$$

TABLE 1.1

Item	Physical	Conjugate Variable	Relation
ω	Angular frequency	t (time)	$\Delta\omega\Delta t \geq \tfrac{1}{2}$
k_x	Propagation along x	x	$\Delta k_x \Delta x \geq \tfrac{1}{2}$
k_y	Propagation along y	y	$\Delta k_y \Delta y \geq \tfrac{1}{2}$
k_z	Propagation along z	z	$\Delta k_z \Delta z \geq \tfrac{1}{2}$
E	$\hbar\omega$ = energy	t	$\Delta E\Delta t \geq h/4\pi$
p_x	Momentum along x	x	$\Delta p_x \Delta x \geq h/4\pi$
p_y	Momentum along y	y	$\Delta p_y \Delta y \geq h/4\pi$
p_z	Momentum along z	z	$\Delta p_z \Delta z \geq h/4\pi$

*Whether the factor in (1.6.1) should be 1, $\tfrac{1}{2}$, or some other number close to 1 depends on how $\Delta\omega$ and Δt are defined.

then the average location of the field is at $y = 0$ and the "uncertainty" Δy is found from

$$(\Delta y)^2 = \frac{\int_{-\infty}^{+\infty} (y - 0)^2 E^2(y) dy}{\int_{-\infty}^{+\infty} E^2(y) dy} \quad (1.6.5)$$

In other words, the mathematical formula for the field can also be interpreted as a probability function. The Fourier transform (in k_y space) is given by

$$E(k_y) = \sqrt{\pi} w_0 E_0 \exp\left[-\left(\frac{k_y w_0}{2}\right)^2\right] \quad (1.6.6)$$

Thus there is a distribution of k_y wave vectors around $k_y = 0$ and thus the "uncertainty" of k_y is

$$(\Delta k_y)^2 = \frac{\int_{-\infty}^{+\infty} (k_y - 0)^2 E^2(k_y) dk_y}{\int_{-\infty}^{+\infty} E^2(k_y) dk_y} \quad (1.6.7)$$

It is left for a problem to show that this particular field distribution has the minimum value permitted: $\Delta y \cdot \Delta k_y = 1/2$.

1.7 SPREADING OF AN ELECTROMAGNETIC BEAM

Let us use the uncertainty relationships to predict the spread of a beam of light energy. Now we know that this beam is traveling more or less at the velocity of light, c; hence, the wave vector k_z is very well defined at $k_z = \omega/c$ (and sure enough, the beam is almost everywhere along the z axis!). But if this is a "beam," its extent in the transverse dimension is limited to the beam diameter, as shown in Fig. 1.2.

If we assume that this "beam" has a smooth "Gaussian-like" spatial extent in the y direction of the form given by (1.6.4):

$$E(y) = E_0 \exp\left[\left(-\frac{y}{w_0}\right)^2\right]$$

then we must also allow for a spread in wave vectors centered around $k_y = 0$:

$$E(k_y) = \pi^{1/2} w_0 E_0 \exp\left[-\left(\frac{k_y w_0}{2}\right)^2\right]$$

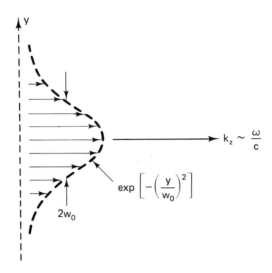

Figure 1.2 Beam of light of diameter $2w_0$ passing the surface $z = 0$. We will use the uncertainty relations to predict the beam diameter along the propagation path.

This interrelationship is sketched in Fig. 1.3. Although a Gaussian spatial envelope is unique in the sense that it is also a Gaussian in k space, the conclusions are the same irrespective of what is chosen for $E(y)$.

Thus, we can construct a diagram for the propagation vectors k_y and k_z as shown in Fig 1.4. It is obvious that the angle $\theta_0/2$ is given by

$$\left. \begin{array}{c} \dfrac{\theta_0}{2} = \dfrac{\Delta k_y}{k_z} = \dfrac{\lambda}{\pi w_0} \\[1em] \text{or} \\[1em] \theta_0 = \dfrac{2\lambda}{\pi w_0} \end{array} \right\} \quad (1.7.1)$$

Thus, a large beam does not spread—indeed, a uniform plane wave (one with

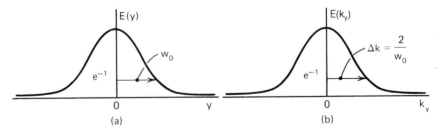

Figure 1.3 Interrelationship between (a) the spatial extent of a beam and (b) the wave number k_y.

Sec. 1.8 Wave Propagation in Anisotropic Media

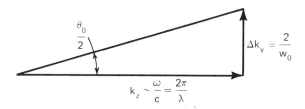

Figure 1.4 Vector addition of k_z and Δk_y to estimate the beam spread.

$w_0 = \infty$) has a zero spread, in accordance with every elementary text on electromagnetic theory. (It has no place to go!)

It is instructive to consider some numbers here. Let $\lambda = 694.3$ nm and $2w_0 = 0.1$ cm; then θ_0 is 8.8×10^{-4} rad. To achieve the same beam spread at 10-cm wavelength would require an antenna aperture $2w_0$ of 144 m. Such a small divergence of an optical beam justifies the simple ray-tracing approach of Chapter 2.

1.8 WAVE PROPAGATION IN ANISOTROPIC MEDIA

Materials that are anisotropic to electromagnetic waves have many uses in optical electronics; modulation, sensing, and harmonic generation are just a few examples. Indeed, most crystalline materials are anisotropic and even some of the amorphorous ones, such as glass, become so when subjected to an electric field, a magnetic field, or mechanical stress. This section introduces the formalism for handling such cases.

We limit our attention to uniaxial media whose dielectric "constant" depends on the direction of the electric field, and thus the displacement vector **D** is described by a matrix multiplication of ϵ with the electric field **E**.

$$\begin{bmatrix} D_x \\ D_y \\ D_z \end{bmatrix} = \epsilon_0 \begin{bmatrix} \epsilon_1 & 0 & 0 \\ 0 & \epsilon_1 & 0 \\ 0 & 0 & \epsilon_2 \end{bmatrix} \cdot \begin{bmatrix} E_x \\ E_y \\ E_z \end{bmatrix} \quad (1.8.1)$$

Our goal is to predict the value of the wave vector **k** as the wave propagates at an angle θ with respect to the z axis (the optical axis) as shown in Fig. 1.5.

From the algebraic form of Maxwell's equations, we know that the wave vector **k** is perpendicular to **D** in any and all cases—anisotropy or no anisotropy!

$$\mathbf{k} \times \mathbf{h} = -\omega \mathbf{D}$$
$$\mathbf{k} \cdot (\mathbf{k} \times \mathbf{H}) \equiv 0 = -\omega \mathbf{k} \cdot \mathbf{D} \quad (1.8.2)$$

Hence there is one orientation of the electric field where we *know* the answer for the orientation of the fields with respect to **k**. This is shown in Fig. 1.5(b), and since the case is so "ordinary," it is given that name. Note that if **k** is constrained

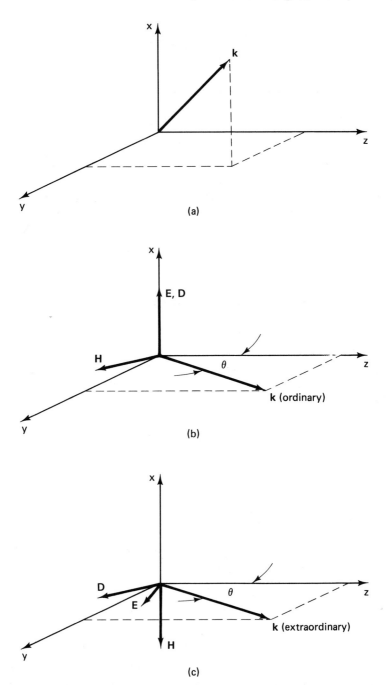

Figure 1.5 Orientation of **k**, **E**, and **D** for a uniaxial crystal. (a) The general problem. (b) The ordinary wave. (c) The extraordinary wave.

Sec. 1.8 Wave Propagation in Anisotropic Media

to the yz plane, then **D** is always in the x direction, and thus $\mathbf{E} = E_x \mathbf{a}_x$. The same argument can be applied to the case where the displacement vector is perpendicular to the plane containing **k** and the z axis—the so-called optic axis. For such cases, the propagation constant is given by

$$k^2 = \omega^2 \mu_0 \epsilon_0 \epsilon_1$$

or
$$\frac{k_0^2}{k^2} = \frac{1}{\epsilon_1} = \frac{1}{n_1^2} \quad \text{(independent of } \theta\text{)} \qquad (1.8.3)$$

If, however, **D** is not perpendicular to the plane containing **k** and the optic axis (i.e., $[\mathbf{a}_z \times \mathbf{k}] \cdot \mathbf{D} = 0$) as shown in Fig. 1.5(c), we have a problem. **D** is still perpendicular to **k** ($\mathbf{D} \cdot \mathbf{k} = 0$), but **E** is not! Hence we can expect a mixture of ϵ_1 and ϵ_2 in the expression for the propagation constant, and a somewhat "extraordinary" behavior as a function of θ—a task to which we turn.

For this polarization shown in Fig. 1.5(c), **k** and **D** can be expressed as

$$\mathbf{k} = k(\cos\theta\, \mathbf{a}_z + \sin\theta\, \mathbf{a}_y) \qquad (1.8.4a)$$

$$\mathbf{D} = D(-\cos\theta\, \mathbf{a}_y + \sin\theta\, \mathbf{a}_z) \qquad (1.8.4b)$$

(Note that $\mathbf{k} \cdot \mathbf{D} \equiv 0$.)
We use (1.8.4b) in conjunction with (1.8.1) to find **E**:

$$E_y = \frac{D}{\epsilon_0 \epsilon_1}[-\cos\theta] \qquad (1.8.5a)$$

$$E_z = \frac{D}{\epsilon_0 \epsilon_2}[\sin\theta] \qquad (1.8.5b)$$

Now it is a straightforward exercise in vector analysis to show (see Problem 1.3) that

$$k^2 = \omega^2 \mu_0 \frac{\mathbf{D} \cdot \mathbf{D}}{\mathbf{E} \cdot \mathbf{D}} \qquad (1.8.6a)$$

or
$$\left(\frac{k_0}{k}\right)^2 = \frac{1}{n_{\text{eff}}^2} = \epsilon_0 \frac{\mathbf{E} \cdot \mathbf{D}}{\mathbf{D} \cdot \mathbf{D}} \qquad (1.8.6b)$$

where the effective index is defined by $k/k_0 = n_{\text{eff}}$. Combining (1.8.6b) with (1.8.5) yields

$$\frac{1}{n_{\text{eff}}^2} = \frac{\cos^2\theta}{n_1^2} + \frac{\sin^2\theta}{n_2^2} \qquad (1.8.7)$$

The forms of normalized propagation vector (k/k_0) expressed by (1.8.3) and (1.8.7) are conveniently shown on a graph called the index surface (see Fig. 1.6. Equation (1.8.3) states that the effective index for the ordinary wave is independent of the angle θ—hence it is shown as a circle. The effective index for extraordinary wave does depend on θ in the form of an ellipse.

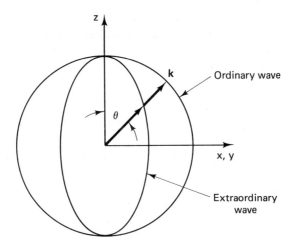

Figure 1.6 The index ellipsoid for a uniaxial crystal.

It is apparent from Fig. 1.6 and (1.8.3) and (1.8.7) that the phase constants for the ordinary and extraordinary waves are not equal for $\theta \neq 0$. This fact plays a critical role in nonlinear optics where it is crucial that the phase constants of, for example, the fundamental wave and any harmonic or intermodulation terms, *must* be synchronized. Fortunately, the dielectric constants are not constant with frequency (i.e., λ) and thus it is possible to choose a phase matching angle θ_m such that the effective index for the fundamental frequency ω, when propagated as an ordinary (extraordinary) wave, equals the effective index for the second (third, etc.) harmonic when it is propagated as an extraordinary (ordinary) wave.

1.9 ELEMENTARY BOUNDARY VALUE PROBLEMS IN OPTICS

The propagation of electromagnetic waves is determined by Maxwell's equations, but these are incomplete without a specification of boundary conditions. After all, they are partial differential equations which presume that all field variables and material properties are continuous functions of the coordinates. However, we will have many occasions to consider abrupt junctions between different materials (windows, mirrors, etc.) where the electrical parameters are different and as a consequence, the field variables change discontinuously.

Most elementary texts derive the relationship between the tangential and normal components of the field at each side of an abrupt interface:

$$\mathbf{a}_n \times (\mathbf{E}_1 - \mathbf{E}_2) = 0 \quad (1.9.1a)$$

$$\mathbf{a}_n \cdot (\mathbf{D}_1 - \mathbf{D}_2) = \rho_{s2} \quad (1.9.1b)$$

Sec. 1.9 Elementary Boundary Value Problems in Optics

$$\mathbf{a}_n \times (\mathbf{H}_1 - \mathbf{H}_2) = \mathbf{J}_{s2} \qquad (1.9.1c)$$

$$\mathbf{a}_n \cdot (\mathbf{B}_1 - \mathbf{B}_2) = 0 \qquad (1.9.1d)$$

where $\mathbf{a}_n$ is a unit vector from 2 to 1 and perpendicular to the interface. The concept of a surface charge, ρ_s, and surface current, $\mathbf{J}_s$, both existing in zero depth in medium 2, are useful approximations at low frequencies, $\nu < 10^{12}$ Hz, but those approximations are almost never utilized in the optical domain. Hence, we will let the right-hand side of (1.9.1b) and (1.9.1c) be zero.

The formal method of handling the interface problem is to first solve Maxwell's equations in the two media and then match the fields at the boundary with (1.9.1a) and (1.9.1c). It is sufficient to match tangential components only, because the normal components will then be matched automatically, provided the fields in the respective media obey Maxwell's equations.

Many times, one can sidestep a lot of dull mathematics implied by the above by applying some elementary physical reasoning. Some very important examples of this approach are shown below.

1.9.1 Snell's Law

Consider a *u*niform *p*lane *w*ave (upw) impinging on the interface shown in Fig. 1.7 making an angle θ_1 with respect to the normal to the surface; the discontinuity generates a second wave at an angle θ_2 and a reflected wave. One could grit one's teeth and match field components at the interface and solve the problem completely. This procedure is necessary if the amplitude and phase of the *transmitted* and *reflected* waves are desired. However, if only the *direction* is desired, the procedure can be greatly simplified.

The point to be remembered is that the incident wave is the *source*, and the transmitted and reflected waves are the *responses*. Hence the phases of both responses, whatever they are, must be synchronized with respect to the source along the boundary where the responses are generated.

The relative phase of the source along the interface is

$$\phi = (\omega/c)\, n_1 \sin \theta_1 \qquad (1.9.2)$$

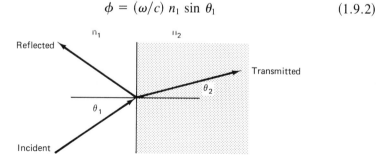

Figure 1.7 Geometry for Snell's law.

and this must be the phase of both responses as measured along the interface. If medium 1 is isotropic, this fact forces the incident and reflected waves to make the same angle with respect to the normal.

For the transmitted field, we force the phases along the boundary to be the same:

$$(\omega/c)\, n_1 \sin \theta_1 = (\omega/c)\, n_2 \sin \theta_2 \tag{1.9.3a}$$

or
$$n_1 \sin \theta_1 = n_2 \sin \theta_2 \tag{1.9.3b}$$

For an anisotropic medium for 2, the incident wave can generate two transmitted waves, but both must remain tied to the phase of incident wave along the interface.

1.9.2 Brewster's Angle

Windows oriented at Brewster's angle are commonly used on gas lasers because, in principle, they transmit waves *without* reflection for one polarization of the electric field. The geometry of the electromagnetic problem is shown in Fig. 1.8 for two possible polarizations of the incident field. In both cases, Snell's law is applicable and thus the wave vector, **k**, is bent toward the normal in the window material.

There are some artifacts added to Fig. 1.8 to help visualize the physical situation: the orientations of the induced dipoles in the dielectric material are shown, for it is their reradiation that generates the reflected wave.

Now every elementary text in electromagnetic theory shows that electric dipoles radiate perpendicular to their axis—*not* along it. Thus for the TE orientation, there is no problem in generating a reflected wave. However, for the TM case and a particular angle of the incident wave, the reflected wave would try to come off the ends of the dipole, which is impossible. Hence there is *no* reflected wave when the angles $\theta_1 + \theta_2 = \pi/2$. Combining this fact with Snell's law yields an expression for the angle of zero reflection:

$$\theta_1 + \theta_2 = \frac{\pi}{2}$$

$$n_1 \sin \theta_1 = n_2 \sin \theta_2 \quad \text{(Snell's law)} \tag{1.9.4}$$

Hence
$$n_1 \sin \theta_1 = n_2 \sin(\pi/2 - \theta_1) = n_2 \cos \theta_1$$

Therefore
$$\tan \theta_1 = \frac{n_2}{n_1} \quad \text{(Brewster's angle)} \tag{1.9.5}$$

It should be emphasized that mathematics involved in matching fields across an interface will lead to the same result, but one should appreciate the physical reasoning presented above also.

Sec. 1.10   Coherent Electromagnetic Radiation   21

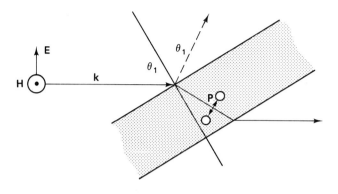

(a) TM or "p" polarized

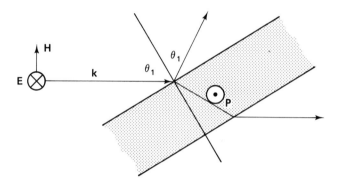

(b) TE or "s" polarized

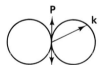

(c) Dipole radiation

Figure 1.8   Brewster's angle windows.

1.10 COHERENT ELECTROMAGNETIC RADIATION

Let us reiterate the goals of this book: to understand the physical basis for the *generation, transmission,* and *detection* of electromagnetic radiation in the "optical" portion of the spectrum. But we should be more precise and focus our attention on a specific characteristic that distinguishes the laser from a simple lamp.

The distinguishing characteristic is the generation of *coherent* electromagnetic radiation. Now, the topic of coherence is most involved and complex to describe with precision, but it is relatively easy to understand the first-order consequences.

Most people who have had electronic experience at low frequencies, say less than 30 GHz, with classical generators never address this subject, because most of our generators had a long coherence time or length. In other words, they are almost perfectly coherent. But what does this mean, and how would one measure either coherence time or length?

In a loose sort of way, coherence time is the net delay one can insert in a wave train and still obtain interference. Since electromagnetic waves travel with a velocity of c, the longitudinal coherence length is simply c times the coherence time. Note that the key word is "interference." Let us illustrate these ideas with a "thought" experiment taken from low-frequency electronics and compare it with a similar experiment at optical frequencies (visible wavelengths).

Consider a simple transmission-line measurement of the standing-wave ratio on a short-circuited transmission system as shown in Fig. 1.9. To make the conventional "slotted-line" measurement of the "voltage" standing-wave ratio (VSWR), we move a short dipole antenna and a rectifying diode along the z axis. The output of the detector is proportional to the square of the electric field (usually); hence, the relative output of the detector would be as shown in Fig. 1.10. The VSWR, V_{max}/V_{min}, is very large—for all practical purposes it is infinity. This is precisely what we observe in a normal laboratory.* Even elementary theory would predict this result, as demonstrated below.

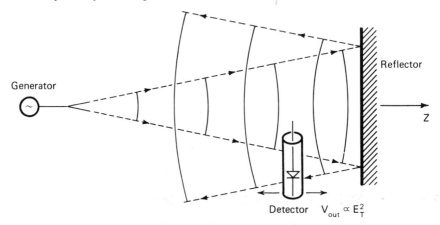

Figure 1.9 Simple interference experiment.

*In fact, Fig. 1.9 bears a close resemblance to the original experiments of Hertz, who demonstrated the equivalence of light and low-frequency waves as predicted by Maxwell's theory.

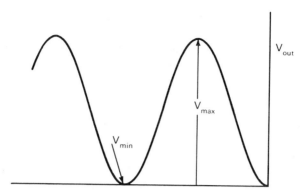

Figure 1.10 Measurement of the VSWR. (NOTE: Most detectors produce an output [i.e., voltage] proportional to the power sampled by the antenna. Consequently, the quantity V_{max}/V_{min} would correspond to the power standing-wave ratio.)

The electric field traveling to the right is given by

$$E^+ = E_0 \exp(-jkz), \quad z < 0, \quad k = \frac{2\pi}{\lambda} \qquad (1.10.1)$$

with the time factor, $\exp(j\omega t)$, suppressed. The reflected wave is given by

$$E^- = -E_0 \exp(+jkz) \qquad (1.10.2)$$

Hence, the output of the detector is given by

$$V_{out} \propto E_T E_T^* = 4E_0^2 \sin^2 kz \qquad (E_T = E^+ + E^-) \qquad (1.10.3)$$

Although this analysis is quite adequate for normal laboratory experiments at low frequencies, we have made the serious assumption of a perfectly coherent source. Such a device does not exist!

We have assumed that the phase of the incoming wave at a point z is predictable from the phase of the wave that crossed this point at a time $2z/c$ seconds earlier. But, of course, it is not tied perfectly to this earlier waveform; its phase could have "wandered" in the time it took the initial wave to traverse the distance from the observation point to the reflector and back. Thus, we should modify (1.10.1) to read

$$E^+ = E_0 \exp\{-j[kz + \Delta\phi(t)]\} \qquad (1.10.1a)$$

where $\Delta\phi(t)$ is a random variable, characteristic of the source.* Thus the output of the detector changes to

$$V_{out} \propto E_T^2 = 4E_0^2 \sin^2\left[kz + \frac{\Delta\phi(t)}{2}\right] \qquad (1.10.4)$$

In this case, the minimum (or maximum) is *not* where we think it should be, and worse yet, it wanders in time according to the whims of $\phi(t)$. In it almost as if

---

*$\Delta\phi$ is the amount by which the phase can change in the round-trip delay time $2z/c$.

the standing-wave pattern is "jittering" back and forth in a random fashion, as indicated in Fig. 1.11. Normally, the time rate of change of ϕ is small compared to the angular frequency, ω, and this fact explains why we never see this effect at low frequencies from any "decent" source.

Example

Suppose that the maximum value of $d\phi/dt$ was 10^{-4} of the angular frequency, ω_0, of the source (a rather poor one, but let's use it). Let the nominal frequency of the source be 1 GHz. If the observation point z of our detector were a "room-like" distance away from the reflector, say 3 m, the time interval between the passage of the first wave train and its return is only

$$\Delta t = \frac{2z}{c} = \frac{2 \times 3}{3 \times 10^8} = 20 \text{ ns}$$

and the phase could, at most change by

$$\Delta\phi = \left.\frac{d\phi}{dt}\right|_{max} \Delta t = 10^{-4} \times 2\pi \times 10^{+9} \times 20 \times 10^{-9} = 0.004\pi \Rightarrow 0.72°$$

In other words, the position of the minimum is only jittering by $0.72°/360° = 0.2\%$ of a wavelength (30 cm) or $\Delta L = 0.6$ mm (probably smaller than the wire used for the dipole antenna).

However, the numbers and the effects change considerably if one performs the same type of interference experiment at optical frequencies. Since most components and detectors are huge compared to optical wavelengths, the techniques are slightly different—but not in their essential function.

Consider the Michelson interferometer shown in Fig. 1.12. Collimated light is divided and passed around the two arms of the interferometer in the

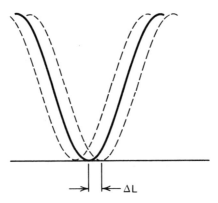

Figure 1.11 "Jittering" of the minimum position due to the random jumps in phase of the later portion of the wave.

Sec. 1.10   Coherent Electromagnetic Radiation

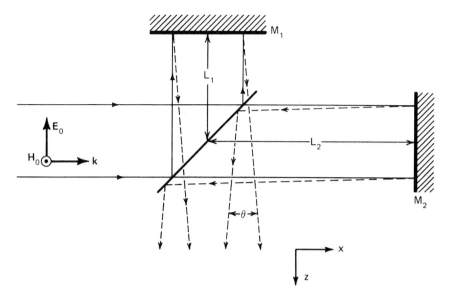

Figure 1.12 Michelson interferometer.

manner indicated in Fig. 1.12. Obviously, the radiation that went the M_2 route is retarded in time by $2(L_2 - L_1)/c$ with respect to that returning from M_1. Shown also is the probable situation of the two beams propagating at a slight angle with respect to each other. Thus the respective electric fields at the plane of the detector are given by

$$E_1 = \frac{E_0}{\sqrt{2}} \exp\left[-j\left(k \cos \frac{\theta}{2} z + k \sin \frac{\theta}{2} x\right)\right] \exp(-j2kL_1) \exp(-j\Delta\phi)$$

$$E_2 = \frac{E_0}{\sqrt{2}} \exp\left[-j\left(k \cos \frac{\theta}{2} z - k \sin \frac{\theta}{2} x\right)\right] \exp(-j2kL_2) \quad (1.10.5)$$

In (1.10.5), we have allowed for the phase of one wave to wander with respect to the other. Again, we must remember that $\Delta\phi(t)$ is the change in phase of the reference during the time that the other signal is delayed. If the two arms of the interferometer are exactly equal, the phase term disappears. Optical detectors (eyes, photographic emulsions, and photoelectric devices) respond to the intensity of the radiation; hence, the relative intensity along the x direction for a fixed z is given by

$$I(x, z = \text{constant}) = \frac{(\mathbf{E}_1 + \mathbf{E}_2) \cdot (\mathbf{E}_1^* + \mathbf{E}_2^*)}{2\eta_0}$$

$$= 2\left(\frac{E_0^2}{2\eta_0}\right) \cos^2\left[\frac{k\theta x}{2} + k(L_2 - L_1) + \frac{\Delta\phi}{2}\right] \quad (1.10.6)$$

where we have assumed that θ is small. Thus the detector—say our eyes—would see a series of bright and dark bands in the manner indicated in Fig. 1.13.

As indicated in Fig. 1.13, θ must be very small for Δx to be a reasonable size. For instance, if $\lambda = 0.5$ μm (green-blue) and Δx is to be greater than 0.5 cm, then $2\theta \leq 10^{-4}$ rad. Note, too, that the position of the minimum in intensity depends on the difference in optical path length. Hence, if this difference changes a fraction of a wavelength (due to room vibrations), the fringe position will jitter accordingly.

Even if these significant mechanical stability problems are overcome and the alignment is achieved, the fringe position will still change, owing to the random wandering of $\phi(t)$. Whereas in the microwave "experiment," we could conceive of measuring the jitter in the fringes, our optical detector would average the intensity over its time constant. [For instance, the eye retentivity (or time constant) is on the order of $\frac{1}{10}$ sec.] This leads to a degradation of the fringe visibility, defined as

$$V = \frac{\langle I_{\max}\rangle - \langle I_{\min}\rangle}{\langle I_{\max}\rangle + \langle I_{\min}\rangle} \quad (1.10.7)$$

where the brackets $\langle\ \rangle$ indicate time averages.

Again, let us take some typical numbers for a spectral line. Let us assume that $d\phi/dt$ is at most 10^{-5} of $\omega = 2\pi c/\lambda$ and solve for the difference in lengths that keeps the fringe position within $\Delta x/2$ of its predicted position (an interchange between "bright" and "dark" bands). Thus $\Delta \phi/2 = \pi/2$ radians.

$$\Delta\phi = \left.\frac{d\phi}{dt}\right|_{\max} \Delta t = 10^{-5} \times \frac{2\pi c}{\lambda} \frac{2(L_2 - L_1)}{c} = \pi$$

or

$$L_2 - L_1 = 10^5 \times \frac{\lambda}{4}$$

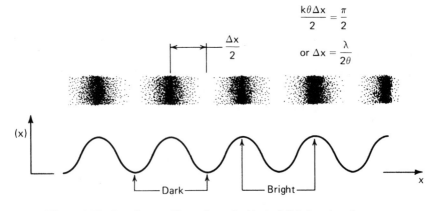

Figure 1.13 Interference fringes formed with the Michelson interferometer.

For a wavelength of 5000 Å, this yields a coherence length, $2(L_2 - L_1)$, of 2.5 cm (1 in.).

Another way of looking at this phenomenon and, at the same time, gaining some insight into the origin of the changes in the phase is to change to the spectral representation of the field. The instantaneous frequency of the source is given by the time rate of change of the phase:

$$\omega = \frac{d\phi}{dt} = \frac{d}{dt}[\omega_0 t + \phi(t)]$$
$$= \omega_0 + \frac{d\phi}{dt}$$
(1.10.8)

If we consider the source as being a collection of identical oscillators being turned on and off randomly,* each contributing a part of the field, then the distribution of frequencies will be centered about ω_0 but containing a smaller amplitude on either side, as shown in Fig. 1.14. This will be discussed in more detail when the line shape is considered.

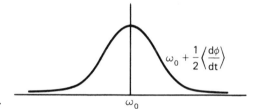

Figure 1.14 Profile of an emission line.

1.11 EXAMPLE OF COHERENCE EFFECTS

Let us consider a problem that emphasizes the role of the random jumps in phase due to the finite spectral width of the source. Assume that we are measuring the interference between the two waves by means of a camera and a photographic film, as shown in Fig. 1.15. We can choose the exposure time by the shutter and thus control the "density" of the blackening of the film. This density is proportional to the time-integrated optical power density incident on a point of the film.

We assume that the two waves have the same amplitude, polarization, and nominal frequency, ω_0. However, we allow the phase of 1 to jump discontinuously by $\Delta\phi$ every T seconds according to the following random prescription.

Use three different denominations of coins—say, a penny, a nickel, and a dime—flip the coins every T seconds, and assign the jump in phase according

---

*As we shall see, the atoms are the oscillators. In a laser, the field stimulates the atoms to give up their energy in a predictable manner, and thus the coherence time is much larger.

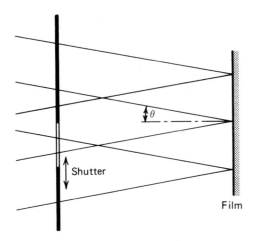

Figure 1.15 Paper experiment to demonstrate coherence.

to Table 1.2. Now let us compute the fringe visibility by determining the density of the exposure of the film assuming that the shutter is open for a $t < T$ seconds and $t = NT$ seconds. In the short-exposure case, the integrated exposure yields perfect fringe contrast according to (1.10.6). But for a long-exposure time, we have

$$D_N(x) = \frac{1}{NT}\left[I_{max} T \cos^2\left(\frac{k\theta x + \phi_0}{2}\right) \right.$$
$$+ I_{max} T \cos^2\left(\frac{k\theta x + \phi_0 + \phi_1}{2}\right)$$
$$+ I_{max} T \cos^2\left(\frac{k\theta x + \phi_0 + \phi_1 + \phi_2}{2}\right) + \cdots$$
$$\left. + I_{max} T \cos^2\left(\frac{k\theta x + \phi_0 + \phi_1 + \cdots + \phi_{N-1}}{2}\right) \right]$$

TABLE 1.2

Penny	Nickel	Dime	Jump in phase $\Delta\phi$
H	H	H	0
H	H	T	+45°
H	T	H	+90°
T	H	H	+135°
T	T	T	+180°
T	T	H	−45°
T	H	T	−90°
H	T	T	−135°

Sec. 1.11 Example of Coherence Effects

TABLE 1.3 RESULTS OF THE "TOSS OF COINS"

Coin Toss	1	2	3	4	5	6	7	8
Penny	H	H	T	H	T	T	T	H
Nickel	T	T	H	H	H	T	T	H
Dime	H	H	H	T	T	H	T	H
$\Delta\phi = +90$	+90	+135	+45	$-90°$	$-45°$	$180°$	0
$\phi_0 = 0, \phi = +90$	180	-45	0	$-90°$	$-135°$	$+45°$	$45°$

Obviously the results depend on the toss of the coin; the author obtained the results given in Table 1.3 (you should generate your own sequence). A plot of the relative film density for various values of N is shown in Fig. 1.16. If one is able to photograph the fringes in a time interval short compared to the phase change, very sharp fringes result. However, if the shutter is open for many phase changes, the visibility decreases more-or-less exponentially with exposure time. For very long periods of time, the film would be exposed more or less uniformly.

With this loose understanding of longitudinal coherence, let us turn to the problem of "transverse" coherence length. Here we inquire whether the phase changes along a transverse coordinate in a smooth and slow fashion, as wave 1

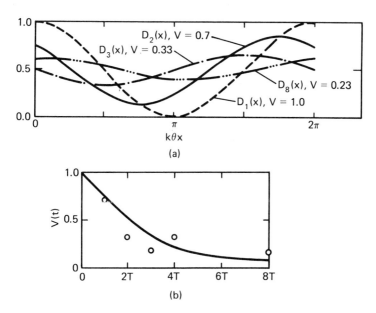

Figure 1.16 Relative density (D) or blackening of the film is shown in (a) along with fringe visibility V. The latter is plotted in (b) as a function of exposure time.

of Fig. 1.17, or rapidly in space and time and in an unpredictable fashion, as wave 2. We could have the same power in the two waves, and as far as visual observation is concerned, the beam size at $z = 0$ would be the same, but there would be a remarkable difference in the divergence of the beam in the two cases. This is easily shown by applying the previous discussion of the uncertainty principle considering the fundamentals of wave propagation.

If the phase is to be changing in the x direction, there must be a component of the wave vector in that direction, even though the wave is traveling primarily in the z direction. The magnitude of this phase change can be estimated by using the mean change in phase $\overline{\Delta \phi}$ divided by the distance over which this occurs:

$$\Delta k_x \sim \frac{\overline{\Delta \phi}}{L} \qquad (1.11.1)$$

Thus for the two waves shown in Fig. 1.17, $\Delta k_{x1} = \overline{\Delta \phi_1}/L_1$ and $\Delta k_{x2} = \overline{\Delta \phi_2}/L_2$. Hence, the diagrams for the transverse and longitudinal wave vectors are as shown in Fig. 1.18. Thus the far-field divergence angle is given by

$$\theta \sim \frac{\Delta k_x}{k_z}, \qquad \theta_1 \sim \frac{\overline{\Delta \phi_1}}{L_1} \frac{\lambda}{2\pi}, \qquad \theta_2 = \frac{\overline{\Delta \phi_2}}{L_2} \frac{\lambda}{2\pi} \qquad (1.11.2)$$

Obviously, $\theta_2 \gg \theta_1$, for the situation shown in Fig. 1.14, and the divergence of the second beam is much greater than the first. Incidentally, (1.11.2) is much more restrictive than one that is usually specified in terms of the amplitude of the wave [i.e., (1.7.1)].

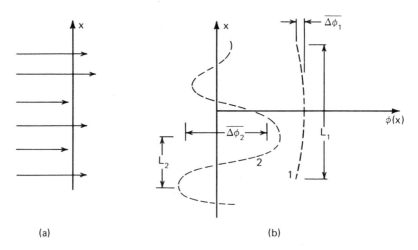

Figure 1.17 Two beams of the same size but with radically different variations of phase in the transverse direction.

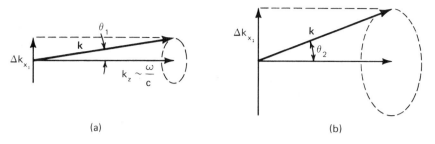

Figure 1.18 Beam spreads for the two beams of Fig. 1.17.

PROBLEMS

1.1. Why is the factor 2 present in the expression for Poynting vector (i.e., $S = E \times H^*/2$)?

1.2. Assume that the electromagnetic fields vary as $\exp(-j\mathbf{k}\cdot\mathbf{r})$ and use the rules for the curl, gradient, and divergence to derive the algebraic form of Maxwell's equations (1.4.2).

1.3. The algebraic forms for Maxwell's equations for a linear homogeneous anisotropic medium are

$$\mathbf{k} \times \mathbf{H} = -\omega \mathbf{D}$$
$$\mathbf{k} \times \mathbf{E} = \omega \mathbf{B}$$

where $\mathbf{B}$ is related to $\mathbf{H}$ and $\mathbf{D}$ to $\mathbf{E}$ by

$$\mathbf{B} = \mu_0(\mathbf{H} + \mathbf{M})$$
$$\mathbf{D} = \epsilon_0 \mathbf{E} + \mathbf{P}$$

For many materials, the polarization vector $\mathbf{P}$ is not collinear with $\mathbf{E}$; hence, $\mathbf{D}$ is not collinear with $\mathbf{E}$ either. The same comments apply to $\mathbf{B}$, $\mathbf{M}$, and $\mathbf{H}$. Assume a dielectric medium with $\mathbf{M} = 0$ but with no restrictions placed on $\mathbf{D}$ and $\mathbf{E}$.
(a) Show that $\mathbf{k} \cdot \mathbf{D} \equiv 0$.
(b) Show that the wave vector $\mathbf{k}$ always points in the direction of $\mathbf{D} \times \mathbf{B}$.
(c) Show that the amplitude of the wave vector $\mathbf{k}$ is given by

$$k^2 = \omega^2 \mu_0 \frac{\mathbf{D} \cdot \mathbf{D}}{\mathbf{E} \cdot \mathbf{D}}$$

(d) Show that the Poynting vector, $\mathbf{S} = \mathbf{E} \times \mathbf{H}^*/2$, can point in a direction other than that of the wave vector $\mathbf{k}$.

1.4. Suppose that we are using an optical beam of diameter D to monitor the particle content of a column of gas. For many applications we would prefer to sample as small a volume as possible, and consequently we would first choose a very small beam. But if the path length is long, a very small beam would diverge quickly and thus sample a larger cross-sectional area of the gas column. Use the uncertainty relations to derive an expression for the beam diameter to minimize the volume of gas sampled. Assume a helium/neon probing laser ($\lambda = 632.8$ nm) and a simple cone describing the convergence and divergence of the beam envelope so as to evaluate for a gas column 10 m long.

1.5. There are various ways to specify the frequency and photon energy of a laser: some use the energy in eV; some specify the wavelength in angstrom units (10^{-10} m), in nanometers, (10^{-9} m), or in micrometers (10^{-6} m); others use the wave number $\bar{\nu}$, the number of wavelengths that will fit inside a centimeter of vacuum; and still others specify cycles (Hz). Convert the specification of photons from common coherent sources to the other units.

Source	eV	λ(Å)	λ(nm)	ν(Hz)	$\bar{\nu}$(cm^{-1})
GaAs	1.47				
Ar$^+$		5145			
He:Ne			632.8		
CO_2					943
ISM band		(in meters)		13.56 MHz	
KrF			249		

1.6. Repeat the coin-flipping routine of Sec. 1.8 and find the fringe visibility as a function of exposure time.

1.7. Show that the Fourier transform of the field given by

$$E(y) = E_0 \exp\left[-\left(\frac{y}{w_0}\right)^2\right]$$

is

$$E(k_y) = \pi^{1/2} w_0 E_0 \exp\left[-\left(\frac{k_y w_0}{2}\right)^2\right]$$

1.8. The TEM$_{0,0}$ Gaussian beam has the smallest value of the product $\Delta x \, \Delta k_x = 1/2$ allowed by the uncertainty relationship. (The meaning of the terminology TEM$_{0,0}$ will be covered in Chapter 3.) The quantities Δx and Δk_x are to be interpreted as

$$\Delta x^2 = \int x^2 |E(x)|^2 \, dx \Big/ \int |E(x)|^2 \, dx$$

$$\Delta k_x^2 = \int k_x^2 |E(k_x)|^2 \, dk_x \Big/ \int |E(k_x)|^2 \, dx$$

with $E(x)$ and $E(k_x)$ being related by the Fourier transform.

(a) What are the values for Δx and Δk_x for $E(x) = E_0 \exp\left[-\left(\dfrac{x}{w_0}\right)^2\right]$ (i.e., $\text{TEM}_{0,0}$)?

(b) What is the uncertainty product for a field given by
$$E_{10} = (\sqrt{2}\, x/w) \exp -[(x^2 + y^2)/w_0^2]$$

(c) Sketch the intensity $(E \cdot E^*)$ as a function of x.

1.9. Show that the factor of 2 belongs in (1.10.4).

1.10. Quartz windows oriented at Brewster's angle are commonly used for He:Ne lasers in the manner indicated in Fig. 0.1. What is the angle measured from the axis of the cavity? ($n[\text{quartz}] = 1.43$)

REFERENCES AND SUGGESTED READINGS

1. E. C. Jordan and K. G. Balmain, *Electromagnetic Waves and Radiating Systems*, 2nd ed. (Englewood Cliffs, N.J.: Prentice-Hall, Inc., 1968), Chaps. 1–7.
2. S. Ramo, J. R. Whinnery, and T. Van Duzer, *Fields and Waves in Communication Electronics* (New York: John Wiley & Sons, Inc., 1965), Chaps. 1–6.
3. A. Nussbaum and R. Phillips, *Contemporary Optics for Scientists and Engineers* (Englewood Cliffs, N.J.: Prentice-Hall Inc., 1976), Chap. 6.
4. R. M. Eisberg, *Fundamentals of Modern Physics* (New York: John Wiley & Sons, Inc., 1961), Chap. 6.
5. F. A. Jenkins and H. E. White, *Fundamentals of Optics*, 3rd ed. (New York: McGraw-Hill Book Company, 1957). An excellent introduction to optics.
6. W. H. Hayt, Jr., *Engineering Electromagnetics* New York: McGraw-Hill Company, 1974).
7. D. K. Cheng, *Field and Wave Electromagnetics* (Reading, Mass.: Addison-Wesley Publishing Company, Inc., 1983).
8. N. N. Rao, *Elements of Engineering Electromagnetics*, 2nd Ed. (Englewood Cliffs, N.J.: Prentice-Hall, Inc., 1987).

2

Ray Tracing in an Optical System

2.1 INTRODUCTION

In Chapter 1 we found that a physically small beam of electromagnetic energy at very high frequencies does not spread to any degree. This allows us to follow our intuition and follow rays of light as they traverse an optical path. Let us first define a "ray" as the *path* that the center of a very slowly diverging electromagnetic beam would take as it goes through the system. Note that the word "path" was emphasized, for a ray does not have an amplitude. Furthermore, we require that the spatial extent of this beam in the transverse direction be small compared to the size of the optical components.

2.2 RAY MATRIX

In order to describe the "gyrations" of a ray in an optical system consisting of lenses, lengths of free space, interfaces between different dielectric media, mirrors, and so on, we should first ask: What is necessary to specify everything about it? The answer is deceptively simple:

1. Where is it with respect to some arbitrarily chosen axis?
2. In what direction is it heading?

We can answer these questions by inspection for the most important building block of an optical system—a length of free space (see Fig. 2.1). Obviously, if

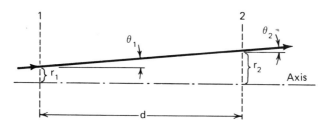

Figure 2.1 Ray in a homogeneous dielectric of length d.

we know where the ray is at the first plane (1) and know its slope with respect to the axis, then we know where the ray emerges and where it is going at the exit plane (2). Indeed, the example is so trivial that it could be boring, but since it is so simple, we use it to introduce the notation.

We assume here and in everything to follow that all rays are paraxial, so that $\tan \theta \approx \sin \theta \sim \theta$ for all angles measured with respect to the optic axis, and thus the angle θ equals the slope of the ray, r'. The output parameters are related to the input parameters by

$$r_2 = 1 \cdot r_1 + d \cdot r_1'$$
$$r_2' = 0 \cdot r_1 + 1 \cdot r_1' \qquad (2.2.1)$$

or we could (and will) write this in matrix form as

$$\begin{bmatrix} r_2 \\ r_2' \end{bmatrix} = \begin{bmatrix} 1 & d \\ 0 & 1 \end{bmatrix} \begin{bmatrix} r_1 \\ r_1' \end{bmatrix} \qquad (2.2.2)$$

In general, the relation between the output and input parameters of a general optical system is given by the *ABCD* matrix of the form

$$\begin{bmatrix} r_{\text{out}} \\ r_{\text{out}}' \end{bmatrix} = \begin{bmatrix} A & B \\ C & D \end{bmatrix} \begin{bmatrix} r_{\text{in}} \\ r_{\text{in}}' \end{bmatrix} \qquad (2.2.3)$$

Thus ray tracing through a sequence of optical components is reduced to simple 2×2 matrix multiplication. For instance, if we consider two lengths of free space (see Fig. 2.2), the output of one optical component is the input to the other. Thus we have

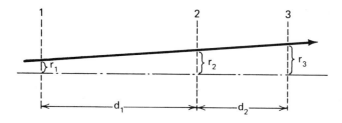

Figure 2.2 Example of two lengths of space.

$$\begin{bmatrix} r_3 \\ r_3' \end{bmatrix} = \begin{bmatrix} 1 & d_2 \\ 0 & 1 \end{bmatrix} \begin{bmatrix} r_2 \\ r_2' \end{bmatrix}; \quad \begin{bmatrix} r_2 \\ r_2' \end{bmatrix} = \begin{bmatrix} 1 & d_1 \\ 0 & 1 \end{bmatrix} \begin{bmatrix} r_1 \\ r_1' \end{bmatrix}$$

or (2.2.4)

$$\begin{bmatrix} r_3 \\ r_3' \end{bmatrix} = \begin{bmatrix} 1 & d_2 \\ 0 & 1 \end{bmatrix} \begin{bmatrix} 1 & d_1 \\ 0 & 1 \end{bmatrix} \begin{bmatrix} r_1 \\ r_1' \end{bmatrix} = \begin{bmatrix} 1 & d_1 + d_2 \\ 0 & 1 \end{bmatrix} \begin{bmatrix} r_1 \\ r_1' \end{bmatrix}$$

Equation (2.2.4) is a very complex way of obtaining an obvious result; if (2.2.2) is correct, then the length $d_1 + d_2$ is substituted for d for the cascade.

Some may start to recognize the analogy between the formalism used here and that used for the cascade of electrical networks (see Fig. 2.3). The analogy is excellent (and intentional). The notation is the same, $ABCD$, and even some of the special details are the same. For instance, the determinant of coefficients of T is unitary for a bilateral electrical network:

$$AD - BC = 1 \tag{2.2.5}$$

Note that this holds for the matrices in (2.2.2) and (2.2.4). For optical rays this is true provided that the index of refraction at the exit plane is the same as the entrance plane.

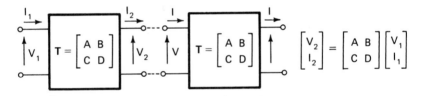

Figure 2.3 Correspondence with electrical network theory.

2.3 SOME COMMON RAY MATRICES

The most important ray matrix in optical systems is that of a length of free space; the next most important element is a thin lens of focal length f. Let us use the formalism of the preceding section and some rather obvious facts about lenses to obtain its ray matrix (see Fig. 2.4). Here we assume that the lens is so thin that there is negligible distance between the entrance (1) and exit (2) planes. Thus no matter what the slope of the incoming ray, the output position is always equal to the input position, or

$$r_2 = r_1$$

Therefore $A = 1$ and $B = 0$.

Now consider the circumstances provided by ray α in the diagram. For this special case, the input slope $r_1' = 0$, yet it is obvious that the output slope is $-r_1/f$.

Sec. 2.3 Some Common Ray Matrices

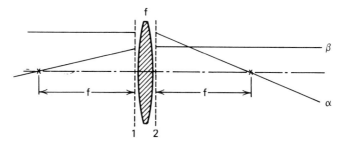

Figure 2.4 Paper experiment with a "thin" lens.

$$r'_{2\alpha} = Cr_{1\alpha} + Dr'_{1\alpha} = -\frac{r_{1\alpha}}{f} = Cr_{1\alpha} + D \cdot 0$$

or

$$C = -\frac{1}{f}$$

In the other case, ray β comes in with a slope of $+r_1/f$ and obviously exits parallel to the axis. Thus $r'_2 = 0$.

$$r'_{2\beta} = 0 = -\frac{1}{f}r_{1\beta} + Dr'_{1\beta}, \qquad r'_{1\beta} = \frac{r_{1\beta}}{f}$$

Therefore

$$D = 1$$

Hence the ray matrix of a thin lens is given by

$$\mathbf{T} = \begin{bmatrix} 1 & 0 \\ -\dfrac{1}{f} & 1 \end{bmatrix} \tag{2.3.1}$$

Again note that the determinant of **T** is unitary.

Fig. 2.5 combines the two elements in cascade. The transmission matrix between 1 and 3 is given by

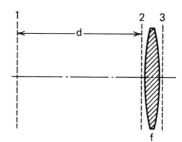

Figure 2.5 Combination of a lens plus free space.

$$\mathbf{T} = \begin{bmatrix} 1 & 0 \\ -\dfrac{1}{f} & 1 \end{bmatrix} \begin{bmatrix} 1 & d \\ 0 & 1 \end{bmatrix} = \begin{bmatrix} 1 & d \\ -\dfrac{1}{f} & 1 - \dfrac{d}{f} \end{bmatrix} \quad (2.3.2)$$

Note that the matrices appear in reverse order to that encountered as the ray goes through the system. (This is because the output of the first component is the input to the second.) Again, note that the determinant is unitary.

Another important component is a mirror (see Fig. 2.6). Here we run into a bit of a problem, because the element causes a major redirection of the ray. In other words, the entrance and exit planes are on the same side of the surface of the mirror. If we ride *with* the ray, the effect of the mirror is to slightly redirect its path according to the laws of geometric optics. To an observer riding with the ray, the effect of the spherical mirror of radius R is to direct the ray toward the axis just like a thin lens. It is very easy to show that the focal length of a spherical mirror is just one-half of the radius of curvature. Thus the transmission matrix is given by

$$\mathbf{T} = \begin{bmatrix} 1 & 0 \\ -\dfrac{2}{R} & 1 \end{bmatrix} \quad (2.3.3)$$

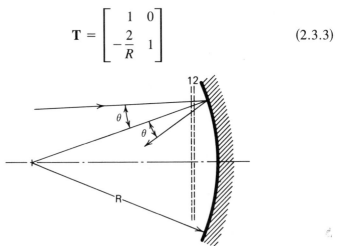

Figure 2.6 Mirror. Note that the entrance (1) and exit (2) planes are on the same side of the mirror.

2.4 APPLICATIONS OF RAY TRACING—OPTICAL CAVITIES

One of the most important uses of ray tracing is in the analysis of optical cavities such as the very simple but very important one shown in Fig. 2.7. Obviously, if a ray gets started inside the cavity, it will bounce back and forth between the two mirrors, being redirected and focused each time it hits the surface.

If the ray position stays "close" to the optical axis even after many transits

Sec. 2.4 Applications of Ray Tracing—Optical Cavities

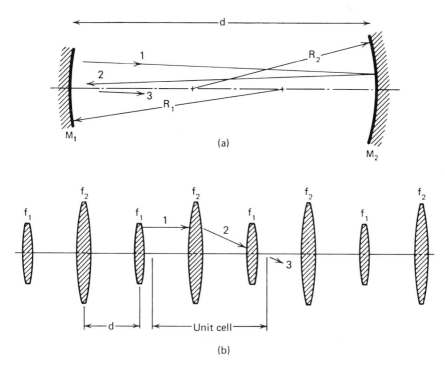

Figure 2.7 (a) Optical cavity showing a ray bouncing back and forth between the mirrors; (b) lens-waveguide equivalent to the mirror system shown in (a).

between the mirror, the system is *stable;* if the ray naturally "walks off" one of the mirror surfaces, it is *unstable:* and if the mirrors must be perfectly aligned to keep the ray near the axis, it is *conditionally stable*.

To analyze the stability of such a cavity, we construct an equivalent lens waveguide that would redirect the beam in the same manner as the mirrors. For instance, if we follow the rays shown in Fig. 2.7(a), we first encounter a length d, then a focusing element M_2, then another length d, and finally a focusing element M_1; then we start all over again. The infinite sequence of lenses shown in Fig. 2.7(b) would redirect the ray in the same manner as the cavity and is called the equivalent-lens waveguide.

Given the information of preceding sections, it is easy to compute the transmission matrix of the unit cell shown; it is just the product of two matrices of the form given by (2.3.2).

$$\mathbf{T} = \begin{bmatrix} 1 & d \\ -\dfrac{1}{f_1} & \left(1 - \dfrac{d}{f_1}\right) \end{bmatrix} \begin{bmatrix} 1 & d \\ -\dfrac{1}{f_2} & 1 - \dfrac{d}{f_2} \end{bmatrix}$$

(2.4.1)

$$\mathbf{T} = \begin{bmatrix} 1 - \dfrac{d}{f_2} & d + d\left(1 - \dfrac{d}{f_2}\right) \\ -\dfrac{1}{f_1} - \dfrac{1}{f_2}\left(1 - \dfrac{d}{f_1}\right) & \left(1 - \dfrac{d}{f_1}\right)\left(1 - \dfrac{d}{f_2}\right) - \dfrac{d}{f_1} \end{bmatrix}$$

This matrix is sufficiently messy to entice us to use the general letter symbols *ABCD* for it and thereby cover all unit cells for all cavities at one time. After we finish, we can return to (2.4.1) to examine the details about this particular cavity.

Let us find a second-order difference equation for the ray as it passes the various planes of the succeeding unit cells which corresponds to observing a ray as it makes successive round-trips through the cavity. We do this by eliminating the slope from these equations:

$$r_{s+1} = A r_s + B r'_s \quad \text{or} \quad r'_s = \frac{1}{B}(r_{s+1} - A r_s)$$

and

$$r'_{s+1} = \frac{1}{B}(r_{s+2} - A r_{s+1}) = C r_s + D r'_s$$

Substituting r'_s from the first line, we obtain

$$\frac{1}{B}(r_{s+2} - A r_{s+1}) = C r_s + \frac{D}{B}(r_{s+1} - A r_s)$$

Combining terms and remembering that $AD - BC = 1$ for a complete round-trip in *any* cavity leads to

$$r_{s+2} - 2\left(\frac{A + D}{2}\right) r_{s+1} + r_s = 0 \qquad (2.4.2)$$

We now ask the question, *Does (2.4.2) have solutions in which the magnitude of r is less than some maximum value?*

It is most important to realize why we should ask the question. If the answer to it is yes, then the position of the ray from the axis undulates as it propagates along the lens waveguide (or between the two mirrors). If the ray's position is not bounded, it will eventually become so big that it will "miss" one of the components and thus walk off the mirror. These possibilities are shown in Fig. 2.8. The solid circles represent a typical position variation of a stable situation; the open dots represent an unstable resonator. The points are connected by a dotted line to help in visualization, but it must be emphasized that the *ABCD* matrix relates the two planes, $s + 1$ to s, but tells you nothing about the trajectory of the ray between the two planes. For complicated cavities it could make some wondrous gyrations between the two planes.

Sec. 2.4   Applications of Ray Tracing—Optical Cavities

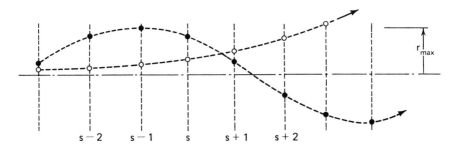

Figure 2.8 Example of the ray's position at the various planes of the lens waveguide.

However, the visualization provided by the dotted curve leads us to attempt a trigonometric solution. Let us assume that the solution to (2.4.2) has the following form:

$$r_s = r_0(e^{j\theta})^s = r_0 e^{js\theta} \tag{2.4.3}$$

Now, of course, the position r must be real; hence we anticipate that we will obtain two solutions (to the second-order difference equation) whose combination yields a real answer. Substituting this guess* into (2.4.2), we obtain

$$r_0 e^{js\theta}\left[e^{j2\theta} - 2\left(\frac{A+D}{2}\right)e^{j\theta} + 1\right] = 0 \tag{2.4.4}$$

Now r_0 is specified by the initial conditions; it cannot be set equal to zero for the general case. The exponential term is not zero; hence, the second factor must be zero. That is a quadratic equation in $\exp(j\theta)$; the two solutions are

$$e^{j\theta} = \frac{A+D}{2} \pm j\left[1 - \left(\frac{A+D}{2}\right)^2\right]^{1/2} \tag{2.4.5}$$

Note that we did obtain two solutions, and, if all quantities in (2.4.5) are real, the solutions are complex conjugates. Thus the general solution to (2.4.2) is a linear combination of the form

or
$$\left.\begin{array}{c} r_s = r_0 e^{js\theta} + r_0^* e^{-js\theta} \\ \\ r_s = r_{\max} \sin(s\theta + \alpha) \end{array}\right\} \tag{2.4.6}$$

This last form emphasizes the fact that the position must be real.

---

*Guessing is an acceptable method of solving difference *and* differential equations—provided that it is successful!

2.5 STABILITY—STABILITY DIAGRAM

Equations (2.4.5) and (2.4.6) only make sense provided that all quantities are real. If they are complex, the physical interpretation is considerably different and (2.4.6) must be changed. It is sufficient for $(A + D)/2$ to be less than ± 1 so that θ be real and thus have a bounded solution implied by (2.4.6). This is the general condition for stability:

$$-1 \leq \left(\cos \theta = \frac{A + D}{2}\right) \leq 1 \qquad (2.5.1)$$

By adding 1 to this equation and then dividing by 2, we obtain

$$0 \leq \frac{A + D + 2}{4} \leq 1 \qquad \text{(stable)} \qquad (2.5.2)$$

Before we proceed further, let us return to the specific case of the two-spherical-mirror cavity. Equation (2.4.1) gives the elements of the ray matrix for the unit cell. Thus the stability condition becomes

$$\frac{A + D + 2}{4} = \frac{1}{4}\left[1 - \frac{d}{f_2} - \frac{d}{f_1} + \left(1 - \frac{d}{f_2}\right)\left(1 - \frac{d}{f_1}\right) + 2\right]$$

$$= 1 - \frac{d}{2f_1} - \frac{d}{2f_2} + \frac{d^2}{4f_1 f_2} \qquad (2.5.3a)$$

$$= \left(1 - \frac{d}{2f_1}\right)\left(1 - \frac{d}{2f_2}\right)$$

Since $2f_1 = R_1$ and $2f_2 = R_2$, the stability condition can be written in a compact and simple form:

$$0 \leq g_1 g_2 \leq 1 \qquad (2.5.3b)$$

where

$$g_{1,2} = 1 - \frac{d}{R_{1,2}} \qquad (2.5.4)$$

Even though this is a trivial equation, it is so important that it is graphed in Fig. 2.9. One can tell by a glance at this graph whether the system is stable. If the dimensions of the cavity are such that product of the g parameters is inside the crosshatched region, the cavity is stable; if outside, it is unstable; and if on the border, it is conditionally stable, requiring perfect alignment.

One of the surprises that this diagram provides is that the confocal geometry, consisting of two identical mirrors separated by the sum of the two focal lengths, is on the borderline of stability ($g_1 = g_2 = 0$). One would think that this

Sec. 2.6   The Unstable Region   43

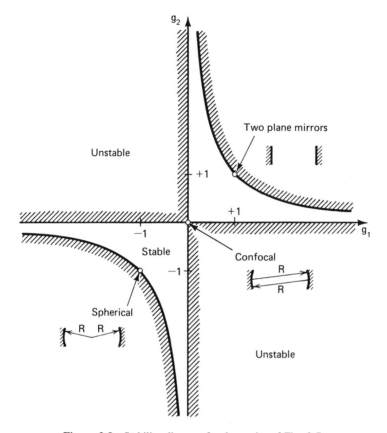

Figure 2.9 Stability diagram for the cavity of Fig. 2.7.

should be the most stable of all cavities—indeed, it was one of the first cavities analyzed "exactly" with analytic techniques. (In fact, it is the only one that yields an analytical answer. See Sec. 13.3.3.) Nevertheless, it is on that borderline, and most lasers are made to avoid that situation by increasing or decreasing d slightly.

2.6 THE UNSTABLE REGION

The unstable region is described by the condition

$$\left(\frac{A + D}{2}\right)^2 > 1 \qquad \text{(unstable)} \qquad (2.6.1)$$

and is shown by the crosshatched region in Fig. 2.9. It is a natural human tendency to avoid the "unstable" region, because of its name. However, resonators operating in the unstable region have become very useful for high-gain laser systems. In that case, the rays that walk off the mirrors constitute the output!

A very crude example illustrates why we should not let personal misgivings about these resonators interfere with scientific judgment. Suppose that we had a small pencil-like beam of light (i.e., a ray) starting out in the previous cavity and that after 10 reflections this beam missed one of the mirrors. Assume further that the medium inside the cavity increases the power to the beam by a factor of 5 each time the beam passes through the medium. Since it makes 10 passes, the emerging beam that misses a mirror is amplified 10 times and thus contains much more power than the original one by a factor of $5^{10} = 9.76 \times 10^6$. If the initial beam contained only 1 mW of power, the emerging beam has nearly 10 kW of power. Although the example is crude, it illustrates that unstable resonators have their place. We have more to say about this later.

2.7 EXAMPLE OF RAY TRACING IN A STABLE CAVITY

There are many applications of ray tracing—deciding whether a cavity is stable or unstable is just one case. To show some of the possibilities afforded by the formalism introduced here, let us consider the cavity shown in Fig. 2.10.

We imagine a situation where the incoming position and slope are specified in the manner illustrated. We take the specific case first and then generalize the result.

We assume that the incoming ray is perpendicular to the flat mirror ($R_1 = \infty$) and displaced from the axis by a specified amount $-r_0$. Thus this ray is directed toward the focal point of M_2, reflects from the flat mirror, and heads toward the spherical mirror along a radius. Once this ray reaches M_2, it starts a retrace of its initial path and forevermore stays along paths indicated. This is an example of a repetitive ray path.

Let us now apply the formalism developed here to this case to establish

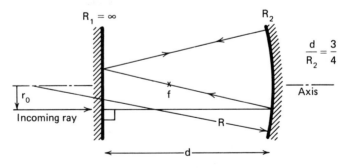

Figure 2.10 Ray tracing in a stable cavity.

Sec. 2.7 Example of Ray Tracing in a Stable Cavity

general conditions for discrete beam paths in a cavity and to apply our previous analysis. First, note that this cavity is stable because $g_1 = 1 - d/R_1 = 1$, and $g_2 = 1 - d/R_2 = \frac{1}{4}$; hence, $g_1 g_2 = \frac{1}{4} < 1$. Note, too, that the beam stays within a maximum displacement of $2r_0$ from the axis—consistent with the ideas of stability developed earlier. Let us now consider the lens-waveguide equivalent to this cavity, shown in Fig. 2.11. The transmission matrix for the first unit cell is given by

$$\mathbf{T}_1 = \begin{bmatrix} 1 - \dfrac{d}{f} & d + d\left(1 - \dfrac{d}{f}\right) \\ -\dfrac{1}{f} & 1 - \dfrac{d}{f} \end{bmatrix} \tag{2.7.1}$$

From (2.4.5), we have

$$e^{j\theta} = \frac{A+D}{2} + j\left[1 - \left(\frac{A+D}{2}\right)^2\right]^{1/2} = \cos\theta + j\sin\theta$$

or

$$\cos\theta = \frac{A+D}{2} = 1 - \frac{d}{f}$$

for this case. Since $f = R/2$, we find that $\cos\theta = 1 - \frac{3}{2} = -\frac{1}{2}$ or $\theta = 2\pi/3$ (120°).

Let us also try to apply the second form of the solution to the difference equation and follow the real ray position as it leaves (and enters) unit cell 1. For this case, (2.4.6) becomes

$$r = +r_0 \sin\left(s\frac{2\pi}{3} - \frac{\pi}{2}\right) \tag{2.7.2}$$

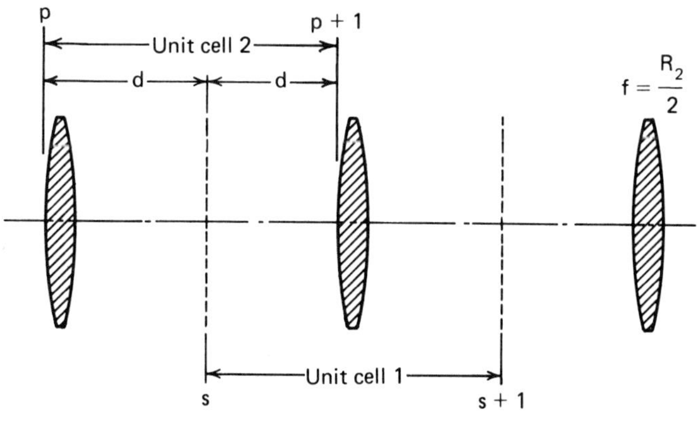

Figure 2.11 Lens-waveguide equivalent to the cavity of Fig. 2.10.

where the phase angle, $-\pi/2$, expresses the fact that the ray is $-r_0$ below the axis at the start of the first round-trip (call this plane $s = 0$). At the end of this round-trip (and one unit cell later), $s = 1$, and the ray's position is given by

$$r = -r_0 \sin\left(\frac{2\pi}{3} + \frac{\pi}{2}\right) = -r_0 \sin 210° = +\frac{r_0}{2}$$

After another unit cell and one more round-trip, $s = 2$ and (2.7.2) predicts that

$$r = -r_0 \sin\left(\frac{4\pi}{3} + \frac{\pi}{2}\right) = -r_0 \sin 330° = +\frac{r_0}{2}$$

and then finally, after three round-trips, $s = 3$, we have

$$r = -r_0 \sin\left(\frac{6\pi}{3} + \frac{\pi}{2}\right) = -r_0 \sin 90° = -r_0$$

and we start all over again.

There appears to be something wrong here. According to (2.7.2), the maximum displacement is r_0, whereas the simple walk through the cavity at the beginning of Sec. 2.6 indicated that the maximum displacement was $2r_0$. Why the difference?

There is nothing wrong; indeed, unit cell 1 was chosen to illustrate this particular point. The *ABCD* matrix for a unit cell relates the position and slope of the ray as it enters and leaves the reference planes. It tells us nothing about what happens in between. If we need to know about the position between the reference planes of the unit cell (say at the spherical mirror), we must apply the individual ray matrix to obtain that information. For the choice of reference plane indicated for unit cell 1, the maximum displacement at those terminals is r_0.

If we had chosen the other unit cell, say cell 2, of Fig. 2.11, the ray matrix is different but the angle θ is the same.

$$\mathbf{T}_2 = \begin{bmatrix} 1 & 2d \\ 0 & 1 \end{bmatrix} \begin{bmatrix} 1 & 0 \\ -\frac{1}{f} & 1 \end{bmatrix} = \begin{bmatrix} 1 - \frac{2d}{f} & 2d \\ -\frac{1}{f} & 1 \end{bmatrix}$$

$$\cos\theta = \frac{A+D}{2} = \frac{1}{2}\left(2 - \frac{2d}{f}\right) = 1 - \frac{d}{f} = 1 - \frac{2d}{R} = -\frac{1}{2}, \quad (2.7.3)$$

$$\theta = \frac{2\pi}{3}, \; 120°$$

Thus the displacement can be related to the *p*th reference plane of this unit cell by

$$r = 2r_0 \sin\left(p\frac{2\pi}{3} - \frac{\pi}{6}\right)$$

At the start of this first unit cell, $p = 0$, and the ray is impinging on the spherical mirror at $r = r_0$ below the axis. After one round-trip, $p = 1$, and the ray's position is $2r_0 \sin(120° - 30°) = 2r_0$. At the end of the second round-trip, $p = 2$, and we find that

$$r = 2r_0 \sin\left(2 \times \frac{2\pi}{3} - \frac{\pi}{6}\right) = 2r_0 \sin 210° = -r_0$$

Note that this last choice of the unit cell tells us nothing about the trials and tribulations of the ray at the flat mirror.

2.8 REPETITIVE RAY PATHS

The preceding example illustrates a case whereby the beam retraces its path after a discrete number of round-trips. In the particular case shown, the ray went three complete round-trips and then was in the same position, with the same slope, as it was when it started. We can generalize this result for any cavity.

Equation (2.4.6) indicates that the ray position could be found from

$$r_s = r_{max} \sin(s\theta + \alpha)$$

where (2.4.6)

$$\theta = \cos^{-1}\left(\frac{A + D}{2}\right)$$

If s is increased by m units, corresponding to m round-trips, the ray returns to its original position after these m round-trips, when θ satisfies

$$m\theta = 2n\pi \quad \text{and} \quad n < \frac{m}{2} \qquad (2.8.1)$$

when n and m are integers. For the case analyzed in the preceding section, $n = 1$ and $m = 3$. Note that inequality guarantees that θ is always less than π, in accordance with the principal value of $\cos \theta = (A + D)/2$.

2.9 INITIAL CONDITIONS—STABLE CAVITIES

The example given in Sec. 2.7 was transparent enough to make the evaluation of the constants in (2.4.6) easy—obtained almost by inspection. Let us now attempt to formalize the procedure for a more complex cavity. Let us suppose that we have unfolded a complex cavity and have found the transmission matrix for a unit cell, and that the ray's position, a, and slope, m, are known at a given reference plane (call it $s = 0$).

If a cavity is stable, we can find the angle θ from the transmission matrix of the unit cell:

$$\theta = \cos^{-1}\left(\frac{A + D}{2}\right) \tag{2.9.1}$$

The initial conditions $r_0 = a$ and $r_0' = m$ yield the first equation involving r_{max} and α:

$$a = |r_{max}| \sin(0 \cdot \theta + \alpha) = |r_{max}| \sin \alpha \tag{2.9.2}$$

The ray matrix also provides us with a second equation for the ray position after traversing one unit cell (or round-trip):

$$r_1 = Ar_0 + Br_0' = |r_{max}| \sin(1 \cdot \theta + \alpha)$$
$$Aa + Bm = |r_{max}| \sin(\theta + \alpha) \tag{2.9.3}$$

Expanding the $\sin(\theta + \alpha)$ and recognizing that $|r_{max}| \sin \alpha = a$ and $\cos \theta = (A + D)/2$ yields

$$|r_{max}| \cos \alpha = \frac{1}{\sin \theta}\left[a\left(\frac{A - D}{2}\right) + Bm\right] \tag{2.9.4}$$

Thus the angle α is given by

$$\alpha = \tan^{-1}\left[\frac{a\left(1 - \left(\frac{A + D}{2}\right)^2\right)^{1/2}}{a\left(\frac{A - D}{2}\right) + Bm}\right] \tag{2.9.5}$$

2.10 INITIAL CONDITIONS—UNSTABLE CAVITIES

If the cavity is unstable, the trigonometric solution (2.4.6) is confusing, misleading, and in any case more trouble than it is worth. It is best to return to the difference equation and reinterpret the conclusions that followed.

We can still assume the same functional form for a solution to (2.4.2), although we give it a slightly different "name." Let

$$r_s = r_0(F)^s \tag{2.10.1}$$

Then (2.4.4) becomes

$$rF^s\left[F^2 - 2\left(\frac{A + D}{2}\right)F + 1\right] = 0 \tag{2.10.2}$$

Thus there are two solutions (again), but now they are both real:

$$F_1 = \frac{A+D}{2} + \left[\left(\frac{A+D}{2}\right)^2 - 1\right]^{1/2} \quad (2.10.3a)$$

$$F_2 = \frac{A+D}{2} - \left[\left(\frac{A+D}{2}\right)^2 - 1\right]^{1/2} \quad (2.10.3b)$$

The general solution becomes

$$r_s = r_a(F_1)^s + r_b(F_2)^s \quad (2.10.4)$$

For unstable cavities $(A+D)/2$ is either greater than 1 or less than -1; in either case, one of the solutions has a magnitude greater than 1. Thus after a few steps through the unit cell, the ray's position is dominated by the larger quantity raised to the sth power:

$$r_s \sim r_0(F_>)^s$$

and is farther and farther away from the axis of the system.

The constants, r_a and r_b, are evaluated in the same manner as in Sec. 2.9. At the plane where the position, a, and slope, m, are specified, $s = 0$ and (2.10.4) yields

$$a = r_a + r_b \quad (2.10.5)$$

After one round-trip, we have

$$r_1 = aA + Bm = r_a F_1 + r_b F_2 \quad (2.10.6)$$

Solving for r_a and r_b yields

$$r_b = \frac{1}{F_1 - F_2}[a(F_1 - A) - Bm] \quad (2.10.7)$$

$$r_a = \frac{1}{F_2 - F_1}[a(F_2 - A) - Bm] \quad (2.10.8)$$

2.11 ASTIGMATISM*

When a material body is placed in the path of a ray and is tilted in the manner shown in Fig. 2.12, one must account for the change in the optical path in two orthogonal directions. For instance, windows placed at the Brewster angle are very common in gas lasers. The angles ϕ_x and ϕ_y can be considered as small and thus paraxial to the optic axis. However, if the dielectric material is a Brewster's angle window on a gas laser tube, then $\theta = \tan^{-1} n$, and for quartz with

---

*Some of the consequences of astigmatism for lasers are discussed in Ref. 8.

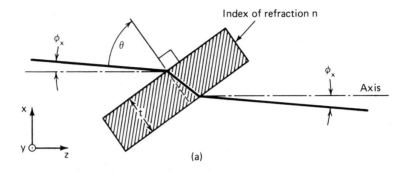

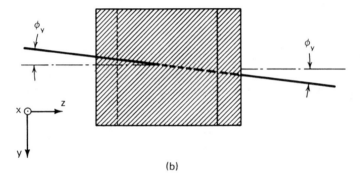

Figure 2.12 Astigmatism of a window. (a) Side view. (b) Top view.

$n = 1.45845$,*$\theta = 55.56°$. Obviously, the angle $\theta - \phi_x$ is not small, and one must account for bending and resulting displacement of the beam in the xz plane as shown in Fig. 2.12(a). The paraxial approximation is quite adequate for rays in the yz plane. It is left as a problem to show that optical paths traversed by the two rays through a Brewster's angle window are given by

$$d_y = t \frac{(n^2 + 1)^{1/2}}{n^2} \tag{2.11.1a}$$

$$d_x = t \frac{(n^2 + 1)^{1/2}}{n^4} \tag{2.11.1b}$$

when $\tan \theta_B = n$. The fact that these distances are not equal gives rise to astigmatism.

A curved mirror is often used in ring laser cavities in the manner shown in Fig. 2.13. It is a tiring exercise in geometry to show that the mirror focuses

*From the *Handbook of Chemistry and Physics* (Chemical Rubber Co., Publisher) at $\lambda = 589.29$ nm.

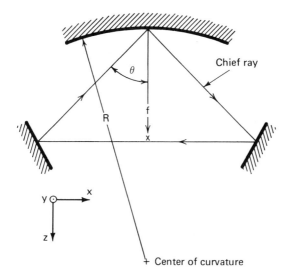

Figure 2.13 Astigmatic laser cavity.

parallel rays in the two planes at different locations, leading to different effective focal lengths in the xy and xz planes. Thus, for rays that are paraxial to the "chief ray path," we use an effective focal length given by

$$f_x = f \cos \theta \tag{2.11.2a}$$

$$f_y = \frac{f}{\cos \theta} \tag{2.11.2b}$$

Astigmatism leads to elliptical beams in ring lasers and plays a critical role in dye-laser cavities.

2.12 CONTINUOUS LENS–LIKE MEDIA

Consider the case in which a ray is propagating in a medium in which the index of refraction is nonuniform in the transverse direction. We assume that the medium is not a function of the axial (i.e., z) coordinate.

Two contrasting examples of this type of a medium are shown in Figs. 2.14 and 2.15. The first case is that of an optical fiber used as a waveguide for communication purposes. The manufacturing process involves the drawing of different glasses with different dopants and different coatings through a die to attain a radial variation in the index of refraction.

The variation of the index of refraction, $n(r) = [\epsilon(r)]^{1/2}$, is shown with typical numbers to emphasize that the change is usually quite small. As we shall see, this small change is sufficient to guide the electromagnetic energy efficiently over long distances.

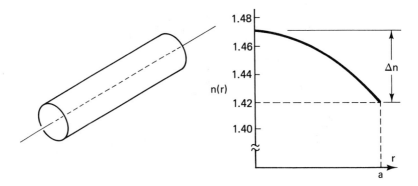

Figure 2.14 Optical fiber with an index of refraction depending on r.

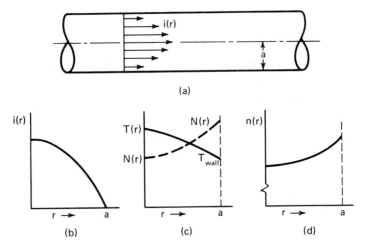

Figure 2.15 Variation of the index of refraction of a gas due to local heating by the current density distributed as in (b).

The second case is that of a gas discharge used to pump some of the common low-power lasers such as helium:neon (632.8 nm), argon ion (488.8 nm, 514.5 nm), or CO_2 (10.6 μm) systems. It will be shown much later that the current is *not* uniform throughout the bore of the tube shown but rather is distributed nonuniformly in the manner sketched. Since the current is maximum at the center, the temperature of the gas is highest there also and decreases to the value of the walls. To a first approximation, the pressure is constant across the bore of the tube, and since the temperature is a function of radius, the density of neutral gas atoms must be the inverse function

$$N(r)kT(r) = \text{constant} \tag{2.12.1}$$

Sec. 2.12 Continuous Lens—Like Media

Inasmuch as the index of refraction of the gas is directly related to the density by

$$n - 1 = 2\pi\alpha N(r) \tag{2.12.2}$$

where $2\pi\alpha$ is the specific refractivity of the gas, it follows that the index of refraction is also a function of r, as shown in Fig. 2.15(d). In both cases the index of refraction is a function of r.

2.12.1 Propagation of a Ray in an Inhomogeneous Medium

We again assume that the paraxial ray is propagating primarily in the z direction, as shown in Fig. 2.16. We shall apply Snell's law to the interface shown on the diagram:

$$\frac{\omega}{c}n(r) \cos \theta_1 = \frac{\omega}{c}n(r + \Delta r) \cos (\theta_1 + \Delta\theta) \tag{2.12.3}$$

Using the first two terms of a Taylor series expansion for $n(r)$ and the expansion of $\cos (\theta_1 + \Delta\theta)$ yields

$$n(r) \cos \theta_1 = \left[n(r) + \frac{\partial n}{\partial r}\Delta r \right] (\cos \theta_1 \cos \Delta\theta - \sin \theta_1 \sin \Delta\theta)$$

or

$$\frac{\partial n}{\partial r} = n(r) \tan \theta_1 \left(\frac{\Delta\theta}{\Delta r} \right) \tag{2.12.4}$$

Within the context of the paraxial ray approximation, $\tan \theta_1 = \Delta r/\Delta z$ and it follows that

$$\tan \theta_1 \cdot \frac{\Delta\theta}{\Delta r} = \frac{1}{n} \cdot \frac{\partial n}{\partial r} = \frac{\Delta r}{\Delta z} \cdot \frac{\Delta\theta}{\Delta r} = \frac{\Delta\theta}{\Delta z} \longrightarrow \frac{\partial^2 r}{\partial z^2} \tag{2.12.5}$$

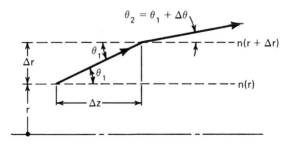

Figure 2.16 Propagation of a beam in an inhomogeneous medium.

The combination of (2.12.4) and (2.12.5) yields the basic equation for the propagation of a ray in an inhomogeneous dielectric medium:

$$\frac{d^2r}{dz^2} = \frac{1}{n(r)} \frac{dn(r)}{dr} \qquad (2.12.6)$$

2.12.2 Ray Matrix for a Continuous Lens

A simple thin positive lens bulges at $r = 0$, indicating more mass and a longer optical path there than at the edges. Similarly, a positive continuous lens has an index of refraction that decreases with r. For a cylindrically symmetric positive lens, the first two terms of Taylor series expansion for $n(r)$ would have the following format:

$$n(r) = n_0 \left[1 - \frac{1}{2l^2} r^2 + O(r^4) \cdots \right] \qquad (2.12.7)$$

We assume that the second term, $r^2/2l^2$, is much less than 1 for all values of the radius of concern to us. It should also be noted that the parameter l in (2.12.7) is simply a scale factor which indicates how fast n varies with r. For instance, if $a = 50$ μm for the radius of the fiber in Fig. 2.14 and $n(a) = 1.42$ with $n_0 = 1.47$ then $\Delta n = 0.05$. The relationship between Δn and l is given by simple arithmetic:

$$\Delta n = n_0 - n(a) = \frac{n_0 a^2}{2l^2} \text{ or } l = a \cdot \sqrt{\frac{n_0}{2\Delta n}}$$

For the numbers quoted above, $l = 191.7$ μm, which is large compared to the radius of the fiber. Most fibers have a change in index much less than the value chosen here, and hence l is even larger.

Because of the inequality we can use the first term of (2.12.7) for n in (2.12.6) and the derivative of the second for dn/dr:

$$\frac{d^2r}{dz^2} = -\frac{r}{l^2} \qquad (2.12.8)$$

The solution is straightforward. Assume that $z = 0$ is the input plane to this optical component where the position r_1 and slope r_1' are known:

$$r(z) = r_1 \left[\cos\left(\frac{z}{l}\right) \right] + r_1' \left[l \sin\left(\frac{z}{l}\right) \right] \qquad (2.12.9)$$

Then, by differentiation, we obtain the slope at any position z:

$$r'(z) = r_1 \left[-\frac{1}{l} \sin\left(\frac{z}{l}\right) \right] + r_1' \left[\cos\left(\frac{z}{l}\right) \right] \qquad (2.12.10)$$

Sec. 2.12   Continuous Lens—Like Media

Thus the ray matrix for a length $z = d$ is

$$\mathbf{T} = \begin{bmatrix} \cos\left(\dfrac{d}{l}\right) & l \sin\left(\dfrac{d}{l}\right) \\ -\dfrac{1}{l}\sin\left(\dfrac{d}{l}\right) & \cos\left(\dfrac{d}{l}\right) \end{bmatrix} \quad (2.12.11)$$

for

$$n(r) = n_0\left(1 - \dfrac{r^2}{2l^2}\right)$$

Note that if $l \to \infty$, the medium is uniform and (2.12.11) reproduces the ray matrix given by (2.2.2).

If the index of refraction increases with r (as in the case of a gas discharge), the geometric fudge factor l must take on imaginary values to predict $n(r)$ from (2.12.7). If we let $l = jL$ and use the fact that $\cos j\theta = \cosh \theta$ and $\sin j\theta = j \sinh \theta$, we obtain the ray matrix for a negative lens directly from (2.12.11):

$$\mathbf{T} = \begin{bmatrix} \cosh\left(\dfrac{d}{L}\right) & L \sinh\left(\dfrac{d}{L}\right) \\ \dfrac{1}{L}\sinh\left(\dfrac{d}{L}\right) & \cosh\left(\dfrac{d}{L}\right) \end{bmatrix} \quad (2.12.12)$$

for

$$n(r) = n_0\left(1 + \dfrac{r^2}{2L^2}\right) \quad (2.12.13)$$

You are cautioned that the matrices (2.12.11) and (2.12.12) describe the ray propagation within the medium. The matrices do *not* describe the changes that might occur rather abruptly at the entrance and exit planes. For instance, if n_0 is significantly different from 1, as indicated by Fig. 2.14 for a glass fiber, the slope of a ray changes in passing from the air into the fiber (and conversely).

Let us examine (2.12.11) in a very imprecise manner so as to anticipate a result of a later section. The C term in this matrix is the negative of the focal length of this lens:

$$C = -\dfrac{1}{f} = -\dfrac{1}{l}\sin\left(\dfrac{d}{l}\right)$$

Thus the focal length is positive (i.e., converging) provided that the argument of the sine function is less that π radians. If d is long enough—as in a fiber-optic communication system—the focal length is alternately positive and negative, corresponding to a converging and diverging system. Thus we could anticipate the "beam" undulating as shown in Fig. 2.17 as it propagates down the fiber. Going one step further, anticipate a situation where the natural divergence of a small beam is continuously counteracted by convergence of the medium. Such fibers exist and are used in fiber-optic communication links.

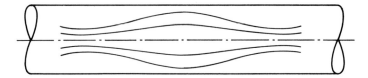

Figure 2.17 Propagation of a beam within a fiber.

2.13 WAVE TRANSFORMATION BY A LENS

Let us now broaden our viewpoint and consider the effect of a lens on a limited extent uniform plane wave. We can consider each of the horizontal lines in Fig. 2.18 to be a ray, and the bundle of rays to be a beam. Furthermore, we assume that the phase *surface* of the incident beam is more-or-less planar, as indicated by the vertical dashed lines.

The question naturally arises: What is the shape of the phase front after passing through the lens? Obviously, the phases of A and B are equal at the left of the lens *and* at the focal point at the right. (This is sometimes referred to as the Fermat principle—the light always takes the path that makes the transit time a minimum.) But if the waves are in phase at f, then A' and B' are not in phase (for equal z coordinates). The ray at B' has farther to travel; hence, its phase must be ahead of that of A'. This extra distance is

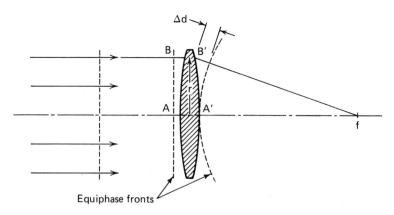

Figure 2.18 Beam transformation by a lens.

$$\Delta d = (r^2 + f^2)^{1/2} - f \doteq f\left(1 + \frac{1}{2}\frac{r^2}{f^2}\right) - f$$

or (2.13.1)

$$\Delta d \doteq \frac{1}{2}\frac{r^2}{f}$$

Thus we have an approximate expression for the field just to the right of the lens:

$$E = E_0 \exp\left(+j\frac{kr^2}{2f}\right) \qquad (2.13.2)$$

A moment's consideration indicates that the wave front is now curved (toward the focal point). The more exact analysis of Chapter 3 shows that this is correct.

PROBLEMS

2.1. Derive the ray matrix for a ray entering a spherical dielectric interface.

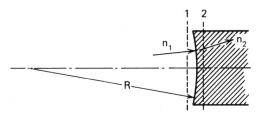

2.2. Derive the ray matrix for the plane dielectric interface.

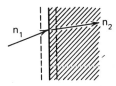

2.3. Derive the ray matrix for the plane dielectric slab of thickness d.

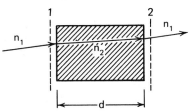

2.4. Combine the results of problems 2.1 and 2.2 to derive the ray matrix for the negative lens. (Assume that $R \gg d$.)

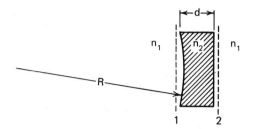

2.5. The GRIN lens shown in the diagram below consists of a *graded index* fiber with $n(r) = n_0[1 - (r^2/2l^2)]$ of length d.
  (a) Find the *ABCD* matrix for this lens. Do not ignore the air-index boundary (use the ray matrix found in Problem 2.2 for that operation). Keep d arbitrary.
  (b) Evaluate for $d = \pi l/2$.
  (c) Show that this lens is equivalent to the system shown at the right.
  (d) If $n_0 = 1.53$, $n(a) = 1.525$, and $a = 2$ mm, find f.

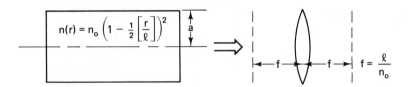

2.6. Find the *ABCD* matrix for the lens combination shown below.

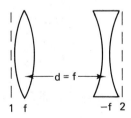

2.7. Consider the ring laser cavity shown in the accompanying diagram.
  (a) Show an equivalent-lens waveguide for this cavity and identify a unit cell starting just after the lens and proceeding counterclockwise around the triangle.

(b) What is the transmission matrix for this unit cell? (Demonstrate that you have the component matrices in proper order.)
(c) What are the values of d/f that make this a stable cavity?

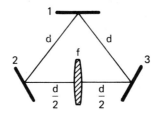

2.8. Consider the cavity shown in the accompanying diagram.
  (a) Construct an equivalent-lens waveguide.
  (b) Indicate a unit cell starting at a flat mirror.
  (c) Find the ray matrix for the unit cell of (b).
  (d) Discuss the stability of this cavity by constructing a diagram similar to Fig. 2.9.

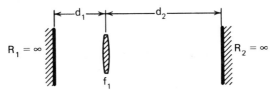

2.9. Sketch the ray pattern shown in Fig. 2.10 if $d/R = 1/2$. Assume the initial ray is a distance r_0 below the axis.
  (a) What is the maximum excursion of the path from the axis?
  (b) Fig. 2.10 shows three round-trips to obtain a repetition. How many are required here?

2.10. Consider the obviously unstable cavity shown in the accompanying diagram. Suppose that a ray starts out at the flat mirror with a zero slope and at a position that is one tenth of the radius of the mirror.
  (a) Show the equivalent-lens waveguide for the cavity. Identify two unit cells, one starting at the flat mirror and the second starting just before the spherical mirror.
  (b) Since the cavity is unstable, one should return to the second-order difference equation to obtain a solution of the form (use unit cell 1)

$$r_s = r_a(F_+)^s - r_b(F_-)^s$$

  (1) What are the values of F_+ and F_-?
  (2) What are the values of r_a and r_b?
  (c) How many passes through the cavity does the ray make before it misses the flat mirror?

(d) Where does the ray leave the cavity, at the flat mirror or at the curved one? How many passes does it make? [To answer this, you will have to repeat (b) and (c) for the second unit cell and then decide.]

(e) If the beam associated with this ray started with 1 μW of power and the power gain per pass $G = 5$, what is the power leaving the cavity? (Assume that the mirrors are perfectly reflecting.)

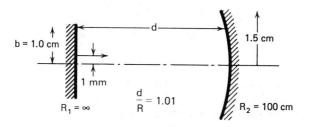

2.11. In the analysis of the cavity shown in Fig. 2.13, one must keep track of displacements from the chief ray in the plane of and perpendicular to the paper. If the chief ray forms an equilateral triangle with sides d and the spherical mirror has a radius of curvature R, what is the limit of d/R to ensure stability?

2.12. Consider the accompanying diagram of an optical cavity that is to be used for a gas discharge laser. The empty cavity is on the borderline of stability. As discussed in Sec. 2.12, the nonuniform gas heating by the discharge current creates a *negative* gas lens. Does this fact create an unstable situation? Assume that the parameter $d/L \ll 1$ and estimate the amount by which the stability margin is improved or degraded by finding the first *nonvanishing* term of a Taylor series expansion of $(A + D + 2)/4$.

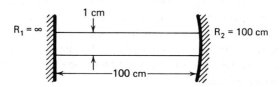

2.13. **(a)** Suppose a cylindrical gas discharge tube of Problem 2.12 was filled with helium to a pressure of 1 Torr at 23°C (i.e., 1/760 of an atmosphere). It was then sealed and detached from the vacuum system. Suppose that the current through the tube heated the gas such that the temperature on axis were 300°C decreasing in a parabolic fashion to 50°C at the wall. The pressure increases because of the higher temperature but must be constant independent of radius, that is, $p = N(r)kT(r) \neq f(r)$. Estimate the value of the parameter d/L. At standard temperature and pressure (STP), $n(\text{helium}) = 1.000036$. (Remember that the total number of atoms cannot change in a sealed-off tube.)

    (b) Repeat (a) if the gas were CO_2 and the initial fill was to 100 torr. At STP $n(CO_2) = 1.000449$.

2.14. Typical parameters of a glass fiber are radius = 20 μm, $n_0 = 1.5$, and $\Delta n = 8 \times 10^{-3}$. How many times would a ray cross the axis of a fiber 1 km long?

2.15. In the derivation of (2.4.2), use was made of $AD - BC = 1$. However, here is a restriction on this relationship. Is it possible for $AD - BC \neq 1$ for a unit cell of an optical cavity?

2.16. Read the paper by H. Kogelnik and T. Li, "Laser Beams and Resonators," Appl. Opt. 9, 1550-1566, Oct. 1966. Verify ray matrices 4, 5, and 6 listed in Table I of that paper. (CAUTION: System 5 uses a different notation from that used in Sec. 2.11.)

REFERENCES AND SUGGESTED READINGS

There are many excellent references to matrix methods in optical design. However, there is a sequence of classic papers which bear directly on the subject matter of this and the following 4 chapters. It is recommended that the student read the following papers:

1. J. R. Pierce, "Modes in Sequences of Lenses," Proc. Natl. Acad. Sci. *47*, 1808–1813, Nov. 1961.
2. A. G. Fox and T. Li, "Resonant Modes in a Maser Interferometer," Bell Syst. Tech. J. *40*, 453–488, Mar. 1961.
3. G. D. Boyd and J. P. Gordon, "Confocal Multimode Resonator for Millimeter through Optical Wavelength Masers," Bell Syst. Tech. J. *40*, 489–508, Mar. 1961.
4. G. D. Boyd and H. Kogelnik, "Generalized Confocal Resonator Theory," Bell Syst. Tech. J. *41*, 1347–1369, July 1962.
5. H. Kogelnik, "Imaging of Optical Modes—Resonators with Internal Lenses," Bell Syst. Tech. J. *44*, 455–494, Mar. 1963.
6. H. Kogelnik and T. Li, "Laser Beams and Resonators," appl. Opt. 5, 1550–1556, Oct. 1966. This article contains an additional 44 references.
7. F. A. Jenkins and H. E. White, *Fundamentals of Optics*, 3rd ed. (New York: McGraw-Hill Book Company, 1957).
8. H. W. Kogelnik, E. P. Ippen, A. Dienes, and C. V. Shank, "Astigmatically Compensated Cavities for CW Dye Lasers," IEEE J. Quant. Electron. *QE-8*, 373, Mar. 1972.
9. A. Yariv, *Introduction to Optical Electronics* (New York: Holt, Rinehart & Winston, 1971) Chap. 2.
10. A. Yariv, *Quantum Electronics* (New York: John Wiley & Sons, Inc., 1975), Chap. 6.
11. A. Siegman, *Lasers* (Mill Valley, Calif.: University Science Books, 1986), Chap. 15.

3

Gaussian Beams

3.1 INTRODUCTION

Although the concept of ray tracing is intuitively satisfying (and simple), it is important to realize its limitations. A ray is a path; it is not a field; it does not have an amplitude, a phase, or a spatial extent. Yet this information is vital to all phases of electronics.

Consequently, it is necessary for us to obtain a more complete wave description of the beams produced by the laser. This will be accomplished by an approximate analytic solution to the wave equation yielding fields which are completely specified at all points in space. After a bit of mathematics, it will be shown that there exist a complete set of "modes" which are naturally generated by stable laser cavities.

It is most important to continually question the significance of the mathematical results so that one can appreciate *why* these modes are generated by lasers. The reason, in the simplest terms, is that these modes *match* the mirror surfaces—they fit!

3.2 PRELIMINARY IDEAS—TEM WAVES

We should first realize that most optical beams propagating in free space are almost pure TEM (transverse electric and magnetic). In other words, the field components lie in the plane perpendicular to the direction of propagation. This is easy to show by a simple application of the divergence equation

$$\nabla \cdot \mathbf{E} = 0 \quad \text{(for free space)} \qquad (3.2.1a)$$

Sec. 3.2   Preliminary Ideas—TEM Waves

Obviously, this same comment applies to **H** since $\nabla \cdot \mathbf{H} = 0$. The divergence operation can be broken up into transverse divergence (∇_t) of the transverse field (E_t) and the longitudinal derivative of the z component of the field:

$$\nabla \cdot \mathbf{E} = \nabla_t \cdot \mathbf{E}_t + \frac{\partial}{\partial z} E_z = 0 \qquad (3.2.1b)$$

Here some physical insight can save hours of arithmetic. The wave is obviously propagating with a velocity of approximately c. Hence, the *major* variation of the field with z is a term of the approximate form $\exp(-jkz)$ with $k \sim \omega n/c = 2\pi n/\lambda_0$. At optical frequencies of $>10^{13}$ Hz, λ_0 is quite small, and hence k is a large number. Thus

$$\frac{\partial E_z}{\partial z} \sim -j \frac{2\pi n}{\lambda_0} E_z \qquad (3.2.2)$$

Furthermore, a beam has a finite diameter D which is typically 1 cm or so. Thus we can *approximate* the transverse divergence by

$$\nabla_t \cdot \mathbf{E}_t \sim \frac{E_t}{D} \qquad (3.2.3)$$

Using (3.2.2) and (3.2.3) in (3.2.1) yields a comparison of the magnitude of the z component of the field in terms of the transverse components:

$$|E_z| \sim \frac{\lambda_0}{2\pi n D} |E_t| \qquad (3.2.4)$$

This ratio is exceedingly small for visible wavelengths (0.5 μm) and reasonable beam diameters (~ 1 cm). But a word of caution is in order here. If one is dealing with a focal spot of lens, D can be quite small, making some of these conclusions questionable. Nevertheless, the assumption of a TEM wave for optical beams is usually quite good.

Let us now return to the point made earlier—that $k = 2\pi n/\lambda_0$ is a very large number. This means that the complete description of the fields must involve functions that change very rapidly with z. If one were to attempt to solve the wave equation on a digital computer, a numerical nightmare would result because of the rapid variation with z.

Thus it behooves us to eliminate this rapid variation with z from our equation. After all, we know that the fields propagate with a velocity of roughly c/n. Thus we look for solutions of the form

$$E(x, y, z) = E_0 \psi(x, y, z) e^{-jkz} \qquad (3.2.5)$$

It is most important to realize the physical significance of (3.2.5). The factor E_0 is the customary amplitude factor expressing the intensity of the wave; the factor $\exp(-jkz)$ expresses our feelings that the wave should propagate more or less as a uniform plane wave; and the factor ψ measures how the beam deviates from a uniform plane wave.

We substitute (3.2.5) into the time-independent wave equation to derive another equation involving ψ alone. In other words, we say to ourselves: "We know about uniform plane waves; what about the deviation?"

$$\nabla^2 E + \frac{\omega^2}{c^2} n^2 E = 0$$

or

$$\nabla_t^2 E + \frac{\partial^2 E}{\partial z^2} + \frac{\omega^2}{c^2} n^2 E = 0 \tag{3.2.6}$$

Let

$$E = E_0 \psi(x, y, z) \exp(-jkz)$$

where

$$k = \frac{\omega}{c} n$$

In (3.2.6) we presume a uniform index of refraction; if n is a function of r, the derivation must be repeated.

The following derivatives are necessary:

$$\nabla_t^2 E = (\nabla_t^2 \psi) \exp(-jkz)$$

$$\frac{\partial E}{\partial z} = \left(-jk\psi + \frac{\partial \psi}{\partial z}\right) \exp(-jkz)$$

$$\frac{\partial^2 E}{\partial z^2} = \left(-k^2 \psi - j2k\frac{\partial \psi}{\partial z} + \frac{\partial^2 \psi}{\partial z^2}\right) \exp(-jkz)$$

When these derivatives are substituted into (3.2.6), the term with k^2 cancels that with $(\omega n/c)^2$, and, of course, the common factor $\exp(-jkz)$ cancels out of all terms:

$$\nabla_t^2 \psi - j2k\frac{\partial \psi}{\partial z} + \frac{\partial^2 \psi}{\partial z^2} = 0 \tag{3.2.7}$$

Equation (3.2.7) is exact; it is just a different representation of the wave equation. But now the first approximation is used; the second derivative term is neglected, with the following justification.

One can anticipate that the "beam" parameters contained in ψ do change with z leading to nonzero values for both $d\psi/dz$ and $d^2\psi/dz^2$. But the first derivative is multiplied by this very large number, k, whereas unity is the coefficient for the last term. Thus it is neglected, to yield

$$\nabla_t^2 \psi - j2k\frac{\partial \psi}{\partial z} = 0 \qquad (3.2.8)$$

Equation (3.2.8) is the central equation for Gaussian beams. (Incidentally, it has the same form as the time-dependent Shrödinger equation.)

3.3 LOWEST-ORDER TEM$_{0,0}$ MODE

To keep the arithmetic to a minimum, we look for a solution that is cylindrically symmetric. Equation (3.2.8) becomes

$$\frac{1}{r}\frac{\partial}{\partial r}\left(r\frac{\partial \psi}{\partial r}\right) - j2k\frac{\partial \psi}{\partial z} = 0 \qquad (3.3.1)$$

As is typical with the solution to differential equations, we "guess" at the functional form of the solution and then force the unknown coefficients or functions to fit the equation. We choose

$$\psi_0 = \exp\left\{-j\left[P(z) + \frac{kr^2}{2q(z)}\right]\right\} \qquad (3.3.2)$$

where the subscript 0 indicates the fundamental lowest-order TEM$_{0,0}$ mode (more about the meaning of the subscripts in Sec. 3.5). Our goal is to find ψ_0 by reducing the partial differential equation (3.3.1) to ordinary differential equations for the unknown functions $P(z)$ and $q(z)$. Thus the following derivatives are necessary:

$$-j2k\frac{\partial \psi_0}{\partial z} = \left[-2kP'(z) + \frac{k^2r^2q'(z)}{q^2(z)}\right]\psi_0$$

$$\frac{\partial \psi_0}{\partial r} = -j\frac{kr}{q(z)}\psi_0 \qquad \therefore r\frac{\partial \psi_0}{\partial r} = -j\frac{kr^2}{q(z)}\psi_0$$

$$\frac{\partial}{\partial r}\left(r\frac{\partial \psi_0}{\partial r}\right) = -j\frac{kr^2}{q(z)}\left[-j\frac{kr}{q(z)}\right]\psi_0 - j2\frac{kr}{q(z)}\psi_0$$

$$\frac{1}{r}\frac{\partial}{\partial r}\left(r\frac{\partial \psi_0}{\partial r}\right) = \left[-\frac{k^2r^2}{q^2(z)} - j\frac{2k}{q(z)}\right]\psi_0$$

These functions are substituted into (3.3.1), and the terms with equal powers of r are grouped together:

$$\left\{\left[\frac{k^2}{q^2(z)}(q'(z) - 1)\right]r^2 - 2k\left[P'(z) + \frac{j}{q(z)}\right]r^0\right\}\psi_0 = 0 \qquad (3.3.3)$$

For the assumed form, (3.3.2), to be a solution, every factor of a power of r must be equal to zero. This yields two simple ordinary differential equations:

$$q'(z) = 1 \quad (3.3.4a)$$

$$P'(z) = -\frac{j}{q(z)} \quad (3.3.4b)$$

Since (3.3.4a) is decoupled from the other, its solution and implications will be discussed first. The solution is trivial:

$$q(z) = q_0 + z \quad (3.3.5)$$

where q_0 is the value of q at $z = 0$. (Where is $z = 0$? We have to face up to this question later.) Now, it is obvious the dimensions of q_0 must be the same as z—length—so one is tempted to use the letter symbol z_0 for it. But is q_0 real?

To answer this, we must refer back to the initial form involving $q(z)$:

$$\psi_0 = \exp\left[-j\frac{kr^2}{2q(z)}\right] \exp[-jP(z)] \quad (3.3.2)$$

If $q(z)$ were always real then $|\exp[-jkr^2/2q(z)]| = 1$ for all values of r. This would mean that the phase is changing faster and faster with r, with the amplitude remaining a constant. That doesn't describe a beam; it is an absurdity! A beam has most of its energy concentrated in a certain location in the transverse plane and thus the possibility of q being pure real is not interesting. If $q(z)$ has an imaginary part, then the mathematical exercise becomes worthwhile.

Assume that $q(z)$ is complex. Now z is obviously real in (3.3.5), and any real part of q_0 just corresponds to a shift in the spatial coordinate. Consequently, we might as well start with the proper $z = 0$ axis to absorb this real part and let q_0 be imaginary (i.e., $q_0 = jz_0$):

$$q(z) = z + jz_0 \quad (3.3.6)$$

where z_0 is a constant to be determined (and interpreted). If (3.3.6) is substituted into (3.3.2)—say at $z = 0$—we obtain a very satisfying physical picture of part of ψ_0. At $z = 0$,

$$q(z) = jz_0$$

and

$$\psi_0(z = 0) = \exp\left(-\frac{kr^2}{2z_0}\right) \exp[-jP(z = 0)]$$

Note that the exponential term is real and thus the amplitude drops off quite rapidly with r, being down from its peak value of 1 at $r = 0$ to 0.368 at $r = (2z_0/k)^{1/2}$. Obviously, this last quantity is a *scale length* for this beam.

$$w_0^2 = \frac{2z_0}{k} = \frac{\lambda_0 z_0}{n\pi} \quad \text{or} \quad z_0 = \frac{\pi n w_0^2}{\lambda_0} \quad (3.3.7)$$

Sec. 3.3 Lowest-Order TEM$_{0,0}$ Mode

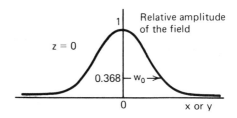

Figure 3.1 Variation of the field in the transverse plane.

Thus the field varies as $\exp(-r^2/w_0^2)$ at the plane $z = 0$ (Fig. 3.1). Since our eyes respond to the square of the field, we would observe a single "dot" with a major fraction of the power contained in a beam of radius $\sim w_0$. Because of this, w_0 is called the "spot size." (Actually, it is the minimum spot size, as will be shown below.)

At any point z, the value of q changes according to (3.3.6). Since we are interested in the inverse of q, let us compute that factor and examine the imaginary part.

$$\frac{1}{q(z)} = \frac{1}{z + jz_0} = \frac{z}{z^2 + z_0^2} - j\frac{z}{z^2 + z_0^2} = \frac{1}{R(z)} - j\frac{\lambda_0}{\pi w^2(z)} \qquad (3.3.8)$$

Again, we return to (3.3.2) for a physical interpretation of this bit of manipulation.

$$\psi_0 = \left\{\exp\left[-\frac{kz_0 r^2}{2(z^2 + z_0^2)}\right]\right\}\left\{\exp\left[-\frac{jkzr^2}{2(z^2 + z_0^2)}\right]\right\}\{\exp[-jP(z)]\} \qquad (3.3.9)$$

Now, as we move away from the axis, the phase changes faster and faster with r, but the amplitude becomes insignificant at large r. Thus the previous absurdity is avoided.

We again realize that the term multiplying r^2 in the first exponential factor is a scale length, and we name it the spot size of the beam, which is now a function of z:

$$w^2(z) = \frac{2}{kz_0}(z_0^2 + z^2) = \frac{2z_0}{k}\left[1 + \left(\frac{z}{z_0}\right)^2\right]$$

or using the previous definition of w_0, (3.3.7), we obtain

$$w^2(z) = w_0^2\left[1 + \left(\frac{\lambda_0 z}{\pi w_0^2}\right)^2\right] \qquad (3.3.10)$$

Let us also abbreviate the terms in the second exponential factor by $R(z)$:

$$R(z) = \frac{1}{z}(z^2 + z_0^2) = z\left[1 + \left(\frac{z_0}{z}\right)^2\right]$$

$$= z\left[1 + \left(\frac{\pi n w_0^2}{\lambda_0 z}\right)^2\right] \qquad (3.3.11)$$

Both (3.3.10) and (3.3.11) require considerable discussion to appreciate the physical interpretation. This will be done after we have disposed of the remaining bit of mathematics—namely, to find the function $P(z)$. This is related to $q(z)$ by

$$P'(z) = \frac{-j}{q(z)} = \frac{-j}{z + jz_0} \qquad (3.3.4b)$$

or

$$jP(z) = \int_0^z \frac{dz'}{z' + jz_0} = \ln(z' + jz_0)\Big|_0^z$$

$$= \ln(z + jz_0) - \ln(jz_0)$$

$$jP(z) = \ln\left[1 - j\left(\frac{z}{z_0}\right)\right]$$

Now we use the fact that

$$1 - j\left(\frac{z}{z_0}\right) = \left[1 + \left(\frac{z}{z_0}\right)^2\right]^{1/2} \exp\left[-j \tan^{-1}\left(\frac{z}{z_0}\right)\right]$$

to find the real and imaginary parts of $P(z)$:

$$jP(z) = \ln\left[1 + \left(\frac{z}{z_0}\right)^2\right]^{1/2} - j \tan^{-1}\left(\frac{z}{z_0}\right) \qquad (3.3.12)$$

We need $\exp[-jP(z)]$:

$$e^{-jP(z)} = \frac{1}{[1 + (z/z_0)^2]^{1/2}} e^{+j \tan^{-1}(z/z_0)} \qquad (3.3.13)$$

Thus the complete expression for the fundamental or lowest-order $TEM_{0,0}$ mode is found from the various definitions made previously:

$$\frac{E(x, y, z)}{E_0} = \left\{\frac{w_0}{w(z)} \exp\left[-\frac{r^2}{w^2(z)}\right]\right\} \qquad \text{amplitude factor}$$

$$\times \exp\left\{-j\left[kz - \tan^{-1}\left(\frac{z}{z_0}\right)\right]\right\} \qquad \text{longitudinal phase} \qquad (3.3.14)$$

$$\times \exp\left[-j\frac{kr^2}{2R(z)}\right] \qquad \text{radial phase}$$

Sec. 3.4   Physical Description of TEM$_{0,0}$ Mode

where

$$w^2(z) = w_0^2\left[1 + \left(\frac{\lambda_0 z}{\pi w_0^2}\right)^2\right] = w_0^2\left[1 + \left(\frac{z}{z_0}\right)^2\right] \quad (3.3.10)$$

$$R(z) = z\left[1 + \left(\frac{\pi w_0^2}{\lambda_0 z}\right)\right] = z\left[1 + \left(\frac{z_0}{z}\right)^2\right] \quad (3.3.11)$$

$$z_0 = \frac{\pi w_0^2}{\lambda_0} \quad (3.3.7)$$

3.4 PHYSICAL DESCRIPTION OF TEM$_{0,0}$ MODE

The physical interpretation of (3.3.14), together with various definitions, is the subject of this section. Hopefully, we will be able to penetrate the mathematical maze of the preceding section so as to appreciate the physical simplicity of the answer.

3.4.1 Amplitude of the Field

The first term in (3.3.14) describes the amplitude of the field as a function of the radial coordinate and how this changes as the beam propagates along z. It was noted in Sec. 3.3 that at $r = w$, the field is down by e^{-1} of its peak value at $r = 0$.

$$\left|\frac{E(x, y, z)}{E_0}\right| = \frac{w_0}{w(z)} \exp\left[-\left(\frac{r}{w}\right)^2\right] \quad (3.4.1)$$

In Fig. 3.2, the $1/e$ point of the field is sketched as a function of the z coordinate. As the beam propagates along z, the spot size, w, becomes larger; hence, the $1/e$ points become farther from the axis. It is obvious from this figure that the beam expands from its minimum value of w_0 by a factor of $\sqrt{2}$ when $z = z_0$.*

$$w^2(z) = w_0^2\left[1 + \left(\frac{z}{z_0}\right)^2\right] \quad (3.3.10)$$

---

*It will become obvious in Chapter 5 that $2 \times z_0$ is the spacing for a confocal mirror arrangement. Hence, $2z_0$ is also called the *confocal parameter*.

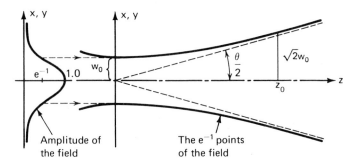

Figure 3.2 Spreading of a TEM$_{0,0}$ mode.

Furthermore, the beam has its minimum spot size at a certain point along the z axis (the definition of $z = 0$!).

For large z, the spot size is asymptotic to the dashed lines described by

$$w(z \gg z_0) = \frac{w_0 z}{z_0} = \frac{\lambda_0 z}{\pi n w_0}$$

Thus the expansion angle of the beam is

$$\frac{\theta}{2} = \frac{dw}{dz} = \frac{\lambda_0}{\pi n w_0}$$

$$\theta = \frac{2\lambda_0}{\pi n w_0} \tag{3.4.2}$$

This is the minimum spread that a beam of diameter $2w_0$ can have. (See the discussion on the uncertainty principle in Chapter 1.)

There is a perfectly logical and simple reason why the factor $w_0/w(z)$ appears in the amplitude of the field. To appreciate this, we compute the total power crossing an arbitrary plane z which must be equal to a constant.

$$P = \frac{1}{2} \iint \frac{EE^*}{\eta} dA = \frac{1}{2} \frac{E_0^2}{\eta} \frac{w_0^2}{w^2(z)} \int_0^{2\pi}\!\!\int_0^\infty \exp\left[-\frac{2r^2}{w^2(z)}\right] r\, dr\, d\phi$$
$$= \frac{1}{2} \frac{E_0^2}{\eta} \left(\frac{\pi w_0^2}{2}\right) \tag{3.4.3}$$

Note that the power carried by the beam is constant independent of z—a most logical result. It would be embarrassing if it were not true, for we have assumed a lossless medium. Thus the factor $w_0/w(z)$ reduces the peak amplitude of the field in response to the divergence of the beam.

3.4.2 Longitudinal Phase Factor

The second factor in (3.3.14) expresses the change in phase of the wave in the direction of propagation,

$$\phi = kz - \tan^{-1}\left(\frac{z}{z_0}\right) \quad (3.4.4)$$

where k is the customary wave number of a uniform plane wave, $\omega n/c$. Thus the phase velocity of this Gaussian beam is close to, but slightly greater than, the velocity of light (in the uniform medium):

$$v_p = \left(\frac{\phi}{\omega z}\right)^{-1} = \frac{c/n}{1 - (\lambda_0/2\pi nz)\tan^{-1}(z/z_0)} \quad (3.4.5)$$

Even though the propagation velocity is close to c/n, the difference can be measured and does play a role in resonator theory.

The fact that the wave number is so close to k of a uniform plane wave means that we obtain the magnetic field intensity by the simple relationship

$$\mathbf{H} = \frac{\mathbf{k} \times \mathbf{E}}{\omega \mu_0} = \mathbf{a}_z \times \frac{\mathbf{E}}{\eta} \quad (3.4.6)$$

where η is the wave impedance for the dielectric $(\mu_0/\epsilon)^{1/2} = \eta_0/n$.

3.4.3 Radial Phase Factor

The final factor in (3.3.14),

$$\exp\left[-j\frac{kr^2}{2R(z)}\right]$$

indicates that the plane z = constant is *not* an equiphase surface. As one moves off the axis, the local field lags that at $r = 0$ [assuming $R(z)$ is positive]. Obviously, if the phase front is not planar, it is curved. Because the letter symbol is used, $R(z)$, we can anticipate that the equiphase surfaces are spherical with a curvature given by $R(z)$.

This is easily shown by considering a limited extent spherical wave as shown in Fig. 3.3 and finding the phase of the wave close to the axis. The field for such a wave would be

$$E \sim \frac{1}{R}\exp(-jkR)$$

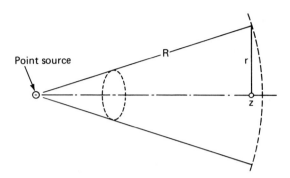

Figure 3.3 Origin of the phase front curvature.

where $R = (r^2 + z^2)^{1/2}$. We place ourselves at a large distance from the origin where $R \sim z \gg r$. Thus the binomial theorem is used in a judicious fashion:

$$R = z\left(1 + \frac{r^2}{z^2}\right)^{1/2} \sim z + \frac{1}{2}\frac{r^2}{z} \sim z + \frac{1}{2}\frac{r^2}{R}$$

since $R \sim z$. Hence the phase of field close to the z axis varies in the following manner:

$$E \sim \frac{1}{R} \exp(-jkz) \exp\left(-j\frac{kr^2}{2R}\right)$$

The last term has the same functional form of the last factor of (3.3.14)—hence its name.

But in the case of a Gaussian beam, the apparent center for the curved wave front changes. Recall the relation for $R(z)$:

$$R(z) = z\left[1 + \left(\frac{z_0}{z}\right)^2\right] \qquad (3.3.11)$$

Only when z is much larger than z_0 does the beam appear to originate at $z = 0$. As one gets closer to the point $z = 0$, the center recedes until at $z = 0$ the "center" of curvature is at infinity and now the wave front is planar. Indeed, it is easy to show that the equiphase surfaces are orthogonal to the beam expansion curves shown in Fig. 3.2.

Thus there are two alternative but equivalent definitions for the plane $z = 0$:

1. Where the spot size is a minimum
2. Where the wave front is planar

Sec. 3.5 Higher-Order Modes

Let us now repeat (3.3.14) and assign a brief physical interpretation to each term.

$$E(x, y, z)$$ 	The electric field
$$\|$$
$$E_0$$ 	Amplitude at $z = 0$
$$\times$$
$$\frac{w_0}{w(z)} \exp\left[-\frac{r^2}{w^2(z)}\right]$$ 	Variation of the amplitude with r
$$\times$$
$$\exp\left\{-j\left[kz - \tan^{-1}\left(\frac{z}{z_0}\right)\right]\right\}$$ 	Longitudinal phase factor
$$\times$$
$$\exp\left[-j\frac{kr^2}{2R(z)}\right]$$ 	Radial phase factor

3.5 HIGHER-ORDER MODES

In the previous work, the simplifying assumption of cylindrical symmetry was made. Although this may make the mathematics simple, the laser does not particularly fear (or know about) the complexity. It has no trouble in solving (3.2.8).

Indeed, if one considers the simple gas laser shown in Fig. 3.4, there are trivial reasons why it would not oscillate in the lowest-order mode. For instance, suppose that the window had a streak of "dirt" or "lint" right at the center of the tube, or suppose that the center of the mirror was absent.* If the electric field is as described by (3.3.14), there would be considerable scattering losses due to the lint and a major coupling loss due to the hole. Later, it will be seen that a

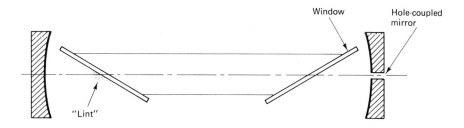

Figure 3.4 Simple laser.

*This is called "hole coupling."

laser will oscillate in that mode with the highest gain-to-loss ratio. Hence, we must consider possibilities that are not cylindrically symmetric.

One can choose to work in cylindrical (r, ϕ, z) or Cartesian coordinates (x, y, z), allowing for variations in ϕ in the former and different variations in the x and y coordinates in the latter. Different mode descriptions apply for each coordinate system. For the simple system shown in Fig. 3.4, and for most lasers, the Cartesian coordinate system is most appropriate. The reason is that the windows provide a "bias" that discriminates against the purely cylindrical modes.

It can be verified by direct substitution* that the following functions satisfy (3.2.8):

$$\frac{E(x, y, z)}{E_{m,p}} = H_m\left[\frac{2^{1/2}x}{w(z)}\right] H_p\left[\frac{2^{1/2}y}{w(z)}\right]$$

$$\times \frac{w_0}{w(z)} \exp\left[-\frac{x^2 + y^2}{w^2(z)}\right]$$

$$\times \exp\left\{-j\left[kz - (1 + m + p)\tan^{-1}\left(\frac{z}{z_0}\right)\right]\right\}$$

$$\times \exp\left[-j\frac{kr^2}{2R(z)}\right] \quad (3.5.1)$$

where all symbols, $w(z)$, w_0, z_0, and $R(z)$ are as defined and interpreted previously. The symbol $H_m(u)$, stands for the Hermite polynomial of order m and argument u and is defined by

$$H_m(u) = (-1)^m e^{u^2} \frac{d^m e^{-u^2}}{du^m} \quad (3.5.2)$$

One notices a great deal of similarity between (3.5.1) and (3.3.14); the radial-phase factor is the same, the exponential variation with $r^2 = x^2 + y^2$ is the same, and the multiplying factor $w_0/w(z)$ is the same; but there are differences.

First note that the phase shift in the z direction depends on the mode numbers m and p. This will play a role in the oscillation frequency of the laser.

The major change in visible appearance is due to the Hermite polynomials. It is instructive to consider a few of the lower-order ones:

$$H_0(u) = 1 \Rightarrow 1$$
$$H_1(u) = 2(u) \Rightarrow u$$
$$H_2(u) = (2u^2 - 1)2 \Rightarrow 2u^2 - 1$$

---

*A very long and painful exercise in arithmetic!

Sec. 3.5   Higher-Order Modes

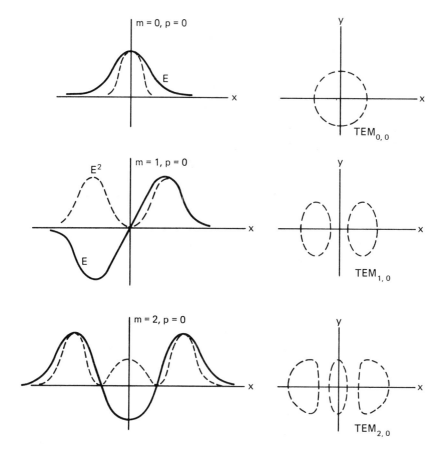

Figure 3.5 The field E, intensity E^2, and "dot" pattern of various modes.

where the arrow indicates that we can absorb common numerical factors into an amplitude factor $E_{m,p}$. Thus the field has a more spectacular variation in the transverse plane, as shown in Fig. 3.5. Note that for large x (or y), the exponential behavior still dominates and thus the "beam" is tightly bound to the z axis. But at small x, the field is modified considerably by the polynominals being forced to zero at a finite number of points. The number of times that the field goes to zero, other than at ∞, is the mode number m. If the laser is visible, there will be $m + 1$ dots encountered going across the beam in the x direction. The same considerations apply to the y direction; hence there will be $(m + 1)(p + 1)$ "dots" in a TEM$_{m,p}$ mode.

At this point we should be cautioned that there is a built-in difficulty with the name "spot size" $w(z)$. The spot size $w(z)$ is the same for all three of the modes illustrated, but the field occupies a bigger "spot" on the paper as the mode number gets larger.

It is a natural tendency to associate the words "spot size" with the radial extent of the beam, but it is wrong to do so if you also associate the same words with $w(z)$. This quantity, $w(z)$, is a scale length for measuring the variation of field in the transverse direction. *All* $\text{TEM}_{m,p}$ modes have this same scale length $w(z)$, but the higher-order modes use a larger transverse area.

3.6 DIVERGENCE OF THE HIGHER-ORDER MODES—SPATIAL COHERENCE

An example will illustrate why one should be very careful about the use of the term "spot size." Let us ask: What is the divergence of a beam consisting of one or more $\text{TEM}_{m,p}$ modes? For instance, one can obtain a rather large physical "spot" by a linear combination of these modes.

The answer is that *all* Hermite-Gaussian beams have *exactly* the same divergence (or far-field angle), given by

$$\theta = \frac{2\lambda_0}{\pi n w_0} \qquad (3.6.1)$$

where w_0 is the minimum spot size for the $\text{TEM}_{0,0}$ mode.

Thus the controlling factor on the beam spread is the characteristic dimension w_0, not the physical spot seen on the wall. If the beam spread is a factor in the application of the laser,* one attempts to ensure that oscillation takes place in the $\text{TEM}_{0,0}$ mode. Thus this mode has the greatest intensity (power/area) for the minimum beam spread as compared to all other modes or field distributions.

Note that the $\text{TEM}_{0,0}$ mode has a uniphase surface—albeit curved but still the field is in phase on this spherical surface. The term "spatial coherence" is used to describe this fact; that is, the field has one common phase on this spherical surface. For contrast, note that the field of the $\text{TEM}_{1,0}$ mode reverses direction for negative x; and for the higher-order modes, the field reverses direction many times. Within each dot, the equiphase surface has the same spherical curvature as the $\text{TEM}_{0,0}$ mode.

This also explains why a "flashlight" beam spreads so much faster than a laser beam, even with the parabolic reflector and large aperture on the former. The atoms in the heated filament of the tungsten wire radiate an incoherent wave; that is, the phase from one group of atoms bears little, if any, relationship to another group. Consequently, we cannot, by any stretch of imagination, identify the spot from a flashlight as being a uniphase surface. The characteristic dimen-

---

*The beam spread is always a consideration. For a laser transit, laser radar, and laser communications with free-space transmission, one desires a minimum beam spread. But these same considerations all apply to focusing. The smallest spot size achievable by a lens is also controlled by w_0. Thus beam spread is a factor in raw power applications.

PROBLEMS

3.1. The following questions are intended as a review and to test your understanding and appreciation of the Hermite-Gaussian beam modes. Answer these questions with a sketch, some simple mathematics, or a few sentences:
   (a) What is the physical significance of the *distance* z_0?
   (b) If $z = z_0$ and $r^2 = w^2(z_0)$ by how much does the phase of the field lead or lag that at $r = 0$?
   (c) Which factor expresses the idea that the beams are not plane waves and the phase velocity is greater than c?

3.2. (a) A certain commercial helium:neon laser is advertised to have a far-field divergence angle of 1 milliradian at $\lambda_0 = 632.8$ nm. What is the spot size, w_0?
   (b) The power emitted by this laser is 5 mW. What is the peak electric field in volts per centimeter at $r = z = 0$?
   (c) How many photons per second are emitted by this laser beam?
   (d) Electromagnetic energy can only come in packages of $h\nu$. If one more photon per second were emitted by this laser, what is the new power specification? (The point of this part of the problem is to recognize that there is a time and a place for making the distinction between a classical field and a photon: should we start here?)

3.3. Given a 1-W $TEM_{0,0}$ beam of $\lambda_0 = 514.5$ nm from an argon ion laser with a minimum spot size of $w_0 = 2$ mm located at $z = 0$:
   (a) How far will this beam propagate before the spot size is 1 cm?
   (b) What is the radius of curvature of the phase front at this distance?
   (c) What is the amplitude of the electric field at $r = 0$ and $z = 0$?

3.4. A 10-W argon ion laser oscillating at 4880 Å has a minimum spot size of 2 mm.
   (a) How far will this beam travel before the spot size is 4 mm?
   (b) What fraction of the 10 W is contained in a hole of *diameter* $2w(z)$?
   (c) Express the frequency/wavelength of this laser in eV, nm, μm, ν(Hz), and $\bar{\nu}$(cm^{-1}).
   (d) What is the amplitude of the electric field when $w = 1$ cm?

3.5. Sketch the variation of the intensity with $x(y = 0)$ of a beam containing 1 W of power in a $TEM_{0,0}$ mode and $\frac{1}{2}$ W in the $TEM_{1,0}$ mode (i.e., total power = 1.5 W).

3.6. Consider a linear combination of two equal amplitude $\text{TEM}_{m,p}$ modes given by:

$$\mathbf{E} = E_0\{(\text{TEM}_{1,0})\mathbf{a}_x \pm j(\text{TEM}_{0,1})\mathbf{a}_y\}$$

(a) Sketch the "dot" pattern or equal intensity contours for each component (i.e., $\mathbf{a}_x$ or $\mathbf{a}_y$). Indicate the direction of the electric field.

(b) Sketch the pattern for the linear combination.

(c) Label the positions where the intensity is a maximum and a minimum. (This is sometimes referred to as the "donut mode" or $\text{TEM}^*_{0,1}$.)

3.7. The intensity of a laser has the following visual appearance when projected on a surface.

(a) Name the mode. (i.e., $\text{TEM}_{m,p}$; $m = ?$; $p = ?$)

(b) A plot of the relative *intensity* of *another* mode as a function of x (for $y = 0$) is shown below at the right. The variation with respect to y is a simple bell-shape curve. What is the "spot size," w?

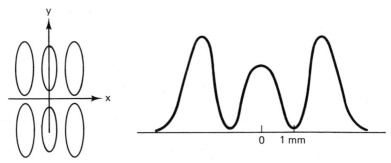

3.8. Suppose that a $\text{TEM}_{m,p}$ mode impinged on a perfectly absorbing plate with a hole of radius a centered on the axis of the beam. Plot the transmission coefficient of this hole as a function of the ratio of a/w for the $(0,0)$, $(0,1)$, and $(1,1)$ modes assuming that the fields are not affected by the plate.

3.9. Show that the Hermite-Gaussian beam modes are orthogonal in the following sense:

$$\text{Re}\left[\int (\mathbf{E}_{m,n} \times \mathbf{H}^*_{p,q}) \cdot d\mathbf{S}\right] = 0$$

3.10. Repeat the analysis from (3.2.6) to (3.3.14) for the case where the index of refraction is nonuniform and is given by

$$n(r) = n_0(1 - r^2/2l^2)$$

3.11. The same arguments advanced for the derivation of (3.2.4) can be used for the magnetic field intensity H. Check the accuracy of this equation by considering a dominant $\text{TE}_{1,0}$ mode in a rectangular waveguide of width a and height b and computing the ratio of H_z/H_x.

3.12. The news media has shown the astronauts placing laser retroreflectors on the moon. Use the expansion law for Gaussian beams to predict the diameter of a laser beam when it hits the moon. Consider two cases:
  (a) A laser rod of 2 cm diameter
  (b) This same laser sent through a telescope backward so that the beam starts with a diameter of 2 meters
  (c) Eye damage intensities are in the range of 10 μW/cm^2. If the laser on earth produced a pulse power of 10 MW, was there danger to the astronauts from the optical radiation?

3.13. A quadrant detector in the configuration shown below is used to detect the axis of an optical beam in the manner shown on the diagram below. Each part converts the photons intercepted by its active area into a proportional amount of electrical current. (Assume that each photon produces one electron, which flows through the resistors on the input of the amplifiers.)
  (a) Estimate the error signal (in volts) produced by a 1 mW 6328 Å TEM$_{0,0}$ laser beam slightly misaligned in the manner indicated below (assume $a/w = 2$ and infinite input impedances for all amplifiers).
  (b) Should the ratio a/w be big or small to optimize the ability to sense the axis?
  (c) The circuit shown below produces an ambiguous error signal in the sense that a displacement of the beam along x is sensed, but not along y. Can you suggest another configuration, possibly using more amplifiers to avoid this ambiguity?

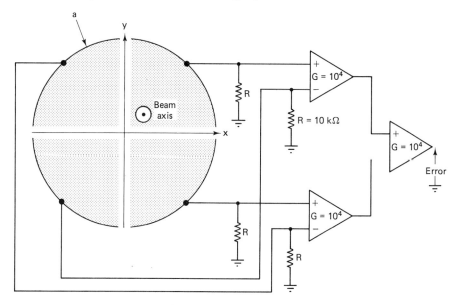

$G = 10^4$ = voltage gain of the amplifiers
R = resistance

REFERENCES AND SUGGESTED READINGS

1. G. D. Boyd and J. P. Gordon, "Confocal Multimode Resonator for Millimeter through Optical Wavelength Masers," Bell Syst. Tech. J. *40*, 489–508, Mar. 1961.
2. G. D. Boyd and H. Kogelnik, "Generalized Confocal Resonator Theory," Bell Syst. Tech. J. *41*, 1347–1369, July 1962.
3. H. Kogelnik, "Imaging of Optical Modes—Resonators with Internal Lenses," Bell Syst. Tech. J. *44*, 455–494, Mar. 1963.
4. H. Kogelnik and T. Li, "Laser Beams and Resonators," Appl. Opt. 5, 1550–1556, Oct. 1966.
5. A. Yariv, *Introduction to Optical Electronics*, 2nd ed. (New York: Holt, Rinehart and Winston, 1976), Chap. 3.
6. D. R. Herriott, "Applications of Laser Light," Sci. Am. *219*, 141–156, Sept. 1965.
7. A. Maitland and M. H. Dunn, *Laser Physics* (Amsterdam: North-Holland Publishing Company, 1969), Chaps. 4–7.
8. H. A. Haus, *Waves and Fields in Optoelectronics* (Englewood Cliffs, N.J.: Prentice-Hall Inc., 1984).
9. A. E. Siegman, *Lasers* (Mill Valley, Calif.: University Science Books, 1986), Chaps. 16–21.

4

Gaussian Beams in Continuous Media

4.1 INTRODUCTION

In Chapter 3 we discussed the characteristics of the Hermite-Gaussian beam modes in a uniform dielectric medium such as that provided by free space. As we shall see in Chapter 5, such modes "fit" into laser cavities in a very natural manner and thus are generated by the process of stimulated emission.

Even though free space has the economic advantage of being free, it is subject to local weather conditions in which dust, sleet, snow, rain, and temperature gradients would render the propagation path useless for communication purposes. For that purpose, one turns to optical waveguides in the form of a glass fiber (say 50 μm in diameter). Such fibers can be produced cheaply and are immune to the difficulties listed above.

Although the main purpose of this book is to provide a simple understanding of the laser, in the future these fibers are destined to be the dominant propagation media for optical communications and control. Furthermore, the fibers mate in a natural manner to the very important semiconductor laser and detector technology. Thus a brief introduction to beam propagation in dielectric waveguides is given in this chapter.

However, one of the topics, the *ABCD* law, is very useful in problems involving free space and/or discrete components in addition to simplifying the analysis of fiber-optic propagation.

4.2 THE DIELECTRIC SLAB WAVEGUIDE

Wave propagation in a nonuniform dielectric fiber can be quite complicated if one insists on an exact solution to Maxwell's equations. However, if one is willing to use a bit of imagination, one can understand the essential features with a minimum of arithmetic detail.

For instance, consider the situation sketched in Fig. 4.1. We assume that a beam from a laser is focused onto the core of the fiber by a lens; that the beam is propagating primarily in the z direction (so that the angle θ is small), and that the nonuniformity in the dielectric consists of discrete jumps at $x = \pm a$. Furthermore, we treat the fields as being approximated by uniform plane waves and neglect any variation in the y direction. Our goal is to find a field configuration or mode that is mostly confined within the core region, with negligible amounts penetrating out to the sheath.

To show that such modes exist, it is sufficient to demonstrate that a wave experiences total internal reflection if θ_2 is small enough. This is easily shown by matching the tangential phase constants of the waves in the core and the cladding (a rederivation of Snell's law):

or
$$\left.\begin{array}{c} \dfrac{\omega}{c} n_2 \cos \theta_2 = \dfrac{\omega}{c} n_1 \cos \theta_1 \\ \\ \cos \theta_1 = \dfrac{n_2}{n_1} \cos \theta_2 \end{array}\right\} \quad (4.2.1)$$

Obviously, if θ_2 is small enough, $\cos \theta_2 \sim 1$, and if $n_2 > n_1$, then $\cos \theta_1 > 1$. What does such an apparent mathematical absurdity mean?

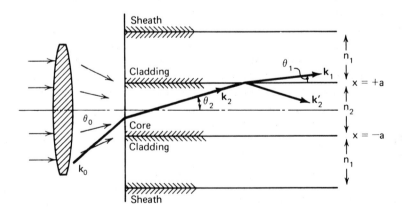

Figure 4.1 Dielectric-slab waveguide.

Sec. 4.2 The Dielectric Slab Waveguide

It means that wave is entirely confined to the core and that there is an attenuation of the field in the x direction. To demonstrate this last statement, consider a wave propagating in the θ_1 direction in the cladding material:

$$E_1 = E_0 \exp(-j\mathbf{k}_1 \cdot \mathbf{r})$$

where

$$\mathbf{k}_1 = \frac{\omega}{c} n_1 (\cos\theta_1 \, \mathbf{a}_z + \sin\theta_1 \, \mathbf{a}_x)$$

(4.2.2)

One uses (4.2.1) to express k_1 in terms of the angle θ_2. Now, since $\sin^2\theta + \cos^2\theta = 1$ for all angles—even complex ones—we find that $\sin\theta_1$ must be an imaginary number.

$$\sin\theta_1 = (1 - \cos^2\theta_1)^{1/2} = \pm j \left[\left(\frac{n_2}{n_1}\cos\theta_2\right)^2 - 1 \right]^{1/2} \quad (4.2.3)$$

Are we compounding the absurdity by insisting on an angle that yields an imaginary value for the sine? No, the angle, θ_1, is only useful insofar as it describes the field, and the field is perfectly well behaved and not at all mysterious.

$$E_1 = E_0 \exp\left\{ -\frac{\omega}{c} n_1 \left[\left(\frac{n_2}{n_1}\cos\theta_2\right)^2 - 1 \right]^{1/2} (x-a) \right\}$$
$$\times \exp\left[-j\frac{\omega}{c} n_1 \left(\frac{n_2}{n_1}\cos\theta_2\right) z \right]$$

(4.2.4)

Thus the field in region $x > a$ decays exponentially away from the interface. However, there is no power flow in the x direction, and hence the energy must be guided by the central core, with a minimal amount contained in the cladding and almost nothing in the sheath. Thus we have achieved the goal of guiding the beam by the slab.

A sketch of the variation of the index of refraction and the field for this slab waveguide is shown by the dashed curves in Fig. 4.2. Note that the field "looks" like that of the Gaussian beam mode of Chapter 3. The fact that the mathematical expression is trigonometric for $|x| < a$, exponentially decaying for $|x| > a$, and not Gaussian [$\sim \exp(-x^2)$] is irrelevant—the field looks like that of a beam.

It should be remembered, however, that this guidance of the beam was caused by the dielectric discontinuity at $x = \pm a$, in particular the decrease of $n(r)$ there. Thus if one goes one step further and makes the index a continuously decreasing function of r [as shown by the solid curve in Fig. 4.2(a)], one can anticipate guidance there also. In fact, one should not be surprised if the field configuration turned out to be a Hermite-Gaussian beam mode. The following sections show that this is indeed possible.

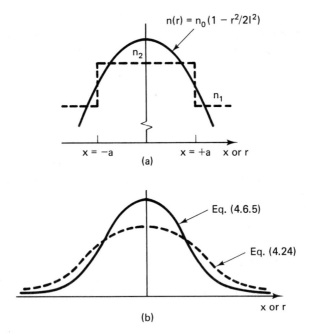

Figure 4.2 Variation of (a) the index and (b) the field within a fiber. The dashed curves correspond to each other, as do the solid curves.

4.3 NUMERICAL APERTURE

Let us return to the slab fiber of Fig. 4.1 and examine the input conditions. If, for instance, the angle θ_2 is too big, then the angle θ_1 is real, and the wave merely propagates outward through the cladding to be absorbed in the sheath or radiated (to $x = \pm\infty$). Such waves are not guided, and the whole purpose of this fiber construction is defeated.

The angle θ_2 is, of course, determined by the air-core interface problem, which is shown in greatly exaggerated form in Fig. 4.1. In order to have a *guided* wave, we need the sine of the angle θ_1 to be imaginary and $\cos \theta_1 > 1$ in accordance with (4.2.3). However, this is only true provided θ_2 is small enough.

$$\cos \theta_2 < \frac{n_1}{n_2} \qquad (4.3.1)$$

On the other hand, θ_2 is controlled by the angle θ_0 and applying Snell's law (again) for the air—n_2 interface yields

$$\frac{\omega}{c} \sin \theta_0 = \frac{\omega n_2}{c} \sin \theta_2 \qquad (4.3.2)$$

Combining (4.3.1) and (4.3.2) yields the maximum angle over which the fiber will collect and *guide* the electromagnetic radiation:

$$\sin \theta_0 < n_2[\sin \theta_2 = (1 - \cos \theta_2)^{1/2}]$$

or

$$\sin \theta_0 < (n_2^2 - n_1^2)^{1/2} = NA \quad (4.3.3)$$

The angle θ_0 is the acceptance angle of the fiber and, because $n_2 - n_1$ is quite small for most fibers, the angle is also quite small. The quantity $[n_2^2 - n_1^2]^{1/2}$ is usually referred to as the numerical aperture (NA) in analogy to the corresponding $f\#$ associated with a lens.

While Fig. 4.1 implies that one could use a large-diameter lens to collect a lot of light and focus it onto the core of the fiber, it is quite futile (useless) to do so. That fraction of the radiation contained in angles beyond the limit specified by (4.3.3) is not guided by the fiber.

4.4 MODES IN A STEP-INDEX FIBER (OR A HETEROJUNCTION LASER)

Even though most fibers are circular in cross section, we restrict our attention to the symmetric slab geometry shown in Fig. 4.1. This ploy enables us to obtain a formal solution to Maxwell's equations without endless haranguing about the marvels of Bessel functions that arise naturally in cylindrical coordinates. All of the mathematical steps and many of the conclusions of the slab are directly applicable to the round fiber. Furthermore, such a model is a good representation of the active region of a heterojunction* laser. It is a fortuitous fact of nature that the substitution of aluminum for gallium in a GaAs crystal, indicated by $Al_xGa_{1-x}As$, leads to a material with an *increased* band gap, a *decreased* index of refraction, and, most importantly, almost *identical* lattice constants. Thus the central region of Fig. 4.1 can be assigned to GaAs, and the outer regions to $Al_xGa_{1-x}As$ of a *p-n* junction laser. There will be much more on the physics of such lasers in Chapter 11, but for now, our focus is on the electromagnetic problem of guiding waves along this slab.

If one solves the wave equation for the "z" components, E_z or H_z, then the transverse components for any guide can be found from:

$$\mathbf{E}_t = \frac{1}{\beta^2 - \left(\frac{\omega n}{c}\right)^2} \{j\beta \nabla_t E_z - j\omega\mu_0 \mathbf{a}_z \times \nabla_t H_z\} \quad (4.4.1a)$$

---

*A heterojunction uses different materials with different bandgaps in contrast to the junctions in elementary semiconductors silicon or germanium. The latter are called homojunctions.

$$\mathbf{H}_t = \frac{1}{\beta^2 - \left(\frac{\omega n}{c}\right)^2} \{j\omega\epsilon_0 n^2 \mathbf{a}_z \times \nabla_t E_z + j\beta \nabla_t H_z\} \quad (4.4.1b)$$

where the fields are assumed to be propagating along z as $\exp(-j\beta z)$. From (4.4.1), we see that there is a natural classification of the types of modes depending on where H_z or E_z is zero for TE or TM, respectively.* Both may not be zero, or a TEM mode, unless the denominator of (4.4.1) vanishes. Sometimes the boundary conditions require the presence of both E_z and H_z and these describe the hybrid EH or HE mode depending on which component is dominant. In any case these longitudinal components obey the scalar wave equation:

$$\nabla_t^2 E_z + \left[\left(\frac{\omega n}{c}\right)^2 - \beta^2\right] E_z = 0 \quad (4.4.2a)$$

$$\nabla_t^2 H_z + \left[\left(\frac{\omega n}{c}\right)^2 - \beta^2\right] H_z = 0 \quad (4.4.2b)$$

Note that the character of the solution to (4.4.2) changes depending on whether $[(\omega n/c)^2 - \beta^2]$ is greater than or less than zero. If the former is true, then the equation "looks" like that of a simple harmonic oscillator and we would anticipate a trigonometric or "standing" wave type of solution in the transverse plane. If $[(\omega n/c)^2 - \beta^2] < 0$, then an exponential solution decaying away from the central slab is appropriate. This is, of course, exactly what was found in Sec. 4.2 where uniform plane waves were reflected from the interfaces.

This idea can be made more formal by considering the Pythagorean relation for components of the wave vector as shown in Fig. 4.3. The wave vector is always related to the material properties of the medium by $\mathbf{k} \cdot \mathbf{k} = (\omega n/c)^2$ which holds even if the transverse projection is imaginary.

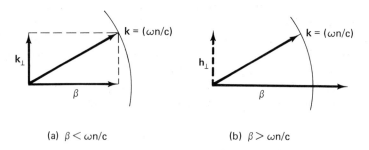

(a) $\beta < \omega n/c$ (b) $\beta > \omega n/c$

Figure 4.3 The Pythagorean relation for the components of the wave vector.

---

*Some authors name the mode after the nonvanishing component of the axial field. Thus a TE mode is also referred to as an H mode and a TM as an E mode.

Sec. 4.4 Modes in a Step-Index Fiber (or a Heterojunction Laser)

For the case illustrated in Fig. 4.3(a), $\beta < \omega n/c$ and thus $k_\perp$ can be real but constrained to

$$\mathbf{k} \cdot \mathbf{k} = (\omega n/c)^2 = [\beta \mathbf{a}_z + k_\perp \mathbf{a}_t] \cdot [\beta \mathbf{a}_z + k_\perp \mathbf{a}_t] = \beta^2 + k_\perp^2$$

$$k_\perp^2 = (\omega n/c)^2 - \beta^2$$

If β is larger than $\omega n/c$ as indicated in Fig. 4.3(b), then the component of $\mathbf{k}$ in the transverse direction must be imaginary (exponentially growing or decaying) in order for the Pythagorean relation to hold:

$$\mathbf{k} \cdot \mathbf{k} = (\omega n/c)^2 = [\beta \mathbf{a}_z \pm jh_\perp \mathbf{a}_t] \cdot [\beta \mathbf{a}_z \pm jh_\perp \mathbf{a}_t] = \beta^2 - h_\perp^2$$

$$h_\perp^2 = \beta^2 - (\omega n/c)^2$$

The analysis of Sec. 4.2 incorporated the above ideas, and it is useful to keep it in mind for the material below.

4.4.1 TE Mode ($E_z = 0$)

Consider the symmetric slab index fiber whose index of refraction variation with respect to x is sketched in Fig. 4.4. To keep the mathematics to a minimum we assume a symmetrical slab with $n_1 = n_3$ and $n_1 < n_2$. (Thus the sketch would correspond to a radial cut through a round fiber.) We search for fields in the three regions which are "guided" by the dielectric step discontinuity at $x = \pm a$.

Let us first agree on what is meant by "guided"; it merely means that most of the field is in the central core, $-a < x < a$, and becomes very small for $|x - a|$ sufficiently large. Furthermore, all would agree that such solutions must obey the wave equation (4.4.2) and the appropriate boundary conditions. For the TE modes for the slab fiber, $\partial/\partial y = 0$ and (4.4.2a) becomes

$$\frac{\partial^2 H_z^{(1,2,3)}}{\partial x^2} + \left[\left(\frac{\omega n_{1,2,3}}{c}\right)^2 - \beta^2\right] H_z^{(1,2,3)} = 0 \qquad (4.4.3)$$

There is a very important point to be noted about (4.4.3): There are three

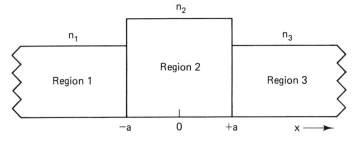

Figure 4.4 Index of refraction in a heterojunction laser or a slab fiber.

different fields corresponding to the three different regions, but there is only one phase constant β which is common to all three regions. This is what is meant by a guided *mode*—a field configuration which retains its proportionality (i.e., shape) along x but which travels with a constant phase velocity independent of x.

If the mode amplitudes are to vanish at sufficient distances away from the central slab, one must hope for an exponential solution in regions 1 and 3 as discussed previously:

$$H_z^{(1)} = B_1 \exp[h(x+a)] + A_1 \exp[-h(x+a)] \quad (4.4.4a)$$

$$H_z^{(3)} = A_3 \exp[h(x-a)] + B_3 \exp[-h(x-a)] \quad (4.4.4b)$$

with

$$h^2 = \beta^2 - \left(\frac{\omega n_{1,3}}{c}\right)^2 \quad (4.4.4c)$$

In order to keep the field finite at $x = -\infty$, $A_1 = 0$; and $A_3 = 0$ to eliminate a similar embarrassment at $x = +\infty$.

The analysis of Sec. 4.2 suggests using a trigonometric form of the solution of (4.4.3) for the central core:

$$H_z^{(2)} = A_2 \cos k_\perp x + B_2 \sin k_\perp x \quad (4.4.5a)$$

with

$$k_\perp^2 = (\omega n/c)^2 - \beta^2 \quad (4.4.5b)$$

Equation (4.4.5a) suggests that there is a natural classification of the modes into ones which are symmetric and ones that are antisymmetric about the plane $x = 0$. Our interest lies primarily with the transverse field components E_t and H_t, since they carry the optical power; and (4.4.1) indicates that their symmetry is opposite of H_z. Thus, for example, the symmetric transverse field components of the TE modes involve only the sine functions for H_z in (4.4.5) with the cosine functions describing antisymmetric ones. Let us focus on the symmetric case ($A_2 = 0$) and collect our equations together:

$$H_z^{(1)} = B_1 \exp[h(x+a)] \quad (4.4.6)$$

$$H_z^{(3)} = B_3 \exp[-h(x-a)] \quad (4.4.7)$$

$$H_z^{(2)} = B_2 \sin k_\perp x \quad (4.4.8)$$

These fields must be continuous at $x = \pm a$; hence,

$$B_1 = -B_2 \sin k_\perp a \quad (4.4.9a)$$

$$B_3 = +B_2 \sin k_\perp a \quad (4.4.9b)$$

$$\therefore B_1 = -B_3 \quad (4.4.9c)$$

Sec. 4.4 Modes in a Step-Index Fiber (or a Heterojunction Laser)

The transverse fields that are parallel to the slab are found by applying (4.4.1) to (4.4.6) through (4.4.8) and using (4.4.9a–c) for amplitude of the fields in the three regions.

$$E_y^{(1)} = \frac{j\omega\mu_0}{h} B_3 \exp[h(x+a)] \qquad (4.4.10)$$

$$E_y^{(2)} = \frac{j\omega\mu_0}{k_\perp} B_3 \cos k_\perp x \qquad (4.4.11)$$

$$E_y^{(3)} = \frac{j\omega\mu_0}{h} B_3 \exp[-h(x-a)] \qquad (4.4.12)$$

where the symmetry equation (4.4.9c) was used in (4.4.10). Again, these fields must be continuous at $x = \pm a$, and, because of symmetry, it is only necessary to work with one boundary condition.

$$\frac{j\omega\mu_0}{h} B_3 = \frac{j\omega\mu_0}{k_\perp} B_2 \cos k_\perp a \qquad (4.4.13)$$

Combining (4.4.13) with (4.4.9) yields a single implicit equation for the propagation constant β, which is contained in the quantities $k_\perp$ and h.

$$\frac{ha}{k_\perp a} = \tan k_\perp a \qquad (4.4.14a)$$

If one follows the same procedure outlined above for the antisymmetric TE modes, one obtains

$$\frac{ha}{k_\perp a} = -\cot k_\perp a \qquad (4.4.14b)$$

We will return to (4.4.14) after addressing similar issues for the TM modes.

4.4.2 TM Modes ($H_z = 0$)

The procedure for the TM case parallels the above except that now it is E_z which is the parent field. The variation of it in the various regions is precisely that given by (4.4.6) to (4.4.8) (after changing H_z to E_z on the left-hand side), the symmetry arguments are the same, and the relationship between the coefficients expressed by (4.4.9) is also the same. However, the transverse fields analogous to (4.4.10) to (4.4.12) are magnetic ones and are found by applying (4.4.1) to E_z in the various regions. For the symmetric TM case, we have

$$H_y^{(1)} = \frac{j\omega\epsilon_0 n_1^2}{h} B_1 \exp[h(x+a)] \qquad (4.4.15)$$

$$H_y^{(2)} = \frac{j\omega\epsilon_0 n_2^2}{k_\perp} B_2 \cos k_\perp x \qquad (4.4.16)$$

$$H_y^{(3)} = \frac{j\omega\epsilon_0 n_1^2}{h} B_3 \exp[-h(x-a)] \qquad (4.4.17)$$

For the symmetric mode, $B_1 = B_3$ as before, and, after matching the fields at the boundary, we obtain another implicit equation for the propagation constant, β, for the TM case:

$$\frac{ha}{k_\perp a} = \frac{n_1^2}{n_2^2} \tan k_\perp a \qquad \text{(Symmetric TM)} \qquad (4.4.18a)$$

For the antisymmetric modes

$$\frac{ha}{k_\perp a} = -\frac{n_1^2}{n_2^2} \cot k_\perp a \qquad \text{(Antisymmetric TM)} \qquad (4.4.18b)$$

4.4.3 Graphic Solution for the Propagation Constant

There is a very convenient and transparent graphic solution procedure for equations (4.4.14) and (4.4.18). Let us define dimensionless variables $X = k_\perp a$ and $Y = ha$ and recall the definitions of $k_\perp$ and h from (4.4.4c) and (4.4.5b). With those definitions, the quantity $R^2 = X^2 + Y^2$ is given by

$$R^2 = (\beta a)^2 - \left(\frac{\omega n_1 a}{c}\right)^2 + \left(\frac{\omega n_2 a}{c}\right)^2 - (\beta a)^2 = X^2 + Y^2$$

or
$$R = \left(\frac{\omega a}{c}\right)[n_2^2 - n_1^2]^{1/2} \qquad (4.4.19)$$

Thus $R = (2\pi a/\lambda_0)NA$

Thus the R number of the slab fiber is $2\pi/\lambda_0$ times the dimension a times the numerical aperture of the fiber. (In round fibers, a is the core radius, and the customary name for R is the "V" number.)

Thus (4.4.14) and (4.4.18) can be rewritten in terms of X and Y and coupled to R:

$$\begin{aligned} Y &= X \tan X & &\text{Symmetric TE} \\ Y &= -X \cot X & &\text{Antisymmetric TE} \\ Y &= (n_1/n_2)^2 X \tan X & &\text{Symmetric TM} \\ Y &= -(n_1/n_2)^2 X \cot X & &\text{Antisymmetric TM} \end{aligned} \qquad (4.4.20)$$

with
$$R^2 = X^2 + Y^2$$

Sec. 4.5 The ABCD Law for Gaussian Beams

The graphic solution is shown in Fig. 4.5 in the form suggested by the choice of variables X, Y, and R. Note that for $R < \pi/2$ only the lowest-order symmetric TE or TM modes can propagate.

Note too that ha represents the exponential decay rate of the fields away from the core of the fiber—a larger value of ha means that the field is more tightly confined to the central region. In this sense then, the TE mode experiences greater confinement to the central region and thus can be considered the dominant mode. This plays a significant role in semiconductor lasers where the region $|x| > a$ is not excited and thus is lossy. The graphic solution above shows that the parameter ha is smaller for the TM mode compared to the TE for the same value of R and thus the field extends further into region 1 or 3 for the TM case. Such regions are lossy for heterojunction lasers, and hence they tend to oscillate in the TE orientation.

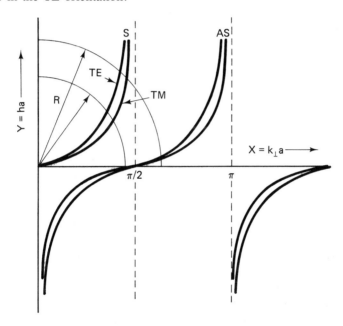

Figure 4.5 Graphic solution for the propagation constant in a slab fiber. The quantity (n_1/n_2) is assumed to be < 1 (as usual) for this graph.

4.5 THE ABCD LAW FOR GAUSSIAN BEAMS

The ABCD law is very important and yet very simple: it states that the complex beam parameter, q, of a Gaussian beam changes as it propagates through an optical system according to

$$q_2 = \frac{Aq_1 + B}{Cq_1 + D} \quad (4.5.1)$$

where A, B, C, and D are the terms of the ray matrix for that optical system and q_1 and q_2 the complex beam parameters at the entrance (1) and exit (2) planes, respectively.

The law is truly amazing in the sense that we have no right to expect such a simple connection between the simple ray theory of Chapter 2 and the Gaussian beams of Chapter 3. Yet that connection is made by (4.5.1).

It is very difficult to prove this law once and for all for all cases, but it is very easy to consider all known special cases—one at a time—and verify its validity. For instance, the differential equations of Chapter 3 predicted that for free space, q varies as

$$q(z) = q_0 + z \qquad (3.3.5)$$

For free space of length z, $A = 1$, $B = z$, $C = 0$, $D = 1$, and (4.5.1) yields precisely the same answer. The complex beam parameter, q, is most easily interpreted in terms of its reciprocal:

$$\frac{1}{q} = \frac{1}{R} - j\frac{\lambda_0}{\pi n w^2} \qquad (3.38) \rightarrow (4.5.2)$$

We can manipulate (4.5.1) to accept and present the information in that format:

$$\frac{1}{q_2} = \frac{C + D(1/q_1)}{A + B(1/q_1)} \qquad (4.5.3)$$

If one assumes a beam with a minimum spot size w_0 and a planar wave front at $z = 0$ and utilizes the $ABCD$ parameters for free space, one recovers the expansion law for a Gaussian beam:

$$\frac{1}{q_2(z)} = \frac{0 + 1 \cdot (-j\lambda/\pi w_0^2)}{1 + z \cdot (-j\lambda/\pi w_0^2)} \overset{\Delta}{=} \frac{1}{R(z)} - j\frac{\lambda}{\pi w^2(z)} \qquad (4.5.4)$$

If one separates (4.5.4) into its real and imaginary parts, one recovers the expansion laws (3.3.10) and (3.3.11) directly.

As another example of the veracity and utility of the $ABCD$ law, let us reconsider the beam transformation by a thin lens (as discussed in Sec. 2.13). Assume that a large-diameter Gaussian beam with a *planar wave front* impinges on a thin lens in the manner shown in Fig. 4.6. Equation (4.5.1) would indicate that beam parameter q' to the right of the lens is given by (4.5.3) with the appropriate values for $ABCD$:

$$\frac{1}{q'} = \frac{-1/f + 1 \cdot (1/q_1)}{1 + 0 \cdot (1/q_1)} = -\frac{1}{f} + \frac{1}{q_1} \qquad (4.5.5)$$

For air, $n = 1$, and

$$\frac{1}{q_1} = \frac{1}{\infty} - j\frac{\lambda_0}{\pi w_{01}^2} \qquad (4.5.6)$$

Sec. 4.5 The ABCD Law for Gaussian Beams

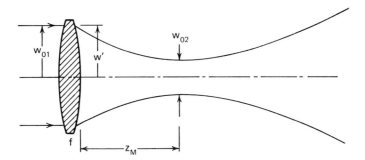

Figure 4.6 Focusing of a Gaussian beam by a lens.

Hence,

$$\left(\frac{1}{q'} \triangleq \frac{1}{R'} - j\frac{\lambda}{\pi w'^2}\right) = -\frac{1}{f} - j\frac{\lambda_0}{\pi w_{01}^2} \quad (4.5.7)$$

Equation (4.5.7) states that the spot size just to the right of the lens equals that at the left, and the beam appears to be converging toward the focal point f. This latter point is precisely the conclusion of Sec. 2.13, and the equality of spot sizes indicates that power is conserved—a most logical result.

As an example of the utility of the law, we can use it to predict the minimum focal spot size achievable with a lens. The transmission matrix for a lens plus a length of free space z is

$$\mathbf{T} = \begin{bmatrix} 1 - \dfrac{z}{f} & z \\ -\dfrac{1}{f} & 1 \end{bmatrix} \quad (4.5.8)$$

Thus the beam parameters at any point z away from the lens are given by

$$\frac{1}{R(z)} = \frac{-1/f + z(1/f^2 + 1/z_{01}^2)}{(1 - z/f)^2 + (z/z_{01})^2} \quad (4.5.9)$$

$$\frac{\lambda_0}{\pi w^2(z)} = \frac{1/z_{01}}{(1 - z/f)^2 + (z/z_{01})^2} \quad (4.5.10)$$

where $z_{01} = \pi w_{01}^2/\lambda_0$ (as usual).

From (4.5.10), we find, much to our surprise, that the spot size does not minimize at $z = f$, but at z_M, given by

$$z_M = \frac{f}{1 + (f/z_{01})^2} \quad (4.5.11)$$

However, for reasonable values of the input beam, $z_{01} \gg f$ and $z_M \sim f$, in accordance with our intuition.

We can use (4.5.11) in (4.5.10) to predict the "focal" spot size:

$$\frac{\lambda_0}{\pi w_{02}^2} = \frac{1/z_{01}}{(1 + z_M/f)^2 + (z_M/z_{01})^2}$$

or

$$w_{02} \simeq \frac{2\lambda_0}{\pi} \frac{f}{2w_{01}}$$

for

$$z_{01} = \frac{\pi w_{01}^2}{\lambda_0} \gg f$$

(4.5.12)

Thus the *smallest* focal spot occurs when the *largest* Gaussian beam impinges on the lens. Obviously, the diameter of the lens must be at least $2w_{01}$ to capture most of the beam. Hence, (4.5.8) becomes

$$w_{02} \simeq \frac{2\lambda_0}{\pi}(f^\#) \qquad (4.5.13)$$

where the f number ($f^\#$) of the lens is equal to the focal length divided by the diameter.

Let us remove the assumption that the incoming beam had a planar wave front (i.e., $R_1 = \infty$), and examine the beam just as it emerges from the lens. According to the *ABCD* law, we have:

$$\frac{1}{q_2} = \frac{C + D(1/q)}{A + B(1/q)} \qquad (4.5.3)$$

But

$$\frac{1}{q_1} = \frac{1}{R_1} - j\frac{\lambda_0}{\pi w_1^2}; \quad C = -1/f; \; A = 1; \; B = 0; \; D = 1$$

Thus

$$\frac{1}{q_2} = \frac{-1}{f} + \frac{1}{R_1} - j\frac{\lambda_0}{\pi w_1^2} \qquad (4.5.14)$$

Thus the thin lens keeps the spot size the same and therefore conserves power (thank heavens) and changes the radius of curvature of the incoming beam to

$$\frac{1}{R_2} = \frac{1}{R_1} - \frac{1}{f} \qquad (4.5.15)$$

Note that if $1/R_1 > 1/f$, the lens does *not* focus the beam!

Let us leave the thin lens and turn to the last type of special case for the *ABCD* law—that of a continuous lens with a parabolic index of refraction $n(r)$. This is reserved for a problem at the end of the chapter; only the procedure is indicated here.

Sec. 4.6 Fiber-Optic Waveguide—An Approximate Annalysis

We use the square of (2.12.7) in the wave equation,

$$\nabla_t^2 E + \frac{\partial^2 E}{\partial z^2} + \frac{\omega^2}{c^2} n^2(r) E = 0 \qquad (4.5.16a)$$

or

$$\nabla_t^2 E + \frac{\partial^2 E}{\partial z^2} + \frac{\omega^2}{c^2} n_0^2 \left(1 - \frac{r^2}{l^2}\right) E = 0 \qquad (4.5.16b)$$

and then proceed to rederive every equation of Chapter 3 from (3.2.6) to (3.3.14). Unless you lose your mind in the mathematical maze of this operation, you will have verified the ABCD law for a continuous lens.

Thus (4.5.1) is a simple and compact way of describing the evolution of a Gaussian beam in an optical system. We will see other examples of its usefulness in the following section and in Chapter 5.

4.6 FIBER-OPTIC WAVEGUIDE— AN APPROXIMATE ANALYSIS

Let us now address the issue brought up in Sec. 4.2: namely, the guidance of electromagnetic energy by a fiber whose index of refraction continuously decreases with r. The fact that the field configuration for the simple slab geometry resembled a Gaussian beam encourages us to postulate that a Gaussian beam is a normal mode of the fiber-optic waveguide with a parabolic index of refraction. We verify this postulate by demonstrating self-consistency.

If Gaussian beams are the normal modes of the waveguide, it must be described by a complex beam parameter q. We do not know the value of q, only that it exists (by the postulate). However, we also know that q will transform according to the ABCD law, from (4.5.1):

$$q(z) = \frac{A(z)q(0) + B(z)}{C(z)q(0) + D(z)} \qquad (4.6.1)$$

where **T** is given by

$$\mathbf{T} = \begin{bmatrix} \cos\left(\frac{z}{l}\right) & l \sin\left(\frac{z}{l}\right) \\ -\frac{1}{l} \sin\left(\frac{z}{l}\right) & \cos\left(\frac{z}{l}\right) \end{bmatrix} \qquad (2.12.11)$$

Now, of course, the fiber does not know of our arbitrary choice of the axis and hence does not know the location of $z = 0$. Therefore the only way for the postulated beams to be the normal modes is for the unknown complex beam parameter q to be independent of z (and our choice of the origin).

$$q(z) = q_0 \quad \text{(i.e., } q \text{ to be independent of } z\text{)} \qquad (4.6.2)$$

Combining (4.6.1) and (4.6.2) leads to a single equation for $1/q_0$:

$$\frac{1}{q_0} = -\frac{A-D}{2b} - j\left[1 - \left(\frac{A+D}{2}\right)^2\right]^{1/2}\bigg/ B \qquad (4.6.3)$$

or

$$\frac{1}{q_0} = 0 - j\frac{[1 - \cos^2(z/l)]^{1/2}}{l \sin(z/l)} = -\frac{j}{l} \qquad (4.6.4)$$

If we substitute this answer for $1/q$ into (3.3.2), we have the variation of the field with a minimum of fuss.

$$E = \exp\left(-j\frac{kr^2}{2q_0}\right) = \exp\left(-\frac{kr^2}{2l}\right) \qquad (3.3.2) \rightarrow (4.6.5)$$

One can define a characteristic spot size for this newly found Gaussian beam by

$$w^2 = \frac{2l}{k} \qquad (4.6.6)$$

A simple numerical example will convince us that this beam will fit nicely within a fiber.

Example

1. Diameter of fiber: 50 μm = 0.05 mm.
2. Index of refraction at center of core: $n_0 = 1.52$.
3. Index of refraction at $r = 25$ μm: $n = 1.52 - 0.008 = 1.512$.
4. Wavelength region of interest: $\lambda_0 = 1.06$ μm.

Combining specifications (3) and (4) leads to the evaluation of the quantity l [see (2.12.7)]:

$$\frac{n_0 a^2}{2l^2} = \Delta n \quad \text{or} \quad l = \left(\frac{n_0 a^2}{2\Delta n}\right)^{1/2} = 2.44 \times 10^{-2} \text{ cm}$$

$$k = \frac{2\pi n_0}{\lambda_0} = 9.01 \times 10^4 \text{ cm}^{-1}$$

Therefore

$$w = \left(\frac{2l}{k}\right)^{1/2} = 7.36 \times 10^{-4} \text{ cm}$$

Obviously, this spot size is much smaller than the fiber diameter, and thus there is plenty of room for higher-order modes. It is also obvious that the field at the edge of this fiber ($r = 25$ μm) is insignificant, being reduced by a factor of $\exp[-(25/7.36)^2] \simeq 9.75 \times 10^{-6}$ of that at the center.

The fact that the spot size in this Gaussian beam does *not* change with z is due to the focusing properties of the medium. As anticipated in Chapter 2, the natural divergence of the beam is exactly balanced by the convergence of the medium and thus leads to a constant beam size in the fiber.

The fact that q is independent of z leads to a very simple solution for the parameter $P(z)$ of (3.3.2). If one repeats the derivation of (3.3.4a) and (3.3.4b) for a quadratic index variation given by (2.12.7), one obtains

$$P'(z) = -j\frac{1}{q(z)} \qquad (3.3.2) \to (4.6.7)$$

But since $1/q(z) = -j(1/l)$ by (4.6.4) and is independent of z, the solution is trivial.

$$P(z) = -\frac{z}{l} \qquad (4.6.8)$$

Hence, the electric field has the following approximate form:

$$E(r, z) = E_0 \exp\left(-\frac{kr^2}{2l}\right) \exp\left[-j\left(k - \frac{1}{l}\right)z\right] \qquad (4.6.9)$$

Thus we have shown that the postulate of a Gaussian beam within this fiber is self-consistent. Note, however, that this beam is somewhat different than those encountered in Chapter 3. For instance, freely propagating beams expand with z, whereas here the spot size remains independent of z. In free space, the longitudinal-phase factor [see (3-22)] depends on z, whereas the corresponding term in (4.6.9) is independent of z. Those differences are due to the fact that the beam is being continuously focused to exactly balance the natural tendency to spread.

4.7 FIBER-OPTIC WAVEGUIDE— AN EXACT ANALYSIS

We can show that the Hermite-Gaussian beams are the normal modes of the fiber with the parabolic index by solving the wave equation directly. Although this involves considerably more mathematics, we need not go through mental gyrations (or chicanery) to arrive at the conclusions. As a bonus, we will be able to compare the approximate analysis with an exact solution and thus gain confidence in the validity of the *ABCD* law (and appreciate its simplicity).

We assume that the square of the index of refraction, the relative dielectric constant, has the following functional form:

$$\epsilon(r) = n^2(r) = n_0^2\left(1 - \frac{r^2}{l^2}\right) \qquad (4.7.1)$$

Thus the wave equation becomes

$$\nabla^2 E + \frac{\omega^2}{c^2} n_0^2 \left(1 - \frac{r^2}{l^2}\right) E = 0 = \nabla^2 E + k^2 \left(1 - \frac{r^2}{l^2}\right) E \qquad (4.7.2)$$

where

$$k = \frac{\omega}{c} n_0$$

The fact that the approximate solution, as represented by (4.6.9), implied a phase constant that was independent of z leads us to assume that fact from the start. Namely, let $E(x, y, z) = E(x, y) \exp(-j\beta z)$, where β is a constant to be determined.* Substituting that form into (4.7.2) leads to

$$\frac{\partial^2 E}{\partial x^2} + \frac{\partial^2 E}{\partial y^2} + \left(k^2 - \beta^2 - \frac{k^2 r^2}{l^2}\right) E = 0 \qquad (4.7.3)$$

with

$$r^2 = x^2 + y^2$$

Now we proceed with the standard method of solving second-order partial differential equations. Assume that $E(x, y)$ is a product function[†] $X(x) Y(y)$. Substitute and differentiate, divide by the product, and rearrange the debris such that all functions of x are on one side of the equation and all functions of y are on the other.

$$\frac{Y''}{Y} + k^2 - \beta^2 - \frac{k^2}{l^2} y^2 = -\frac{X''}{X} + \frac{k^2}{l^2} x^2 = T \qquad (4.7.4)$$

The only way the equality demanded by (4.7.4) can be satisfied for all values of x and y is for the right- and left-hand sides to be equal to a common constant—call that separation constant T. Thus we have two ordinary separated differential equations[‡]:

$$X'' + \left(T - \frac{k^2}{l^2} x^2\right) X = 0 \qquad (4.7.5)$$

$$Y'' + \left(k^2 - \beta^2 - T - \frac{k^2}{l^2} y^2\right) Y = 0 \qquad (4.7.6)$$

---

*If our previous approach had been exact, then (4.6.9) would indicate that $\beta = k - 1/l$. As we will see, that is a reasonable answer, but still, it is an approximation.

† Most fibers are round; hence, cylindrical coordinates might be used. However, there is nothing wrong with staying with more familiar Cartesian coordinates.

‡ The equations have the same appearance as the Schrödinger equation for a harmonic oscillator.

Sec. 4.7 Fiber-Optic Waveguide—An Exact Annalysis

After a frantic search of mathematical tables, one finds that the Hermite-Gaussian functions of the form $H_m(u) \exp(-u^2/2)$ solve these differential equations provided the variable u and the constant terms, T and $k^2 - \beta^2 - T$, are chosen correctly. If we substitute u for $(k/l)^{1/x}x$, (4.7.5) becomes

$$\frac{d^2X}{du^2} + \left(\frac{Tl}{k} - u^2\right)X = 0 \qquad (4.7.7a)$$

where

$$\left(\frac{k}{l}\right)^{1/2} x \triangleq u \qquad (4.7.7b)$$

The Hermite polynomial of degree m times a Gaussian is a solution provided that the constant in (4.7.7a) is an odd integer, $2m + 1$,

$$X(x) = H_m\left[\left(\frac{k}{l}\right)^{1/2} x\right] \exp\left(-\frac{kx^2}{2l}\right) \qquad (4.7.8)$$

if

$$\frac{Tl}{k} = 2m + 1 \implies T = (2m + 1)\frac{k}{l} \qquad (4.7.9)$$

We repeat the same manipulations on (4.7.6) to find that

$$Y(y) = H_p\left[\left(\frac{k}{l}\right)^{1/2} y\right] \exp\left(-\frac{ky^2}{2l}\right) \qquad (4.7.10)$$

if

$$\frac{l}{k}(k^2 - \beta^2 - T) = 2p + 1$$

or

$$\beta^2 = k^2 - T - (2p + 1)\frac{k}{l} \qquad (4.7.11)$$

where $2p + 1$ is a different odd integer.

Now one combines (4.7.9) and (4.7.11) to solve for the "unknown" propagation constant β:

$$\beta^2 = k^2 - \frac{2k}{l}(1 + m + p) \qquad (4.7.12)$$

or

$$\beta = k\left[1 - \frac{2}{kl}(1 + m + p)\right]^{1/2} \qquad (4.7.13)$$

If we define a characteristic spot size w by the relationship $w^2 = 2l/k$, which is the same one we used before [see (4.6.6)], we obtain a familiar-appearing expression,

$$E(x, y, z) = E_0 H_m\left(\frac{\sqrt{2}x}{w}\right) H_p\left(\frac{\sqrt{2}y}{w}\right) \exp\left(-\frac{r^2}{w^2}\right)$$
$$\times \exp\left\{-jk\left[1 - \frac{2}{kl}(1 + m + p)\right]^{1/2} z\right\} \quad (4.7.14)$$

Note that there are differences between this "exact" answer and the approximate one [see (4.6.9)]. The presence of the Hermite polynominals could be expected based on the corresponding terms in the free-space modes. The only unexpected difference lies in the longitudinal-phase constant and even that is minor. Compare for $m = p = 0$:

$$\beta = k\left[1 - \frac{2}{kl}(1 + m + p)\right]^{1/2} \iff \beta = k - \frac{1}{l}$$

"Exact" eq. (4.7.13)   "Approx." eq. (4.6.9)

If we expand the exact answer, with the assumption of kl being large (for the example of the last section, $kl = 2.2 \times 10^3$), we obtain the approximate answer given by (4.6.9). The fact that one obtains such a close answer with the approximate analysis should inspire confidence in the theoretical answer obtained by the simple application of the *ABCD* law.

4.8 HIGHER-ORDER TEM MODES

The exact analysis does yield the variation of the field in the transverse direction in a natural fashion. Furthermore, this mathematical analysis does allow us to *estimate* the number of transverse modes that can exist in a given fiber. For instance, in the prior example of Section 4.6, the spot size of the (0, 0) mode was 7.4 μm, whereas the radius of the parabolic fiber is 25 μm. How many higher-order $\text{TEM}_{m,p}$ modes can be guided? In answering, we will gain some mathematical insight into some of the characteristics and peculiarities of Hermite-Gaussian functions.

Consider (4.7.7a) again and insert the value of the constant given by (4.7.9):

$$\frac{d^2X}{du^2} + [(2m + 1) - u^2]X = 0 \quad (4.8.1)$$

Let us look for an *approximate* solution to this equation.

If m is large, then this is almost a simple harmonic oscillator equation for u small. Thus we could expect a solution that looks like a trigonometric function with a varying frequency:

$$X \sim \sin\left[(2m+1) - u^2\right]^{1/2} u \quad \text{for } u \text{ small} \tag{4.8.2}$$

But at large enough values of $u = (k/l)^{1/2} x$, the sign of the second term changes and we would expect an exponential-like decay:

$$X \sim \exp\{-[u^2 - (2m+1)]^{1/2} u\} \quad \text{for } u \text{ large} \tag{4.8.3}$$

This crude line of reasoning provides a reasonably good picture of the Hermite-Gaussian functions and also allows us to estimate the maximum mode number for a given fiber.

If a mode is to be guided by the fiber, the field in the cladding must be negligible compared to that in the core. Consequently, the mode number m must be sufficiently small so that the field would be described by (4.8.3) at the core-cladding interface. Thus, at x equal to the radius, a, of the core, $2m + 1$ must be less than u^2, or

$$2m + 1 < u^2 \triangleq \frac{kx^2}{l} \tag{4.8.4}$$

Since $w^2 = (2l/k)$, then $k/l = 2/w^2$ and we obtain a simple estimate:

$$2m + 1 < 2\left(\frac{a}{w}\right)^2 \tag{4.8.5}$$

In the numerical example discussed earlier, the fiber radius was 25 μm, and w was found to be 7.4 μm. Thus the mode index m is restricted to be less than ~ 11. Similar considerations apply to the y variation.

Thus, if we do not let minor inaccuracies bother us, we can estimate that there are $\simeq m \cdot p \simeq 100$ different modes that can be guided in this one fiber. Furthermore, those modes with different $m + p$ values have different phase and group velocities and hence transport energy at a different rate. This has serious implications for optical data communications.

4.9 DISPERSION IN FIBERS

Let us ask a very simple question: How long does it take for a short pulse to travel the length of a fiber? We can give a first-order answer very quickly: we divide the length by the velocity of propagation, c/n, and obtain the transit time. Although this is a useful first-order answer, it is not accurate enough for sophisticated communication applications.

Some of the pitfalls with this line of reasoning are as follows. The time delay is given by $l \div (c/n)$ *provided* that the electromagnetic energy is in the form of a uniform plane wave *and* the index of refraction is independent of frequency *and/or* we are dealing with a single-frequency source. Seldom are any of these provisos true; the fiber does *not* propagate a uniform plane wave; the index of refraction varies with wavelength; and a typical source for fiber-optic communications might be a multimode semiconductor laser or light-emitting diode (LED), both of which generate optical signals spread out over a band of wavelengths. Consequently, we must be more precise in the calculation of this time delay.

Electromagnetic energy travels at the group velocity v_g defined by

$$v_g = \left(\frac{d\beta}{d\omega}\right)^{-1} \tag{4.9.1}$$

Hence, we return to (4.7.13) for the phase constant of the TEM$_{m,p}$ mode:

$$\beta = k\left[1 - \frac{2(1 + m + p)}{kl}\right]^{1/2} \tag{4.7.13}$$

where

$$k = \frac{\omega}{c} n(\omega) \tag{4.9.2}$$

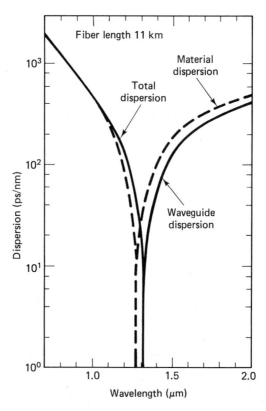

Figure 4.7 The delay dispersion of an 11 km fiber. (Data from Ref. 19.)

Chap. 4   Dispersion in Fibers   **103**

There are two causes of dispersion. As is obvious from (4.7.13), even the phase constant of the TEM$_{m,p}$ mode depends on the mode number. Thus it should come as no surprise that a multimode fiber—one permitting many values of m and p—suffers from modal dispersion. In other words, the energy in the different (m, p) modes travels with different group velocities.

One can use a single-mode fiber and eliminate that source of dispersion. However, there still remains the problem that the index of refraction of material is frequency (or wavelength) dependent. This is called material dispersion and is most important. If we ignore the second term in the radical of (4.7.13) as being small, then

$$\frac{d\beta}{d\omega} = \frac{dk}{d\omega} = \frac{n(\omega)}{c} + \frac{\omega}{c} \cdot \frac{dn(\omega)}{d\omega} \qquad (4.9.3a)$$

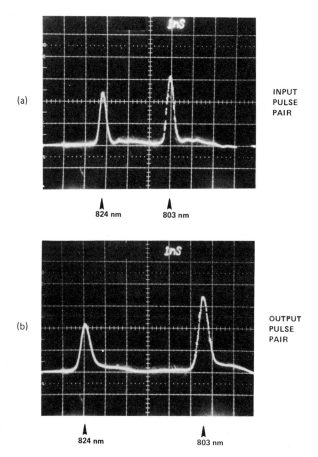

Figure 4.8 Dispersion effects in an optical fiber: (a) input pulse pair; (b) output pulse pair. (From Ref. 11.)

$$= \frac{n(\lambda)}{c} \cdot [1 - \frac{\lambda_0}{n(\lambda)} \cdot \frac{dn(\lambda)}{d\lambda}] \quad (4.9.3b)$$

Thus the group velocity is given by

$$v_g = c/n_g \quad (4.9.4a)$$

where

$$n_g = n(\lambda) \cdot [1 - \frac{\lambda_0}{n(\lambda_0)} \cdot \frac{dn(\lambda)}{d\lambda}] \quad (4.9.4b)$$

The parameter n_g is called the group index, and it too is wavelength dependent. This leads to a difference in the arrival time of different parts of a pulse on a fiber and thus to a spreading of the waveform. Fig. 4.7 is an example of the added arrival time expressed as picoseconds per nanometer of band width used by the pulse for an 11 km length of fiber. Note that there is a minimum near 1.3 μm, and this fact accounts for the frantic research effort on semiconductor lasers in that wavelength region.

Figure 4.8 shows the effect of this dispersion in a simple graphic manner. In Fig. 4.8(a), the two input pulses are shown separated in time by ~3.2 ns. After propagating for 1 km, the pulses are separated by 5.6 ns and are obviously broadened in the pulse width. Such factors must be accounted for in the design of an optical communication link.

Such considerations would take us far afield. However, a few problems are given next to further identify the issue and possibly inspire you to pursue the subject elsewhere.

PROBLEMS

4.1. Verify the *ABCD* law for a continuous lens by starting with (4.5.16b) and following the analysis of Sec. 3.1 through 3.3.

4.2. A convenient, if oversimplified definition of a focal length of a lens, is that it converges a parallel beam of light to a point. But if the spot size at the focus were zero, as implied by a point, the expansion of the beam would be infinitely fast and by symmetry would also correspond to its convergence—both statements being obvious contradictions. Use a simple geometric argument based on the convergence (and expansion) to estimate the minimum spot size in the focal region of a lens. Compare with the exact answer.

4.3. Suppose that a Gaussian beam with $w = 2$ cm and a planar wave front impinges on a lens of focal length $f = 4$ cm ($\lambda_0 = 1.0$ μm).
   (a) If $z = 0$ is the location of the lens, where does the output beam reach its minimum spot size?
   (b) What is the far-field expansion angle?

4.4. A typical fiber with a quadratic variation of index $n(r) = n_0 - \Delta n(r/a)^2$ has the following parameters: $n_0 = 1.5$, $\Delta n = 8 \times 10^{-3}$, and $a = 20$ μm. Assume that all modes with numbers $(m, p \leq 20)$ are excited more-or-less uniformly at $z = 0$ and with a common phase. Calculate the phase shift of the (m, p) mode relative to $(0, 0)$ mode at $z = 1$ km. Estimate the distribution of phases of the various modes relative to $m = p = 0$.

4.5. Consider an optical fiber whose core has a parabolic index variation in the manner shown below. The index of refraction on the axis, $n(r = 0) = n_0 = 1.52$; the diameter of the core is 40 μm; and the change in index is $\Delta n = 0.006$. Assume $\lambda_0 = 1.315$ μm.

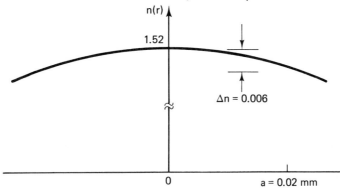

(a) What is the scale length l in the formula for the index?
(b) What is the numerical value of the spot size in this fiber?
(c) Suppose the $TEM_{0,0}$ and $TEM_{0,1}$ modes were excited with equal powers and with a common phase at $z = 0$. Find the distance, z_1, where the two modes are 180° out of phase.
(d) Make a careful sketch of the total intensity, $E \cdot E^*/2\eta_0$, as a function of $(x, y = 0, z = 0)$ and for $(x, y = 0, z = z_1)$. Assume that both modes have the same sense of polarization.
(e) What is the value of the index at $r = w$?

4.6. Find the normalized coordinates $(x/w, y/w)$ at which the electric field of the $TEM_{1,0}$ mode is a maximum.

4.7. The complex beam parameter in a parabolic index fiber is given by
$$\frac{1}{q} = -\frac{(A - D)}{2B} - j\frac{\{1 - [(A + D)/2]^2\}^{1/2}}{B}$$
which was derived from the $ABCD$ matrix via a quadratic equation. However the $(\pm)$ sign in front of the imaginary part of this definition has been changed to a simple $(-)$ sign. For what physical reason or reasons can we drop the positive portion of the $(\pm)$ sign before the imaginary part of $1/q$? (Do not just say that the $(+)$ sign is physically impossible. Explain why it is impossible.)

4.8. Repeat the analysis of Sec. 3.1 and 3.3 for a medium in which the dielectric constant is complex and depends on r in the following manner:

$$\epsilon(r) = \epsilon_0 \left[\epsilon' + j\epsilon'' \left(1 - \frac{r^2}{l^2} \right) \right]$$

The term ϵ'' can be positive or negative corresponding to gain or loss, but in any case, it is much less than ϵ' and the scale length l is much larger than r.

(a) If $\epsilon'' > 0$, does the medium have gain or loss?

(b) With gain or loss, the amplitude of the field does not remain constant with z. Hence, one assumes a solution of the form

$$E = E_0 \psi(x, y, z) \exp \left[\frac{\gamma}{2} - j\frac{\omega}{c}(\epsilon')^{1/2} \right] z$$

(1) What is a logical choice for the relationship between the gain coefficient γ and $\epsilon''(r)$?

(2) What is the differential equation for the wave function ψ?

(c) It is possible to obtain an amplified beam profile in such a medium. Find that beam and relate it to the parameters of the media.

4.9. Compare, by means of a graph, the exact field distribution for a $TEM_{3,0}$ mode given by (4.5.14) and the approximate solution given by (4.8.1) through (4.8.3).

4.10. A fiber is excited in the manner shown below (obviously not drawn to scale).

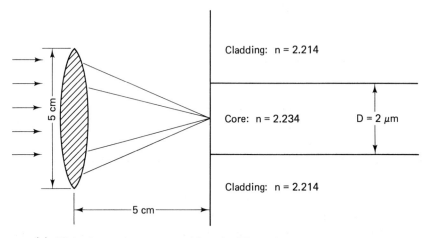

(a) If the incoming wave uniformly fills a circular lens, what fraction of the power can be captured and guided by the fiber? (Assume a round fiber and use geometric optics for this part of the problem.) If this were a slab fiber excited by a cylindrical lens, what fraction of the incident power would be captured and guided?

(b) What is the R# for this fiber?

(c) If only a single TE (or TM) mode is to be guided by the slab fiber, what is the shortest wavelength (i.e., highest frequency) that can be used?

(d) How thick must the cladding be in order for the field at the outside radius to be 10^{-6} of that at the edge of the core?

(e) What is the phase constant β (in cm^{-1}) for the TE$_0$ mode?

4.11. The laser cavity shown below produces a TEM$_{0,0}$ mode with $z = 0$ located at the flat mirror, and its output impinges on a lens of focal length f_3. Assume w_0 is known (0.5 mm), $\lambda_0 = 6328$ Å, $d_3 = 1$ m, and $f_3 = 0.25$ m.

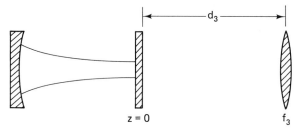

(a) What are the spot size and radius of the curvature of the wave impinging on f_3?

(b) What is the radius of curvature after passage through f_3.

4.12. Is the cavity shown below stable? You must demonstrate the logic of your answer by (a) constructing a unit cell starting at the flat mirror; (b) finding the ABCD matrix for that cell; and (c) applying the stability criteria. (d) What are the circumstances under which the quantity $[AD - BC]$ can be different from 1? Why is $AD - BC$ always equal to 1 for a cavity?

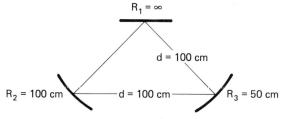

4.13. The ABCD matrix for a flat mirror with uniform reflectivity is trivial with $A = D = 1$ and $B = C = 0$. This problem concerns a nonuniform flat mirror with a reflectivity that is "tapered" with radius

$$\rho^2(r) = \exp\left[-(tr)^2\right]$$

Assume that a TEM$_{0,0}$ Gaussian beam impinges on such a mirror (with the axis of the beam corresponding to the axis of the mirror). Find a new ABCD matrix for this tapered mirror such that the ABCD law still applies for this element.

$$q_2 = \frac{Aq_1 + B}{Cq_1 + D}$$

where $q_{1,2}$ is the complex beam parameter of the incident and the reflected wave, respectively. (NOTE: Your answer must reduce to the usual one when $t = 0$. Do not waste your time tracing rays.)

REFERENCES AND SUGGESTED READINGS

1. A. G. Fox and T. Li, "Resonant Modes in a Maser Interferometer," Bell Syst. Tech. J. *40*, 453–488, Mar. 1961.
2. H. Kogelnik, "Imaging of Optical Modes—Resonators with Internal Lenses," Bell Syst. Tech. J. *44*, 455–494, Mar. 1963.
3. D. Marcuse, *Light Transmission Optics* (Princeton, N.J.: D. Van Nostrand Company, 1972).
4. D. Marcuse and S. E. Miller, "Analysis of a Tubular Gas Lens," Bell Syst. Tech. J. *43*, 1759–1787, July 1964.
5. D. Marcuse, "Theory of a Thermal Gradient Gas Lens," IEEE Trans. *MTT-13*, 734–739, Nov. 1965.
6. E. A. J. Marcateli and R. A. Schmeltzer, "Hollow Metallic and Dielectric Waveguides," Bell Syst. Tech. J. *43*, 1783–1809, July 1964.
7. P. K. Tien, "Integrated Optics and New Wave Phenomena in Optical Waveguides," Rev. Mod. Phys. *49*, 361–420, Apr. 1977.
8. Special Issue on Light Wave Communications, Phys. Today *29*(5), May 1976.
9. W. S. Boyle, "Light Wave Communications," Sci. Am., *237*, Aug. 1977.
10. D. Botez and G. J. Herskowitz, "Components for Optical Communications Systems: A Review," Proc. of IEEE, *68*, 689–731, June 1980. *Note:* The bibliography of this reference contains 640 references.
11. D. L. Franzen and G. W. Day, "Measurement of Propagation Constants Related to Material Properties in High-Bandwidth Optical Fibers," IEEE J. Quant. Electron. *QE-15*, 1409–1414, Dec. 1979.
12. K. A. Jones, *Introduction to Optical Electronics* (New York: Harper and Row, Publishers Inc., 1987).
13. R. G. Hunsperger, *Integrated Optics: Theory and Technology*, Springer Series in Optical Sciences (New York: Springer-Verlag, 1984).
14. H. A. Haus, *Waves and Fields in Optoelectronics* (Englewood Cliffs, N.J.: Prentice-Hall, Inc., 1984), Chap. 6.
15. G. Keiser, *Optical Fiber Communications* (New York: McGraw-Hill Book Company, 1983.)
16. A. H. Cherin, *An Introduction to Optical Fibers* (New York: McGraw-Hill Book Company, 1983).
17. D. Marcuse, *Theory of Dielectric Optical Waveguides* (New York: Academic Press, 1974).
18. E. A. Lacy, *Fiber Optics* (Englewood Cliffs, N.J.: Prentice-Hall, Inc.).
19. J. Yamada, "High Speed Optical Pulse Transmission at 1.29 μm Wavelength using Low-loss Single Mode Fibers." IEEE J. Quant. Electron. *QE-14*, 791, 1978.

5

Optical Resonators

5.1 INTRODUCTION

A most important part of any laser is the feedback system—in the simplest case, this consists of two mirrors with the active medium located between them. It is the purpose of this chapter to relate the field parameters of the characteristic cavity modes to the geometry of the feedback structures.

5.2 GAUSSIAN BEAMS IN SIMPLE STABLE RESONATORS

In Chapter 3 Gaussian beams were discussed in exhaustive detail. We may ask: How are the parameters of the Gaussian beam determined by a real-life cavity?

To answer this question, look at Fig. 5.1, showing the expansion of the

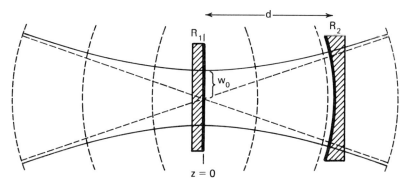

Figure 5.1 Expansion of a Gaussian beam.

TEM$_{0,0}$ mode [i.e., the variation of $w(z)$] and also the curvature of the phase front, $R(z)$. We choose our coordinate system such that the minimum spot size occurs at $z = 0$, where the wave front has an infinite radius of curvature. At this stage we do *not* know the value of w_0—only that it exists. But if we did know w_0, we *could* predict everything about the beam.

Now suppose that we choose a set of mirrors and adjust their positions and curvature so that their surfaces exactly *match* the surfaces of constant phase of the beam. For instance, suppose that we choose a flat mirror $R_1 = \infty$ and a curved mirror of radius R_2. Since the rays associated with this Gaussian beam impinge perpendicular to the mirror surface, they will be reflected back on themselves and return to the other. Thus we have obtained a self-consistent description of a normal mode of this cavity. We need three formulas from Chapter 3 to relate the beam parameters to the mirror specifications:

$$R(z) = z\left[1 + \left(\frac{z_0}{z}\right)^2\right] \qquad (3.3.11)$$

$$w^2(z) = w_0^2\left[1 + \left(\frac{z}{z_0}\right)^2\right] \qquad (3.3.10)$$

with

$$z_0 = \frac{\pi n w_0^2}{\lambda_0} \qquad (3.3.7)$$

Thus we choose the value of w_0 such that the equiphase surfaces do coincide with our choice of mirrors. For the example shown in Fig. 5.1, the flat mirror obviously *matches* the phase surface at $z = 0$. We now *force* the phase surface to match the curved mirror at $z = d$.

$$R(z) = z\left[1 + \left(\frac{z_0}{z}\right)^2\right]$$

or $\qquad (5.2.1)$

$$R(d) = R_2 = d\left[1 + \left(\frac{z_0}{d}\right)^2\right]$$

Solving this equation for the unknown z_0 is trivial:

$$z_0 = \frac{\pi w_0^2}{\lambda} = (dR_2)^{1/2}\left(1 - \frac{d}{R_2}\right)^{1/2} \qquad (5.2.2)$$

In obtaining this last result, we have used (3.3.7) to relate w_0 to z_0 and have assumed $n = 1$.

Note that z_0 and w_0 are real quantities provided that $0 \leq d/R_2 \leq 1$. Outside this region, (5.2.2) yields nonsensical answers—or, to put it bluntly, it is impossible to choose, in a self-consistent manner, the parameters of a Gaussian beam to match the resonator surfaces.

Within these limits we have achieved a solution and we now know the value of the minimum spot size w_0, (5.2.2), and where it is located—at the flat mirror. From this information we can predict what the Gaussian beam parameters are at any point in space. For instance, the spot size at the spherical mirror is found by

$$\frac{\pi w^2(d)}{\lambda_0} = \frac{\pi w_0^2}{\lambda_0}\left[1 + \left(\frac{d}{z_0}\right)^2\right] \qquad (5.2.3)$$

Substituting (5.2.2) into (5.2.3) yields the desired information.

$$\begin{aligned}\frac{\pi w^2(d)}{\lambda_0} &= (dR_2)^{1/2}\left(1 - \frac{d}{R_2}\right)^{1/2}\left[1 + \frac{d^2}{dR_2(1 - d/R_2)}\right] \\ &= \frac{(dR_2)^{1/2}}{(1 - d/R_2)^{1/2}}\end{aligned} \qquad (5.2.4)$$

We could have started with a more complicated problem with two curved mirrors, as shown in Fig. 5.2. In this problem we must adjust the position of the plane $z = 0$ so that the radius of curvature of the beam going to the right matches the surface of R_2, and simultaneously let this same beam expand to the left to mate with R_1. Here we run into a slight notation problem: the wave front on the left of $z = 0$ has a (mathematical) negative radius of curvature, but we know that the mirror R_1 was ordered from the optics company with positive (focusing) properties. Thus we must be sensible and wide awake in our application of the mathematics. We treat all distances z_1, z_2 as positive numbers and let the radii of curvature of the mirrors carry their own sign.

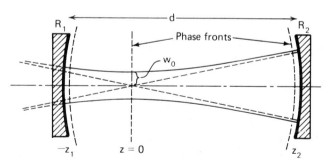

Figure 5.2 Simple stable cavity.

Since we do not know where $z = 0$ is located, we also do not know the z coordinates of the mirrors. But we do know that the sum $z_1 + z_2 = d$. Thus the following equations are pertinent:

$$z_1 + z_2 = d \qquad (5.2.5)$$

$$R(z_2) = R_2 = z_2\left[1 + \left(\frac{z_0}{z_2}\right)^2\right] \qquad (5.2.6)$$

$$R(z_1) = -R_1 = -z_1\left[1 + \left(\frac{z_0}{z_1}\right)^2\right] \qquad (5.2.7)$$

We now have the unpleasant and uninteresting task of solving three nonlinear equations in three unknowns. The solutions for the desired quantities are as follows:

$$z_0^2 = \left(\frac{\pi w_0^2}{\lambda_0}\right)^2 = \frac{d(R_1 - d)(R_2 - d)(R_1 + R_2 - d)}{(R_1 + R_2 - 2d)^2} \qquad (5.2.8)$$

$$z_1 = \frac{d(R_2 - d)}{R_1 + R_2 - 2d}$$

$$z_2 = \frac{d(R_1 - d)}{R_1 + R_2 - 2d} \qquad (5.2.9)$$

This procedure of forcing the phase fronts to coincide with the mirror surface is always successful for stable cavities but may become somewhat tedious for complex systems. Fortunately, there is a very straightforward way of handling all stable cavities, complex or simple. This is shown in the next section, where the *ABCD* law of Sec. 4.5 is used.

5.3 APPLICATION OF THE *ABCD* LAW TO CAVITIES

We can find the characteristic modes of an optical cavity by the application of the *ABCD* law, after we have established a definition of a characteristic mode in a cavity.

> A cavity mode is a field distribution that reproduces itself in relative shape and in relative phase after a round trip through the system.

This definition is so basic and so obvious that its importance is sometimes overlooked. It applies equally well to low-frequency resonators, *LC* circuits, microwave cavities, and optical cavities.

In Fig. 5.3(a), a simple short-circuited coaxial cable is used to construct a resonator. The dominant TEM mode in this system has a $1/r$ field, which propagates down to the short, reflects, comes back to the input port, reflects, and starts all over again. Because of the losses in the cable, the amplitude of the field starting the second trip is smaller than the first. But its relative amplitude $(1/r)$ shape is the same, and the relative phase is the same.

The same ideas are there for the $TE_{1,0}$ microwave cavity mode shown in Fig. 5.3(b). If one launches a $TE_{1,0}$ mode down the waveguide, it propagates to the short, back to the other short, and back to the start. Again, the relative shape and phase of the field are the same after a round trip. There is an apparent

Sec. 5.3 Application of the *ABCD* Law to Cavities   113

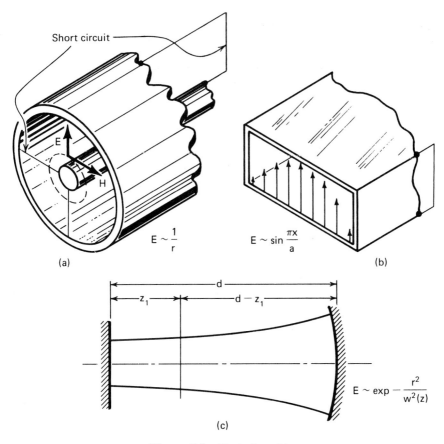

Figure 5.3 Typical cavities.

difference between the optical cavity in Fig. 5.3(c) and that shown in part (a) or (b). The fields in the optical system change in "size" as they traverse the cavity, whereas those in the low-frequency cavities do not. This is merely a consequence of the *guided* waves in Fig. 5.3(a) and (b) as opposed to a freely propagating wave in (c). All three fields fit the definition of a cavity mode.

Indeed, in their classic work, Fox and Li[1] utilized this definition to let a computer pick out a normal mode of a cavity. The mathematical detail used by them is more involved than is justified here, and the computer programing is out of place; however, the physics of their reasoning is perfect for our consideration.

They started with an *assumed* field distribution on the mirror surface M_1. That distribution was probably wrong, but as we shall see, the procedure is self-correcting. They then used Huygen's principle to predict the field distribution at mirror M_2. This involves a messy integral from an analytic standpoint but is straightforward numerically and involves merely computer time to find the field distribution at M_2. Then Huygen's principle is used once again to find the field at M_1, which is most probably different than what was started at the beginning of this paragraph.

Then they began the whole process all over again with the field that survived one complete round trip. After many such round trips through the cavity (and the DO loop), the field closely resembles that which was started. At that point, they claim they had achieved a solution for the characteristic mode in the cavity.

Rather than using a computer, we will use the *ABCD* law to follow a Gaussian beam through a round trip. In particular:

1. We *assume* that the Hermite-Gaussian beams are the characteristic modes of the optical cavity. For this to be true:
2. We *force* the complex beam parameter to repeat itself after a round trip.

Consider the cavity shown in Fig. 5.3 (analyzed previously in Sec. 5.2). We assume that there is a complex beam parameter $q(z_1)$ which describes the field on an arbitrary plane inside the cavity. We *do not* know its value at this stage—only that it exists. (We hope!) We determine the value of *q at the plane* z_1 by forcing the q to transform into *itself* after a round trip. This will be true if

$$q(z_1 + \text{round trip}) = q(z_1) \tag{5.3.1}$$

Since the left-hand side can be found from the *ABCD* law, we find that

$$q(z_1) = \frac{Aq(z_1) + B}{Cq(z_1) + D} \tag{5.3.2}$$

where A, B, C, and D are the elements of the transmission matrix for the unit cell chosen for the problem.

We now solve (5.3.2) for q—or rather $1/q(z_1)$:

$$Cq^2 + Dq = Aq + B$$

or $$\tag{5.3.3}$$

$$B\left(\frac{1}{q}\right)^2 + 2\left(\frac{A-D}{2}\right)\left(\frac{1}{q}\right) - C = 0$$

Therefore

$$\frac{1}{q} = -\frac{A-D}{2B} \pm \frac{1}{B}\left[\left(\frac{A-D}{2}\right)^2 + BC\right]^{1/2} \tag{5.3.4}$$

Now we use the fact that $AD - BC = 1$ and manipulate (5.3.4) to conform to the definition of $1/q$.

Sec. 5.3   Application of the ABCD Law to Cavities

$$\frac{1}{q(z_1)} = -\frac{A-D}{2B} - j\frac{\left[1-\left(\frac{A+D}{2}\right)^2\right]^{1/2}}{B} \tag{5.3.5}$$

Thus the radius of curvature of the beam and the spot size at the plane z_1 are given by

$$R(z_1) = -\frac{2B}{A-D} \tag{5.3.6}$$

$$\frac{\pi n w^2(z_1)}{\lambda_0} = \frac{B}{\left[1-\left(\frac{A+D}{2}\right)^2\right]^{1/2}} \tag{5.3.7}$$

It is most important to recognize that these beam parameters are found *at the plane* z_1, where the unit cell starts (and stops). The unit cell for the case shown in Fig. 5.3(c) is shown in Fig. 5.4. The transmission matrix of the unit cell shown is

$$\mathbf{T} = \begin{bmatrix} 1 & d+z_1 \\ 0 & 1 \end{bmatrix} \cdot \begin{bmatrix} 1 & d-z_1 \\ -\frac{1}{f} & 1-\frac{d-z_1}{f} \end{bmatrix} \tag{5.3.8}$$

Note that if the unit cell is started at the flat mirror (i.e., $z_1 = 0$), it is symmetric and $A = D$. Thus we have automatically found the position of the beam waist ($z_1 = 0$) where the phase front is planar or $R = \infty$. This, of course, mates to the mirror surface in the same manner as was found in Section 5.2. The beam waist

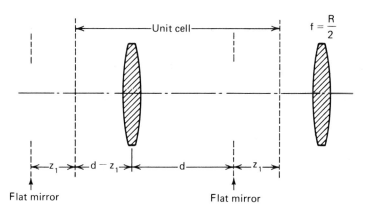

Figure 5.4 Equivalent-lens waveguide for the cavity shown in Fig. 5.3(c).

is

$$\frac{\pi w_0^2}{\lambda_0} = \frac{d + d(1 - d/f)}{[1 - (1 - d/f)^2]^{1/2}} = \frac{2d(1 - d/R)}{[4(d/R)(1 - d/R)]^{1/2}}$$

or

$$\frac{\pi w_0^2}{\lambda_0} = (dR)^{1/2}\left(1 - \frac{d}{R}\right)^{1/2}$$

(5.3.9)

which is the same as (5.2.2).

It is important to recognize that any and all cavities can be analyzed by the method outlined above. The essential steps in the analysis are:

1. Postulate that Hermite-Gaussian beams are the normal modes of the cavity.
2. Formulate an equivalent transmission system for this cavity showing *at least* one complete round trip.
3. Identify a unit cell. Is the cavity stable?
   (a) The starting point is arbitrary. However, the parameters determined from (5.3.6) and (5.3.7) correspond to R and w at the corresponding point in the cavity.
   (b) Considerable arithmetic can be avoided by an intelligent choice of a unit cell.
4. Force the complex beam parameter to transform into itself after a round trip by use of the *ABCD* law.
5. Evaluate R and w from (5.3.6) and (5.3.7).

Although the foregoing steps are logical and straightforward, nonsense can result from a blind application of (5.3.6) and (5.3.7). For instance, if $(A + D)/2$ is greater than 1, these formulas predict an imaginary radius of curvature and an imaginary spot size—which is ridiculous. Thus the theory applies for stable cavities only—no information is obtained for unstable systems.

5.4 MODE VOLUME IN STABLE RESONATORS

Let us return to (5.2.2) and (5.2.4) for the spot sizes on the mirrors of the cavity analyzed in Sec. 5.2.

$$\frac{\pi w_0^2}{\lambda_0} = (dR_2)^{1/2}\left(1 - \frac{d}{R_2}\right)^{1/2} \tag{5.2.2}$$

$$\frac{\pi w^2(d)}{\lambda_0} = \frac{(dR_2)^{1/2}}{(1 - d/R_2)^{1/2}} \tag{5.2.4}$$

The relative variation of these spot sizes as a function of mirror separation is shown in Fig. 5.5. A few minutes of arithmetic in evaluating (5.2.2) or (5.2.4)

Sec. 5.4 Mode Volume in Stable Resonators

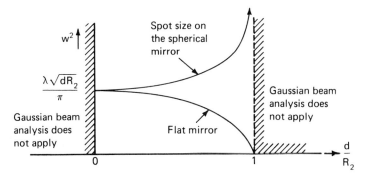

Figure 5.5 Spot sizes in a hemispherical cavity.

should convince you that these spot sizes are quite small for visible wavelengths. For instance, let $\lambda = 632.8$ nm, $d = 1$ m, $R_2 = 20$ m, with $R_1 = \infty$. The spot size w_0 is only 0.94 mm (and R_2 is a reasonably "flat" mirror.).

As we will see later, the active atoms in a laser interact with the square of the electric field; hence, it behooves us to compute the effective mode volume of the Gaussian beam. Knowing this mode volume, one can quickly estimate the *number* of atoms that must be present and radiating to generate a given optical power.

We define the mode volume in terms of the peak electric field in the cavity by

$$E_0^2 V = \int_0^d \int_{-\infty}^{+\infty} \int_{-\infty}^{+\infty} E(x, y, z) E^*(x, y, z) \, dx \, dy \, dz \quad (5.4.1)$$

where E_0 is the peak value of the electric field within the cavity, obviously occurring at the beam waist and on the axis. The complex-conjugate multiplication eliminates all phase factors, leaving only transverse variation of the field described by the Hermite-Gaussian functions and the z variation of $w(z)$. For a single mode, we have

$$E_0^2 V_{m,n} = E_0^2 \int_0^d \frac{w_0^2}{w^2(z)} \int_{-\infty}^{+\infty} \int_{-\infty}^{+\infty} H_n^2\left(\frac{2^{1/2}x}{w}\right) e^{-2x^2/w^2} H_m^2\left(\frac{2^{1/2}y}{w}\right) e^{-2y^2/w^2} \, dy \, dx \, dz$$
$$(5.4.2)$$

This equation can be rearranged and put in a more conventional form by the substitution

$$u = \frac{2^{1/2}x}{w} \quad \text{or} \quad \frac{2^{1/2}y}{w}$$

$$V_{m,n} = \int_0^d \frac{w_0^2}{2} dz \left[\int_{-\infty}^{+\infty} H_m^2(u) e^{-u^2} du\right] \left[\int_{-\infty}^{+\infty} H_n^2(u) e^{-u^2} du\right] \quad (5.4.3)$$

Now

$$\int_{-\infty}^{+\infty} e^{-u^2} H_m^2(u)\, du = 2^n n! \pi^{1/2}$$

Hence, the mode volume is given by

$$V_{m,n} = \frac{\pi w_0^2}{2} d(m!n!2^{n+m}) \quad (\text{Note: } 0! = 1) \quad (5.4.4)$$

The first factor in (5.4.4) has the satisfying interpretation: $\sim$area($\pi w_0^2/2$) $\times$ length (d); the last factor is the modification of this basic volume for the higher-order modes.

If we take the example considered previously and restrict our attention to the $TEM_{0,0}$ mode, the mode volume is quite small:

$$w_0 = 0.94 \text{ mm} \qquad R_2 = 20 \text{ m}$$
$$d = 1 \text{ m} \qquad R_1 = \infty$$

Therefore

$$V_{0,0} = 1.38 \text{ cm}^3$$

With this number one can quickly estimate the number of atoms that can possibly interact with the mode and thus contribute to the laser output power. For instance, suppose that we had a pressure of 0.1 torr for neon for this example, with each atom being excited (on the average) of 10 times per second (by the gas discharge) and thus producing a photon at 632.8 nm. What is the maximum power that we could expect from this laser?

$$\text{Energy per photon} = h\nu = \frac{hc}{\lambda_0} = 3.14 \times 10^{-19} \text{ J} = 1.96 \text{ eV}$$

$\times$

Number of neon atoms = $0.1(3.54 \times 10^{16})V_{0,0} = 4.88 \times 10^{15}$

$\times$

$$\left\{\begin{array}{l}\text{Average excitation rate per atom} \\ \| \\ \text{Average emission rate per atom}\end{array}\right\} = 10 \text{ sec}^{-1}$$

$\|$

Power $= 15.3$ mW

This is typical for a laser. There are only a couple ways to increase this power. One could increase the length, d, to make the mode volume large. But there is a practical limit; a 10-m-long laser would be most unwieldy. If one could excite the atoms at a faster rate, the power would be higher. But as we will see later, this is already optimistic. Thus we are left with changing the mode volume.

PROBLEMS

5.1. Consider the optical cavity shown in the accompanying diagram. Assume that it is stable.

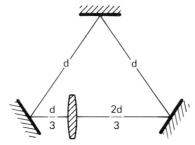

(a) Sketch an equivalent-lens waveguide.
(b) Compute the minimum spot size for a Gaussian beam and identify the place in the cavity where the beam achieves this minimum.
   (1) Identify the appropriate unit cell that makes the *ABCD* matrix symmetric.
   (2) Identify the plane $z = 0$ in the cavity.
   (3) What is the *ABCD* matrix for the unit cell?
(c) What is the formula for the minimum spot size?
(d) Show that this formula is valid for stable cavities only.

5.2. Find the spot size and the radius of curvature on the lens for the cavity shown below. The following procedure *must* be followed:
(a) Show an equivalent-lens waveguide with a unit cell starting just after the lens and proceeding in a counterclockwise fashion.
(b) For what values of d/f is this cavity stable?
(c) If $d = 20$ cm, $f = 40$ cm, and $\lambda_0 = 6000$ Å, find R and w (at the lens).
(d) Identify the plane in the cavity where the spot size is a minimum.

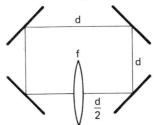

5.3. The GRIN lens shown on the diagram below consists of a *graded index* fiber with $n(r) = n_0(1 - r^2/2a^2)$ of length $d = \pi a/2$. ($a \gg r$)

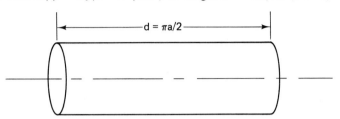

(a) Find the *ABCD* matrix of this lens. (CAUTION: Do not ignore the air-dielectric interfaces.)
(b) If we consider this short section of the fiber a cavity, what are the spot sizes at the entrance and exit planes?

5.4. (a) Construct a graph similar to Fig. 5.5 for the cavity of Fig. 5.2 ($R_1 = R_2$).
(b) Use the expansion law for a Gaussian beam to find the spot sizes on the mirrors.
(c) If $R_1 = R_2$ in Fig. 5.2, find the distance that maximizes the mode volume.

5.5 (a) If $d/2$ of the space adjacent to the flat mirror of Fig. 5.1 were filled with a negative gas lens [see (2.12.12)], show that the cavity is stable for $d = R$.
(b) Find the spot size on the mirrors with $d = R$, and the parameter L of (2.12.12) equal to 100 m.

5.6. In the stable optical cavity shown in the diagram below, the plane $z = 0$ occurs at a distance 25 cm to the left of M_1 with the beam parameter $z_0 = 125$ cm. The distance between the two mirrors is 75 cm.

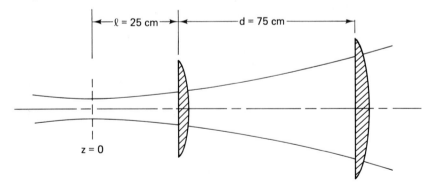

(a) Find a formula for the resonant frequency of the TEM$_{m,p,q}$ mode.
(b) Find the difference between the resonance frequency of the TEM$_{1,2,q}$ and TEM$_{0,0,q}$ modes.
(c) Find the radius of curvature for the mirrors M_1 and M_2.

5.7. Consider the optical cavity consisting of two flat mirrors with a converging lens as shown in the accompanying diagram.
  (a) What are the stability limits for this cavity? Express your answer in the form of an inequality involving the ratio of d_1/f and d_2/f.
  (b) Construct a stability diagram expressing this inequality.

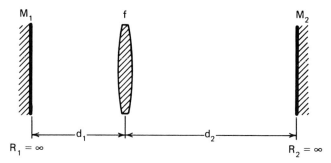

5.8. Find the spot sizes at the mirrors M_1 and M_2 of the cavity shown in Problem 5.7.

5.9. (a) Sketch the e^{-1} points of the field for the $TEM_{0,0}$ mode for the cavity of Problem 5.7.
  (b) Find the far-field divergence angle of the radiation emerging from the two mirrors. (Label this angle on the sketch.)

5.10. Find a formula for the spot size of a $TEM_{0,0,q}$ mode at the spherical mirror of Fig. 5.1 by following the procedure below.
  (a) Show an equivalent-lens waveguide for this cavity and identify a unit call such that the ABCD law will yield the spot size on the spherical mirror directly.
  (b) Use the ABCD law to find a formula for $\pi w_s^2/\lambda_0$.

5.11. The following relationship was derived for cavities:

$$\frac{1}{q} = -\frac{(A-D)}{2B} - j\frac{\{1 - [(A+D)/2]\}^{1/2}}{B}$$

It is a solution to the quadratic equation for $1/q$. What happened to the ($\pm$) sign in front of the imaginary part of $1/q$? For what physical reason or reasons can the (+) sign be dropped?

REFERENCES AND SUGGESTED READINGS

1. A. G. Fox and T. Li, "Resonant Modes in a Maser Interferometer," Bell Syst. Tech. J. *40*, 453–488, Mar. 1961.
2. H. Kogelnik and T. Li, "Laser Beams and Resonators," Appl. Opt. 5, 1550–1567, Oct. 1966.

3. A. Yariv, *Introduction to Optical Electronics* (New York: Holt, Rinehart & Winston, 1971), Chap. 4.
4. A. Yariv, *Quantum Electronics,* 2nd ed. (New York: John Wiley & Sons, Inc., 1975), Chap. 7.
5. A. E. Siegman, *Lasers* (Mill Valley, Calif.: University Science Books, 1986), Chaps. 16–21.

6

Resonant Optical Cavities

6.1 GENERAL CAVITY CONCEPTS

Until now we have implied that the cavity is an integral part of any laser because of the feedback that it provides. This is indeed true, but there is more than just raw feedback involved. As we see in this chapter, the cavity provides the ultimate in frequency-determining properties of the laser. It is most important to understand the *classical electromagnetic problem* of a cavity so as to appreciate why a laser always oscillates at a cavity resonance, why the *photon lifetime* is so important to the threshold of oscillation condition, and why the cavity "*filters*" the spontaneous emission from the atoms in the cavity so that stimulated emission takes place to generate a coherent electromagnetic wave.

6.2 RESONANCE

Resonance of an electromagnetic wave at optical frequencies is no different than resonance of any other system, be it mechanical or electrical. There is always an interchange of energy in such a system between potential and kinetic forms in the case of a mechanical system, with attendant friction losses, or between electric and magnetic energy with resistive losses in an electromagnetic problem.

Quite often, the phenomenon of resonance gets lost in the mathematics when analyzing a low-frequency system; fortunately, a much simpler physical picture emerges when we consider systems where the wavelength is much less than the physical dimensions of the components.

In order to make the problem as familiar as possible, we consider the excitation of the cavity shown in Fig. 6.1 by an external source, such as a tunable laser or a variable-frequency oscillator. To keep the arithmetic to a minimum, we consider all waves, incident on the cavity from the left, inside the cavity, or transmitted through it to the right to be uniform plane waves of limited spatial extent transverse to the direction of propagation. Our task is to relate the fields, running wave intensities, and stored energy on the inside of the cavity to those quantities that we can measure on the outside.

Let us follow a wave as it bounces back and forth between the two mirrors. Consider the initial field at the plane just to the right of M_1, labelled by $\mathbf{E}_0$. It propagates to M_2 and back to the starting plane and experiences an amplitude change of $\rho_1 \cdot \rho_2$ and a phase factor $\exp[-jk2d]$ as it travels that round trip and thus generates the field labelled $\mathbf{E}_1^+$, which experiences the same trials and tribulations as $\mathbf{E}_0$, and it in turn generates $\mathbf{E}_2^+$, and so on. At every point along the path from M_1 to M_2, the fields $\mathbf{E}_1^+$, $\mathbf{E}_2^+$, and so on are to be added to $\mathbf{E}_0$ to which we assign the reference phase of 0°. This phasorial addition is shown in Fig. 6.2 where, because there is an assumed lagging phase angle, we have assumed that the round-trip phase shift (RTPS), $2\theta = 2kd$, is almost but not quite an integral multiple of 2π radians. That deficiency is labelled by ϕ and is related to kd by

$$2\theta = 2kd = q2\pi - \phi \tag{6.2.1}$$

Note that if the angle ϕ is significant, the total field propagating to the right inside the cavity is merely the difference between the origin and the spiral of phasors—quite similar to the straight-line distance between the beginning and

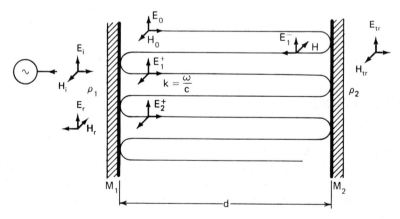

Figure 6.1 Optical cavity.

Sec. 6.2   Resonance

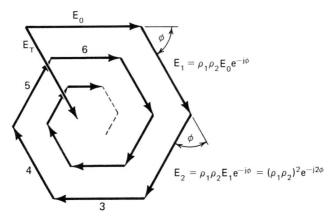

Figure 6.2 Phasor diagram.

end of a coiled rope. Not much at all! But, if that rope is uncoiled, the distance becomes much larger.

In a similar fashion, the *total* field $\mathbf{E}_T$ will be many times the initial value $\mathbf{E}_0$ if $\rho_{1,2}$ are close to 1 and $\phi = 0$. This point is most important and needs repeating many times over. The following quantities are all maximized by the simple equation $\phi = 0$: the total field travelling to the right (and thus to the left also); the magnetic fields associated with $\mathbf{E}$; the intensities of the waves; the number of photons bouncing back and forth; and the stored energy. These physical facts are characteristic of *resonance*, which is defined by

Round Trip Phase Shift   (RTPS) = $2kd = q2\pi$ (6.2.2)

This simple equation is most important and should be committed to memory for instant recall.

Equation (6.2.2) contains a lot of information that can be discussed in complementary ways. Since $k = \omega n/c = 2\pi/\lambda$, we can use one of the equalities to find the resonant wavelengths:

$$k \cdot 2d = \frac{\omega n \cdot 2d}{c} = \frac{2\pi \cdot 2d}{\lambda} = q \cdot 2\pi \qquad (6.2.3a)$$

or

$$d = \frac{q \cdot \lambda}{2} \qquad (6.2.3b)$$

where $\lambda = \lambda_0/n$. This view of resonance states that there has to be an integral number of half wavelengths between the two mirrors. This implies that the integer q is a very large number for optical frequencies and reasonable size cavities as the following examples illustrate.

Example 6.1: A semiconductor laser cavity

Material   Gallium arsenide (GaAs)
Index of refraction   3.6
Length of cavity d 100 μm = 0.01 cm = 3.94 × 10^{-4} in.
Wavelength region of interest   8000 Å

Thus

$$q = \frac{nd}{(\lambda_0/2)} = \frac{3.6 \cdot 10^{-2} \text{ cm}}{0.4 \cdot 10^{-4} \text{ cm}} = 900$$

We can turn this problem around and ask, What is the wavelength for $q = 899$ and 901?

$$\frac{\lambda_0}{2} = \frac{nd}{899} \quad \text{or} \quad \lambda_0 = 8008.898 \text{ Å for } q = 899$$

$$\frac{\lambda_0}{2} = \frac{nd}{901} \quad \text{or} \quad \lambda_0 = 7991.121 \text{ Å for } q = 901$$

Example 6.2: A gas laser

Material   Helium-neon gas at 1 torr
Index of refraction   1.000+
Length of cavity   20 cm = 8 in.
Wavelength region of interest   6328 Å

Thus $q = 632{,}111$ for $\lambda_0 = 6328.0025$ Å

and $q = 632{,}110$ for $\lambda_0 = 6328.0125$ Å

Note that there is only 0.010 Å difference between two adjacent wavelengths for the large gas laser, a difference which is beyond the resolution capabilities of most monochromators. Obviously, it was somewhat silly to carry out the calculation of wavelength to eight significant digits when the distance is only given to two. However, the point is that the wavelengths are very close together even for the small semiconductor and extremely so for a large gas laser.

Equation (6.2.2) can also be interpreted in terms of frequency ν:

$$k \cdot 2d = \omega \cdot \frac{2nd}{c} = 2\pi\nu \cdot \frac{2nd}{c} = q(2\pi)$$

$$\nu = q \cdot \frac{c}{2nd}$$

(6.2.4)

Sec. 6.3　Sharpness of the Resonance: Q, Finesse, Photon Lifetime　　127

Because q is restricted to integer values, there are only discrete frequencies which obey the resonance condition. The separation between those frequencies is given by

$$\nu_{q+1} - \nu_q = \frac{c}{2nd} \tag{6.2.5}$$

For reasonable-size cavities, this separation yields numbers that should be comfortable to most persons. For the helium-neon laser cavity of example 2, this difference is 750 MHz. Although the frequency difference is important (it is called the free spectral range), do not lose sight of the resonant frequency given by $\nu = q(c/2d) = 473.7553$ THz for $\lambda_0 = 6328.002$ Å, which is very high compared to those "comfortable" values.

Stimulated emission, which is the key issue with any laser, is always proportional to the energy (E^2); hence that stimulation will always be a maximum at a cavity resonance.

6.3 SHARPNESS OF THE RESONANCE: Q, FINESSE, PHOTON LIFETIME

It is obvious from Fig. 6.2 that this magic doesn't happen only if $\phi = 0$, but rather that it should be small. Our question is: How small? To answer, we need to describe the variation of the fields for arbitrary frequencies and thus phase shifts for a round-trip. The total field propagating to the right is given by

$$\mathbf{E}_T^+ = \sum_0^\infty E_n^+ = E_0[1 + \underbrace{\rho_1\rho_2 e^{-jk2d}}_{\text{one}} + \underbrace{(\rho_1\rho_2 e^{-jk2d})^2}_{\text{two}} + \underbrace{(\rho_1\rho_2 e^{-jk2d})^3}_{\text{three round-trips}} + \text{etc.}$$

$$= \frac{E_0}{1 - \rho_1\rho_2 e^{-j2\theta}} \tag{6.3.1}$$

where θ = electrical length of the cavity = $\omega nd/c$

The field returning from M_2 is just ρ_2 times the round-trip phase factor, $\exp[-jk2d]$, multiplying the wave going to the right.

$$\mathbf{E}_T^- = \sum E_n^- = \rho_2 \cdot e^{-j2\theta} \cdot \mathbf{E}_T^+$$

$$= E_0 \left[\frac{\rho_2 e^{-j2\theta}}{1 - \rho_1\rho_2 e^{-j2\theta}} \right] \tag{6.3.2}$$

What we need now is a prescription for the field E_0 in terms of the external source exciting the system in Fig. 6.1. This is most conveniently done by using the scattering matrix formalism (stolen from microwave electronics*), which merely relates the fields going *away* from an element to those impinging on it. Thus, with almost no information whatsoever about the construction of the mirrors, we can set down the scattering matrix for each of the mirrors:

$$\begin{bmatrix} E_r \\ \sum E_n^+ \end{bmatrix} = \begin{bmatrix} \rho_1 & jt_1 \\ jt_1 & \rho_1 \end{bmatrix} \begin{bmatrix} E_i \\ \sum E_n^- \end{bmatrix} \quad (6.3.3a)$$

where, for an assumed lossless mirror, we have the conservation of energy equation

$$\rho^2 + t^2 = 1$$

or

$$R + T = 1; \quad R = |\rho|^2, \quad T = |t|^2 \quad (6.3.3b)$$

The summations in (6.3.3a) merely represent the waves impinging on and receding from the right side of M_1. Actually, the reflected wave $\mathbf{E}_r$ is also a summation of the partial transmission through the mirror from right to left. The bottom line of this matrix equation yields a relation between the field $\mathbf{E}_0$ and that from the external source:

$$\sum E_n^+ = jt_1 E_i + \rho_1 \sum E_n^-$$

Combining (6.3.1) and (6.3.2) with this equation, we find $\mathbf{E}_0$:

$$\mathbf{E}_0 = jt_1 \mathbf{E}_i \quad (6.3.4)$$

The total field reflected from the cavity can be found from the top line of (6.3.3a):

$$E_r = \rho_1 E_i + jt_1 \sum E_n^-$$

Using $\mathbf{E}_0 = jt_1 \mathbf{E}_i$ by (6.3.4), we find that the reflected field is given by

$$\frac{E_r}{E_i} = \frac{\rho_1 - \rho_2 e^{-j2\theta}}{1 - \rho_1 \rho_2 e^{-j2\theta}} \quad (6.3.5)$$

To find the field transmitted through M_2, we propagate $\Sigma \mathbf{E}^+$ from M_1 to M_2 by the normal propagator, $\exp[-jkd]$, and then apply the bottom line of (6.3.3a) with subscripts changed to 2 for M_2. (Note that there is *nothing* incident on M_2 from the right in Fig. 6.1.)

---

*See the book by H. A. Haus[9] for an elegant and rigorous discussion. We need it to avoid solving the boundary value problem in detail.

Sec. 6.3 Sharpness of the Resonance: Q, Finesse, Photon Lifetime

$$\frac{E_{tr}}{E_i} = jt_2 e^{-j\theta} \cdot \frac{\sum E_n^+}{E_i} = \frac{-t_1 t_2 e^{-j\theta}}{1 - \rho_1 \rho_2 e^{-j2\theta}} \quad (6.3.6)$$

At optical frequencies, we measure the photon flux or Poynting vector rather than the field. This quantity is found from the power transmission coefficient *through both mirrors* obtained by multiplying (6.3.6) by its complex conjugate:

$$T = \left|\frac{E_{tr}}{E_i}\right|^2 = \frac{t_1^2 \cdot t_2^2}{1 - 2\rho_1 \rho_2 \cdot \left[\frac{e^{j2\theta} + e^{-j2\theta}}{2}\right] + \rho_1^2 \rho_2^2} \quad (6.3.7a)$$

Now $\cos 2\theta = 1 - 2\sin^2\theta$ and $1 - 2\rho_1\rho_2 + (\rho_1\rho_2)^2 = [1 - \sqrt{R_1 R_2}]^2$ where the power reflectivity $R = \rho^2$ and $|t|^2 = 1 - R$:

$$T = \frac{(1 - R_1) \cdot (1 - R_2)}{(1 - \sqrt{R_1 R_2})^2 + 4\sqrt{R_1 R_2} \sin^2\theta} \quad (6.3.7b)$$

The power reflection coefficient of the cavity $|E_r/E_i|^2$ is given by

$$R_{net} = \left[\frac{E_r}{E_i}\right]^2 = \frac{(\sqrt{R_1} - \sqrt{R_2})^2 + 4\sqrt{R_1 R_2} \sin^2\theta}{(1 - \sqrt{R_1 R_2})^2 + 4\sqrt{R_1 R_2} \sin^2\theta} \quad (6.3.7c)$$

There is considerable physics tied up in (6.3.7) so let us dissect it in every way possible.

It is quite instructive to plot this transmission coefficient as a function of the electrical length $\theta = \omega n d/c$ as in Fig. 6.3. Note that the transmission peaks at resonance $\theta = q\pi$, as predicted by (6.2.2), with a maximum value of

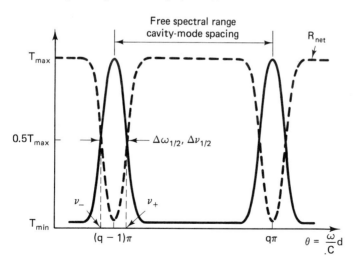

Figure 6.3 Transmission or bandpass characteristics of a Fabry-Perot cavity.

$$T_{\max} = \frac{(1-R_1)\cdot(1-R_2)}{(1-\sqrt{R_1 R_2})^2} \tag{6.3.8a}$$

The fact that it is a maximum should surprise no one since the field is a maximum, which maximizes the wave incident on and through M_2. However, its numerical value may shock some so let us go slowly.

If the mirrors are identical such that $R_1 = R_2 = R$, then $T_{\max} = 1$; that is, *all* of the power incident on the cavity from the left is passed through the cavity. Why the fuss? Check some numbers and think about the implications. Suppose $R = 0.995$ and 1 watt was incident on the cavity from the source. Then the arithmetic states that 1 watt emerges from M_2, which only has a power transmission coefficient T of 0.005. Thus the power incident on M_2 inside the cavity amounts to 200 watts! Where did this extra power come from, or have we violated conservation of power? Not guilty!

If one thinks of the source as being a sequence of very short pulses of photons with a repetition rate equal to the round-trip time in the cavity, the origin of this large running wave power can be understood. For the first pulse, only a small fraction of the incident photons enters the cavity (0.5% for the numbers chosen above). Most of those that do enter the cavity survive a round trip with a high probability, $R_1 R_2 = 0.99$; indeed, it takes nearly 100 round-trips for these photons to decay to 36.8% of the initial value. For $d = 20$ cm, this is a significant time interval of $100 \times (2nd/c) = 133$ ns for the cavity of example 2. During this time, more photons are being added by successive pulses from the source at the correct time and, because we assume a coherent source, *at the same phase* as the initial input. We are adding the same amount per unit of time, but we lose only a fraction of what is present. Thus the energy accumulates until the rate of input equals the losses with the interior fields being much greater than the external ones. This, of course, happens in a simple R-L-C circuit with voltage across L at resonance being much larger than the source.

Note also that the reflected power is zero if $R_1 = R_2$ by (6.3.7c). This is also due to the very large internal fields at resonance. The sum of the transmissions through M_1 of the partial waves traveling to the left in the cavity exactly cancels the large component of $\mathbf{E}_i$ reflected by M_1.

At antiresonance, the transmission through the cavity is virtually zero with a value given by

$$T_{\min} = \left[\frac{1-R}{1+R}\right]^2 \tag{6.3.8b}$$

For the numbers chosen above, $T_{\min} = 6 \times 10^{-6}$ and thus the interior power incident on M_2 is only 1.25×10^{-3} of that of the source. The ratio between the fields at resonance to that at antiresonance is given by

$$\frac{E^+(\text{resonance})}{E^+(\text{antiresonance})} = \left(\frac{T_{\max}}{T_{\min}}\right)^{1/2} = \frac{1+(R_1 R_2)^{1/2}}{1-(R_1 R_2)^{1/2}} \tag{6.3.9}$$

Sec. 6.3 Sharpness of the Resonance: Q, Finesse, Photon Lifetime

For highly reflecting mirrors, this is a large number. Thus the cavity expresses a *loud* and *definite preference* for fields at the resonant frequency of the cavity.

Of course, there is a narrow band of frequencies about the resonant one, where the fields are appreciable. A measure of the sharpness of the resonance is the Q (for quality) of the cavity:

$$Q = \frac{\nu_0}{\Delta \nu_{1/2}} \tag{6.3.10}$$

where $\Delta \nu_{1/2}$ is the full width at half maximum (FWHM) (see Fig. 6.3) and ν_0 is the resonant frequency of the cavity $\nu_0 = q(c/2nd)$.

Equation (6.3.7) is used to solve for the frequencies ν_+ and ν_-, which makes the response decrease to one half of its maximum value. For highly reflecting mirrors, the electrical angle will deviate only slightly from $q\pi$, and one can use the small-angle approximation for $\sin(q\pi + \Delta\theta) \sim \Delta\theta$.

$$4(R_1 R_2)^{1/2} \sin^2 \frac{\omega_\pm nd}{c} = [1 - (R_1 R_2)^{1/2}]^2$$

or

$$\sin \frac{\omega_\pm nd}{c} = \pm \frac{1 - (R_1 R_2)^{1/2}}{2(R_1 R_2)^{1/4}}$$

$$\nu_\pm = q \frac{c}{2nd} \pm \frac{c}{2nd} \frac{1 - (R_1 R_2)^{1/2}}{2\pi (R_1 R_2)^{1/4}} \tag{6.3.11}$$

Therefore

$$\Delta \nu_{1/2} = \nu_+ - \nu_- = \frac{c}{2nd} \frac{1 - (R_1 R_2)^{1/2}}{\pi (R_1 R_2)^{1/4}}$$

Thus the cavity Q is given by $q(c/2nd)$ divided by $\Delta \nu_{1/2}$. One should recognize that q is merely the number of half wavelengths between the two mirrors; thus

$$Q = \frac{q(c/2nd)}{\Delta \nu_{1/2}} = \frac{2\pi nd}{\lambda_0} \frac{(R_1 R_2)^{1/4}}{1 - (R_1 R_2)^{1/2}} \tag{6.3.12}$$

since

$$q = \frac{nd}{\lambda_0/2}$$

Those who were introduced to Q at radio and/or microwave frequencies should be comfortable with the concept of Q as a measure of the sharpness of the resonance. Unfortunately, the numerical values of Q at optical frequencies are astronomical, primarily because of the smallness of λ_0. For instance, suppose that $d = 1$ m, R_1, $R_2 = 0.99$, and $\lambda = 632.8$ nm; then $Q = 9.88 \times 10^8$. But, of course, the resonant frequency is also astronomical—$\nu = c/\lambda = 4.74 \times 10^{14}$. Only the half-width $\Delta\nu_{1/2}$ is a familiar quantity, 480 kHz.

To avoid such large numbers and yet provide a measure of the filtering properties of the cavity, one uses another term—the Finesse (F).*

or

$$F = \frac{\text{free spectral range}}{\text{full width at half maximum}} = \frac{c/2nd}{\Delta\nu_{1/2}}$$

$$F = \frac{\pi(R_1R_2)^{1/4}}{1 - (R_1R_2)^{1/2}}$$

(6.3.13)

For the numbers used above, this yields a more reasonable value of $F = 313$.

6.4 PHOTON LIFETIME

It was mentioned earlier that the cavity stores the input in the field between the mirrors, and this fact allows the running wave power to be much larger than the external source. Obviously, it must take time to build up the energy to the high value. By the same line of reasoning, once the field is established inside the cavity, it takes time for the energy to decay down to zero after the external supply is shut off. To compute this time, we follow a package of N_p photons around the cavity as shown in Fig. 6.4.

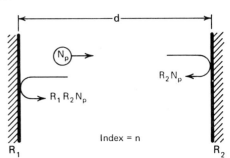

Figure 6.4 Decay of photons in a cavity.

After one round-trip the photon number has decreased to $R_1R_2N_p$, and the loss $(1 - R_1R_2)N_p$ has occurred in a time interval $2nd/c$. Thus the rate of change of photons inside the cavity is

$$\frac{dN_p}{dt} = \frac{1 - R_1R_2}{2nd/c}N_p$$

(6.4.1)

---

*Historically, the use of the term came from those using optical instruments, whereas the Q concept came from the lower-frequency domain. At optical frequencies, the mirror system is referred to as a Fabry-Perot cavity. It is quite often used as a bandpass filter to examine very narrow portions of a spectrum.

Sec. 6.4 Photon Lifetime

which has the simple solution given by

$$N_p(t) = N_{p0} \exp\left(-\frac{t}{\tau_p}\right)$$

where

$$\tau_p = \frac{2nd/c}{1 - R_1 R_2} \quad (6.4.2)$$

$$= \frac{\text{time for a round-trip}}{\text{fraction lost per round-trip}}$$

and N_{p0} is the initial number of photons in the cavity at $t = 0$. The quantity τ_p is called the *photon lifetime*.

One can relate this quantity to Q (and thus F) by using the theoretical definition* of Q.

$$Q = 2\pi \left(\frac{\text{energy stored in the system at resonance}}{\text{energy lost in a cycle of oscillation}}\right) \quad (6.4.3a)$$

$$= \omega_0 \left(\frac{\text{energy stored in the system at resonance}}{\text{average power lost}}\right) \quad (6.4.3b)$$

Since the average power lost from a system is the (decreasing) time rate of change of energy, we have

$$Q = \omega_0 \frac{W}{-dW/dt} \quad (6.4.3c)$$

where W is the stored energy. The solution to (6.4.3c) is again trivial and provides the connection among Q, F, and τ_p:

$$W(t) = W_0 \exp\left(-\frac{\omega_0}{Q} t\right) \quad (6.4.4)$$

where W_0 is the initial value of stored energy and equal to the initial number of photons times the energy per photon $h\nu$. The connection between τ_p and Q is obvious:

$$\tau_p = \frac{Q}{\omega_0} \quad (6.4.5)$$

Since $\omega_0/Q = \Delta\omega_{1/2}$, the full width at half-maximum in radian frequency units, we obtain an "uncertainty" relation,

$$\Delta\omega_{1/2} \tau_p = 1 \quad (6.4.6)$$

---

*The previous definition of $Q = \nu_0/\Delta\nu_{1/2}$ is most likely to be used by experimentalists in the laboratory, whereas (6.4.3) is most useful for theoretical calculations.

This should surprise nobody!

If one uses (6.4.5) and (6.4.2) to solve for Q, one obtains

$$Q = 2\pi\nu_0 \frac{2nd}{c} \frac{1}{1 - R_1 R_2} = \frac{4\pi nd}{\lambda_0} \frac{1}{1 - R_1 R_2} \qquad (6.4.7)$$

which is slighty *different* from that presented in (6.3.12). However, for highly reflecting mirrors, both lead to the same approximate answer. Using the same numbers as before, we obtain $Q = 9.99 \times 10^8$, as opposed to 9.88×10^8 using (6.3.12). This difference is of minor importance.

It is the resonance of the fields inside the cavity that is of major importance. As we will see in Chapter 7, the atoms are stimulated by the square of the electric field; hence, only the fields near resonance participate in the interaction to any extent. It is this fact that makes the cavity so important in the operation of a laser.

6.5 RESONANCE OF THE HERMITE-GAUSSIAN MODES

In the analysis to this point in the chapter we have assumed a limited-extent uniform plane wave for the fields, whereas the more exact description leads to the Hermite-Gaussian modes. The concepts introduced previously are still applicable, but there are refinements.

Resonance is still determined by (6.2.1); the round-trip phase must equal 2π radians. However, the phase constant for the Hermite-Gaussian modes is not $\omega n/c$ but is much more complicated. The phase shift experienced by a TEM$_{m,p}$ mode in propagating from $z = 0$ to a point $z = d$ is given by

$$\phi(0 \longrightarrow d) = kd - (1 + m + p) \tan^{-1}\left(\frac{d}{z_0}\right) \qquad (3.5.1)$$

Thus the equation for resonance for the mirror system shown in Fig. 6.5 is as follows:

$$kd - (1 + m + p) \tan^{-1}\left(\frac{d}{z_0}\right) = q\pi \qquad (6.5.1)$$

If we recall (5.2.2), which relates z_0 to the radius of curvature, R_2,

$$z_0 = (dR_2)^{1/2}\left(1 - \frac{d}{R_2}\right)^{1/2} \qquad (5.2.2)$$

and perform some painful, yet trivial manipulations on (6.5.1), we obtain a more easily interpreted formula for the resonant frequencies of the TEM$_{m,p,q}$ modes.

$$\nu_{m,n,q} = \frac{c}{2nd}\left\{q + \frac{1 + m + p}{\pi} \tan^{-1}\left[\frac{(d/R_2)^{1/2}}{(1 - d/R_2)^{1/2}}\right]\right\} \qquad (6.5.2)$$

Sec. 6.5 Resonance of the Hermite-Guassian Modes

or

$$\nu_{m,n,q} = \frac{c}{2nd}\left[q + \frac{1+m+p}{\pi}\cos^{-1}\left(1 - \frac{d}{R_2}\right)^{1/2}\right] \quad (6.5.3)$$

If M_1 has a finite radius of curvature R_1, the arithmetic is most painful, but the answer is quite similar to (6.5.3):

$$\nu_{m,n,q} = \frac{c}{2nd}\left[q + \frac{1+m+p}{\pi}\cos^{-1}(g_1 g_2)^{1/2}\right] \quad (6.5.4)$$

where

$$g_{1,2} = 1 - \frac{d}{R_{1,2}}$$

Note that the frequency separation between modes with the same transverse numbers is still $c/2nd$ as in the previous section, but there is a difference between modes with different m and p values.

For example, let $d/R_2 = \frac{1}{2}$ in Fig. 6.5; hence, the arc cosine term in (6.5.3) is $\pi/4$.

$$\nu_{m,n,q} = \frac{c}{2nd}\left(q + \frac{1+m+p}{4}\right) \quad (6.5.5)$$

Thus, as m or p increases by 1, the frequency changes by one fourth of the fundamental spacing $c/2nd$. Furthermore, the (0, 0) mode is shifted from that predicted by the plane-wave analysis by this same value. These effects are shown in Fig. 6.6.

Also shown in Fig. 6.6 is the fact that two modes, with different values of m, p, and q, can have the same resonant frequency. This is an example of frequency degeneracy.

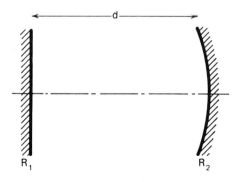

Figure 6.5 Optical cavity for the Hermite-Gaussian beam modes.

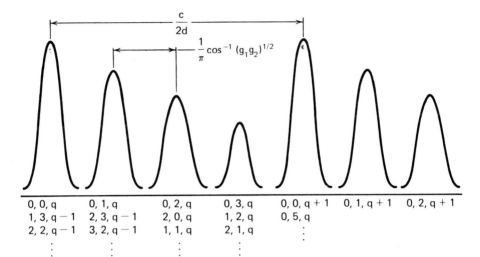

Figure 6.6 Frequency degeneracy in an optical cavity.

6.6 DIFFRACTION LOSSES

The Hermite-Gaussian beam modes are the characteristic values for stable infinite aperture mirrors, whereas any practical system has a finite size. Furthermore, the transverse dimensions of the active median of the laser may be considerably smaller than that of the mirrors—indeed, this is the most logical case.

In either event, the beam modes are only approximate solutions for the cavity—fortunately, excellent ones for most laser cavities. But one does have to account for the fact that the modes extend to a considerable distance from the axis and may *not* hit a reflecting surface. The fraction of the incident power that is not reflected by the mirrors is called diffraction losses.

An exact field description is quite complicated and is beyond the scope of the treatment here. However, one can generate an approximate technique by using some elementary physical reasoning, as shown in Fig. 6.7.

Here we show a Hermite-Gaussian beam impinging on an iris, an absorbing plate with a hole in it. (Quite often, such a device is used to restrict oscillation to the dominant TEM$_{0,0}$ mode in a cavity.) If we let our imagination loose, we would estimate that this aperture would transmit only the amount of power contained in the cross-sectional area of the hole. Thus the transmission coefficient is given by

$$T = \frac{\int_{r=0}^{a} \int_{0}^{2\pi} |E(r, \phi, z)|^2 r \, dr \, d\phi}{\int_{0}^{\infty} \int_{0}^{2\pi} |E(r, \phi, z)|^2 r \, dr \, d\phi} \tag{6.6.1}$$

Sec. 6.7 Cavity with Gain—An Example

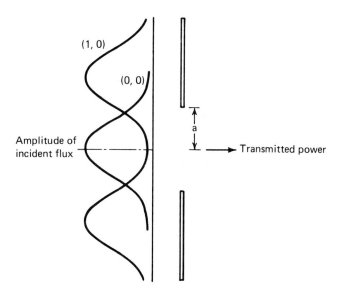

Figure 6.7 Transmission through an aperture.

For the two modes illustrated in Fig. 6.7, one obtains the following:

$$T_{(0,0)} = 1 - \exp\left[-2\left(\frac{a}{w}\right)^2\right] \tag{6.6.2a}$$

$$T_{(1,0)} = 1 - \left[1 + 2\left(\frac{a}{w}\right)^2\right]\exp\left[-2\left(\frac{a}{w}\right)^2\right] \tag{6.6.2b}$$

By inspection of (a) and (b), one can see that this simple iris transmits more of the (0, 0) mode than of the (1, 0) mode.

One should not attempt to carry this approximate analysis too far, for it is only a zeroth-order approximation. Only when the loss is low (or the reflectivity high) does this physical approach lead to the precise numerical value. However, it is close and provides guidelines.

6.7 CAVITY WITH GAIN—AN EXAMPLE

Let us redo the material of Sec. 6.3 but now include a medium with a power gain G_0 between the two mirrors. (Thus the field is amplified by $G_0^{1/2}$.) The serious student should repeat each step of the development from (6.3.1) to (6.3.8), which amounts to replacing the round-trip propagator $1 \cdot \exp[-jk \cdot 2d]$ by $G_0 \exp[-jk \cdot 2d]$, to find formulas for the power transmission and reflection coefficients. The answers are

$$T = \frac{G_0(1 - R_1) \cdot (1 - R_2)}{(1 - G_0\sqrt{R_1 R_2})^2 + 4\sqrt{R_1 R_2}\, G_0 \sin^2\theta} \tag{6.7.1}$$

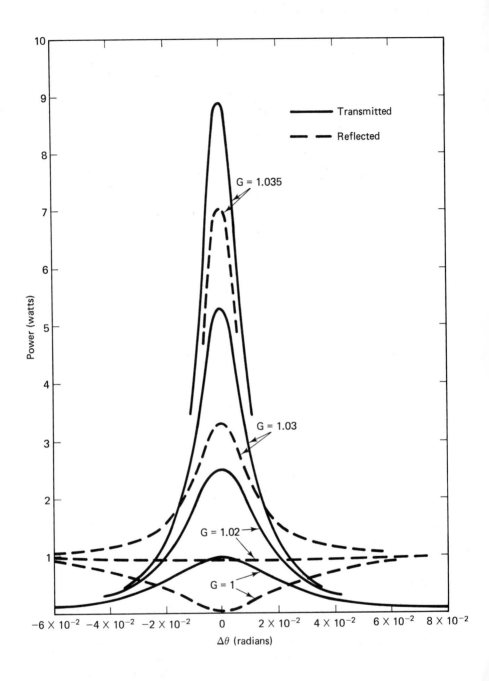

Figure 6.8 Response of an active cavity. The parameter G is the single pass power amplification factor.

$$R_{\text{net}} = \left[\frac{E_r}{E_i}\right]^2 = \frac{(\sqrt{R_1} - \sqrt{R_2})^2 + 4G_0\sqrt{R_1 R_2}\sin^2\theta}{(1 - G_0\sqrt{R_1 R_2})^2 + 4G_0\sqrt{R_1 R_2}\sin^2\theta} \quad (6.7.2)$$

These functions are sketched for various values of G_0 in Fig. 6.8 assuming that the wave impinging on the cavity contains 1 watt of optical power. Note that the reflected power does not go to zero at resonance for $G_0 > 1$; indeed, it peaks along with the transmitted wave, both being considerably larger than the incident power. All the "action" takes place near resonance, and it, too, is critically dependent on the single pass gain G_0.

The full width at half maximum of the resonance is

$$\text{FWHM} = 2\Delta\theta_{1/2} = \frac{1 - G_0\sqrt{R_1 R_2}}{G_0^{1/2}\sqrt[4]{R_1 R_2}} \quad (6.7.3)$$

If G_0 becomes bigger than $1/R_1 R_2$, the mathematics indicates an infinite (!) output with a zero-line width. What this absurdity means is that we have an oscillator—we can shut off the external source. Thus it is essential to discuss the physics leading to the gain, G_0, and this is the subject of the remainder of the book.

PROBLEMS

6.1. The following questions refer to the optical cavity shown in Fig. 6.5 with $d = (3/4)R_2$, $\rho_1^2 = 0.99$, and $\rho_2^2 = 0.97$.

(a) Find an expression for the resonant frequencies of the $\text{TEM}_{0,0}$ modes of the cavity.

(b) If the radius of curvature is 2.0 m and the wavelength region of interest is 5000 Å, compute the following quantities:
  (1) Free spectral range in MHz and in Å units
  (2) Cavity Q
  (3) Photon lifetime in nsec
  (4) Finesse

Problems 6.2 through 6.4 refer to the optical cavity shown in the accompanying diagram.

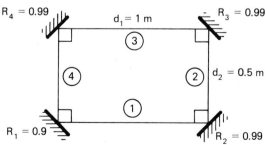

6.2. If the optical paths 1 through 4 are lossless, what is the photon lifetime of this cavity? (Ans. 78.9 nsec.)

6.3. What is the cavity Q (assume that the wavelength region of interest is 5000 Å)? (Ans. 2.97×10^8.)

6.4. (a) Suppose that path 1 has a transmission coefficient of 0.85 rather than 1 as in Problem 6.2. What is the new photon lifetime? (Ans. 38.8 nsec.)

  (b) Suppose that path 1 had a power gain of 1.1. What is the new photon lifetime?

  (c) If one blindly plugs into the formulas, τ_p becomes negative for G sufficiently large. What is the meaning of this apparent absurdity?

Problems 6.5 through 6.10 refer to the optical cavity in the accompanying diagram. It is excited by a variable-frequency source, and the detected intensity is as shown.

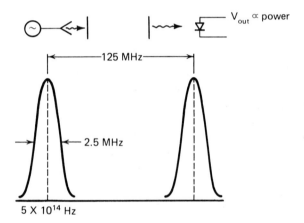

6.5. What is the nominal wavelength of the source?

6.6. How long is the cavity?

6.7. What is the finesse?

6.8. What is the Q?

6.9. What is the photon lifetime?

6.10. Suppose that the cavity is filled with an active medium with a single-pass gain of G. How large should G be to obtain oscillation?

6.11. The following two questions refer to the laser cavity shown in the accompanying diagram.

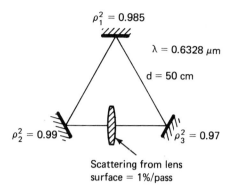

(a) What is the photon lifetime? (Ans. = 78.66 n.)
(b) What is the cavity Q? (Ans. 2.35×10^8.)

6.12. Drawn to scale on the graph below is the relative power transmission through a Fabry-Perot cavity when the distance d is increased slightly. The source is a He:Ne laser at $\lambda_0 = 6328$ Å.

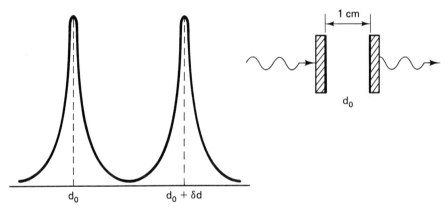

(a) What is the distance δd?
(b) What is the finesse of the cavity?
(c) What is the cavity Q?

6.13. Drawn to scale in the graph below is the relative power transmitted through the cavity as the distance d is increased from its initial value of 2 cm to 2 cm + 0.5 μm. The source is a single-mode laser of wavelength λ_0.
(a) What is the wavelength of the source?
(b) What is the finesse?
(c) Find the full width at half maximum (FWHM) of the resonance and express your answer in MHz.

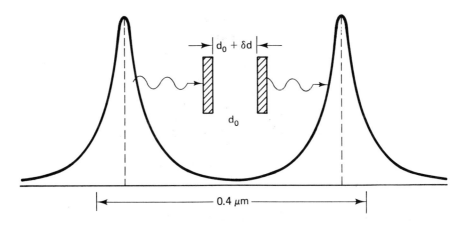

(d) What is the cavity Q?
(e) Find the photon lifetime.

6.14. Consider a cavity constructed by terminating a transmission line of length d with an inductance L at one end and a capacitance C at the other. Assume that the characteristic impedance of the line is Z_0 and its phase velocity is c.

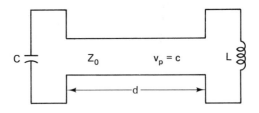

(a) Find a transcendental equation that determines the resonant frequency of this cavity by applying the condition that the round-trip phase shift is an integral number times 2π radians.
(b) Show that the equation derived for (a) reduces to $\omega_0^2 = 1/LC$ if d is sufficiently small.

[HINT: From transmission line theory, the voltage reflection coefficient is given by $\rho = (Z - Z_0)/(Z + Z_0)$, and thus there is a phase shift associated with ρ for complex terminations.]

6.15. Consider the optical cavity shown in the diagram below in which the variation of the spot size $w(z)$ is also shown. Note that the beam waist occurs at a distance d to the left of M_1 and that the mirrors have curvatures of opposite signs. (Assume that M_1 is so thin that it does not focus or otherwise affect the beam passing through it.)

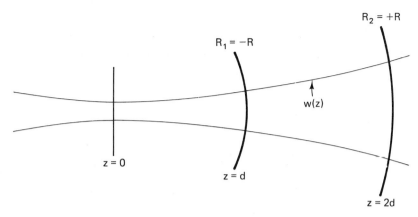

(a) Assume that the cavity is stable. Solve for the ratio d/R and evaluate the parameter z_0^2.

(b) If $d = 100$ cm and $z_0 = 100\sqrt{2}$ cm, what is the difference between the $\text{TEM}_{0,0,q}$ and $\text{TEM}_{1,0,q}$ resonant frequencies (in MHz)?

6.16. (a) Find an expression for the difference in the resonant frequencies of the $\text{TEM}_{0,0,q}$ and the $\text{TEM}_{m,n,q}$ modes of the cavity shown below. You may assume that the cavity is stable and the parameters z_{01} and z_{02} are known. Express your answer in terms of d_1, d_2, z_{01}, and z_{02}—do not evaluate.

(b) What is the photon lifetime and Q of the passive cavity if $\lambda_0 = 500$ nm? (A numerical answer is required.)

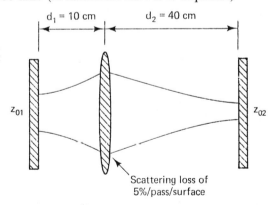

6.17. Make a careful sketch of the peaks (and valleys) of the transmission through a Fabry-Perot cavity as a function of frequency around $\lambda_0 = 6000$ Å with a finesse of 10 and free spectral range of 20 GHz.

(a) Find a numerical value for FWHM in GHz, Å, and cm^{-1}

(b) What is the cavity Q?

(c) What is the photon lifetime?
(d) What is the free spectral range in GHz, Å, and cm^{-1}?

6.18. Consider the following laser cavity and assume that the parameters z_{01} and z_{02} are known. Assume also that the cavity is stable.

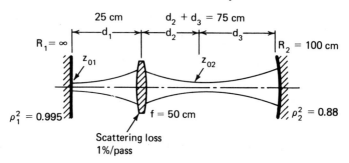

(a) What is the radius of curvature of the phase at the spherical mirror? (Ans. 100 cm.)
(b) What is the photon lifetime? (Ans. 47.01 nsec.)
(c) Derive a formula for the resonant frequency of the TEM$_{m,p,q}$ mode.

$$\left(\text{Ans: } \nu = \frac{c}{2(d_1 + d_2 + d_3)}\left\{q + \frac{1 + m + p}{\pi}\left[\tan^{-1}\left(\frac{d_1}{z_{01}}\right) + \tan^{-1}\left(\frac{d_2}{z_{02}}\right) + \tan^{-1}\left(\frac{d_3}{z_{03}}\right)\right]\right\}.\right)$$

6.19. Consider the laser shown in the accompanying diagram.
(a) Is this cavity stable?

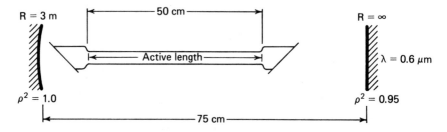

(b) What would be the frequency difference between the TEM$_{0,0,q}$ mode and the TEM$_{1,0,q}$ mode? (Ans. = 33.3 MHz.)
(c) What should be the bore size of the laser tube so that less than 0.1% of the TEM$_{0,0,q}$ mode intercepts the tube wall? (Ans. 2.14 mm diameter.)
(d) What is the minimum gain *coefficient* of the laser tube to sustain oscillation? (Ans. 5×10^{-4} cm^{-1}.)

6.20. A laser cavity was excited by a 1 nsec pulse from an external source at $\lambda = 5577$ Å. When the medium was not pumped, the detected transmission was as shown in the upper part of the diagram. When the medium was irradiated by an intense electron beam, the lower part of the diagram resulted.
(a) How long is the cavity? (Ans. 75 cm.)
(b) What is the photon lifetime? (Ans. From graph, 75 nsec.)
(c) What is the cavity Q? (Ans. 2.5×10^8.)
(d) What is the gain coefficient under e-beam excitation? (Ans. 1.89×10^{-4} cm^{-1}.)
(e) What is the cold-cavity finesse? (Ans. $F = 94.2$.)

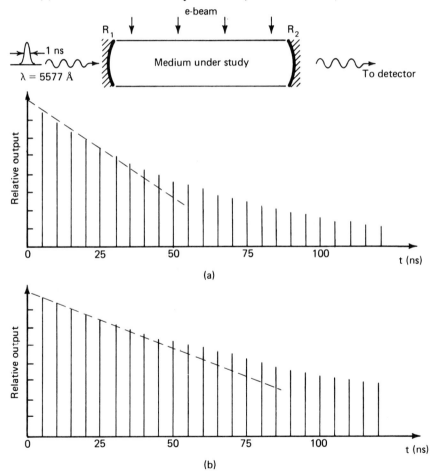

6.21. Consider the accompanying diagram of a cavity designed to be utilized with a helium/neon laser at $\lambda_0 = 632.8$ nm.

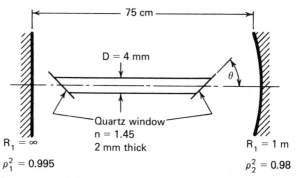

(a) Is the cavity stable?
(b) What is the spot size of the beam at the flat mirror?
(c) What is the spot size of the beam at the spherical mirror?
(d) The windows are cemented to the tube at Brewster's angle. What is the angle θ as shown on the sketch?
(e) Assuming that the tube bore is centered with respect to the axis of the $\text{TEM}_{0,0}$ mode, compute the loss introduced by the aperturing action of the tube walls. (Zero is *not* an acceptable answer.)
(f) What is the formula for the resonant frequency of the $\text{TEM}_{m,p,q}$ mode?

6.22. Consider a single $\text{TEM}_{0,0,q}$ mode of the laser shown in Problem 6.21. Because of room vibrations, sound waves, and temperature variations, the distance, $d = 75$ cm, varies slightly about its nominal value. If the optical frequency of a mode is to be held constant to 1 kHz, what is the maximum allowable variation in d? The answer should disturb you, especially when you consider that atoms are *spaced* about 4 Å apart. Nevertheless, such frequency control is possible.

6.23. A variable-frequency, constant-amplitude dye laser irradiates the active Fabry-Perot cavity containing a medium with a single-pass power gain of G. Even though there is gain, it is not sufficient to support oscillation (i.e., $G^2 R_1 R_2 < 1$). Assume $R_1 = R_2 = 0.90$.
(a) Find the power magnification factor of this cavity if the source is tuned to the center of the cavity resonance, i.e., find
$$M = \frac{P_{\text{ref}} + P_{\text{trans}}}{P_{\text{inc}}}$$
NOTE: If $G = 1$ (i.e., a passive cavity), then $M = 1$ since this is a lossless system.
(b) Plot the transmission through the cavity as the frequency of dye laser is changed for $G^2 R_1 R_2 = 0.95$.
(c) Compute and plot the line width of the cavity as the gain is varied from 1 to 99.9% of its maximum value (i.e., $G^2 = 1/R_1 R_2$).

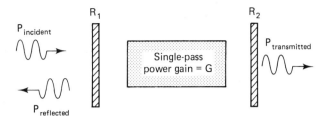

6.24. One of the means for "tuning" a Fabry-Perot cavity is by changing the index of refraction of the medium between the two mirrors, which is accomplished by varying a gas density in the enclosure. The index of refraction scales as $n = 1 + kp$, where p is the pressure in atmospheres. Suppose the etalon spacing is 0.2 cm, the finesse was 20 at 6328 Å, and the gas constant in the index equation is 8×10^{-4} atmospheres^{-1}. Make a careful sketch of the relative transmission through the etalon as the pressure is varied.

(a) What change in pressure is required to scan the etalon over one free spectral range?
(b) What is the spectral range (in GHz)?
(c) What is the photon lifetime and Q of this cavity?

6.25. For the cavity shown below, the Hermite-Gaussian beam parameters are given by $z_{01} = \pi w_{01}^2/\lambda_0 = 6.45$ cm and $z_{02} = 38.7$ cm with $d_1 = 25$ cm, $d_2 = 50$ cm, $R_1 = 0.98$, $R_2 = 0.93$, and a transmission through the lens of 95%. The wavelength region of interest is 5145 Å.

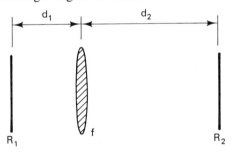

(a) Find a formula for the resonant frequencies of the TEM$_{m,p,q}$ modes in terms of the distances $d_{1,2}$ and the parameters z_{01} and z_{02}.
(b) What is the photon lifetime?
(c) Evaluate the passive Q.
(d) If an active medium were incorporated within the cavity and it amplified the intensity by a factor of 1.13 per pass, find the new photon lifetime and line width of the cavity response.

6.26. A GRIN waveguide is constructed from a material described in problem 5.3. There is a reflection at each end from the air-dielectric mismatch or from intentional coating, and thus a length can be considered as a cavity with Gaussian beams as its characteristic modes.

(a) Use the *ABCD* law to find the complex beam parameter, q, of a Gaussian beam whose spot size does *not* change along the z axis. Evaluate this spot size for $n_0 = 1.87$; $\lambda_0 = 1$ μm; and $a = 250$ μm.

(b) The phase constant of the Gaussian beam found in (a) is given by $\beta \approx \left(k - \dfrac{1}{a}\right)$ where $k = \omega n_0/c$ as usual. Find an expression for the resonant frequencies of the $TEM_{0,0,q}$ modes for this cavity.

(c) The power reflectivities at the planes $z = 0$ and $z = d$ can be approximated by the usual plane wave or transmission line formula: $R = \left[\dfrac{n_0 - 1}{n_0 + 1}\right]^2$. Justify the use of this formula, given that the spot size found in (a) is much less than the scale length a.

(d) If the ends were coated such that the reflectivities were 0.5, $d = 10$ cm, $n_0 = 1.87$; and $a = 250$ μm, find the photon lifetime and cavity finesse.

REFERENCES AND SUGGESTED READINGS

It should be obvious to all by now that Chapter 6 is the culmination of the previous 5 chapters. Much of any "new" material contained in this chapter can be found in any book on elementary electrical circuits, possibly under a different name, but still the same ideas.

1. A. G. Fox and T. Li, "Resonant Modes in a Maser Interferometer," Bell Syst. Tech. J. *40*, 453–488, Mar. 1961.
2. H. Kogelnik and T. Li, "Laser Beams and Resonators," Appl. Opt. *5*, 1550–1567, Oct. 1966.
3. A. E. Siegman, "Unstable Optical Resonators for Laser Applications," Proc. IEEE *53*, 227–287, Mar. 1965.
4. A. E. Seigman, *Introductions to Lasers and Masers* (New York: McGraw-Hill Book Company, 1968).
5. A. Maitland and M. H. Dunn, *Laser Physics* (Amsterdam: North-Holland Publishing Company, 1969).
6. S. Ramo, J. R. Whinnery, and T. VanDuzer, *Fields and Waves in Communication Electronics* (New York: John Wiley & Sons, Inc., 1965).
7. A. Yariv, *Introduction to Optical Electronics* (New York: Holt-Rinehart and Winston, 1971).
8. A. Yariv, *Quantum Electronics*, 2nd ed. (New York: John Wiley & Sons, Inc., 1975).
9. H. A. Haus, *Waves and Fields in Optoelectronics* (Englewood Cliffs, N. J.: Prentice-Hall, Inc., 1984).

7

Atomic Radiation

7.1 INTRODUCTION AND PRELIMINARY IDEAS

This chapter introduces the elementary concepts leading to gain in an atomic system and thus, with proper feedback, a laser. Since a laser is a quantum device, we first review the necessary concepts from this theory to understand the conditions leading to gain.

These concepts are amazingly simple and are quite palatable, even to those who have not had a formal introduction to quantum mechanics.

1. There are discrete energy levels in an atomic (or molecular, or solid, or semiconductor) system. We represent them by an energy-level diagram as shown in Fig. 7.1.
2. The system can make a *transition* between these two states by the *emission* of a photon of energy $E = E_2 - E_1 = h\nu$, thereby changing an atom

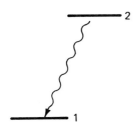

Figure 7.1 Energy-level diagram.

labeled by (2) into one identified with (1). Or to reverse the process, an atom in state 1 can *absorb* a photon of this same energy and be labeled as in state 2.

These two ideas are the only ones we need to "accept"; the rest will follow naturally and with little effort on our part.

In addition to the simple ideas expressed above, we need to understand the origin of the correct description of blackbody radiation theory. (At low frequencies, this radiation is referred to as "white" noise, "Johnson's" noise, or "Nyquist" noise, all being equal to Planck's blackbody radiation formula at those limits.) Indeed, it was this problem that inspired Planck to make the quantum hypothesis—namely that electromagnetic energy at a frequency ν can be present only in discrete multiples of $h\nu$.

Understanding the origin of Planck's development of this formula is essential to appreciate the beauty and *simplicity* of Einstein's description of the role of the atoms in arriving at this relation. The importance of Einstein's approach is that it provides the key to the analysis of systems that are *not* in thermodynamic equilibrium; a laser is a prime example.

Furthermore, Einstein's approach provides us with a connection between *transition rates* between quantum states and quantities that can be measured experimentally. Consequently, we can approach such transitions in a phenomenological manner using the published experimental results, thereby bypassing some very complex and complicated quantum calculations.

The foregoing should not be construed to downgrade the importance of a full quantum description of a laser. There are phenomena which are not predicted by the rate-equation approach and which require a detailed analysis by quantum-theoretical methods. But for the initial understanding, Einstein's rate equations are quite adequate.

We will also have to modify slightly the picture shown in Fig. 7.1 to account for the *uncertainty principle* and "real-life" broadening processes. An atom is never isolated; a gas atom is in helter-skelter motion, it collides with other atoms (with same kind or with different ones), and it may be subjected to external fields. As a consequence, the energy levels are not perfectly sharp, and thus there is a finite band of frequencies emitted by a collection of atoms and a finite band for amplification.

Once these ideas are in hand, the laser gain equation is a trivial application and oscillation is merely a matter of providing feedback.

7.2 BLACKBODY RADIATION THEORY

The starting point for the quantum era was the derivation, from first principles, of a formula that described the radiation emerging from a small hole in a highly polished "cavity" such as that shown in Fig. 7.2(a). The *experimental facts* known from the late 1800s were as follows:

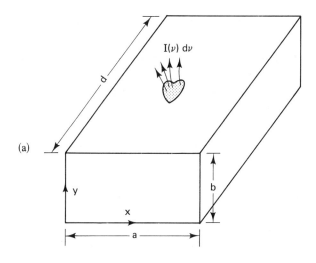

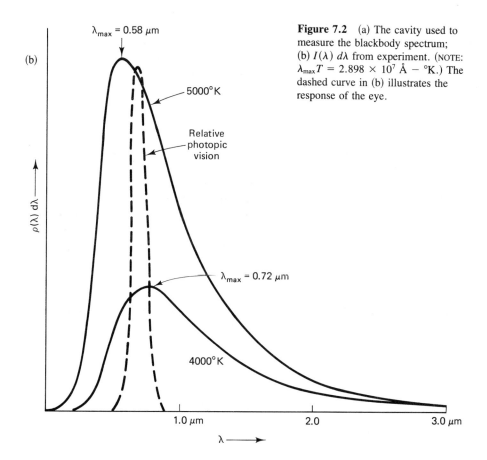

Figure 7.2 (a) The cavity used to measure the blackbody spectrum; (b) $I(\lambda)\,d\lambda$ from experiment. (NOTE: $\lambda_{max}T = 2.898 \times 10^7$ Å $-$ °K.) The dashed curve in (b) illustrates the response of the eye.

1. The hole acted as a nearly perfect blackbody—one whose emission coefficient* as a function of frequency was as high as possible, one (1).
2. Thus the intensity emitted by this hole is directly proportional to the energy density of the electromagnetic radiation inside the cavity.
3. This energy density, in a fixed-frequency interval $d\nu$ (determined by the measurement technique), had the functional form of (7.2.1).

or
$$\rho(\nu) = \frac{a\nu^3}{e^{b\nu/kT} - 1}$$
$$\rho(\lambda)\,d\lambda = \frac{a'}{\lambda^5} \frac{d\lambda}{e^{b'/\lambda T} - 1}$$
(7.2.1)

It is important to realize that these facts were known *before* the advent of quantum mechanics. Even though these facts were known, they were not understood.

One of the triumphs of nineteenth-century physics was the phenomenal success of Maxwell's theory of electromagnetic radiation. Yet here was an electromagnetic problem, apparently coupled to some elementary thermodynamics, which defied explanation. (In fact, Planck, who was ultimately successful, nearly had a nervous breakdown over the dilemma.) To illustrate some of the difficulties, let us follow some of the logic paths used to arrive at the *wrong* answer.

Every electromagnetic phenomenon until that time (and since) had (and has) been explained by Maxwell's theory. Hence it was natural to start there and ask how electromagnetic energy is distributed inside the heated cavity shown in Fig. 7.2(a). (Although a specific geometry is chosen, the results are independent of the shape providing *all* dimensions are large compared to a wavelength in the medium inside cavity.) From Chapter 6, we recognize that only the resonant $TE_{m,p}$ or $TM_{m,p}$ modes of the short-circuited rectangular waveguide can have appreciable energy. Those resonances are determined by the standard formula: RTPS = $q \cdot 2\pi = \beta_{m,p} \cdot 2d$ where $\beta_{m,p}$ is the phase constant of the mode:

$$\beta_{m,p} = \left\{ \left(\frac{\omega n}{c}\right)^2 - \left(\frac{m\pi}{a}\right)^2 - \left(\frac{p\pi}{b}\right)^2 \right\}^{1/2} \quad (7.2.2a)$$

Equation (7.2.2a) applies for either TE or TM modes. Thus the resonant frequencies of the (m, p, q) mode are given by $\beta_{m,p} \cdot 2d = q \cdot 2\pi$ or

$$k^2 = \left(\frac{\omega n}{c}\right)^2 = \left(\frac{m\pi}{a}\right)^2 + \left(\frac{p\pi}{b}\right)^2 + \left(\frac{q\pi}{d}\right)^2 \quad (7.2.2b)$$

---

*Kirchhoff's radiation law states that the emissivity of a body is equal to its absorptivity. Thus the cavity would also absorb all electromagnetic energy entering through the hole and emit as much as possible.

Sec. 7.2 Blackbody Radiation Theory

Equation (7.2.2b) has a nice geometric interpretation: the term on the left is the wave vector, **k**, of a uniform plane wave in the medium; the terms on the right can be considered as the projection of it along the (x, y, z) axes; and, of course, the right-hand side states that the Pythagorean relationship holds. Resonance, then, implies that each component of **k** must be a multiple of a half wavelength. Thus the resonant frequencies are given by

$$\left(\frac{2\pi\nu n}{c}\right)^2 = \left(\frac{m\pi}{a}\right)^2 + \left(\frac{p\pi}{b}\right)^2 + \left(\frac{q\pi}{d}\right)^2 \quad (7.2.2c)$$

To make the formula even simpler, we chose a cube with $b = d = a$ and find that formula for resonance "looks" similar to a spherical coordinate radius vector with the mode numbers being a measure along the Cartesian axis:

$$\nu = \frac{c}{2na}[m^2 + p^2 + q^2]^{1/2} \quad (7.2.3a)$$

This geometric interpretation is shown in Fig. 7.3 where the "points" are allowed mode numbers scaled by the factor $c/2na$ since the density of points in that sphere is exactly 1 (by construction), that is, one mode in the volume $dm = dp = dq = 1$. Thus we can find the number of modes between $\nu = 0$ and ν by finding the volume of one eighth of a sphere with the radius $R = 2na\nu/c$:

$$V_s = \frac{1}{8} \times \frac{4\pi R^3}{3} \quad (7.2.3b)$$

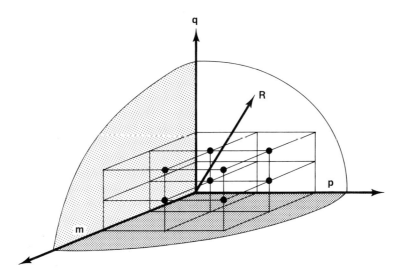

Figure 7.3 The mode diagram for cubic cavity.

Now each point represents two modes, TE and TM; hence the total number of resonances between 0 (i.e., DC) and ν, the frequency of interest, is

$$N = 2 \times \frac{1}{8} \times \frac{4\pi}{3}\left(R \to \frac{2na\nu}{c}\right)^3 = \frac{8\pi n^3 \nu^3}{3c^3} a^3 \qquad (7.2.4a)$$

We are usually interested in the small frequency interval $d\nu$ around ν and also prefer the answer on a per-unit-volume basis to eliminate the geometry. The mode density $p(\nu)$ (volume^{-1} $-$ frequency^{-1}) is given by

$$p(\nu)d\nu = \frac{1}{V}\frac{dN}{d\nu} d\nu = \frac{8\pi n^2 \bar{n}_g}{c^3} \nu^2 d\nu \cong \frac{8\pi n^3}{c^3} \nu^2 d\nu \qquad (7.2.4b)$$

where

$$\bar{n}_g = n + \nu\frac{dn}{d\nu} = n - \lambda\frac{dn}{d\lambda} \qquad (7.2.4c)$$

The quantity $\bar{n}_g$ is called the group index and plays a role in materials, such as semiconductors, where the index of refraction has a significant variation with wavelength. However, to make the following relations less cumbersome, we will use the approximate one, $\bar{n}_g \sim n$.

So much for identifying the electromagnetic modes that couple to that hole in the cavity. Nobody wanted to (or has) changed (7.2.4b); it is *correct*. The problem always arose in assigning an energy to those modes.

Classical physics gets away with assigning an energy $kT/2$ to each "mode" of motion. For instance, free atoms in a gas can move along x, y, or z, and thus the average energy of N atoms is $N(3kT/2)$—in agreement with experiment. If we follow that logic, each electromagnetic mode should have an energy equal to kT since both E and H store energy. Multiplying (7.2.4b) by kT leads to an answer for the energy density in frequency interval $d\nu$; unfortunately, it is wrong—it does not agree with the experiment.

$$\rho(\nu) = \left(\frac{8\pi n^3 \nu^2 d\nu}{c^3}\right) kT \qquad \text{(Wrong!)} \qquad (7.2.5)$$

Equation (7.2.5) is the Rayleigh-Jeans distribution of blackbody radiation, which, as shown in Fig. 7.2(b), did *not* agree with experiment. Although it does agree with experiment at long wavelengths (microwaves and lower frequencies), it predicts the absurd conclusion that there is infinite energy in the system when *all* frequencies are concerned (the ultraviolet catastrophe).

It is hard to argue with the line of reasoning leading to (7.2.5): the number of modes times the average energy per mode. Thus the error must be in the computation of one of these quantities. Planck accepted and retained the mode calculation and turned his attention to computing the average energy per mode.

Planck made the initial quantum hypothesis that electromagnetic energy at

Sec. 7.2 Blackbody Radiation Theory

a frequency ν could only appear as a multiple of the step size $h\nu$. Energies between $h\nu$ and $2\,h\nu$ do *not* occur! This is in stark contrast to classical ideas where all energies are permitted—by allowing the field to have continuously variable amplitudes. Having made this radical departure from accepted concepts, he returned to classical Boltzmann statistics to compute the average energy.

If $\epsilon_1, \epsilon_2, \epsilon_3, \ldots$ are the allowed energies, we must weigh each energy by the relative probability that it can occur and then divide by the sum of all relative probabilities. According to Boltzmann statistics, the relative probability that an energy (ϵ_j) can occur is simply $\exp(-\epsilon_j/kT)$, the Boltzmann factor. Following the recipe above, we obtain

$$\epsilon = nh\nu \quad \text{(the quantum hypothesis)} \tag{7.2.6}$$

$$\langle \epsilon \rangle = \frac{1h\nu e^{-h\nu/kT} + 2h\nu e^{-2h\nu/kT} + 3h\nu e^{-3h\nu/kT} + \cdots}{1 + e^{-h\nu/kT} + e^{-2h\nu/kT} + e^{-3h\nu/kT} + \cdots} \tag{7.2.7}$$

$$= \frac{\sum_{1}^{\infty} nh\nu e^{-nh\nu/kT}}{\sum_{0}^{\infty} e^{-nh\nu/kT}} \tag{7.2.8}$$

where

$$\epsilon_n = nh\nu$$

$n = 0, 1, 2, 3, 4 \cdots$ are the allowed values of the electromagnetic energy

$\exp(-\epsilon_n/kT)$ = relative probability of the energy ϵ_n

$\sum_{0}^{\infty} \exp(-\epsilon_n/kT)$ = the sum of all relative probabilities

One might be tempted to replace the summation by an integral, a procedure that yields an average energy given by (7.2.4), which is wrong. Fortunately, the series given in (7.2.8) can be summed exactly to yield

$$\langle \epsilon \rangle = \frac{h\nu}{e^{h\nu/kT} - 1} \tag{7.2.9}$$

Then the energy density of the electromagnetic field inside the cavity at the center frequency of interest is

$$\rho(\nu) = \left(\frac{8\pi n^3 \nu^2}{c^3}\right) \cdot (h\nu) \frac{1}{e^{h\nu/kT} - 1} = \frac{8\pi n^3 \nu^2}{c^3} \frac{h\nu}{e^{h\nu/kT} - 1} \tag{7.2.10}$$

Equation (7.2.10) is written as a product of three factors to emphasize the origins: the first is a purely classical electromagnetic result, with the second being the quantum value of the average energy of the classical field. Since the package of energy in the field is a multiple of the quantum of energy $h\nu$, the term $1/[\exp(h\nu/kT) - 1]$ is the number of quanta in a cavity mode.

Let us evaluate this number for a reasonable set of circumstances. Let $T = 1200°K$ (an "orange" color temperature) and $\lambda = 6000$ Å (an orange color); then $\nu = 5 \times 10^{14}$ and the average number of photons per mode is $\sim 10^{-9}$.

One must interpret that number correctly. If the energy comes in discrete steps, one cannot have a fraction of a photon in a mode—rather one has a photon in the particular cavity mode for a small—a very small—fraction of the time. This particular point is important and distinguishes a laser that has at least one photon in a particular mode *all* the time.

7.3 EINSTEIN'S APPROACH—THE *A* AND *B* COEFFICIENTS

Einstein was able to arrive at the same functional form of radiation density per frequency interval, $\rho(\nu)$, for a thermodynamic environment by a much simpler and much more transparent line of reasoning. Most important is the fact that the reasoning can be applied to a system which is *not* in thermodynamic equilibrium. He focused the attention on the *atoms* in the cavity walls which were responsible for the generation of the electromagnetic energy inside the cavity.

7.3.1 Definition of Radiative Processes

Einstein accepted the quantum hypothesis that the energy came in discrete packages, $E = h\nu$, and as a consequence there must be two energy states,* E_2 and E_1, associated with those atoms separated by that value as shown in Fig. 7.4. He identified three radiative processes that affect the concentrations of atoms in states 2 and 1. (The fact that there are many other very important processes affecting the concentration need not concern us now. We are limiting our view to the blackbody cavity.)

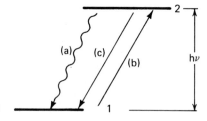

Figure 7.4 Radiative processes in a two-level system.

*As a matter of fact, there is a continuum of states, but this does not change the ideas expressed by a single set of states.

Sec. 7.3 Einstein's Approach—The A and B Coefficients

(a) Spontaneous emission (A_{21}). As the name implies, it appeared as if the atoms in state 2 decayed spontaneously to state 1; in doing so, they added their excess energy to the cavity field in the form of a photon. If the population density in state 2 was N_2, the decay of this state is given by

$$\frac{dN_2}{dt}\bigg|_{\text{spontaneous emission}} = -A_{21}N_2 \qquad (7.3.1)$$

Note that this equation says something very simple and transparent: If no other process took place, the atomic population would run "downhill" with a time constant $\tau = (A_{21})^{-1}$. Obviously, the bottom of the "hill," N_1, must *increase* just as fast as the top decreases.

(b) Absorption (B_{12}). In this process an atom in state 1 *absorbs* a photon from the field and thus converts the atom into one of those in state 2. The rate at which this process takes place must depend on the number of absorbing atoms and the field from which they extract the energy. Thus we have

$$\frac{dN_2}{dt}\bigg|_{\text{absorption}} = +B_{12}N_1\rho(\nu) = -\frac{dN_1}{dt}\bigg|_{\text{absorption}} \qquad (7.3.2)$$

where the string of equalities expresses the obvious idea again that N_1 must decrease if N_2 increases.

(c) Stimulated emission (B_{21}). This process is the reverse of absorption; the atom gives up its excess energy, $h\nu$, to the field, adding coherently to the intensity. Thus the added photon is *at the same frequency, at the same phase, in same sense of polarization, and propagates in the same direction* as the wave that *induced* the atom to undergo this type of transition. Obviously, the rate depends on the number of atoms to be stimulated and the strength of the stimulating field.

$$\frac{dN_2}{dt}\bigg|_{\text{stimulated emission}} = -B_{21}N_2\rho(\nu) = -\frac{dN_1}{dt}\bigg|_{\text{stimulated emission}} \qquad (7.3.3)$$

These three processes are shown in Fig. 7.5, where some artifacts have been used to emphasize various issues.

Note that "nothing" comes into Fig. 7.5(a), but a photon comes out, most likely in a direction different from the one you predict. That is spontaneous emission; the atom can radiate into any of 4π steradians with any sense of polarization. The absorption should be familiar to most; the wave decreases in amplitude and the atom in state 1 is converted to state 2. Obviously, that part of the wave not absorbed continues along its path. The picture for stimulated emission is just the inverse of absorption. If one accepts this inverse relation, much of the "magic" about stimulated emission disappears. One would not question the fact that the transmitted wave in Fig. 7.5(b) is at the same fre-

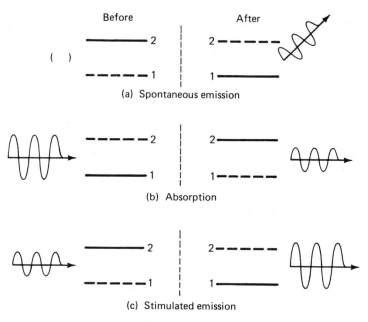

Figure 7.5 Effect of radiation on an atom.

quency, is in the same direction, and has the same polarization as the incident wave. By identifying Fig. 7.5(c) as the inverse process to Fig. 7.5(b), one attributes those same characteristics to the portion of the wave that *adds* to the stimulating wave.

A very important point to remember from Fig. 7.5 is that stimulated emission adds a photon

1. *At the same frequency* of the stimulating wave
2. *In the same polarization* of the stimulating wave
3. *In the same direction* of travel of the stimulating wave
4. *In the same phase* of the stimulating wave

If these characteristics were not true, one would obtain various absurdities such as violating most laws of thermodynamics (see reference 14). With a more detailed view of the field-atom interaction and much more mathematics, these characteristics become obvious.

7.3.2 Relationship Between the Coefficients

By defining these processes, Einstein was able to reproduce the blackbody formula within the framework of thermodynamic equilibrium. The sum of (7.3.1) through (7.3.3) yields the total rate of change of the population density in state 2 (or 1) as a result of the radiative processes.

Sec. 7.3 Einstein's Approach—The A and B Coefficients

At thermodynamic equilibrium, each process going "down" must be balanced exactly by that going "up." (This is the reason why we can ignore any other as yet unidentified processes affecting states 2 and 1; the detail balance must occur between the radiative processes considered here.)

$$\frac{dN_2}{dt} = -A_{21}N_2 + B_{12}N_1\rho(\nu) - B_{21}N_2\rho(\nu) = -\frac{dN_1}{dt} \quad (7.3.4)$$

At *equilibrium*, the time rate of change must be zero.

$$\frac{N_2}{N_1} = \frac{B_{12}\rho(\nu)}{A_{21} + B_{21}\rho(\nu)} \quad (7.3.5)$$

Einstein invoked classic Boltzmann statistics to provide another equation for the ratio of the two populations in states 2 and 1 and set that value equal to (7.3.5):

$$\frac{N_2}{N_1} = \frac{g_2}{g_1}e^{-h\nu/kT} = \frac{B_{12}\rho(\nu)}{A_{21} + B_{21}\rho(\nu)} \quad (7.3.6)$$

where $g_{2(1)}$ = number of ways that an atom can have the energy $E_{2(1)}$. For a simple atom, this quantity is related to the total angular momentum quantum number $J_{2(1)}$ by

$$g_{2(1)} = 2J_{2(1)} + 1 \quad (7.3.7)$$

We need only to solve for $\rho(\nu)$ from (7.3.6), and we manipulate that solution in the following manner:

$$A_{21}\left[\frac{g_2}{g_1}e^{-h\nu/kT}\right] + B_{21}\left[\frac{g_2}{g_1}e^{-h\nu/kT}\right]\rho(\nu) = B_{21}\rho(\nu)$$

or

$$\rho(\nu) = \frac{A_{21}\left[\dfrac{g_2}{g_1}e^{-h\nu/kT}\right]}{B_{21} - B_{21}\left[\dfrac{g_2}{g_1}e^{-h\nu/kT}\right]} \quad (7.3.8)$$

After dividing by $[(g_2/g_1)\exp(-h\nu/kT)]$ and factoring B_{21} out of the denominator, one obtains

$$\rho(\nu) = \frac{A_{21}}{B_{21}} \cdot \frac{1}{\dfrac{B_{12}g_1}{B_{21}g_2}e^{h\nu/kT} - 1} \quad (7.3.9)$$

This is almost the Planck formula of (7.2.10). To make it so, Einstein forced the fit with identification of various interrelationships between the coefficients:

$$\frac{B_{12}g_1}{B_{21}g_2} = 1 \quad \text{or} \quad g_2 B_{21} = g_1 B_{12} \quad (7.3.10a)$$

and

$$\frac{A_{21}}{B_{21}} = \frac{8\pi n^3 h \nu^3}{c^3} \qquad (7.3.10b)$$

With these identifications, (7.3.9) is identical to (7.2.10).

Equations (7.3.10a) and (7.3.10b) are very important because they show a connection between three different radiative processes: spontaneous emission, absorption, and stimulated emission. If one is known, all are known. Although a particular experiment may emphasize one or another coefficient, the result may be applied to a completely different one. For instance, an absorption experiment yields vital information on the stimulated emission coefficient.

It is most important to realize that these coefficients are characteristic of the *atom*. The atom, per se, does not know (or care) whether it is in a thermodynamic equilibrium environment of a heated cavity or in the presence of an intense field (a laser) generated by other atoms. It responds according to the rates indicated by (7.3.1), (7.3.2), and (7.3.3) for electromagnetic radiation.

However, radiation is not the only thing that can affect an excited atom. The atoms can undergo a collision with another atom, an electron, or a lattice vibration (a phonon), which can also cause transitions to take place. Einstein's approach places radiation on an equal footing with these other processes and incorporates the interchange between the states in a natural and straightforward manner. The power of this approach is shown in the next section.

7.4 INTRODUCTION TO THE RATE EQUATIONS—LIFETIME BROADENING

Let us leave the thermodynamic environment inasmuch as the major common device that uses this blackbody radiation is the incandescent light bulb, which while important, is not the most exciting application to be considered. The example shown in Fig. 7.6 will illustrate the transparent simplicity of Einstein's approach and will also serve to introduce the concept of level broadening, which in turn leads to the idea of a line shape of an atomic transition.

Here we imagine a very dilute gas in which a small number, ΔN, is excited to state 2 at $t = 0$. Then all external sources are removed and the system relaxes back toward its equilibrium state, with all atoms in the lowest-energy state. We assume that this number, ΔN, although small, is far in excess of the thermodynamic population predicted by Boltzmann statistics.

It is not necessary to specify how this excitation is provided, but it helps in visualization. Thus, to be definitive in the example, we assume that the gas is irradiated with an energetic electron beam and that the beam current is suddenly clamped to zero at $t = 0$.

Sec. 7.4 Introduction to the Rate Equations—Lifetime Broadening

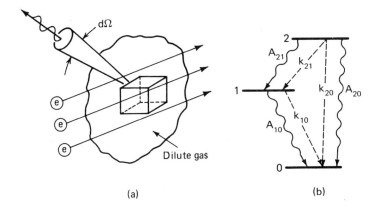

Figure 7.6 (a) Irradiation of a dilute gas by an electron beam and (b) subsequent decay of excitation.

After the beam is turned off, the atoms start to return to their lowest energy state, 0, by emitting a photon, or by colliding with a foreign atom and giving up its internal energy but not yielding a photon. There are literally hundreds of such different processes, but, for now, we bypass the physics of those and just assign rates, k_{21}, to describe the downward trickle caused by the collisions. (This is sometimes described as sweeping our ignorance under the rug.)

Thus state 2 obeys the following differential equation:

$$\frac{dN_2}{dt} = -(A_{20} + A_{21})N_2 - (k_{20} + k_{21})N_2 = -(A_2 + k_2)N_2 \qquad (7.4.1)$$

where

$A_2 = A_{20} + A_{21} + \cdots$ other emissions originating from state 2

$ =$ total decay of state 2 by spontaneous emission

$k_2 = k_{20} + k_{21} + \cdots$ other collisional deactivation paths

$ =$ total quenching of state 2 by collisions

In writing (7.4.1) we have assumed implicitly that the excitation is weak enough so that any resulting radiation is also weak, and hence we can also neglect absorption and stimulated emission. Indeed, we will call this neglect a necessary requirement for the success of this experiment.

The solution to (7.4.1) is straightforward:

$$N_2 = \Delta N_2 \exp\left[-(A_2 + k_2)t\right] = \Delta N_2 \exp\left(-\frac{t}{\tau_2}\right) \qquad (7.4.2)$$

It is obvious from this last equality that the quantity $(A_2 + k_2)^{-1}$ is the lifetime of state 2 under the circumstances of this "Gedanken" experiment.

If we could eliminate all collisional processes (i.e., $k_2 = 0$) in this experiment, there remain the radiative decay processes, and we obtain a slightly different physical interpretation of the Einstein A coefficient.

$$\frac{1}{\tau_r} = \sum_{n<j} A_{jn}$$

τ_r = radiative lifetime of state j (7.4.3)

The point to be made here is that the experiment can be performed with only slightly more sophisticated equipment than that required for *RC* circuit measurements. Thus the Einstein A coefficient should be familiar to everyone.

According to (7.4.2) and Fig. 7.6, state 2 is decaying with a time constant τ_2, with part of this decay proceeding through state 1 and the rest going directly to state 0. The *fraction* of the radiative decay out of state 2 to state 1 is called the *branching ratio* of the transition $2 \to 1$ and is abbreviated by the symbol ϕ_{21}.

$$\phi_{21} = \frac{A_{21}}{\sum_{n<2} A_{2n}} \qquad (7.4.4)$$

Each one of these radiative transitions contributes a photon of energy $h\nu_{21}$, which is spewed out more-or-less equally into 4π steradians. Thus the power at the frequency ν_{21} into the solid angle $d\Omega$ is given by

$$dP = \frac{d\Omega}{4\pi} A_{21} N_2(t) h\nu_{21}$$

or

$$dP(t) = \frac{d\Omega}{4\pi} h\nu_{21} A_{21} \Delta N_2 \exp(-t/\tau_2)$$

$$\frac{1}{\tau_2} = A_2 + k_2 \qquad \text{[from (7.4.2)]}$$

(7.4.5)

The relative intensity of the radiation emerging at the frequency ν_{21} is proportional to that transition probability, A_{21}, and decays with a time constant associated with the upper state.

Even though the experiment could be performed in the manner described, and the results would be as given by (7.4.2) and (7.4.5), there is a philosophical difficulty at this point. We started this chapter by "accepting" the idea of *discrete* energy states in an atom. But if these energy states were perfectly sharp, the uncertainty principle would indicate an infinite indeterminacy in the time that the atom is in one of these states. To put it another way, if the atom were excited to a perfectly well-defined energy state, it would stay there! Yet we have blithely assumed that the excited atoms decay by the emission of a photon.

Sec. 7.4 Introduction to the Rate Equations—Lifetime Broadening

Therefore we are forced to modify our energy-level picture slightly to account for the experimental fact that atoms do radiate. This is accomplished by smearing the energy levels into a sharply peaked band, as shown in Fig. 7.7.

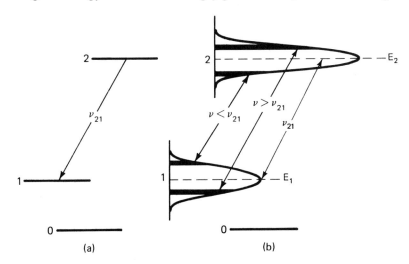

Figure 7.7 (a) Elementary energy-level diagram. (b) Modification of (a) by smearing the energy levels into a sharply peaked band.

Although Fig. 7.7 greatly exaggerates the smearing and broadening, it emphasizes the fact that radiation can and does appear on either side of line center. It also emphasizes the fact that different states have different broadening and that the band of frequencies emitted by the transition $2 \to 1$ will reflect the smearing of *both* upper and lower states.

We can look on these bell-shaped curves as being the relative probability of an atom being found in a band dE_2 around energy E_2, given that the atom is in state 2. Similar considerations apply to state 1 with the subscripts changed. Since transitions can occur between dE_2 and dE_1, the emitted radiation is also smeared, with the most intense part appearing between bands with the greatest occupation probability—obviously at ν_{21}.

We define a function $g(\nu)$, the *line shape*, such that $g(\nu) \, d\nu$ is the probability of emission of a photon with frequency between ν and $\nu + d\nu$. Obviously, if the atom emits a photon, it has to appear somewhere:

$$\int_0^\infty g(\nu) \, d\nu = 1 \tag{7.4.6}$$

(Even though the limits are shown from 0 to ∞, we can anticipate that the main contribution is in a very narrow band about ν_{21}. We will have a more detailed description of the line-shape function in the next section.)

Finally, note that the lifetime of a state is determined by *all* processes affecting the atom. For instance, the lifetime of state 2 is determined by A_{21} and A_{20} *and* by the collisional processes k_{21} and k_{20}, which, as we shall see in the next section, contribute to the "smearing" or "broadening" of the radiation around ν_{21}.

7.5 THE LINE SHAPE

7.5.1 Homogeneous Broadening

One way of obtaining a mathematical description of the line shape for the situation described in Sec. 7.4 is by Fourier-analyzing the spectrum emitted by state 2 in the vicinity of v_0. We return to (7.4.5) and define a classical electric field so as to yield the same power emitted by the atoms.

$$E = E_0 e^{-t/2\tau_2} \cos \omega_0 t \qquad (7.5.1)$$

where E_0 absorbs all constants necessary to predict the correct power. To find the spectral content of this field, one finds the Laplace transform of (7.5.1) and evaluates it on the $s = j\omega$ axis.

$$E(\omega) = \mathcal{L}[E(t)] = \left[\int_0^\infty E(t) e^{-st} dt\right]_{s=j\omega} \qquad (7.5.2)$$

$$= \frac{(1/2\tau_2) + j\omega}{\omega_0^2 - \omega^2 + (1/2\tau_2)^2 + j(\omega/\tau_2)} \cdot E_0$$

To find the spectral distribution of power, we form the product $E(\omega)E^*(\omega) = S(\omega)$.

$$\frac{S(\omega)}{E_0^2} = \frac{(1/2\tau_2)^2 + \omega^2}{[\omega_0^2 - \omega^2 + (1/2\tau_2)^2]^2 + (\omega/\tau_2)^2} \qquad (7.5.3)$$

A few approximations are in order here. We expect the function to be strongly peaked around $\omega = \omega_0$; hence, we set $\omega + \omega_0 = 2\omega_0$ but keep the difference function $\omega_0 - \omega$ explicitly. Furthermore, we neglect $(1/2\tau_2)^2$ compared to ω or ω_0.

$$\frac{S(\omega)}{E_0^2} = \frac{\omega_0^2}{(\omega_0 - \omega)^2(\omega_0 + \omega)^2 + (\omega/\tau_2)^2}$$

$$\doteq \frac{\omega_0^2}{4\omega_0^2[\omega_0 - \omega]^2 + (1/2\tau_2)^2} \qquad (7.5.4)$$

$$= \frac{1}{4} \frac{1}{(\omega_0 - \omega)^2 + (1/2\tau_2)^2}$$

Thus the radiation is not perfectly monochromatic but is distributed over a band of frequencies as shown in Fig. 7.8. Equation (7.5.4) is usually expressed in frequency units:

Sec. 7.5 The Line Shape

$$S(\nu) = \frac{k}{(\nu_0 - \nu)^2 + (\Delta\nu/2)^2} = g(\nu) \quad (7.5.5)$$

To obtain the line shape, we use the normalization condition Eq. 7.4.6 to find the constant k.

$$\int_0^\infty g(\nu)d\nu = \int_0^\infty \frac{kd\nu}{(\nu_0 - \nu)^2 + (\Delta\nu/2)^2} = 1$$

or

$$k = \frac{\Delta\nu}{2\pi}$$

Therefore

$$g(\nu) = \frac{\Delta\nu}{2\pi[(\nu_0 - \nu)^2 + (\Delta\nu/2)^2]} \quad (7.5.6)$$

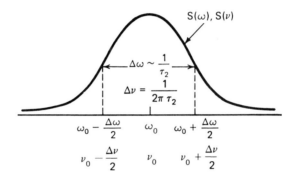

Figure 7.8 Spectral distribution of power radiated by a group of atoms with a common lifetime.

It is important to realize the physical implications of this bit of mathematics. We were faced with the task of measuring the temporal decay of the radiation to determine the lifetime of state 2; now we find that this same information is contained in the spectral distribution of the radiation. Inasmuch as any group of excited atoms decay according to (7.4.2), it is now no longer necessary to clamp the excitation to zero. The line shape is characteristic of this group of atoms and includes the natural decay processes.

In our formulation leading to (7.5.6), the width of the line (2 → 1) is related to the decay rate of state 2 by $\Delta\omega\tau_2 \simeq 1$. In terms of the energy levels, the spread of the energy of the quantum state 2 and its lifetime is related by the uncertainty relation:

$$\Delta\omega\tau_2 = 1 \quad \text{or} \quad (\hbar\Delta\omega)\tau_2 = \hbar \quad \text{or} \quad \Delta E\tau_2 = \hbar \quad (7.5.7)$$

Now these same considerations apply to state 1; it has a nonzero decay rate $\frac{1}{\tau_1}$ and thus a finite spread in energy also.

Consequently, the width of a transition between the two broadened quantum states 1 and 2 is determined by *both* lifetimes,

$$\Delta \nu = \frac{1}{2\pi}\left(\frac{1}{\tau_1} + \frac{1}{\tau_2}\right) \tag{7.5.8}$$

and is given the name "lifetime broadening." *If* all decay processes can be neglected except the radiative ones, we obtain the minimum width of any transition under any circumstance.

$$\Delta \nu_n = \frac{1}{2\pi}(A_2 + A_1) \tag{7.5.9}$$

Unfortunately, this has been called "natural broadening." (There is nothing unnatural about other processes.) It is probably the *least* important broadening mechanism of all.

Probably the most important process is not contained explicitly in the foregoing analysis; it is described by the following *collision* sequence:

$$[2] + [M(\text{anything})] \longrightarrow [2] + [M(\text{anything})] \tag{7.5.10}$$

where $[M]$ represents any atom, molecule, or phonon, with the same sequence applying to atoms in state 1. Equation (7.5.10) states that atoms in state 2 collide with something, $[M]$, exchange very little, if any, energy, and therefore leave the collision partners in the same quantum state as before. Obviously, the frequency of such collisions depends on the density of the M atoms; hence, this is called pressure broadening.

At first glance (7.5.10) might seem to be unimportant—nothing appears to be happening! Although the total energy is conserved by the reaction indicated by (7.5.10), some of the internal potential energy of the quantum state can be converted to kinetic energy along the collision trajectory. As a consequence, the energy levels can be considered to be a function of time, by starting at E_2, changing according to the force law of the "elastic" collision to $E_2(t)$, and ending at the same energy, E_2. Since the time for this interchange occurs on a time scale of 10^{-13} sec or less, far shorter than the mean radiative lifetime, the collisions represent a discontinuous *dephasing* of the wave functions of atoms. (A quantum treatment of the consequences of this dephasing is covered in Sec 14.6.)

A classical picture of this process is that shown in Fig. 7.9 (with great exaggeration). In Fig. 7.9(a), the classical field associated with the radiation from a large number of noninteracting atoms is shown. Fig. 7.9(b) is for atoms undergoing elastic collisions. The field decays at the same rate, indicative of the finite lifetime of the state; however, the phase of the radiation takes discontinuous jumps. This, in turn, shows up as a broadening in the frequency domain.

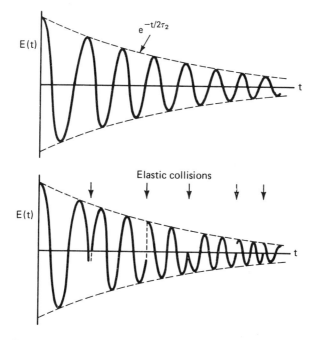

Figure 7.9 Classical picture of the effect of (b) elastic collisions on the decay of radiation shown in (a).

It is a rather unpleasant task in mathematics to account for the random nature of the collision rate around some mean rate, ν_{col}, and then to Fourier-analyze its effect on the spectrum, but the result is somewhat transparent to a physical interpretation. Inasmuch as there are *two* quantum states involved in the transition $2 \rightarrow 1$, each atomic wave function is interrupted ν_{col} times per second; hence, the additional broadening is $2\nu_{col}$. The full width of the transition is given by

$$\Delta \nu = \frac{1}{2\pi}[(A_2 + k_2) + (A_1 + k_1) + 2\nu_{col}] \qquad (7.5.11)$$

where the grouping $(A + k)$ indicates the decay rates of the states.

Whether the factor of 2 belongs in (7.5.11) is somewhat academic, since one must depend on experimental measurements to infer the pressure-broadened line width of a transition. As an example, E.T. Gerry and D.A. Leonard presented the data of Fig. 7.10 on the *absorption* coefficient of 10.6 μm radiation by CO_2 at various pressures. Without becoming unnecessarily complex about the energy levels of CO_2, it should be clear to all that the number of absorbing molecules increases as the pressure is increased. If we followed our

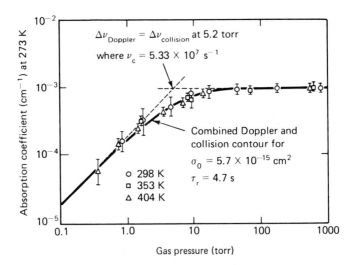

Figure 7.10 Absorption coefficient in CO_2 at 10.6 μm as a function of CO_2 pressure. (After E.T. Gerry and D.A. Leonard, Appl. Phys. Lett. *8*, 227, 1966.)

intuition, we would therefore expect the absorption to increase linearly with the number of molecules. The absorption does increase linearly with pressure at low pressures, but beyond about 10 to 20 torr becomes independent of pressure.

As will be shown later in this chapter, the absorption coefficient is proportional to the number of absorbers (as one would expect) *and* the line-shape function $g(\nu_0)$. For a pressure-broadened line, $g(\nu_0)$ is inversely proportional to the collision rate. Hence, the product becomes independent of pressure, as shown.

The collision rate is related to a cross section (i.e., "area") by

$$\nu_{col} = N_m \langle \sigma v \rangle \tag{7.5.12}$$

where N_m is the density of the background gas (not the excited states N_2 or N_1), σ is the cross section, and v is the relative velocity. The angular brackets $\langle \ \rangle$ indicate an average over all possible collision velocities weighted by the probability that the gas atom will have this velocity. For a Maxwellian distribution specified by a temperature T, with a cross section independent of velocity, (7.5.12) becomes

$$\nu_{col} = N_m \sigma \left[\frac{8kT}{\pi} \left(\frac{1}{M_m} + \frac{1}{M_n} \right) \right]^{1/2}$$

where the M's are the masses of the colliding atoms of type m and n.

Note that every atom has been treated on equal footing in this subsection. We have assumed that every atom is more or less the same as any other one, or

Sec. 7.5 The Line Shape

that there was no distinguishing feature about any one group. This is characteristic of *homogeneous* broadening, and thus we add the subscript "h" to $\Delta\nu$.

In all the cases studied above, the line shape $g(\nu)$ has the same functional form—the Lorentzian.

$$g(\nu) = \frac{\Delta\nu_h}{2\pi[(\nu_0 - \nu_x)^2 + (\Delta\nu_h/2)^2]} \qquad (7.5.6)$$

where

$$\Delta\nu_h = \frac{1}{2\pi}[(A_2 + k_2) + (A_1 + k_1) + 2\nu_{col}] \qquad (7.5.11a)$$

In most practical cases, the last term of 7.5.11a dominates, and the width of the homogeneously broadened line becomes:

$$\Delta\nu_h \simeq \frac{\Delta\nu_{col}}{\pi} = \frac{1}{\pi T_2} \qquad (7.5.11b)$$

where T_2 is the mean time between phase interrupting collisions. If one can distinguish between different groups of atoms under special circumstances, a different functional form of the line shape results.

7.5.2 Inhomogeneous Broadening

Although all atoms exhibit homogeneous broadening as a result of the reasons cited above, there are other processes that contribute to the width of a spectral line. If there is a characteristic that distinguishes one group of atoms from another, then we observe a line shape that reflects the relative probability of occurrence of the two groups. Probably the easiest example to use to illustrate this point is the case of naturally occurring neon, with 80% of it having a mass, 20 atomic mass units, and most of the remainder being of 22 atomic mass units. Because of the isotope effect, the center of the $(3s_2 - 2p_4)$ transition (at $\lambda_0 = 6328$ Å) changes slightly and one obtains an asymmetric line shape, as illustrated in Fig. 7-11.

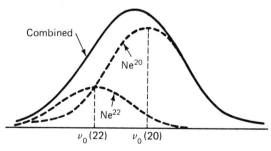

Figure 7.11 Inhomogeneous broadening in neon due to the isotope shift.

This effect plays an important role in a device encountered in everyday living—the common fluorescent lamp. The device works by virtue of the efficient excitation of the 6^3P_1 state of mercury, which is followed by radiation from that state at 254 nm. That radiation excites the phosphor. Mercury has seven isotopes, five with 0 spin S, one with $S = 1/2$, and one with $S = 3/2$. Due to a combination of the isotope shift and the hyperfine interaction, the "line" is split into five significant components, and this allows the radiation to escape from the body of the gas to the phosphor on the walls. Obviously, the composite line shape is not simple, with the details being a prime candidate for a problem (see Chapter 14). However, the point to be made with both examples is that there are distinguishing features—the mass and nuclear spin—that identify or "tag" a particular group of atoms. We, living in the real world, observe a sum of their individual homogeneous line shapes weighted by the probability of their occurrence.

A case of considerable practical importance is due to the Doppler effect from the thermal velocity of the atoms in a gas. If a select group of atoms emit a frequency ν_0 in their rest frame which is moving away from the observer in the laboratory with a velocity v_z, then the observer would measure the Doppler-shifted frequency

$$\nu_0' = \nu_0\left(1 - \frac{v_z}{c}\right)$$

Thus the homogeneous line width radiated by that particular group identified by their velocity is

$$g(v_z, \nu) = \frac{\Delta\nu_h}{2\pi\left[\left(\nu - \nu_0 + \nu_0\frac{v_z}{c}\right)^2 + \left(\frac{\Delta\nu_h}{2}\right)^2\right]} \quad (7.5.13)$$

Now we must multiply (7.5.13) by the probability of occurrence of a group of atoms traveling away from us at a velocity of v_z (within dv_z). This is given by the Maxwell-Boltzmann distribution function.

fraction of the atoms with velocity in the z direction with velocity between v_z and $v_z + dv_z$:

$$\frac{dN}{N} = \left(\frac{M}{2\pi kT}\right)^{1/2} \exp\left(-\frac{Mv_z^2}{2kT}\right) dv_z \quad (7.5.14)$$

By multiplying (7.5.13) by (7.5.14) and integrating over all velocities, we obtain the more realistic line-shape function for these atoms.

$$g(\nu) = \left(\frac{M}{2\pi kT}\right)^{1/2} \int_{-\infty}^{+\infty} \left\{\frac{\Delta\nu_h}{2\pi\left[\left(\nu - \nu_0 + \nu_0\frac{v_z}{c}\right)^2 + \left(\frac{\Delta\nu_h}{2}\right)^2\right]}\right\} \exp\left(-\frac{Mv_z^2}{2kT}\right) dv_z$$

$$(7.5.15)$$

Sec. 7.5 The Line Shape

This function has been tabulated—it is called the Voigt function. One usually resorts to the approximation of replacing the Lorentzian, the quantity in the braces, by a delta function. (We will see later when this approximation is valid.) That is,

$$L(x - x') = \frac{\Delta x}{2\pi[(x - x')^2 + (\Delta x/2)^2]} \longrightarrow \delta(x - x')$$

Thus (7.5.15) can be integrated easily:

$$g(\nu) = \left(\frac{M}{2\pi kT}\right)^{1/2} \int_{-\infty}^{+\infty} \delta\left(\nu - \nu_0 + \frac{\nu_0 v_z}{c}\right) \exp\left(-\frac{Mv_z^2}{2kT}\right) d\left(\frac{v_z \nu_0}{c}\right) \frac{c}{\nu_0}$$

or

$$g(\nu) = \frac{1}{\nu_0}\left(\frac{Mc^2}{2\pi kT}\right)^{1/2} \exp\left(-\frac{Mc^2}{2kT}\right)\left(\frac{\nu - \nu_0}{\nu_0}\right)^2 \quad (7.5.16)$$

This is the "pure" Doppler line shape. The peak value occurs at ν_0 (naturally) and falls to one half of that at frequencies given by

$$\frac{Mc^2}{2kT}\left(\frac{\nu_{+,-} - \nu_0}{\nu_0}\right)^2 = \ln 2$$

Thus the full width of the transition is given by

$$(\nu_+ - \nu_-) = \Delta\nu_D = \left(\frac{8kT \ln 2}{Mc^2}\right)^{1/2} \nu_0 \quad (7.5.17)$$

and (7.5.16) can be reexpressed in the following manner:

$$g(\nu) = \left(\frac{4 \ln 2}{\pi}\right)^{1/2} \frac{1}{\Delta\nu_D} \exp\left[-4 \ln 2 \left(\frac{\nu - \nu_0}{\Delta\nu_D}\right)^2\right] \quad (7.5.18)$$

We can now evaluate the validity of our approximation of replacing the Lorentzian by a delta function. If the Doppler width (7.5.17) is much larger than the homogeneous width, (7.5.18) is valid. If the reverse is true, we should replace the exponential term in (7.5.15) by a delta function centered at $v_z = 0$ and thus recover the homogeneous line shape.

An electronic (i.e., visible) transition in a low-pressure gas tends to be dominated by inhomogeneous or Doppler broadening. If, however, the pressure is high enough, or the center frequency, ν_0, is low enough, homogeneous or pressure broadening will dominate.

7.5.3 General Comments on the Line Shape

There are many other mechanisms for broadening a spectral line. For instance, in a solid, the local crystalline field may split and shift the energy levels as a result of the Stark effect, and lattice vibrations (phonons) can contribute to the breadth of a transition.

The reader should be cautioned against assuming a direct relationship between the amount of mathematics expended here on a broadening mechanism and its relative importance. Although Doppler and pressure-broadening mechanisms are important, they do not overwhelm all other types (indeed, they do not even apply in a solid). In fact, only the *central portion* of some transitions in a gas is adequately described by the theory presented here.

However, the idea of a line shape is most important, quite general, and independent of the maze of mathematics surrounding its development. The *line-shape function,* $g(\nu)\,d\nu$, is the relative probability that

1. A photon emitted by a *spontaneous* transition will appear between ν and $\nu + d\nu$.
2. Radiation in the frequency interval ν to $\nu + d\nu$ can be *absorbed* by atoms in state 1.
3. Radiation in this interval will *stimulate* atoms in state 2 to give up their internal energy.

Obviously, (1) applies to spontaneous emission, (2) to absorption, and (3) to stimulated emission. However, the same line-shape function applies to all three processes.

Many of the real-life line-shape functions are asymmetric and mathematically intractable. However, the atoms have no knowledge of and no trouble with *our* arithmetic. In response, we must be prepared to tolerate and use a real-life line-shape function about which we have imperfect information.

7.6 TRANSITION RATES FOR MONOCHROMATIC WAVES

When the Einstein A and B coefficients were introduced, we were concerned with the interaction of a continuous radiation spectrum, $\rho(\nu)$, with the "discrete" energy levels of a group of atoms. We assumed that the bandwidth of the radiation (i.e., energy spread) was much larger than the band of emission or absorption by the atoms.

The situation is quite often reversed when one is dealing with laser radiation. For many (if not most) cases involving lasers, one has a finite amount of radiant energy (per unit volume) in a bandwidth that is much smaller than the corresponding spread expressed by the line shape of the transition. The arithmetic involved in the rate equations (7.3.2) and (7.3.3) changes accordingly to reflect this difference.

Thus the rate of change of the population of state 2 as a result of a monochromatic wave at a frequency ν with energy density ρ_ν (in joules/m^3) is

$$\frac{dN_2}{dt}\bigg|_{\substack{\text{absorption}\\ \text{stimulated emission}}} = -N_2 B_{21} g(\nu)\rho_\nu + N_1 B_{12} g(\nu)\rho_\nu \qquad (7.6.1)$$

[The simple artifact of dimensional analysis helps to remember this change. The dimensions of a continuous distribution of frequencies, denoted by $\rho(\nu)$, is joules per unit volume per unit frequency. A perfectly monochromatic wave has all its energy at one frequency, but the atom reacts only according to $g(\nu)$, whose dimensions are (frequency)$^{-1}$.]

Rather than use energy density, it is useful to convert to intensity (watts/unit area) by recognizing that electromagentic energy travels at the velocity c/n. Thus

$$I_\nu = \frac{c}{n}\rho_\nu \qquad (7.6.2)$$

and (7.6.1) can be modified accordingly. We can also use (7.3.10a), [i.e., $B_{12} = (g_2/g_1)B_{21}$] and (7.3.10b) [i.e., $B_{21} = (c^3/8\pi n^3 h\nu^3)A_{21}$] in this modification

$$\left. \begin{aligned} \frac{dN_2}{dt}\bigg|_{\substack{\text{absorption} \\ \text{stimulated emission}}} &= -A_{21}\frac{c^3}{8\pi n^3 h\nu^3}\left(N_2 - \frac{g_2}{g_1}N_1\right)g(\nu)\frac{I_\nu}{(c/n)} \\ \text{or} \\ &= -A_{21}\frac{\lambda_0^2}{8\pi n^2}g(\nu)\left(N_2 - \frac{g_2}{g_1}N_1\right)\frac{I_\nu}{h\nu} \end{aligned} \right\} \qquad (7.6.3)$$

Please note that $I_\nu/h\nu$ is just the number of photons streaming past a unit area, the photon flux.

7.7 AMPLIFICATION BY AN ATOMIC SYSTEM

We are now in the position to describe the process of amplification (or attenuation) of electromagnetic energy by its interaction with the atoms. We take Einstein's view of this interaction, although we will *not* assume that the population densities (m^{-3}) of the various states N_2 and N_1 are in thermodynamic equilibrium. We will assume that these densities are being created by a pumping process that is exactly balanced by a loss process—the details of which need not concern us at present.

In Fig. 7.12, we imagine a "Gedanken" experiment in which a slab of these atoms, Δz long, is being irradiated by a polarized electromagnetic wave of intensity I_ν (W/m^2), which after amplification (or attenuation) is received by our detector. We must recognize, however, that a detector system *cannot* distinguish between the various physical processes discussed in this chapter: a photon radiated spontaneously by this slab and reaching the detector causes the same response as does one that has been added by stimulated emission. Obviously, spontaneous emission from this slab contributes "noise" in our experiment.

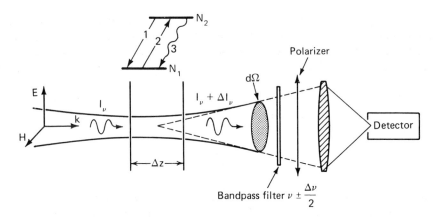

Figure 7.12 Measurement of the gain of an atomic system.

However, if one chooses the source carefully, the noise can be minimized: one limits the optical bandwidth of the detector system by some sort of a filter with a passband $\Delta\nu$ around the frequency ν of the source; one uses a polarizer to reject one half of the spontaneous power orthogonal to the source; and one carefully limits the field of view (FOV) to match the incoming beam. Thus the output consists of the input intensity *plus* that added by the following processes:

1. *Stimulated emission:* the amount of radiation stimulated by the incoming wave. Since this is stimulated emission, the frequency, phase, and direction of the added signal are the same as the incoming wave (this is indicated by process 1 on the energy-level diagram)

minus:

2. *Absorption:* the amount of radiation absorbed by the atoms in state 1

plus:

3. *Spontaneous emission:* the amount of radiation emitted spontaneously by the atoms in state 2 in the direction of the input wave and in the same frequency within bandwidth $\Delta\nu$ of our detector (this is indicated by the wavy line going from 2 to 1).

Now the easiest way to do the bookkeeping on these processes is to arrange the physical factors into vertical columns. We recognize that each transition caused by the foregoing processes contributes (or subtracts) a package of energy $h\nu$ (column 1), which, when multiplied by the rate per atom, is the power contributed by each atom (column 2), but we also recognize that this contribution (or interaction) scales according to the line shape (column 3). Column 4 is the probability that the photon involved has the proper polarization, and 5

Sec. 7.7 Amplification by an Atomic System

is the probability that it is within the solid angle specified by the experimental setup (which is heavily weighted favoring the first two processes).

However, some of the spontaneous power reaches the detector. The radiation from the slab is spread uniformly into 4π steradians, or equivalently into the various electromagnetic modes defined by the beam size and the length Δz. Hence, the detector, with an acceptance cone of $d\Omega$, will only collect a fraction, $d\Omega/4\pi$, of the power radiated by these atoms, and only one half of that fraction has the proper polarization. Finally, column 6 is the number of atoms involved in the interaction. This arrangement is given in (7.7.1).

$$\begin{array}{cccccccc}
 & 1 & 2 & 3 & 4 & 5 & 6 \\
\Delta I_\nu = & h\nu \times & B_{21}\dfrac{I_\nu}{c} \times & g(\nu) \times & 1 \times & 1 & \times N_2\,\Delta z \\
 & -h\nu \times & B_{12}\dfrac{I_\nu}{c} \times & g(\nu) \times & 1 \times & 1 & \times N_1\,\Delta z \\
 & +h\nu \times & A_{21}\,\Delta\nu \times & g(\nu) \times & \dfrac{1}{2} \times & \dfrac{d\Omega}{4\pi} & \times N_2\,\Delta z
\end{array} \quad (7.7.1)$$

Further manipulation yields

$$\frac{\Delta I_\nu}{\Delta z} \longrightarrow \frac{dI_\nu}{dz} = \left[\frac{h\nu}{c}(B_{21}N_2 - B_{12}N_1)g(\nu)\right]I_\nu$$
$$+ \frac{1}{2}\left[h\nu A_{21}N_2 g(\nu)\,\Delta\nu \frac{d\Omega}{4\pi}\right] \quad (7.7.2a)$$

One can readily appreciate that this last term can be called "noise," since this signal is present at the detector without any input I to the slab. Even though it is essential to the laser (to initiate the oscillation), its presence is not required at this time and we neglect it for now.

It is convenient to use some of the relations found in Sec. 7.3 to change the appearance of (7.7.2a). Recall that $B_{12}/B_{21} = g_2/g_1$ and $A_{21}/B_{21} = 8\pi n^3 h\nu^3/c^3$. Thus, the basic equation becomes

$$\frac{dI_\nu}{dz} = \left[A_{21}\frac{\lambda_0^2}{8\pi n^2}g(\nu)\left(N_2 - \frac{g_2}{g_1}N_1\right)\right]I_\nu \triangleq \gamma_0(\nu)I_\nu \quad (7.7.2b)$$

The coefficient $\gamma_0(\nu)$ is called the gain coefficient (m^{-1}) with the subscript 0, indicating that the incoming intensity is sufficiently small as to cause negligible perturbation on the populations N_2 and N_1. It is the central equation of laser theory and should be treated with due respect. (*Memorize it.*)

Some prefer to specify the factors in the expression for the gain coefficient in terms of a stimulated emission or absorption cross section (i.e., "area")

$$\sigma_{SE} = A_{21} \frac{\lambda_0^2}{8\pi n^2} g(\nu_0), \qquad \sigma_{AB} = A_{21} \frac{\lambda_0^2}{8\pi n^2} g(\nu_0) \frac{g_2}{g_1} \qquad (7.7.3)$$

Thus the gain coefficient can be written as the product of the stimulated emission cross section and the inversion:

$$\gamma_0(\nu) = \Delta N \, \sigma_{SE}(\nu) \qquad (7.7.4)$$

$$\Delta N = N_2 - \frac{g_2}{g_1} N_1 \qquad (7.7.5)$$

This is most convenient, for, given the cross section, it is easy to do the mental arithmetic to ascertain the inversion necessary to obtain a required gain. For instance, if the cross section is 10^{-14} cm^2, we need an inversion of 10^{13} cm^{-3} to obtain a gain coefficient of 0.1 cm^{-1}.

Having found the differential form of the gain, we can integrate (7.7.2) to obtain

$$I_\nu(z) = I_\nu(0) \exp[\gamma_0(\nu) z]$$
$$= G_0(\nu) I_\nu(0) \qquad (7.7.6)$$
$$G_0 = \exp[\gamma_0(\nu) d]$$

where G_0 is the small signal (power) gain of an amplifier of length d.

It is most important to remember that the gain coefficient is frequency dependent; consequently, the gain G_0 is even more so, since $\gamma_0(\nu)$ appears in the exponent. Thus we have a narrow band-pass amplifier. In Chapter 8 we put this narrow band amplifier into a feedback loop to obtain oscillation.

7.8 REVIEW

Before we proceed to laser oscillation, it is appropriate to stand back and review the material in this chapter. Although a lot of arithmetic was used, most of it was used to convert one equation into another form. Consequently, definitions are an important part of this chapter. If one remembers the physical meaning of the quantities defined, the few lines of arithmetic follow most naturally.

For instance, the Einstein A and B coefficients were introduced to describe the *rate* at which the atoms respond to the electromagnetic field. In covering this, we are naturally led to the physical interpretation of A and B and then to the idea of *line shape*. This was the probability of a group of atoms interacting with a wave by stimulated emission or absorption, or in producing the wave by spontaneous emission. We also observed that there are different types of mechanisms leading to different types of line shapes, or broadened transitions. If there is no distinguishing feature about the atoms, we have homogeneous broadening, whereas if two (or more) groups can be identified, we have inhomogeneous broadening.

All of these concepts will play a major role in laser oscillation.

PROBLEMS

7.1. The derivation of the formula for the electromagnetic mode density assumed that the cavity dimensions are very large and $N(\nu)$ was found to be

$N(\nu)$ = number of electromagnetic modes between 0 and ν (per unit of volume of a cavity)

$$= \frac{8\pi \nu^3}{3(c/n)^3} \text{ where } n = \text{index of refraction}$$

Hence, the mode density (per unit of volume per unit of frequency) was found to be

$$n(\nu) = \frac{dN}{d\nu} = \frac{8\pi \nu^2}{(c/n)^3}$$

When the dimensions of the cavity become comparable to a wavelength, the approximation is *not* accurate and one must count the modes. This problem is intended to give you confidence in the formula above and also to indicate the exact procedure. Plot the number of modes between 0 and 10 GHz as a function of frequency for a rectangular cavity of dimensions 2 × 5 × 6 cm using the approximate formula given above and by actually counting the allowed modes and plotting the resulting stair-step function. (CAUTION: Only the m or p index of the TE$_{m,p,q}$ mode may be zero, but not both, and only the q index may be zero for the TM mode.)

7.2. The broadening of a transition reflects the width and shape of both quantum levels involved. If one ignores the slight change in occupation probability with energy, the relative probability of a transition at $h\nu$ is a product of the density of states at E_2 and E_1 summed over all possibilities of $\Delta E = h\nu$. Assume that the density of states is distributed according to a Lorentzian around E_2 with a width ΔE_2 with corresponding quantities E_1 and ΔE_1 for state 1. Find the line shape for the transition. (The line shape is a Lorentzian having a width reflecting the sum $\Delta E_1 + \Delta E_2$.)

7.3. Show that the ratio of stimulated emission to spontaneous emission is equal to the number of photons per mode.

7.4. The spontaneous emission profile from a certain transition can be approximated by the shape shown below.

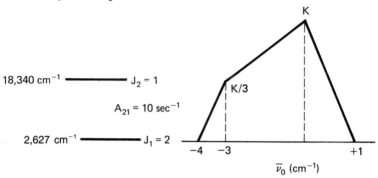

(a) What is the stimulated emission cross section?
(b) What is the absorption cross section?

7.5. Suppose the distribution of center frequencies is a "square" function for those values between $\nu_0 - (\Delta\nu_s)/2 < f < \nu_0 + (\Delta\nu_s)/2$ and zero, otherwise in the manner shown on the diagram below.

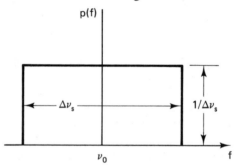

Each group of atoms, $p(f)df$, is characterized by a Lorentzian with a width $\Delta\nu_h$ (i.e., a homogeneous width).

(a) Plot the normalized small signal gain coefficient as a function of frequency for a ratio $\Delta\nu_h/\Delta\nu_s = 1.0$ by dividing it by the value at line center for the "pure" inhomogeneous limit (i.e., $\Delta\nu_h/\Delta\nu_s = 0$). (Since the curve is symmetric about ν_0, only positive values of $\nu - \nu_0 \leq 5\,\Delta\nu_s$ need be plotted.)

(b) Plot the gain coefficient at line center as a function of the normalized intensity I_ν/I_s for $\Delta\nu_h/\Delta\nu_s = 1$. Show, as dashed curves, the saturation behavior for the extreme cases of $\Delta\nu_h/\Delta\nu_s = 0$ and ∞ assuming that the small signal gain coefficient at line center is the same for both cases.

7.6. Consider a transition at 5000 Å with a width of 1 Å and a cavity 2 cm³ in volume ($n = 1$).

(a) Convert this wavelength interval (1 Å) to frequency units (i.e., GHz and cm^{-1}).

(b) How many electromagnetic modes exist in this frequency band for this cavity?

(c) Suppose that the cavity was in the form of a cylinder with a cross-sectional area of 0.1 cm^2 (and thus is 20 cm long). How many TEM$_{0,0,q}$ cavity modes would fit within the frequency band specified by this 1 Å? (Do not forget the two polarizations.)

(d) Combine the results of (b) and (c) to estimate the probability of a spontaneous photon appearing in one of the polarized TEM$_{0,0,q}$ modes.

(e) If the A coefficient for this transition is 10^7 sec^{-1}, what is the stimulated emission cross section?

7.7. Evaluate the Doppler widths of the following helium-neon transition: (a) $3s_2 - 2p_4$ at $\lambda_0 = 6328$ Å; (b) $2s_2 - 2p_4$ at $\lambda_0 = 1.1523$ μm; and (c) $3s_2 - 3p_4$ at $\lambda_0 = 3.39$ μm. Express your answers in GHz and in Å units.

7.8. Note that many of the helium-neon laser transitions share a common upper level (the first-named state) or a common lower level (the second) in Problem 7.7. Use the specifications of the wavelengths to construct a partial energy-level diagram. (NOTE: The answer is given in Chapter 10.)

7.9. Reformulate (7.5.15) using the following substitutions:

$$\omega = \frac{2(\nu - \nu_0)}{\Delta\nu_D}(\ln 2)^{1/2} \qquad a = \frac{\Delta\nu_h}{\Delta\nu_D}(\ln 2)^{1/2} \qquad y = \frac{2v_z}{c}\nu_0\frac{(\ln 2)^{1/2}}{\Delta\nu_D}$$

(a) Show that (7.5.15) becomes

$$g(\omega) = \frac{a}{\pi}\int_{-\infty}^{+\infty}\frac{\exp(-y^2)}{a^2 + (\omega - y)^2}\,dy \qquad \text{(P7-9)}$$

(b) Evaluate this integral numerically for $a = 1$:

	ω	0	1	2	3	4	6	8	10
Ans.	$g(\omega)$	0.428	0.305	0.140	0.066	0.037	0.016	0.009	0.005

(c) Find an analytic expression for $g(\omega)$ for small values of a:

$$\left(\text{Ans.: } g(\omega) = \exp(-\omega^2) - \frac{2a}{\pi^{1/2}}\left(1 - 2\omega e^{-\omega^2}\int_0^\omega e^{x^2}\,dx\right).\right)$$

(d) Show that (P7-9) reduces to a Lorentzian for large values of a.

7.10. On a single sheet of three-cycle semilog graph paper, plot the combined Doppler and homogeneous line shape for the numerical values given in the previous problem (i.e., $a = 1$). The linear frequency scale should be normalized to the Doppler width; that is, $(\nu - \nu_0)/\Delta\nu_D$ and the line shape plotted on the logarithmic axis. On this same graph, plot the "pure" Doppler line shape (i.e., $a = 0$). What is the combined width (expressed as a numerical factor times the Doppler width)?

7.11. In the diagram shown below, the pump P_2 (i.e., electrons, flash lamp, another laser, etc.) excites atoms from state 0 to state 2, nothing to state 1. To make the problem simple and tractable, assume (1) state 0 is not depleted to any significant extent for any time (i.e., $dN_0/dt = 0$); (2) use the simple decay route indicated; (3) neglect stimulated emission; (4) assume $\tau_2 = 1$ μs and $\tau_1 = 2$ μs; and (5) let $P_2 = 10^{20}$ cm^{-3} s^{-1}. Use symbols for (a) and (b) and find numerical values for (c) and (d).

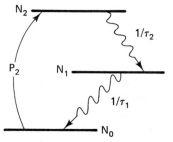

(a) What are the rate equations for states 2 and 1?
(b) Give an expression for the densities $N_{2,1}$ as a function of time.
(c) Over what time interval, δt, is the population difference $N_2 - N_1 > 0$?
(d) What are the steady-state populations in 2 and 1?

7.12. The spontaneous emission profile of a certain laser can be approximated by the triangular shape shown below. If the spontaneous lifetime were 5 nsec, and the gain coefficient were 10 cm^{-1}, find:
(a) The value of the line shape (in sec) at $h\nu/e = 1.476$ eV
(b) The inversion necessary to obtain that gain coefficient

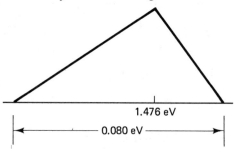

7.13. The following is a tractable representation of the line shape for the helium-neon transition at $\lambda_0 = 6328$ Å ($3s_2 - 2p_4$), which has an A coefficient of 6.56×10^7 sec^{-1}.

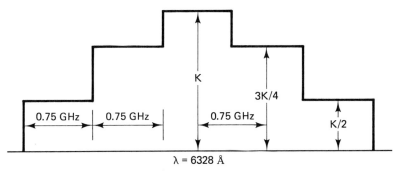

$\lambda = 6328$ Å

(a) What is the value of $g(\nu_0)$?
(b) What is the stimulated emission cross section?
(c) Give a short word description of the physical significance of $g(\nu)$ as it applies to spontaneous emission, absorption, and stimulated emission.

REFERENCES AND SUGGESTED READINGS

1. A. C. G. Mitchell and M. W. Zemansky, *Resonance Radiation and Excited Atoms* (New York: Cambridge University Press, 1971). Chapter 3 is especially germane to the material of this chapter and is highly recommended. Please note: This book was first printed in 1934 and yet the material is pertinent to modern quantum electronic devices. Indeed this book is one of the most quoted references in gas laser theory. The laser gain equation is not new!
2. G. Herzberg, *Atomic Spectra and Atomic Structure*, (Englewood Cliffs, N.J.: Prentice-Hall, Inc., 1937; New York: Dover Publications, Inc., 1944). An excellent introduction to atomic spectra.
3. G. Herzberg, *Molecular Spectra and Molecular Structure*, Vol. 1: *Spectra of Diatomic Molecules* (Princeton, N.J.: D. Van Nostrand Company, 1967).
4. A. E. Siegman, *Introduction to Lasers and Masers* (New York: McGraw-Hill Book Company, 1971), Chap. 3.
5. A. Yariv, *Introduction to Optical Electronics*, 2nd ed. (New York: Holt, Rinehart and Winston, 1976).
6. R. P. Feynman, R. B. Leighton, and M. Sands, *The Feynman Lectures on Physics*, Vol. 3 (Reading, Mass.: Addison-Wesley Publishing Company, Inc., 1965).
7. W. S. C. Chang, *Principles of Quantum Electronics* (Reading, Mass.: Addison-Wesley Publishing Company, Inc., 1969), Chap. 5.
8. A. Maitland and M. H. Dunn, *Laser Physics* (Amsterdam: North-Holland Publishing Company, Inc., 1969), Chaps. 2 and 3.
9. R. M. Eisberg, *Fundamentals of Modern Physics* (New York: John Wiley & Sons, Inc., 1961), Chaps. 2 and 13, 468–471.
10. A. E. Siegman, *Lasers* (Mill Valley, Calif.: University Science Books, 1986), Chaps. 5–6.
11. A. Yariv, *Quantum Electronics* (New York: John Wiley & Sons, Inc., 1975).
12. G. H. B. Thompson, *Physics of Semiconductor Laser Devices* (New York: John Wiley & Sons, Inc., 1980).

13. See the Historical Paper "On the Quantum Theory of Radiation," A. Einstein, reprinted from *The Old Quantum Theory* (Pergamon Press, 1967) in *Laser Theory,* Ed. Frank Barnes (New York: IEEE Press, 1972). There are many other papers of interest in this collection.
14. L. Oster, "Some Applications of Detailed Balancing," Am. J. Phys. *38*, 754–761, 1970.
15. J. H. Van Vleck and D. L. Huber, "Absorption, Emission and Line Breadths: A Semi-historical Perspective," Rev. Mod. Phys. *49*, 939–959, 1977.
16. R. G. Breene, Jr., "Line Shape," Rev. Mod. Phys. *29*, 94–143, 1957.

8

Laser Oscillation and Amplification

8.1 INTRODUCTION—THRESHOLD CONDITION FOR OSCILLATION

It should be obvious from the laser-gain equation

$$\gamma_0(\nu) = A_{21} \frac{\lambda_0^2}{8\pi n^2} g(\nu) \left(N_2 - \frac{g_2}{g_1} N_1 \right) \tag{7.7.26}$$

that it is necessary to have $N_2 > (g_2/g_1) N_1$ in order to have gain. Inasmuch as this is an "abnormal" state of affairs in nature, the population densities are said to be inverted. We will not be concerned here with the specific details as to how this condition is created but with the consequences.

It is obvious that we can construct a narrow-band amplifier. If the length of the medium is l_g, the small-signal power gain is

$$G_0 = \exp\left[\gamma_0(\nu) l_g\right] \tag{8.1.1}$$

To make this into an oscillator, we provide sufficient feedback in the manner illustrated in Fig. 8.1. Threshold for oscillation is determined by the requirement that the round-trip gain exceed 1. If one considers only mirror losses, this implies that the gain coefficient $\gamma_0(\nu)$ must be sufficiently large, so that

$$R_1 R_2 e^{2\gamma_0(\nu) l_g} \geq 1$$

or (8.1.2)

$$\gamma_0(\nu) \geq \frac{1}{2l_g} \ln\left(\frac{1}{R_1 R_2}\right) = \alpha$$

In other words, the gain per unit of length, $\gamma_0(\nu)$, must exceed the loss when that loss is prorated on a per unit of length basis.*

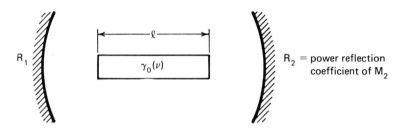

Figure 8.1 A simple laser.

There is considerable physics buried in such a simple equation—so much so that it is appropriate to consider a graphical solution, as shown in Fig. 8.2.

It should be obvious from Fig. 8.2 that "threshold" refers to a situation where the gain coefficient exceeds the loss over a very small band of frequencies. If the gain is made much larger than threshold, there is a considerable band of frequencies over which the inequality (8.1.2) is satisfied.

Which frequency oscillates? What limits the amplitude of oscillation? What physical mechanism starts the laser oscillating?

The remaining sections of this chapter address these questions together with others, as they naturally arise.

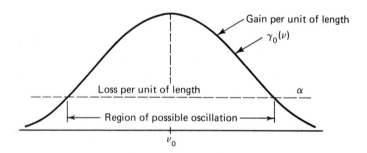

Figure 8.2 Graphical solution of the threshold equation.

*The symbol α will be used to denote a loss per unit of length.

8.2 LASER OSCILLATION AND AMPLIFICATION IN A HOMOGENEOUS BROADENED TRANSITION

Figure 8.2 implies that there is a considerable band of frequencies over which the gain exceeds the loss. This is a very practical—maybe even desirable—situation. For instance, suppose that we are dealing with a homogeneously broadened transition with a width of 1 GHz and that the small-signal gain coefficient at line center ν_0 is N times the loss.

Now the frequency dependence of $\gamma_0(\nu)$ is contained in the line shape, and for a Lorentzian, we have

$$\gamma_0(\nu) = \gamma_0(\nu_0) \frac{(\Delta\nu/2)^2}{(\nu_0 - \nu)^2 + (\Delta\nu/2)^2} \qquad (8.2.1)$$

For $\gamma_0(\nu_0)/\alpha = N$, the gain exceeds the losses over the frequency interval:

$$2|\nu_0 - \nu| \leq (N - 1)^{1/2} \Delta\nu \qquad (8.2.2)$$

For $\gamma_0(\nu_0)/\alpha = 4$, for instance, and $\Delta\nu = 1$ GHz, there is a band 1.7 GHz wide where laser oscillation *can* take place. However, laser oscillation *will* occur at a discrete frequency* dictated by the cavity mode that has the highest gain-to-loss ratio.

To appreciate the logic of this last statement, recall the basic equation associated with stimulated emission:

$$\left.\frac{dN_2}{dt}\right|_{\substack{\text{stimulated}\\\text{emission}}} = -B_{21}N_2\rho_\nu g(\nu) = -B_{21}N_2 g(\nu)\frac{I_\nu}{(c/n)} \qquad (7.6.1)$$

Recall that the coefficient B_{21} represents an integral part of the atom; just because we have decided to build a laser, that atom is not going to change its characteristics. The line-shape function $g(\nu)$ expresses how each atom, on the average, responds to an electromagnetic wave at various frequencies. Although an intense field can affect the line shape, only the characteristic broadening mechanisms affect its shape before oscillation starts. Thus although $g(\nu)$ expresses a preference for frequencies close to line center, it is a mild one, changing only by a factor of 2 for $|\nu - \nu_0| = \Delta\nu/2$. Although we would prefer N_2 to be large, its maximum value is fixed by the external pumping mechanism.

Is there anything in (7.6.1) that makes a sharp distinction as to which frequency or narrow band of frequencies stimulates the atom at the greatest rate? By the process of elimination, we are left with ρ_ν. Is there anything we can do to make any one frequency more favorable than another? If you recall the discussion on resonance from Chapter 6, you know the answer is "yes."

---

*We postpone until later issues such as spatial hole burning and the spectral width of the oscillation.

Recall that the field at a cavity resonance is much larger than those at, say, antiresonance, [see(6.3.9)]. Thus, we can construct the following scenario for the start-up of laser oscillation. We assume that the pumping agent has created the population inversion $N_2 - (g_2/g_1 N_1)$. Even though an inversion exists, spontaneous emission still occurs, spewing out electromagnetic energy into any one of the $(8\pi n^3 \nu^2 \Delta\nu/c^3)V$ modes that are present in the volume of the active medium.

A few numerics are in order here. The number of modes that receive this spontaneous emission is just huge. For instance, suppose that the center wavelength of the transition is 5000 Å ($\nu_0 = 6 \times 10^{14}$ Hz); the width is, as before, 1 GHz; the volume of the active medium is 10 cm³; and $n = 1$. Then there are $(8\pi \nu^2 \Delta\nu/c^3)V = 3.35 \times 10^{10}$ different modes to which an atom can give up its internal energy to this electromagnetic field.

But most of these modes represent waves that are going in the wrong direction—not toward the mirrors but out the side of the laser cell. *But* that part of the spontaneous emission which is in the proper frequency interval to coincide with a cavity resonance (and, of course, is along the axis of the laser) is bounced back and forth between the mirrors, greatly enhancing the standing-wave field.

Thus the initial frequency dependence of the electromagnetic energy density is governed by the frequency dependence of the spontaneous emission [i.e., $g(\nu)$] and by the cavity response. This situation is shown in Fig. 8.3(a), where the fields in the cavity modes are just beginning to be formed by spontaneous emission. Once the energy is present, transitions caused by stimulated emission can take place, adding energy in the proper phase, at the proper frequency, in the proper direction, and in the proper polarization so as to add coherently to the field that stimulated the atoms. Consequently, those resonant fields close to line center are amplified, and in a few, cavity transit times are much greater than their initial values. Those modes in the wings $(0, 0, q - 2)$ and $(0, 0, q + 3)$ are amplified, but much less so.

Now a few round-trips through the cavity are sufficient to make the field quite large. For instance, suppose that the peak intensity of the $(0, 0, q)$ mode in Fig. 8.3(a) was 1 μW/cm² and that the net gain $(R_1 R_2)^{1/2} \exp[\gamma_0(\nu_q)l] = 4$. After just five round-trips, this intensity will have grown by a factor of $4^{10} \sim 1.05 \times 10^6$ provided that the gain coefficient stays constant during this process. The other modes of Fig. 8.3 grow but not nearly as fast. If, for instance, $(R_1 R_2)^{1/2} \exp[+\gamma(\nu_{q+1})l] = 2$ and its initial intensity was 0.5 μW/cm², its value would be 0.5 mW/cm² after five round-trips through the cavity. Clearly, the $(0, 0, q)$ mode is much larger than the rest, but even so, the 0.5 mW/cm² intensity of the $(0, 0, q + 1)$ mode is significant.

It should also be clear that something has to give—the intensity cannot keep growing indefinitely through more and more cavity transit times. Every time one more photon is added to the field inside the cavity, the population inversion must have decreased by 2. (Why not by 1?)

Sec. 8.2 Laser Oscillation and Amplification in a Homogeneous Broadened Transition

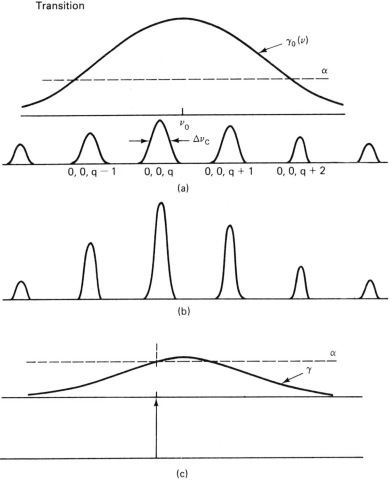

Figure 8.3 Evolution of laser oscillation from spontaneous emission: (a) initial; (b) intermediate; and (c) final.

When the stimulating field is so large as to cause the atoms to give up their energy as fast as they are being pumped up the energy scale, we have reached an equilibrium. Thus the gain of the system must change to a lower value until the rate of *production* of the excess inverted population is balanced by the destruction rate by stimulated emission. (This is called gain saturation and will be analyzed later.)

A moment's consideration of the issues raised in the previous two paragraphs leads to the acceptance of Fig. 8.3(c) for the representation of the final state of the laser gain and spectral content. There are three items to be emphasized and noted.

1. The laser gain coefficient has decreased (or saturated)* to the loss coefficient *at the frequency of laser oscillation*. The relative shape of the gain curve is similar to its initial one, although the peak value at line center is considerably reduced.
2. The spectral shape has changed dramatically from Fig. 8.3(a) or (b), with the central mode much larger than any of the other modes. Indeed, all other modes are now *below* threshold and are invisible in comparison to the laser amplitude. For instance, the gain on the $(q + 1)$ mode at ν_{q+1} is less than the loss at that frequency, and in spite of the initial phase of rapid growth of that mode, dies down to a level sustained only by spontaneous emission.
3. Laser oscillation occurs at the center of the cavity mode with the highest net gain.

The fact that the laser oscillates on only one cavity mode is a consequence of the assumption of *homogeneous broadening*. Recall that the definition of homogeneous broadening is that *all* atoms behave in the same manner. Thus, if any one atom gives up its energy (as a photon) to any field at any frequency, that atom can no longer contribute to the gain at another frequency. Consequently, the gain profile sags while maintaining proportionality over the spectrum.

However, the spectral shape within a cavity-mode resonance changes dramatically. Just as one cavity mode wins the footrace for the energy stored in the population inversion at the expense of other cavity modes, so does the frequency at the peak of the cavity resonance overwhelm the other nearby ones. There is a nonzero width to the oscillation spectrum because of the quantum nature of the generation process, but usually the mechanical, acoustical, or thermal fluctuations in the cavity length contribute a spectral width that is many times that allowed by quantum effects.

8.3 GAIN SATURATION IN A HOMOGENEOUS BROADENED TRANSITION

To describe the final state of the laser, we need a mathematical description of gain saturation. Toward this end, we construct a generalized model of the two atomic states involved in the laser, as shown in Fig. 8.4. In the figure, R represents the rate of producing the appropriate state as a result of all causes other than those indicated on the diagram. For instance, R_1 includes direct excitation from the ground state to state 1 and also any indirect routes such as excitation to a higher state followed by spontaneous emission from the higher

---

*The gain will actually saturate at a value slightly lower than the loss line, with the very slight difference being made up by the spontaneous emission. This leads to a finite spectral width of oscillation. For many, if not most applications, this fine point can be neglected.

Sec. 8.3 Gain Saturation in a Homogeneous Broadened Transition

state back to 1. It does *not* include the spontaneous decay of state 2 into 1 nor the stimulated processes—these will be considered separately. State 2 is pumped directly from ground at a rate R_2 that includes indirect paths to higher levels followed by a decay to 2, or it may represent a transfer from a different gas into the one of interest.

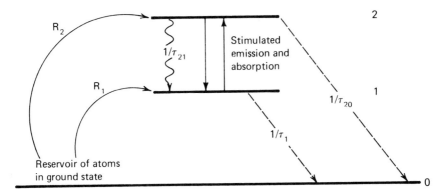

Figure 8.4 Generalized pumping scheme of a laser.

Once the atoms are in the atomic levels, they suffer a variety of fates. For instance, state 2 can radiate a photon of energy $h\nu_{12}$ spontaneously, converting an atom from 2 into an atom named by 1, or some internal collision can effect this conversion, but in either case the rate of *decay* of state 2 is proportional to the density in 2 times a rate constant. Atoms in state 1 suffer similar fates: they can radiate spontaneously to another level, be deactivated by a collision, or be simply swept out of the volume of interest by mass motion (as is done in a flowing dye laser). The stimulated emission rates shown in Fig. 8.4 will be in addition to the above "natural decay" processes. To avoid excessive arithmetic, we assume that the populations in 2 or 1 are very small compared to that in 0 so that we need not worry about the conservation of mass.*

The differential equations describing the dynamics of the populations will be written in terms of the *lifetime* for certain processes to take place. For instance:

1. The sum of the spontaneous emission rate and any other collision process which decreases the population in 2 and simultaneously increases that in 1 is denoted by $1/\tau_{21}$.
2. The rate of loss of state 2 that does not result in an atom in state 1 is denoted by $1/\tau_{20}$.

*In other words, $N_1 + N_2 \ll N_0$ under all circumstances, and thus N_0 is independent of the pumping rates. This is an approximation that will have to be changed for heavily pumped lasers or ones whose lower level is state 0 (see the discussion of ruby lasers in Chapter 10).

3. The total rate of decay of state 2 is the sum of the above rates and defines the "lifetime" of state 2 by $1/\tau_2 = 1/\tau_{21} + 1/\tau_{20}$.
4. The decay rate of state 1 due to any and all causes is denoted by $1/\tau_1$.

The stimulated emission and absorption rates can be expressed in terms of the Einstein coefficients by $B_{21}g(\nu)\rho_\nu$ and $B_{12}g(\nu)\rho_\nu$, respectively, but it is more convenient to use the relationships between them and express the final result in terms of the stimulated emission cross section and the intensity of the radiation. (We assume equal degeneracies, $g_2 = g_1$, so as to allow any serious student of lasers to repeat the following to obtain a more general analysis, which should be done!) Recall that

$$B_{21}g(\nu)\rho_\nu = \frac{c'}{h\nu}\frac{\lambda_0^2}{8\pi n^2}A_{21}g(\nu)\frac{I_\nu}{c'} \qquad (8.3.1)$$

with

$$\rho_\nu = I_\nu/c'$$

and

$$c' = c/n \qquad (8.3.2)$$

Thus the dynamics of the populations involved in the lasing process are described by the following coupled differential equations:

$$\frac{dN_2}{dt} = R_2 - \frac{N_2}{\tau_2} - \frac{\sigma I_\nu}{h\nu}(N_2 - N_1) \qquad (8.3.3a)$$

$$\frac{dN_1}{dt} = R_1 + \frac{N_2}{\tau_{21}} - \frac{N_1}{\tau_1} + \frac{\sigma I_\nu}{h\nu}(N_2 - N_1) \qquad (8.3.3b)$$

These equations are so fundamental and lead to so many important consequences that it is appropriate to take a few special cases as examples.

Case 1. Let us forget about a laser for a moment and assume $I_\nu = 0$. Furthermore, let us assume that the pumping rate of state 2 is in the form of a pulse; that is

$$R_2(t) = R_{20}[u(t) - u(t - T)]$$

where $u(\ )$ represents the Heaviside step function and T is the pulse width, which is assumed to be much longer than τ_2.

Equation (8.3.3a) is decoupled from (8.3.3b) and has a simple solution (similar to the variation of voltage across a leaky capacitor when driven by a current source). For $0 < t < T$, we have

Sec. 8.3 Gain Saturation in a Homogeneous Broadened Transition

$$\frac{dN_2}{dt} + \frac{N_2}{\tau_2} = R_2 \tag{8.3.4a}$$

$$N_2(t) = R_{20}\tau_2(1 - e^{-t/\tau_2}) \tag{8.3.4b}$$

and for $t > T$:

$$N_2(t) = R_{20}\tau_2 e^{-(t-T)/\tau_2} \tag{8.3.4c}$$

Case 2. We modify Case 1 a bit allowing for a finite intensity I_ν, but we assume that state 1 decays at an infinitely fast rate ($\tau_1 = 0$). As we will see, this would be the ideal laser system, but our purpose in making that assumption is to decouple (8.3.3b) from (8.3.3a). The immediate effect of assuming $\tau_1 = 0$ is that $N_1 = 0$, irrespective of anything!

Equation (8.3.3a) is still simple in form and has a simple solution:

$$\frac{dN_2}{dt} + \frac{1}{\tau_2}\left(1 + \frac{\sigma\tau_2}{h\nu}I_\nu\right)N_2 = R_2 \tag{8.3.4d}$$

for $t < T$:

$$N_2(t) = \frac{R_{20}\tau_2}{\left(1 + \frac{\sigma I_\nu \tau_2}{h\nu}\right)}\left\{1 - \exp\left[-\frac{t}{\tau_2}\left(1 + \frac{\sigma I_\nu \tau_2}{h\nu}\right)\right]\right\} \tag{8.3.4e}$$

and for $t > T$:

$$N_2(t) = \frac{R_{20}\tau_2}{\left(1 + \frac{\sigma I_\nu \tau_2}{h\nu}\right)} \exp\left[-\frac{(t-T)}{\tau_2}\left(1 + \frac{\sigma I_\nu \tau_2}{h\nu}\right)\right] \tag{8.3.4f}$$

These two simple solutions are sketched in Fig. 8.5 assuming that the amplitude of the pump R_{20} is the same, and two points should be apparent.

Note that when the source is turned off at $t = T$ in Case 1, the population in state 2 decays with a time constant τ_2 (hence its name—the lifetime of state 2). When there is a stimulating wave present as in Case 2, this time constant *decreases* with I_ν. Indeed, this decrease in lifetime is a classic test for the presence of stimulated emission and provides us with a measure for whether an intensity is strong or weak. If the quantity $\sigma I_\nu \tau_2/h\nu \ll 1$, then (8.3.4d) appears the same as (8.3.4a) and the intensity is considered to be weak; if that quantity is greater than 1, the intensity is strong since it obviously affects both the decay rate of state 2 as well as its magnitude. The quantity $h\nu/\sigma\tau_2$ has the dimensions of intensity and controls this crossover between the two solutions—it is called the *saturation intensity*.

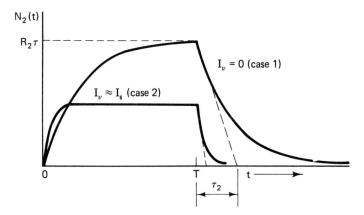

Figure 8.5 Variation of the population in state 2 with time in response to a pulsed excitation in the presence and absence of stimulated emission.

$$I_s = \frac{h\nu}{\sigma \tau_2} \tag{8.3.5}$$

The other point to be noted from Fig. 8.5 is that stimulated emission affects the amplitude of the population N_2 if the intensity is comparable to or greater than I_s. As the intensity gets bigger, the population becomes smaller with increasing intensity. If one anticipates that laser intensity would increase more or less linearly with the pumping R_2 above a threshold rate, then the upper state would first increase with pumping but then be clamped by the simultaneous increase of the denominator of (8.3.4e).

This effect was demonstrated in a convincing fashion by Paoli[20] and the data is presented in Fig. 8.6. The spontaneous emission from the side of a semiconductor laser was measured as a function of the injected current normalized to the threshold value. Now that side radiation is simply related to the population in the upper state—the number of electrons in the conduction band in this case. Note that the upper state as measured by this radiation increases with injected current (i.e., pumping R_2) until threshold for lasing occurs. A further increase of current does *not* increase the upper state; indeed it appears that the upper laser level is clamped to its value *at threshold*. This is a general characteristic of *all* lasers. Even though the above data were chosen from the semiconductor version, they illustrate a very important and general point:

Stimulated emission is never important below threshold for a laser, but it is overwhelmingly important at and above threshold.

Now let us remove another assumption—that of $\tau_1 \neq 0$, the ideal laser system—but invoke steady state.

Sec. 8.3 Gain Saturation in a Homogeneous Broadened Transition

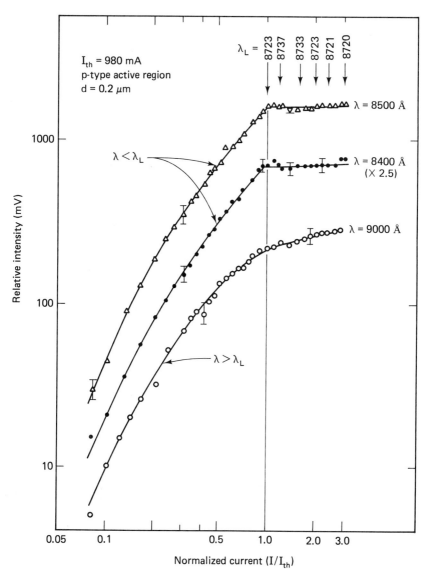

Figure 8.6 Variation of the spontaneous emission from the side of a semiconductor diode as the pumping current is increased. Once the diode starts lasing, stimulated emission uses the carriers as fast as they are injected into the junction. Hence the inversion is clamped at threshold. (Data from T. Paoli. IEEE J. Quant. Electr. QE-9, 267, 1973.)

Case 3. $\partial/\partial t = 0$

Assuming all time derivatives equal to zero reduces the differential equations to a simple algebraic set:

$$\frac{dN_2}{dt} = 0 = R_2 - \frac{N_2}{\tau_2} - \frac{\sigma I}{h\nu}(N_2 - N_1) \tag{8.3.6a}$$

$$\frac{dN_1}{dt} = 0 = R_1 + \frac{N_2}{\tau_{21}} - \frac{N_1}{\tau_1} + \frac{\sigma I}{h\nu}(N_2 - N_1) \tag{8.3.6b}$$

Rewriting these equations in matrix form emphasizes the simplicity (but drudgery) involved with the determination of $N_{2,1}$ as a function of $R_{2,1}$ and I_ν.

$$\begin{bmatrix} \left(\dfrac{1}{\tau_2} + \dfrac{\sigma I}{h\nu}\right) & -\dfrac{\sigma I}{h\nu} \\ -\left(\dfrac{1}{\tau_{21}} + \dfrac{\sigma I}{h\nu}\right) & \left(\dfrac{1}{\tau_1} + \dfrac{\sigma I}{h\nu}\right) \end{bmatrix} \begin{bmatrix} N_2 \\ N_1 \end{bmatrix} = \begin{bmatrix} R_2 \\ R_1 \end{bmatrix} \tag{8.3.6c}$$

Now

$$\Delta = \frac{1}{\tau_1\tau_2} + \left[\frac{1}{\tau_1} + \frac{1}{\tau_2} - \frac{1}{\tau_{21}}\right]\left(\frac{\sigma I}{h\nu}\right)$$

or
$$\Delta = \frac{1}{\tau_1\tau_2}\left[1 + \left(\tau_1 + \tau_2 - \frac{\tau_1\tau_2}{\tau_{21}}\right)\left(\frac{\sigma I}{h\nu}\right)\right] \tag{8.3.7}$$

Using Cramer's rule, we find the populations $N_{2,1}$:

$$N_2 = \left\{R_2\left(\frac{1}{\tau_1} + \frac{\sigma I}{h\nu}\right) + R_1\left(\frac{\sigma I}{h\nu}\right)\right\}/\Delta \tag{8.3.8a}$$

$$N_1 = \left\{R_1\left(\frac{1}{\tau_2} + \frac{\sigma I}{h\nu}\right) + R_2\left(\frac{1}{\tau_{21}} + \frac{\sigma I}{h\nu}\right)\right\}/\Delta \tag{8.3.8b}$$

Now the pertinent parameter insofar as a laser is concerned is the difference in populations, $N_2 - N_1$; hence after some obvioius manipulation, one can express that difference as

$$N_2 - N_1 = \frac{R_2\tau_2\left(1 - \dfrac{\tau_1}{\tau_{21}}\right) - R_1\tau_1}{1 + \left(\tau_1 + \tau_2 - \dfrac{\tau_1\tau_2}{\tau_{21}}\right)\left(\dfrac{\sigma I}{h\nu}\right)} \tag{8.3.8c}$$

Now if (8.3.8c) is multiplied by the stimulated emission cross section, one has

Sec. 8.3 Gain Saturation in a Homogeneous Broadened Transition

an expression for the gain coefficient for any value of the intensity. If the intensity is small enough such that the denominator is approximately 1, then the numerator is an expression for the small-signal gain coefficient in terms of the pumping rates and the lifetimes:

$$\gamma_0(\nu) = \sigma(\nu)\left[R_2\tau_2\left(1 - \frac{\tau_1}{\tau_{21}}\right) - R_1\tau_1\right] \quad (8.3.9a)$$

The rate of increase of intensity with distance divided by the intensity is, by definition, the gain coefficient (for any intensity) and is given by

$$\frac{1}{I_\nu}\frac{dI_\nu}{dz} \triangleq \frac{\gamma_0(\nu)}{1 + \dfrac{I\nu}{I_s(\nu)}} \quad (8.3.9b)$$

where I_s is now given by

$$I_s = \frac{h\nu}{\sigma(\nu_0)\tau_2} \cdot \frac{1}{1 + \dfrac{\tau_1}{\tau_2}\left(1 - \dfrac{\tau_1}{\tau_{21}}\right)} \quad (8.3.9c)$$

Let us focus on (8.3.9c) for a moment: if $\tau_1 \ll \tau_2$, which is referred to as a favorable lifetime ratio; or the quantity $\tau_2/\tau_{21} \sim 1$, which is called a favorable branching ratio; or $g_2/g_1 < 1$, which is called a favorable degeneracy ratio; or any combination of these three "ifs" (which is quite common), then the denominator of (8.3.9c) is approximately 1 and the saturation intensity is identical to that found previously from the temporal analysis (8.3.5).

Thus there are two equivalent definitions of the saturation intensity:

$$I_s = \begin{cases} \text{that intensity which shortens the lifetime of state 2} \\ \text{by a factor of 2 (8.3.5)} \\ \\ \text{that intensity which reduces the gain } \textit{coefficient} \text{ by a} \\ \text{factor of 2 (cf 8.3.9c)} \end{cases} \quad (8.3.10)$$

It must be emphasized that the saturation intensity is just a collection of constants which have the dimensions of intensity and which indicate when a stimulating wave is strong or weak. As we shall see, all lasers operate with an intensity which is more or less equal to the saturation intensity and hence this is the first-order estimation for a laser amplitude inside the cavity. It should also be noted that we have buried the frequency dependence of the stimulated emission cross section under the symbol σ. If one retraces the steps made above, the line shape factor $\overline{g}(\nu)$, normalized to be unity at line center, should multiply σ everywhere it appears. For instance, if the homogeneous line shape were a Lorentzian, then $\overline{g}(\nu)$ is given by

$$\overline{g}(\nu) = \frac{(\Delta\nu/2)^2}{(\nu - \nu_0)^2 + (\Delta\nu/2)^2} \tag{8.3.11}$$

and the saturation law should be written as

$$\frac{1}{I_\nu}\frac{dI_\nu}{dz} = \frac{\gamma_0(\nu)}{1 + \overline{g}(\nu)\dfrac{I_\nu}{I_s}} \tag{8.3.12}$$

For the most part, one is concerned with radiation close to line center where $\overline{g}(\nu) \sim 1$ and most of the following discussion assumes that approximation.

The above analysis is most important and the serious laser student would do well to practice arriving at the above conclusions by as many different routes as possible. It is much more than a dry mathematical description of the formation, decay, and use of atoms by stimulated emission; it also indicates the general physical requirements for achieving a population inversion and gain.

Now let us focus the discussion on (8.3.8c). Obviously, one wants the first term in the numerator of (8.3.8c) to be as large as posible, keeping the second as small as possible. Thus we want the pumping rate of the upper state to be high (R_2 large), keeping R_1 small. Not only should the pumping rates be in the appropriate direction, but the lifetime of state 2 should be long, whereas that of state 1 should be short. However, the limiting lifetime of state 2 is, of course, τ_{21}. For, if τ_2 were infinite, τ_{21} must also be infinite; then A_{21} would be zero and the stimulated emission cross section is also zero. The lifetime of state 1 should be as short as possible—deplete the population N_1 as fast as possible. Indeed, as we will see later, the *depletion* of the lower state and the pumping of the upper state are the rate-limiting processes for a good laser.

Now let us return to (8.3.12), which describes the situation shown in Fig. 8.7, where a signal from an external laser is injected into an amplifier. The intensity of the injected signal is $I_\nu(z = 0)$, and the problem is to determine the output $I_\nu(z = l_g)$. It must be emphasized that the solution to (8.3.12) is *not* given by the simple exponential law with an intensity in the exponent.

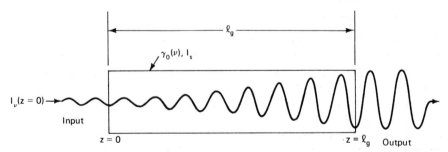

Figure 8.7 An optical amplifier.

Sec. 8.3 Gain Saturation in a Homogeneous Broadened Transition

$$I_\nu(z = l_g) \neq I_\nu(z = 0) \exp\left[\frac{\gamma_0(\nu)l_g}{1 + \bar{g}(\nu)I_\nu/I_s}\right] \quad \text{(No!)}$$

One must solve the differential equation as given by (8.3.12). Bringing all factors involving I_ν to the left side and dz to the right leads to

$$\int_{I_\nu(z=0)}^{I_\nu(z=l_g)} dI_\nu \left[\frac{1}{I_\nu} + \frac{\bar{g}(\nu)}{I_s}\right] = \int_{z=0}^{z=l_g} \gamma_0(\nu)\, dz \quad (8.3.13)$$

thus

$$\ln\frac{I_\nu(l_g)}{I_\nu(0)} + \frac{\bar{g}(\nu)}{I_s}[I_\nu(l_g) - I_\nu(0)] = \gamma_0(\nu)l_g \quad (8.3.14a)$$

or

$$\ln G + \bar{g}(\nu)\frac{I_\nu(0)}{I_s}(G - 1) = \gamma_0(\nu)l_g \quad (8.3.14b)$$

with

$$G \triangleq I_\nu(z = l_g)/I_\nu(z = 0) \quad (8.3.14c)$$

Note that if the output $I_\nu(l_g)$ [and, of course $I_\nu(0)$] is much smaller than I_s, then we can ignore the second term on the left and recover the simple amplification law:

$$\frac{I_\nu(l_g)}{I_\nu(0)} = \exp[\gamma_0(\nu)l_g] = G_0(\nu) \quad \text{(power gain)} \quad (8.3.15)$$

If the input is comparable to the saturation intensity, the rate of increase of intensity is smaller; consequently, the net power gain, G, is smaller. Although (8.3.14) is fairly simple, it is a transcendental equation that must be solved numerically for the output in terms of the input. The general trend of a solution is shown in Fig. 8.8, where the power gain (in dB) is plotted as a function of ratio $I_{\text{in.}}/I_{\text{sat.}}$ (at line center, where $\bar{g} = 1$). It is instructive and quite revealing to go to the extreme limit of $I_\nu(0) \gg I_s$ to find the output intensity directly from (8.3.12).

$$I_\nu(l_g) = I_\nu(0) + \left[\frac{\gamma_0(\nu)I_s}{\bar{g}(\nu)}\right]l_g \quad I_\nu(0) \gg I_s(\nu) \quad (8.3.16)$$

Note that $\gamma_0(\nu)$ contains the same frequency dependence as $\bar{g}(\nu)$, and thus we can only *add* a certain amount of intensity $\gamma_0(\nu_0)I_s$ per unit of length of our amplifier independent of the detuning from line center. Let us reinsert the parameters of $\gamma_0(\nu_0)$ and I_s from (8.3.9a) and (8.3.9c) to show the obvious logic of the result.

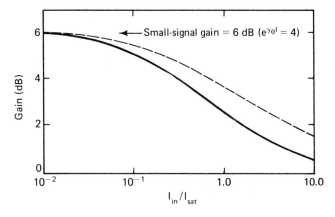

Figure 8.8 Saturation of the gain of an optical amplifier. The solid curve is for a homogeneously broadened system and the dashed curve is for an inhomogeneous one. [See Problem 8.6.]

$$\gamma_0(\nu_0) = \left[R_2 \tau_2 \left(1 - \frac{\tau_1}{\tau_2}\right) - R_1 \tau_1 \right] \sigma(\nu_0) \qquad (8.3.9a)$$

$$I_s = \frac{h\nu}{\sigma(\nu_0)\tau_2 \left[1 + \frac{\tau_1}{\tau_2}\left(1 - \frac{\tau_2}{\tau_{21}}\right)\right]} \qquad (8.3.9c)$$

Thus power added (per unit of length) of this saturated amplifier is

$$\frac{\Delta P}{\text{area} \times \text{length}} = \frac{\gamma_0(\nu)I_s}{\bar{g}(\nu)} = h\nu \left[\frac{R_2\tau_2(1 - \tau_1/\tau_{21}) - R_1\tau_1}{\tau_1 + \tau_2 - \tau_1\tau_2/\tau_{21}}\right] \qquad (8.3.17)$$

This is the *best* we can do! Note that the detailed picture of the photon interacting with the atom has completely disappeared, leaving only lifetimes and pumping rates for our consideration. Indeed, the quantity in the brackets can be considered as an effective pumping rate, each effective pumping event contributing one package of energy, $h\nu$, to the incoming wave.

$$R_{\text{eff}} = \frac{R_2\tau_2(1 - \tau_1/\tau_{21}) - R_1\tau_2}{\tau_1 + \tau_2 - \tau_1\tau_2/\tau_{21}} \qquad (8.3.18)$$

If we again consider the ideal laser system—weak pumping of the lower state (R_1 small), lower state lifetime small compared to the upper one ($\tau_1 \ll \tau_2$) and ($\tau_2 \sim \tau_{21}$)—then R_{eff} reduces to R_2, the pumping rate of the upper state. This means that we can only extract what we put in, a most logical result! Thus the primary job in laser research is to devise a means of pumping the upper state. For that reason, then, we will devote a considerable amount of effort in later chapters to the physical processes that lead to the formation of the upper state.

Before this section is closed, it is well to reexamine the assumptions

Sec. 8.4 Laser Oscillation in an Inhomogeneous System—An Example

implicit in the derivation so as to appreciate the limitations of the results. The major assumption lies in the cavalier manner of writing the rate equations (8.3.6a) and (8.3.6b). We assumed that the radiation at ν interacted with the atoms according to the line shape $g(\nu)$ and implicitly assumed that the line shape for all atoms was the same. Thus, if an atom in state 2 gave up its internal energy to the stimulating radiation field and became part of the population in state 1, the gain coefficient throughout the entire line profile would be reduced. This is shown in Fig. 8.9.

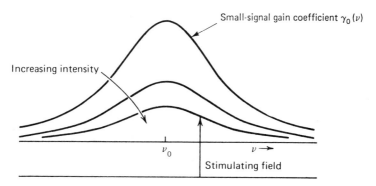

Figure 8.9 Saturation of a homogeneously broadened transition.

8.4 LASER OSCILLATION IN AN INHOMOGENEOUS SYSTEM—AN EXAMPLE

The common He:Ne laser transition is inhomogeneously broadened and some of the conclusions of the previous section need to be modified. However, before we jump too far into the mathematics, it is reasonable to talk our way through a simple and tractable example and use that result to understand common phenomena occurring in such a system. The example is indicated in Fig. 8.10 where we assume a transition consists of five closely spaced lines occurring with a 3 : 4 : 4 : 4 : 3 weight with center frequencies $\nu_1 \ldots \nu_5$ uniformly separated by a common homogeneous line width of each component; thus each line has the same stimulated emission cross section (evaluated at the respective line centers). For instance, these components could represent the hyperfine structure of the atom or the various isotopes.

The key point to recall from Sec. 7.5.2 is that the inhomogeneous line is the *sum of individual homogeneous components*. Because the peak stimulated emission cross section of each individual part, evaluated at the respective centers, is identical (by assumption), we can immediately express the small-signal gain coefficient for this inhomogeneous line:

$$\frac{\gamma_0(\nu)}{\left(N_2 - \frac{g_2}{g_1}N_1\right)\sigma} = \frac{1}{18}\left\{\frac{(\Delta\nu_h/2)^2}{(\nu_1 - \nu)^2 + (\Delta\nu_h/2)^2} \times (3)\right.$$

$$+ \frac{(\Delta\nu_h/2)^2}{(\nu_2 - \nu)^2 + (\Delta\nu_h/2)^2} \times (4) + \frac{(\Delta\nu_h/2)^2}{(\nu_3 - \nu)^2 + (\Delta\nu_h/2)^2} \times (4) \quad (8.4.1)$$

$$+ \frac{(\Delta\nu_h/2)^2}{(\nu_4 - \nu)^2 + (\Delta\nu_h/2)^2} \times (4) + \left.\frac{(\Delta\nu_h/2)^2}{(\nu_5 - \nu)^2 + (\Delta\nu_h/2)^2} \times (3)\right\}$$

where

$$\sigma = A_{21}\left(\frac{\lambda_0^2}{8\pi n^2}\right) \cdot \left(\frac{2}{\pi\Delta\nu_h}\right)$$

and the numeric 18 is the sum of the relative weights. Thus if we had more than five lines, it is a matter requiring tedious attention to detail to extend the number of components. For this example, "stick" heights, equal to the weights of each line, are shown in Fig. 8.10.

Now suppose there is a strong stimulating wave located at $\nu = \nu_0 + 2\Delta\nu_h = \nu_5$. It, of course, will saturate that part of the sum according to the simple saturation law of (8.3.12) with $\bar{g}_4$ given by

$$\bar{g}_4 = \frac{(\Delta\nu_h/2)^2}{(\nu_4 - \nu)^2 + (\Delta\nu_h/2)^2}\bigg|_{\nu=\nu_5=\nu_4+\Delta\nu} = \frac{1}{5}$$

and in a similar fashion we find that $\bar{g}_3 = 1/7, \bar{g}_2 = 1/37$, and $\bar{g}_1 = 1/65$. Thus the stimulating wave interacts most strongly with those atoms labeled by the ν_5 center frequency, less with those labeled by ν_4, considerably less with ν_3 and ν_2, and hardly at all with the ν_1 group.

Fig. 8.10 also illustrates the fate of the laboratory line shape as the strength of this stimulating wave increases. It literally burns a "hole" in the small-signal gain profile with a depth depending on the ratio I/I_s and whose "width" is more or less the homogeneous value plus a contribution caused by I_ν/I_s. Note that the gain available at another frequency, say at $\nu = \nu_2$, is hardly affected by the stimulating wave. These conclusions will be established in a more formal manner in Sec. 8.6 for a Doppler-broadened line.

This "hole burning" phenomenon has some important consequences for lasers operating in an inhomogeneously broadened line. Let us talk our way through a specific case of a low-pressure Doppler-broadened system, such as that found in the He:Ne laser, to emphasize the contrast between it and that described in Sec. 8.2.

Sec. 8.4 Laser Oscillation in an Inhomogeneous System—An Example

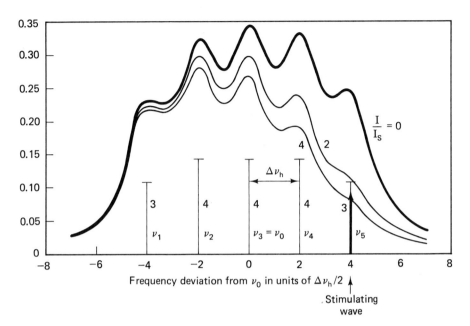

Figure 8.10 An inhomogeneous line consisting of five components in the proportion 3:4:4:4:3 with center frequencies uniformly spaced with respect to the centroid. The graph assumes the same spontaneous emission coefficient, same line width, and same saturation intensity for each component.

Consider Fig. 8.11, which describes the various phases of the buildup of coherent optical power within the laser cavity. Aside from minor differences in the shape of the gain curve, Fig. 8.11(a) and 8.11(b) are identical to Fig. 8.3(a) and 8.3(b). The spontaneous emission goes into the cavity modes more or less like the line shape, and each cavity mode grows according to the gain minus the loss formulation, as before.

The difference becomes startling when saturation occurs. [Compare Figs. 8.11(c) with 8.3(c).] The fact that *we* know that the laboratory line shape is inhomogeneously broadened is quite irrelevant—nobody informed the atoms or the field of this fact. The field interacts with a specific group according to the *homogeneous* line shape of that group in the atom's frame of reference, not the laboratory line shape that we know.

Thus the $(0, 0, q)$ mode interacts most strongly with that group of atoms which "think" that $\nu_{0,0,q}$ is the center frequency of a homogeneous line-shape function $g_h(\nu)$ for the gas. If the homogeneous line width $\Delta\nu_h$ is much less than the Doppler width, this interaction occurs over a very limited portion of the laboratory line shape, leaving the remainder unaffected by the $(0, 0, q)$ mode.

The same considerations apply to the stimulating waves at $\nu_{0,0,q-1}$ and

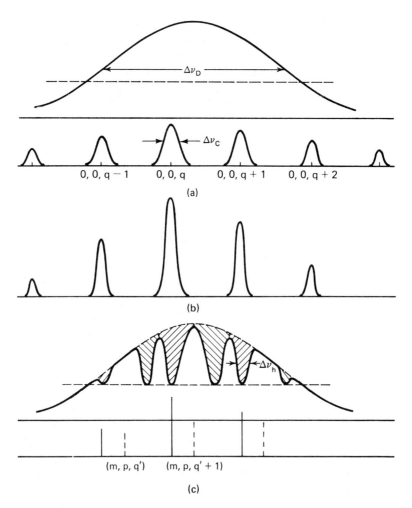

Figure 8.11 Evolution of oscillation in a Doppler-broadened transition.

$\nu_{0,0,q+1}$, which have positive values for gain minus loss. Each will burn a separate "hole" in the gain profile until the gain at each frequency saturates at the loss. As before, the cavity response narrows around the center resonance frequency until the spectral representation is nearly a delta function. In a loose sort of a way, the power in each mode is proportional to the area "burnt" away in forming the "hole."

For Doppler broadening, there is an additional complication that has most interesting consequences. Recall that the electromagnetic energy consists of fields running back and forth between the mirrors, obviously in opposite directions for a simple cavity. If ν_q is less than ν_0 (in the laboratory frame), the wave

Sec. 8.4 Laser Oscillation in an Inhomogeneous System—An Example

that travels in the positive z direction interacts most strongly with atoms that have enough velocity in the opposite (or negative z) direction to make ν_q equal to ν_0 in their frame of reference. When the wave turns around and propagates in the negative z direction, it now interacts most strongly with atoms moving in the positive z direction. There are two distinct groups of atoms that are stimulated by one field, and consequently both contribute their energy to the same laboratory frequency. These two frequencies are images about the line center ν_0.

One of the most dramatic demonstrations of this effect occurs in a very short laser such that only one cavity mode has a gain-to-loss ratio greater than 1. This is shown in Fig. 8.12. By changing d slightly, we can tune the cavity mode across the line. (This is usually done by mounting the mirror on a piezoelectric element and applying a sawtooth voltage to the electrodes. That voltage is shown on the lower trace of the oscilloscope picture. The upper trace is the power out of the laser. The letters at the bottom of the picture correspond to the cavity mode appearing at the corresponding figures above.)

As expected, when the cavity mode is on wings (a and d) of the transition, the power is low; as it is tuned toward the line center—say b—the power increases as the area burned off of the gain curve increases. When, however, the hole at the oscillation frequency overlaps the image hole, the power decreases. It makes no difference to those atoms in the overlap region whether it

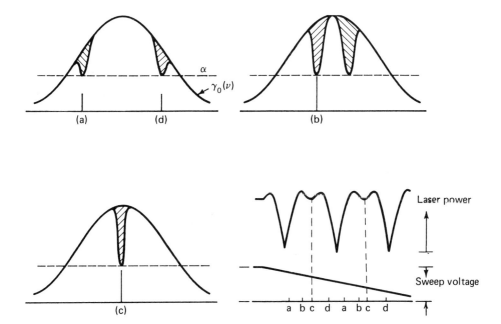

Figure 8.12 Lamb dip. (Courtesy of Spectra-Physics.)

gives up its energy in the form of a photon to the positive or the negative traveling wave. It only has *one* package of energy, $h\nu$, to give up and no more. Consequently the power drops, reaching a minimum at line center. This phenomenon is referred to as the "Lamb dip" after W. E. Lamb[8] who predicted its occurrence on theoretical grounds. It has the very practical application of enabling one to stablize the oscillation frequency to the center of a transition with extreme precision.

Just as atoms have only one unit of energy, $h\nu$, which can be used by stimulated emission, they can also only absorb that same unit from either wave. This absorption can be saturated (i.e., reduced) just as "the amplification" in the discussion above. There is a fortuitous coincidence between the He:Ne transition at 3.39 μm and a vibrational-rotational transition in methane (CH_4). Thus, if a low-pressure methane cell is incorporated inside the laser cavity, there will be

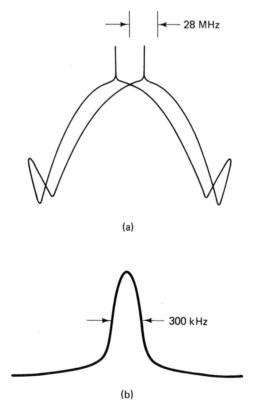

Figure 8.13 (a) Output power of a He:Ne laser at 3.39 μm in the near vicinity of the absorption of CH_4; (b) expanded frequency scale for the peak. (From R. L. Barger and R. L. Hall, Phys. Rev. Lett. 22, 4, 1969.) NOTE: Ref. 21 has identified the *frequency* of the peak to occur at ν = 88.3761816029 THz ± 1.2 kHz.

an inverted Lamb dip when the laser frequency coincides with the center frequency of the methane absorption; that is, the power reaches a maximum at the center. Because of the beautiful symmetry of the methane molecule, the saturated absorption is very narrow—on the order of 100 kHz or so. Such a scheme is easily reproduced and a laser stabilized to this peak has become the international standard for length. An example of the output of a laser as it is tuned across the absorption line, taken from the work of R. L. Barger and J. L. Hall of the National Bureau of Standards,[17] is shown in Fig. 8.13.

8.5 MULTIMODE OSCILLATION

It is obvious from Fig. 8-11 that a laser can oscillate on more than one $\text{TEM}_{0,0,q}$ mode. A few numerical values are in order to illustrate this point. Consider the He:Ne transition at $\lambda_0 = 632.8$ nm, which is Doppler broadened with $\Delta \nu_D \sim 1.5$ GHz. The free-spectral range (the separation between $\nu_{0,0,q}$ and $\nu_{0,0,q-1}$) is $c/2d$, and, for a nominal mirror spacing of $d = 1$ m, is 150 MHz. Thus as many as 10 different frequencies can be oscillating simultaneously on a single transition, assuming that $\gamma_0(\nu_0)/\alpha = 2$.

The situation becomes even more involved when the transverse modes are considered. Not only are the resonant frequencies of the $\text{TEM}_{m,p,q}$ modes different from those of the $(0, 0, q)$ mode—and thus still have net gain at their frequency—but the fields of these higher-order extend over a larger cross section and thus interact with different spatial groups of excited atoms.

This last point is illustrated in Fig. 8.14, where the "dot" patterns of the $(0, 0)$ and the $(1, 1)$ modes are shown. It is obvious that the $(0, 0)$ mode interacts most strongly with the atoms near the axis of the tube, whereas the $(1, 1)$ mode uses the atoms located at a distance away from the axis. Thus even a homogeneously broadened transition can support multimode oscillation by "spatial" hole burning.

One can suggest a situation whereby the field of one (m, p, q') mode competes with the field of another mode for the same group of atoms at the same

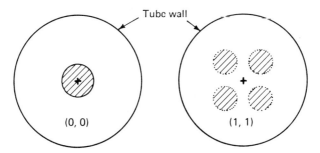

Figure 8.14   Dot patterns of $(0, 0)$ and $(1, 1)$ mode.

location in the gain medium. Furthermore, it is highly probable that the gain medium is *not* uniform in r (or ϕ). To handle this situation—even approximately—is not a trivial task and will be left to more advanced texts.

8.6 GAIN SATURATION IN DOPPLER-BROADENED TRANSTITION—MATHEMATICAL TREATMENT

We have avoided any mathematics in the two preceding sections so as to gain a physical feeling for the saturation process. It is now time to put these feelings on a more quantitative basis.

The key points to recall are (1) the field interacts with the atoms according to the homogeneous line shape in the atom's rest frame, and (2) the inhomogeneous line shape as observed in the laboratory is a weighted sum between the atom's homogeneous line shape and the probability that this group will occur.

Let $p(f)\,df$ be the fraction of the total atom population whose center frequency (in the laboratory frame) coincides with the frequency ν. This group of atoms interacts with the laser frequency at ν according to the homogeneous line-shape function and saturates according to the homogeneous law derived previously (8.3.12). Thus the saturated gain coefficient for an inhomogeneously broadened transition is given by

$$\gamma(\nu, I_\nu) = A_{21}\frac{\lambda_0^2}{8\pi n^2}\left(N_2^0 - \frac{g_2}{g_1}N_1^0\right)\int_0^{+\infty}\frac{p(f)\overline{g}_h(\nu, f)df}{1 + (I_\nu/I_s)\overline{g}_h(\nu, f)} \quad (8.6.1)$$

where

$$\int_0^{+\infty} p(f)\,df = 1 \quad (8.6.2)$$

$$I_s = \frac{h\nu}{(\tau_1 + \tau_2 - \tau_1\tau_2/\tau_{21})\sigma(\nu_0)} \quad (8.3.9c)$$

$\overline{g}_h(\nu, f)$ = the Lorentzian normalized to 1 at $f = \nu$

or

$$\overline{g}_h(\nu, f) = \frac{\pi\Delta\nu_h}{2}g(\nu, f) = \frac{(\Delta\nu_h/2)^2}{(f - \nu)^2 + (\Delta\nu_h/2)^2} \quad (8.6.3)$$

N_2^0, N_1^0 = population of states 2, 1 in the limit of *zero* amplitude of I_ν

To be specific, let us assume a thermal distribution of velocities that translates into a Doppler distribution of frequencies.

Sec. 8.6 Gain Saturation in Doppler-Broadened Transition—Mathematical Treatment

$$p(f) = \left(\frac{4 \ln 2}{\pi}\right)^{1/2} \frac{1}{\Delta \nu_D} \exp\left[-4(\ln 2)\left(\frac{f - \nu_0}{\Delta \nu_D}\right)^2\right] \qquad (7.5.18)$$

Anyone who attempts to substitute (8.6.3), (8.3.9c), and (7.5.18) into (8.6.1) is bound to be overwhelmed by the shear volume of symbols and letters. Let us at least keep $p(f)$ as an integral part of the expression and proceed carefully. If we examine (8.6.1), we see $g_h(\nu, f)$ and $\overline{g}_h(\nu, f)$ appearing in the numerator and denominator; hence, there will be some cancellation of common factors if numerator and denominator are multiplied by $[(\nu - f)^2 + (\Delta \nu/2)^2]$:

$$\gamma(\nu, I_\nu) = A_{21} \frac{\lambda_0^2}{8\pi n^2} \left(N_2^0 - \frac{g_2}{g_1} N_1^0\right) \frac{\Delta \nu_h}{2\pi}$$

$$\times \int_0^\infty \frac{p(f) \, df}{(f - \nu)^2 + (\Delta \nu_h/2)^2 (1 + I_\nu/I_s)} \qquad (8.6.4)$$

Amazingly, the denominator has a Lorentzian frequency dependence with an intensity-dependent line width. Defining this last factor as a "hole width,"

$$\Delta \nu_H^2 = \Delta \nu_h^2 \left(1 + \frac{I_\nu}{I_s}\right) \qquad (8.6.5)$$

(8.6.4) becomes much more palatable with the identification of a Lorentzian with an intensity-dependent width defined by (8.6.5). Rearranging (8.6.4) according to this logic leads us to

$$\gamma(\nu, I_\nu) = A_{21} \frac{\lambda_0^2}{8\pi n^2} \left(N_2^0 - \frac{g_2}{g_1} N_1^0\right) \frac{\Delta \nu_h}{\Delta \nu_H}$$

$$\times \int_0^\infty p(f) \frac{\Delta \nu_H}{2\pi[(f - \nu)^2 + (\Delta \nu_H/2)^2]} \, df \qquad (8.6.6)$$

We cannot go further without some sort of approximation procedure or a resort to a numerical evaluation of this integral. We take the former approach and make the same type of an approximation as was done in Sec. 7.5.2 [see (7.5.15) and (7.5.16)]; namely, we approximate a Lorentzian in the integrand by a delta function:

$$\frac{\Delta \nu_H/2\pi}{(f - \nu)^2 + (\Delta \nu_H/2)^2} \longrightarrow \delta(f - \nu) \qquad (8.6.7)$$

Then the integration of (8.6.6) is trivial:

$$\gamma(\nu, I_\nu) = \left[A_{21} \frac{\lambda^2}{8\pi} \left(N_2^0 - \frac{g_2}{g_1} N_1^0\right) p(\nu)\right] \frac{\Delta \nu_h}{\Delta \nu_H} \qquad (8.6.8)$$

The quantity in the brackets is the small-signal gain coefficient, whereas the last term contains the effect of saturation:

$$\gamma(\nu, I_\nu) = \gamma_0(\nu) \frac{\Delta \nu_h}{\Delta \nu_H} \quad (8.6.9)$$

The ratio $\Delta \nu_h / \Delta \nu_H$ comes from (8.6.5) and can be expressed as

$$\frac{\Delta \nu_h}{\Delta \nu_H} = \frac{1}{(1 + I_\nu/I_s)^{1/2}} \quad (8.6.10)$$

Hence, the gain saturates according to

$$\gamma(\nu, I_\nu) = \frac{\gamma_0(\nu)}{(1 + I_\nu/I_s)^{1/2}} \quad (8.6.11)$$

Note that the saturation intensity as it appears in (8.6.11) is still given by $h\nu/\sigma(\nu_0)\tau_2$, where the line-width factor appearing in $\sigma(\nu_0)$ is $\Delta \nu_h$ (i.e., the homogeneous width) not $\Delta \nu_D$. (8.6.11) predicts that the gain of a Doppler-broadened transition saturates with a functional dependence that is slower than that of a homogeneous one. This is because the width of the hole burnt in the inhomogeneous gain profile is changing simultaneously with the depth of the hole (or gain depression). In a homogeneous line, the width does not change and the profile of the line is maintained while the gain is depressed.

For the Doppler case, the saturation factor $(1 + I/I_s)^{-1/2}$ is frequency independent, where the normalized Lorentzian appears in (8.3.12). Whereas the gain coefficient in a homogeneously broadened transition is hard to saturate by stimulated emission in the wings (where $|\nu - \nu_0| > \Delta \nu/2$), the Doppler-broadened case can be saturated to the same degree with the one intensity over the band. Actually, however, the distinction is minor because seldom is an inhomogeneously broadened medium used as a power amplifier in the manner indicated by Fig. 8.7.

One should be very careful about blindly applying this saturation law for all values of the intensity. If the stimulating field is high enough, the approximation that the intensity-dependent Lorentzian hole width is much narrower than that of the inhomogeneous line, $p(f)$, breaks down. In the intermediate case, the arithmetic complexity is overwhelming, so let us turn to a picture to see the physics of the situation. Fig. 8.15 shows the status.

If the hole width is much larger than the inhomogeneous width $\Delta \nu_D$ because of the intensity or because of the large value of homogeneous width $\Delta \nu_h$, the Lorentzian can be pulled outside the integral in (8.6.6)

$$\gamma(\nu, I_\nu) = A_{21} \frac{\lambda_0^2}{8\pi n^2} \left(N_2^0 - \frac{g_2}{g_1} N_1^0 \right) \frac{\Delta \nu_h}{2\pi[(\nu_0 - \nu)^2 + (\Delta \nu_H/2)^2]} \int_0^\infty p(f)\, df$$

$$(8.6.12)$$

where we have substituted the mean value of $f = \nu_0$ into the Lorentzian function.

Sec. 8.6 Gain Saturation in Doppler-Broadened Transition—Mathematical Treatment

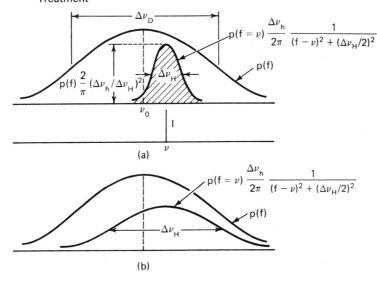

Figure 8-15 Saturation of an inhomogeneously broadened amplifier: (a) intermediate intensity; (b) high intensity.

Since the last integral is unity, we recover the prior formula for gain saturation in a homogeneous medium:

$$\gamma(\nu, I_\nu) = A_{21} \frac{\lambda_0^2}{8\pi n^2} \left(N_2^0 - \frac{g_2}{g_1} N_1^0 \right) \frac{\Delta\nu_h/2\pi}{(\nu - \nu_0)^2 + (\Delta\nu_h/2)^2(1 + I_\nu/I_s)} \quad (8.6.13)$$

Rewriting makes the result more familiar:

$$\gamma(\nu, I_\nu) = \left[A_{21} \frac{\lambda_0^2}{8\pi n^2} \left(N_2^0 - \frac{g_2}{g_1} N_1^0 \right) \frac{\Delta\nu_h/2\pi}{(\nu - \nu_0)^2 + (\Delta\nu_h/2)^2} \right]$$

$$\times \left[\frac{1}{1 + \frac{I_\nu}{I_s} \frac{(\Delta\nu_h/2)^2}{(\nu - \nu_0)^2 + (\Delta\nu_h/2)^2}} \right] \quad (8.6.14)$$

The first set of brackets is just $\gamma_0(\nu)$, whereas the second set can be rewritten using the definition of $\bar{g}_h(\nu)$:

$$\gamma(\nu, I_\nu) = \frac{\gamma_0(\nu)}{1 + (I_\nu/I_s)\bar{g}_h(\nu)} \quad (8.3.12) \to (8.6.15)$$

It is important to realize that physics *demands* this limit in the extreme of very large intensities. As was argued previously [see (8.3.17)], we can extract power from an optical system only at the effective rate that we can pump it. In other words, we can get out only as much as we put in!

8.7 AMPLIFIED SPONTANEOUS EMISSION (ASE)

Although the primary emphasis of this chapter (and most of the book) has been on laser oscillation, it should be pointed out that the energy represented by the population inversion $h\nu[N_2 - (g_2/g_1)N_1]$ can be extracted as broad-band radiation and can be quite intense. This can be a blessing or a curse! As we will see, this enables one to generate intense radiation without an optical cavity, but it also limits the amount of gain that one can design into an amplifier.

Consider the situation shown in Fig. 8.16, with an amplifier being pumped by an external source. Although no externally injected signal is shown there, the atoms in state 2 still radiate spontaneously into the frequency interval that matches the gain profile of the remaining part of the amplifier.

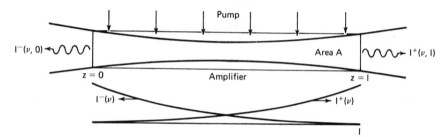

Figure 8-16 Optical amplifier generating broad-band incoherent radiation.

Let us focus on the radiation $I^+(\nu, z)$ traveling in the positive z direction and contained within a constant solid angle $d\Omega/4\pi$. The atoms in N_2 at $z = 0$ radiate part of their energy spontaneously into this solid angle, and this radiation, in turn, is amplified by the inversion between $z = 0$ and $z = l$. Thus spontaneous emission is continuously added to I^+ along z, and simultaneously stimulated emission amplifies the power from previous lengths. (Remember—this is incoherent radiation, and thus powers rather than fields must be added.)

$$\frac{d}{dz}(I^+(\nu, z)\, d\nu) = \gamma_0(\nu)I^+(\nu, z)\, d\nu + h\nu A_{21} N_2 g(\nu)\, d\nu \frac{d\Omega}{4\pi} \quad (8.7.1)$$

A solution to this linear first-order differential equation subject to the obvious boundary condition that $I^+(\nu, z = 0) = 0$ is readily obtained:

$$I^+(\nu, z = l) = \frac{h\nu A_{21} N_2 g(\nu)}{\gamma_0(\nu)}(e^{\gamma_0(\nu)l} - 1)\frac{d\Omega}{4\pi} \quad (8.7.2)$$

If we relate the small-signal gain coefficient $\gamma_0(\nu)$ to the population inversion and the A coefficient, we obtain a very important formula:

$$I^+(\nu, l) = \frac{8\pi n^2 h\nu^3}{c^2} \frac{N_2}{N_2 - (g_2/g_1)N_1}[G_0(\nu) - 1]\frac{d\Omega}{4\pi} \quad (8.7.3a)$$

Sec. 8.7 Amplified Spontaneous Emission (ASE)

where $G_0(\nu) = \exp[\gamma_0(\nu)l]$ is the small-signal gain of the amplifier. This formula applies equally well to an absorptive system with $N_2 < (g_2/g_1)N_1$ and $G_0 < 1$ (i.e., the cell is an attenuator).

Equation (8.7.3a) is very important, and it is worthwhile to digress for a moment to appreciate its significance.

• • •

Case A: An Optically "Thin" Amplifier or Attenuator. If $G_0(\nu)$ is very close to 1, the amplifier (or attenuator) is said to be optically thin, and thus $\gamma_0(\nu)l$ is small. Therefore the Taylor series expansion of $\exp(\gamma_0 l) - 1$ yields $\gamma_0(\nu)l$, and we obtain a most logical result:

$$I^+(\nu, l) = A_{21}h\nu N_2 lg(\nu)\frac{d\Omega}{4\pi} \qquad \text{(optically thin)} \qquad (8.7.3b)$$

This states that the power from $N_2 l$ atoms radiating into $d\Omega/4\pi$ as $g(\nu)$ add their radiation—a result that would be guessed from the start. In other words, each element dz along z contributes an equal amount to the power.

Case B: A Thermal Population. If the atomic populations are such that $N_2 < (g_2/g_1)N_1$, the amplifier is an attenuator and $G_0 < 1$. Furthermore, if N_2/N_1 can be related to a "temperature" by a Boltzmann relation,

$$\frac{N_2}{N_1} = \frac{g_2}{g_1}\exp\left(-\frac{h\nu}{kT}\right)$$

then (8.7.3a) becomes

$$I^+(\nu, l) = \left[\frac{8\pi n^2 h\nu^3}{c^2}\frac{1}{\exp(h\nu/kT) - 1}\right]\frac{d\Omega}{4\pi}(1 - e^{-|\gamma_0(\nu)|l}) \qquad (8.7.3c)$$

One should immediately recognize the quantity in the brackets as the Planck formula for blackbody radiation for $I(\nu) = (c/n)\rho(\nu)$ [see (7.2.10)]. The remaining factor, $1 - \exp[-|\gamma_0(\nu)l|]$ is a measure of the blackbody power available to the outside world; its maximum value is, of course, unity for an optically thick medium (i.e., $\exp[-|\gamma_0(\nu)l|] \ll 1$).

Thus we see that the quantity in parentheses in (8.7.3c) is the *emissivity* of the system for a normal population ratio. Since the factor $1 - \exp[-|\gamma_0(\nu)l|]$ is also the absorption of a wave passing through the attenuator, we have thus derived Kirchhoff's radiation law, which states that a body can emit only as much blackbody radiation at a frequency ν as it can absorb.

If the lower-state population increases with z, as it does, for instance, in a high-pressure sodium lamp (with z treated as the radius), the central part of the spontaneous emission is heavily attenuated, whereas the wings escape more or less unimpeded. Under such circumstances one can easily obtain a *self-reversed line*. (Observe a common sodium vapor street lamp with a small hand-held spectroscope; one sees a broad orange spectrum extending on either side of 589 nm, but a "dark" band at this wavelength, which is the *center* of sodium emission. This is a common example of a self-reversed line.

• • •

Let us now consider some of the consequences of (8.7.3a). First, we should remind ourselves that it was derived under the assumption that the populations N_2 and N_1 were *not* saturated by stimulated emission. Under such circumstances and high gain, the spectral width of the amplified spontaneous emission is narrower than that predicted by the line shape $g(\nu)$ as given, for instance, by (8.7.3b).

A few numbers should convince one of this fact. Suppose that $G_0(\nu_0)$ (i.e., the gain at line center) were 100. Then the factor $G_0(\nu_0) - 1 = 99$. At $\nu = \nu_0 \pm \Delta\nu/2$, $\gamma(\nu) = \gamma_0(\nu_0)/2$, and thus $G_0(\nu) = [G_0(\nu_0)]^{1/2} = 10$. Thus this factor is now equal to 9, a reduction by a factor of 11 in changing the frequency by a mere $\Delta\nu/2$. Thus spectral narrowing is to be expected from a system with high gain. A graphic display of this is shown in Fig. 8.17, where (8.7.3a) is plotted for various values of the gain coefficient (times length) with an assumed Lorentzian line shape for $g(\nu)$.

One cannot carry this analysis to the extreme of letting the intensity become arbitrarily large. If the line is inhomogeneously broadened, too large an intensity at line center will burn a hole at line center but leave the wings unaffected. When this happens, the line width begins to expand back toward its optically thin value, and the arithmetic becomes most unbearable. Equation (8.7.1) used the small-signal gain coefficient, whereas now one would need to include saturation.

A much more serious problem arises in addition to our unhappiness with the mathematics. If the ASE saturates the population inversion, that inversion energy *cannot* be extracted by an externally injected coherent laser signal. Thus the curse of ASE is that it limits the maximum gain that can be built into an amplifier.

To quantify this limit, we generalize the rate equation in Sec. 8.3 to account for a spectral distribution of radiation $I(\nu)$ rather than I_ν in (8.3.3) through (8.3.9b). This amounts to replacing factors of the form $\sigma(\nu)I_\nu$ by $\int \sigma(\nu')I(\nu')\,d\nu'$ (the details are left for a problem). The result is that the gain saturates according to

$$\gamma(\nu) = \frac{\gamma_0(\nu)}{1 + \left[\tau_2 + \left(\frac{g_2}{g_1}\right)\tau_1\left(1 - \frac{\tau_2}{\tau_{21}}\right)\frac{\sigma(\nu_0)}{h\nu}\int \overline{g}(\nu')[I^+(\nu') + I^-(\nu')]\,d\nu'\right]}$$

(8.7.4)

Sec. 8.7 Amplified Spontaneous Emission (ASE)

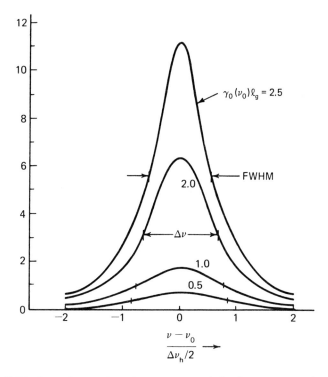

Figure 8.17 Spectral narrowing of spontaneous emission in an unsaturated amplifier.

Next we use the result of the other viewpoint expressed in Sec. 8.3, namely, that the saturation intensity can also be interpreted as that intensity which reduces the lifetime of state 2 by a factor of 2 (for a good laser with $\tau_1 \ll \tau_2$). For a spectral distribution of intensity, the lifetime of state 2 is reduced by a factor of 2 when

$$\frac{\sigma(\nu_0)}{h\nu} \int \overline{g}(\nu')[I^+(\nu', z) + I^-(\nu', z)] d\nu' = \frac{1}{\tau_2} \qquad (8.7.5)$$

Saturation will occur first at $z = l$ or $z = 0$ of the amplifier of Fig. 8.16. If the condition expressed by (8.7.5) holds, further pumping of the amplifier is mostly futile; spontaneous emission extracts the energy from the inversion as fast as the pumping agent can create it.

Let us consider $z = l_g$ where $I^-(\nu, l_g) = 0$ and evaluate (8.7.5) with $I^+(\nu, l_g)$ given by (8.7.3).

$$\frac{\sigma(\nu_0)}{h\nu} \int \overline{g}(\nu') \frac{8\pi n^2 h\nu^3}{c^2} \frac{N_2}{N_2 - \frac{g_2}{g_1}N_1} [e^{\gamma_0(\nu_0) l_g \overline{g}(\nu')} - 1] \frac{d\Omega}{4\pi} d\nu' = \frac{1}{\tau_2} \qquad (8.7.6a)$$

All frequency variations espressed in (8.7.6a) are insignificant compared to the violent variation of $\overline{g}(\nu)$ appearing in the exponential and as a multiplier. Thus the center frequency, $\nu = \nu_0$, is dominant in the exponential where $\overline{g}(\nu_0) = 1$ and (8.7.6a) becomes

$$e^{\gamma_0(\nu_0) l_g} - 1 = \frac{\lambda_0^2}{8\pi n^2} \cdot \frac{1}{\sigma(\nu_0)} \left[\frac{N_2 - (g_2/g_1)N_1}{N_2} \right] \frac{1}{\tau_2} \cdot \left(\frac{d\Omega}{4\pi} \right)^{-1}$$

or

$$G_0(\nu_0) - 1 < \frac{\tau_{21}}{\tau_2} \left(\frac{d\Omega}{4\pi} \right)^{-1} \left(1 - \frac{g_2}{g_1} \frac{N_1}{N_2} \right) \qquad (8.7.6b)$$

where τ_{21} is used instead of A_{21} in expression for $\sigma(\nu_0)$. We can estimate $d\Omega/4\pi$ by the solid angle subtended by the aperture A on a sphere centered at $z = 0$, or $d\Omega/4\pi \sim A/4\pi l_g^2$. For a good laser medium, $N_1 \ll N_2$ and $\tau_2 \sim \tau_{21}$, and thus the maximum permissible gain is dictated by geometry:

$$G_0(\nu_0) - 1 = \frac{4\pi l_g^2}{A} \qquad (8.7.7)$$

Although this derivation suffers from the approximations used, it gives a good physical picture of the limitations that spontaneous emission imposes on the amplifier design. But this limitation is on the small-signal gain of the system; it does not limit the amount of energy that can be extracted. As shown in Sec. 8.3, one can extract as much power as the effective pumping rate can supply with the proviso that ASE does not use it first. Thus, for many high-energy applications, such as laser fusion, one prefers a *low gain* but high energy storage capability in the amplifiers.

8.8 LASER OSCILLATION—A DIFFERENT VIEWPOINT

Let us close this chapter with a review of the basic concepts and simultaneously obtain a slightly different picture of laser oscillation. Incidentally, these considerations also apply to classical oscillators, such as microwave tubes and transistor circuits.

We start with a simple system, such as that shown in Fig. 8.1, with an active medium, characterized by a small-signal gain $G_0 = \exp[\gamma_0(\nu)l]$ and a suitable feedback system in the form of a cavity with mirror reflectivities R_1 and R_2. As discussed previously, the laser oscillation builds up from the spontaneous emission "noise" emitted from the upper state until the coherent photon flux saturates the gain. We can describe this buildup and the necessity of gain saturation with a few lines of transparent mathematics.

The fact that a population inversion exists automatically guarantees a noise source in the form of spontaneous emission into 4π steradians and into a whole band of frequencies in proportion to the line shape $g(\nu) \, d\nu$. What we need is a

Sec. 8.8 Laser Oscillation–a Different Viewpoint

precise description of how these noise photons are allotted to a high-Q cavity mode which will eventually become the intense laser radiation. We obtain this prescription by the following line of reasoning.

Consider the high-Q TEM$_{0,0,q}$ modes of the laser cavity. These modes are separated by a frequency interval $\Delta \nu = c/2nd$, and the fields exist over an effective "area" that overlaps the active medium. (As we will see, it is not necessary to be precise about this area, but one must recognize that it exists.) Thus the number of active atoms in state 2 capable of being stimulated by these modes is $N_2 \times$ area $\times$ length $= N_2 V$. Before stimulated emission becomes important, we must obtain the initial photons from the noise. We note that

1. The rate of generation of spontaneous photons by this group of atoms into all frequencies is $A_{21} N_2 V$.
2. The fraction that appears in the frequency interval $c/2nd$, symmetrically placed around ν_q, is $g(\nu)\Delta\nu$ (with $\Delta\nu = c/2nd$).
3. Only the TEM$_{0,0,q}$ mode has a high Q, but there are $(8\pi n^3 \nu^2/c^3)V \cdot (\Delta\nu = c/2nd)$ electromagnetic modes in that volume and that frequency interval.

Thus the rate of increase of photons in this one TEM$_{0,0,q}$ cavity mode due to spontaneous emission is

$$\left.\frac{dN_p}{dt}\right|_{\text{spont.}} = (A_{21} N_2 V)\left[g(\nu)\frac{c}{2nd}\right]\frac{1 \text{ mode}}{\text{no. modes} = (8\pi n^3 \nu^2/c^3)(c/2nd)V} \quad (8.8.1)$$

where N_p is the number of photons in the cavity mode volume. Canceling common factors of V and $c/2nd$ and recognizing the remaining collection of factors as belonging to the stimulated emission cross section lead to a compact formula:

$$\left.\frac{dN_p}{dt}\right|_{\text{spont.}} = N_2 c \left[A_{21}\frac{\lambda^2}{8\pi}g(\nu)\right] = N_2 c \sigma_{\text{SE}} \quad (8.8.2)$$

Once the photons are in the cavity mode, they bounce back and forth between the two mirrors, being amplified by a factor G per pass, with some of them escaping by transmission through (or absorption by) the mirrors. If N_p starts at an arbitrary point inside the cavity, then $GR_1 GR_2 N_p$ returns after a round trip taking $2nd/c$ seconds. Thus the time rate of change of photon number due to the active cavity is

$$\left.\frac{dN_p}{dt}\right|_{\text{cavity with gain}} = \frac{G^2 R_1 R_2 - 1}{2nd/c} N_p \quad (8.8.3)$$

The addition of (8.8.2) and (8.8.3) yields this simple mathematical model:

$$\frac{dN_p}{dt} = \frac{G^2 R_1 R_2 - 1}{2nd/c} N_p + N_2 c \sigma_{\text{SE}} \quad (8.8.4)$$

If $G^2R_1R_2 - 1 < 0$, we have a very uninteresting situation. The equilibrium value of N_p is enhanced by the active cavity, but even so, the output power is insignificant (see the problem at the end of this section). Now the last term in (8.8.4) is quite small, and a few numbers will convince us of that fact. Let $\sigma = 3 \times 10^{-12}$ cm^2 (a very large stimulated emission cross section), $c = 3 \times 10^{10}$ cm/sec (as usual), and $N_2 = 10^{12}$ cm^{-3} (typical of such a high-gain laser). Then the rate of spontaneous emission into this cavity mode sounds big, $N_2 c \sigma_{SE} = 9 \times 10^{10}$ sec^{-1}, but if each photon has 1 eV of energy, this represents a paltry 14.4 nanowatts!

But if $G^2 > 1/R_1R_2$, the first term on the right-hand side of (8.8.4) is a *positive* coefficient of N_p, and the photon number grows exponentially with time. As long as N_p stays small enough so that G can be considered a constant, the number *grows* exponentially with time.

$$N_p(t) \doteq N_p(0) \exp\left[+ \left(\frac{G^2 R_1 R_2 - 1}{2nd/c} \right) t \right] \tag{8.8.5}$$

But this equation cannot apply for all time—*we* cannot avoid saturation! As long as $G^2R_1R_2 - 1 > 0$, the photon number increases with time and the only way to reverse that behavior is for the sign of the first term in (8.8.4) to change (i.e., revert back to a negative sign). Thus the gain *must* saturate slightly *below* the loss so that the two terms on the right-hand side of (8.8.4) are equal—then we have a steady-state laser!

For any computational purpose, one can safely set $G_{sat.}^2 = 1/R_1R_2$. For instance, suppose that the laser described above produces 10 mW of power, has a length of 50 cm, and has mirrors with reflectivities $R_1 = 1$ and $R_2 = 0.9$. From the specification of the output power, we can compute the power impinging on the mirrors, which, when divided by $h\nu$, yields the number of photons hitting M_1 or M_2 (per second). This number is also equal to the number of photons inside the cavity divided by the round-trip transit time $2nd/c$ (since each of the photons hits each mirror in that time interval).

$$\frac{N_p}{2nd/c} = \frac{P_{out.}}{h\nu} \frac{1}{1 - R_1R_2} \tag{8.8 6}$$

Thus we can compute the saturated gain, $G_{sat.}$, by requiring $dN_p/dt = 0$ in (8.8.4). We must also recognize that the upper-state population is saturated by the laser oscillation and thus one must use the saturated value for N_2.

$$(1 - G_s^2 R_1 R_2) \frac{P_0}{h\nu} \frac{1}{1 - R_1 R_2} = N_2^{(s)} c \sigma_{SE} \tag{8.8.7}$$

For the numbers chosen above, we obtain

$$1 - G_s^2 R_1 R_2 = \frac{h\nu}{P_0}(1 - R_1R_2) N_2^{(s)} c \sigma_{SE}$$

$$< 1.44 \times 10^{-7} \tag{8.8.8}$$

Sec. 8.8 Laser Oscillation–a Different Viewpoint

Surely, then, setting the saturated gain equal to the loss is an excellent approximation!

Even though spontaneous emission is ignorable in the final state of laser oscillation, it is crucial to its initiation. As we will see in Chapter 9, there are some laser media where the spontaneous emission is so low that it takes appreciable time for the laser pulse to develop. Indeed, the spontaneous emission is so low that the population inversion can build up faster than the photon number. This will lead to a "gain-switched" pulse.

The fact that the gain does saturate at a value slightly less than the loss (for this or any oscillator) implies that the laser will have a finite, nonzero, spectral width due to the "noise" contributed by the spontaneous emission. We can compute this width by recognizing that the radiation is distributed in frequency around ν_q according to the Fabry-Perot function for the *active* cavity. Modifying (6.3.1) to account for the saturated gain G_s leads to the following expression for the spectral distribution of power *inside* the laser cavity for $n = 1$.

$$P(\nu) = \frac{K}{[1 - G_s(R_1R_2)^{1/2}]^2 + 4G_s(R_1R_2)^{1/2} \sin^2[2\pi(\nu - \nu_q)d/c]} \quad (8.8.9)$$

We now use a whole sequence of tricks to derive various formulas for the spectral width. The first is a straightforward analysis to obtain the full width at half maximum of the saturated cavity response function from (8.8.9).

$$\Delta\nu_{osc.} = \frac{1 - G_s(R_1R_2)^{1/2}}{\pi[G_s(R_1R_2)^{1/2}]^{1/2}} \frac{c}{2d} \quad (8.8.10a)$$

Now the mathematical manipulations start:

$$1 - G_s^2 R_1 R_2 = [1 - G_s(R_1R_2)^{1/2}][1 + G_s(R_1R_2)^{1/2}] \simeq 2[1 - G_s(R_1R_2)^{1/2}]$$

or

$$1 - G_s(R_1R_2)^{1/2} = \tfrac{1}{2}(1 - G_s^2 R_1 R_2)$$

where the saturated gain is set equal to the loss except when the difference appears. Substituting (8.8.8) for $(1 - G_s^2 R_1 R_2)$ leads to another version of the oscillation bandwidth.

$$\Delta\nu_{osc.} = \frac{h\nu}{P_{out.}}\left[\frac{c}{4\pi d}(1 - R_1R_2)\right]N_2^{(s)} c\sigma_{SE} \quad (8.8.10b)$$

Multiply the quantity in the brackets by ν/ν, convert $c/\nu = \lambda$, and recognize the residue as ν/Q or the passive cavity width $\Delta\nu_{1/2}$—see (6.3.10).

$$\Delta\nu_{osc.} = \frac{h\nu}{P_{out.}} \Delta\nu_{1/2} N_2^{(s)} c\sigma_{SE} \quad (8.8.10c)$$

Now use the one final bit of trickery: we set the saturated gain *coefficient* $(N_2^{(s)} - g_2/g_1 N_1^{(s)})\sigma$, equal to the loss, which is prorated over the length d:

$$N_2^{(s)} \sigma_{SE}\left(1 - \frac{g_2 N_1^s}{g_1 N_2^s}\right) = \frac{1}{2d} \ln\left(\frac{1}{R_1 R_2}\right) = \frac{1}{2d} \ln\left[\frac{1}{1 - (1 - R_1 R_2)}\right]$$

$$\simeq \frac{1 - R_1 R_2}{2d}$$

Therefore
$$N_2^{(s)} \sigma_{SE} \simeq \frac{1 - R_1 R_2}{2d}\left(1 - \frac{g_2}{g_1}\frac{N_1^{(s)}}{N_2^{(s)}}\right)^{-1}$$

Use this expression in (8.8.10c), insert factors of $2\pi/2\pi$ and ν/ν, and again identify some of the factors as $\nu/Q = \Delta\nu_{1/2}$.

$$\Delta\nu_{osc.} = 2\pi \frac{h\nu}{P_{out.}} (\Delta\nu_{1/2})^2 \left(1 - \frac{g_2}{g_1}\frac{N_1^{(s)}}{N_2^{(s)}}\right)^{-1} \qquad (8.8.10d)$$

All forms of (8.8.10) are equivalent—some are more convenient than others for computation purposes. Equation (8.8.10d) is most widely quoted and was originally derived by Schawlow and Townes[12] and also by Gordon.[13]

Let us close by emphasizing that this oscillation bandwidth is extremely small. For the numbers used as an example for this section and assuming an ideal system with $N_1 = 0$ (no lower state), we have $\nu = 2.42 \times 10^{14}$ Hz, $\lambda = 1.24$ μm, $Q = 5.05 \times 10^7$, $\Delta\nu_{1/2} = 4.77$ MHz (for the passive cavity), $P_0 = 10$ mW (an arbitrary, but typical power); thus $\Delta\nu_{osc.} = 2.29 \times 10^{-3}$ Hz! Obviously, we have ample room to allow for nonideal situations ($N_1 \neq 0$), and the limit to oscillation bandwidth is incredibly small. In practice, the wandering of the oscillation frequency caused by very slight perturbations in the mirror separation completely overwhelms the foregoing limit. But in any case, a delta function for the spectral representation of the laser is an excellent approximation.

PROBLEMS

8.1. A homogeneously broadened laser transition at $\lambda = 10.6$ μm (CO_2) has the following characteristics: $A_{21} = 0.34$ sec^{-1}; $J_2 = 21$; $J_1 = 20$; $\Delta\nu_h = 1$ GHz.
  (a) What is the stimulated emission cross section at line center?
  (b) What must be the population inversion density $N_2 - (g_2/g_1)N_1$ to obtain a gain coefficient of 5%/cm. If the lifetime of the upper state is 10 μs and that of the lower state 0.1 μs, what is the saturation intensity?

8.2. An experiment involving a homogeneously broadened optical amplifier is depicted in the diagram below. For an input intensity of 1 W/cm², the gain (output/input) is 10 dB. If the input intensity is doubled to 2 W/cm², the gain is reduced to 9 dB.

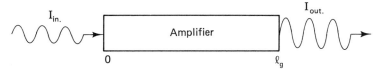

(a) What is the small-signal gain (i.e., $I_{in.} \to 0$) of this amplifier (in dB)?
(b) What is the saturation intensity?
(c) What is the maximum power (per unit area) that can be extracted from this amplifier (in limit of large input intensity)?
(d) What would be the input intensity to extract 50% of this maximum?

8.3. The purpose of this problem is to predict the saturated gain (or transmission) through an amplifier with a small-signal gain coefficient γ_0 that saturates according to the homogeneous law and a loss coefficient α that is not affected by the radiation. Hence the intensity changes with z according to

$$\frac{df}{dz} = \left(\frac{\gamma_0}{1+f} - \alpha\right) \cdot f \qquad 0 \leq z \leq l_g$$

where $f(z) = I(z)/I_s$ and l_g = length of the amplifier.
(a) Assume a small-signal gain $G_0 = \exp(\gamma_0 - \alpha)l = 4$ (i.e., 6 dB) and values of the ratio $\gamma_0/\alpha = 2, 5, 50$. Plot the saturated gain G_s (in dB) as a function of the input intensity (i.e., $I_{in.}/I_s$). Use a semilog graph paper and plot G_s on the linear scale and $I_{in.}/I_s$ on the log scale covering the range from 10^{-2} to 10^3.
(b) If the input intensity is much larger than the saturation intensity, find an analytic expression for the output in terms of the input.

8.4. The model of Sec. 8.2 assumed an atomic system with equal degeneracies $g_1 = g_2$. Use the same logic path as used there to find an expression for the small-signal gain coefficient γ_0 and for the saturation intensity I_s for the case where $g_1 \neq g_2$.

8.5. The ideas of gain saturation are equally applicable to absorption—that is the purpose of this problem. Consider a single-frequency dye laser tuned to the center of the sodium D line at 5889.95 Å and irradiating a heated cell (630°K), 10 cm long, containing a mixture of sodium (Na) vapor at a density of 1.5×10^{15} cm^{-3} and helium (He) gas at a density of 6.53×10^{19} cm^{-3}. The self-broadening of this line, due to collisions between sodium atoms, is 15 MHz for the conditions of this problem. The foreign gas broadening is due to collisions between sodium atoms and helium with a cross section estimated to be 10^{-14} cm^2. The following data for this transition are from NSRDS-NBS (vol. 11), US Department of Commerce, National Bureau of Standards:

N_1 $3^2S_{1/2}$ $g_1 = 2$ $E_1 = 0$
N_2 $3^2P_{3/2}$ $g_2 = 4$ $E_2 = 16,978.07 \text{ cm}^{-1}$; $A_{21} = 6.3 \times 10^7 \text{ sec}^{-1}$

(a) What are all of the pertinent line widths from the various causes ("natural," Doppler, self-broadening, or foreign gas [He] collisions)?

(b) If a "small-signal" laser is tuned to line center and propagates through the 10 cm length, what fraction emerges? Express the attenuation in dB.

(c) Let the input amplitude of the laser be a variable. Use a semilog graph paper and plot the transmission (in percent of the incident value) as a function of the input intensity normalized to a saturation value similar to that shown in Fig. 8.6. To find I_S, follow the procedure of Sec. 8.3, but remember that $N_1 + N_2 = [\text{Na}]$ (i.e., the total number of sodium atoms is conserved).

8.6. (a) The response of an amplifier with a 6 dB small-signal gain to varying input intensities for the case of homogeneous broadening is covered in the text, with the results plotted as the solid curve in Fig. 8.8. Consider this amplifier, but assume that inhomogeneous broadening applies. Show that the equation relating the output to the input is given by

$$\ln\left[\frac{(1 + y_2)^{1/2} - 1(1 + y_1)^{1/2} + 1}{(1 + y_1)^{1/2} - 1(1 + y_2)^{1/2} + 1}\right]$$
$$+ 2((1 + y_2)^{1/2} - (1 + y_1)^{1/2}) = \gamma_0 l_g$$

where

$$y_2 = \frac{I_{\text{out.}}}{I_s} \quad \text{and} \quad y_1 = \frac{I_{\text{in.}}}{I_s}$$

This equation is plotted as the dashed curve in Fig. 8.8.

(b) Show that one recovers the small-signal amplification law for $(y_2, y_1) \ll 1$.

8.7. Consider the ideal laser medium shown below. The pump excites the atoms to state 2 at a rate R_2, which then decays to state 1 at a rate $(\tau_{21})^{-1}$ and back to state 0 at a rate $(\tau_{20})^{-1}$. State 1 decays back to 0 so fast that the approximation $N_1 \approx 0$ is appropriate. The radiative rate for the $2 \rightarrow 1$ transition is $6 \times 10^6 \text{ sec}^{-1}$, and its width is 10 GHz. (Assume a Lorentzian profile and steady state.)

(a) What is the stimulated emission cross section?

(b) What must be the pump rate R_2 in order to obtain a *small*-signal gain coefficient of 1%/cm?

(c) What is the saturation intensity for the $2 \rightarrow 1$ transition?

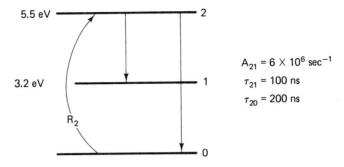

(d) How much power (in W/cm³) is expended in creating the gain coefficient of (b)?
(e) Express the line width in Å units and cm⁻¹ units.

8.8. In the laser system shown below, only state 2 is pumped directly from the ground (or 0) state with a pumping value of R_2 (cm⁻³/sec). State 2 decays to 0 at a rate of 5×10^6 sec⁻¹ and to 1 by spontaneous emission and quenching collisions at a rate of 10^7 sec⁻¹. The lifetime of state 1 is 50 ns. Assume homogeneous broadening; line width $\Delta\nu = 2$ cm⁻¹, $\lambda_0 = 4000$ Å; $A_{21} = 10^6$ sec⁻¹; $M = 40 \times 1.67^{-27}$ kg; $T = 500$ K; and that the medium fills the cavity in the manner shown below. Also assume steady state.

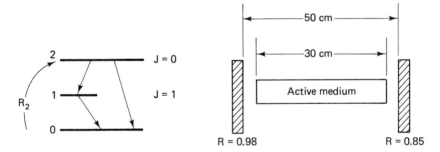

(a) What is the stimulated emission cross section at line center?
(b) What is the pump rate P_2 that brings the laser to threshold?

8.9. The following questions refer to an atomic system with $J_2 = 1$ and $J_1 = 2$.
(a) What is the ratio B_{12}/B_{21}?
(b) What is the formula for the small-signal gain coefficient for the $2 \rightarrow 1$ transition?
(c) If the line shape function could be approximated by the graph shown below, $A_{21} = 10^6$ sec, $\lambda = 6401$ Å, and $N_2 = N_1 - 10^{12}$ cm⁻³, what is the small-signal gain coefficient for the $2 \rightarrow 1$ transmission at $\nu = \nu_0$?

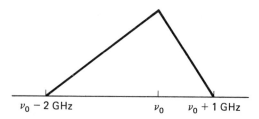

$\nu_0 - 2$ GHz ν_0 $\nu_0 + 1$ GHz

8.10. Consider a homogeneous broadened gain cell of problem 8.2 being irradiated by an external laser of variable amplitude. The small-signal gain of the cell is 10 dB (at the line center), the full width at half maximum (FWHM) of the transition is 1 GHz, and the saturation intensity (at the line center) is 10 W/cm^2.
   (a) Plot the gain (in dB) as a function of input intensity. (Assume that the frequency of the input coincides with the line center.)
   (b) Repeat (a), but assume that the incoming frequency is detuned from the line center by 0.5 GHz. (Show as a dashed curve.)
   (c) In the limit of large input intensities, the gain cell can, at most, add a photon flux to the signal. What is the maximum intensity that can be extracted from this gain cell?
   (d) What should be the input intensity to extract 95% of the maximum found in (c)?

8.11. Consider the energy level diagram shown below, which is representative of lasers such as the molecular nitrogen N_2.

```
            ─── 2        J = 1
         ↑  ⌇              τ₂₁ = 20 ns
      R  │  ─── 1        J = 2
         │  ⌇              τ₁ = 1 μs
            ─── 0
```

   (a) If $R = 10^{20}$ cm^{-3}/sec, what are the equilibrium populations of states 2 and 1?
   (b) Why is this system unsuitable for a CW laser?
   (c) Suppose that the pump had the form of a step function of the amplitude given in (a). Sketch the time variation of the small-signal gain coefficient assuming all populations zero at $t = 0$ and a stimulated emission cross section of 10^{-12} cm^2.

8.12. The purpose of this problem is to point out a simple experimental method for estimating the saturation intensity, $I_{\text{sat.}}$, of a laser. You are given the experimental apparatus shown below, which is made up of a continuously pumped gain medium (small-signal gain coeffieicnt γ_0), two nearly perfect reflecting mirrors, and a photodetector. The photodetector records the side fluorescence power emanating from a small volume of the gain medium. Assume that the laser transition is homogeneously broadened and that the lower laser level population is negligible compared to that in the upper state.

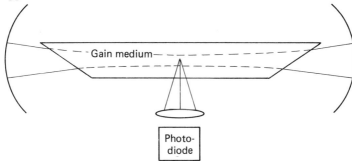

(a) If P_0 is the side fluorescence power (W/cm^{-3}) that is observed with one of the cavity mirrors blocked and P is measured when the laser is operating normally (i.e., mirror unblocked, everything else the same), then derive a simple expression that relates P/P_0 to the saturation intensity of the gain medium.

(b) If the side fluorescence is observed to be suppressed by 50% when the intercavity laser flux is 100 W/cm^2, what is $I_{\text{sat.}}$?

8.13. Consider the laser cavity shown below. The mirrors M_1, M_2 have a power reflectivity of 0.95 and 0.85, respectively, and the Brewster's angle windows transmit 98% of the power for the proper polarization (or minimum loss).

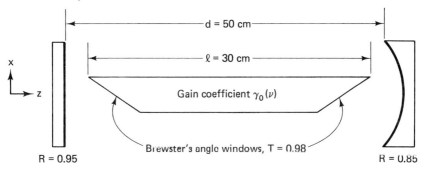

(a) If $\gamma_0 = 0$, find the photon lifetime of the passive cavity. Assume a Brewster's angle polarization that yields a minimum loss.

(b) If $\gamma_0 = 4 \times 10^{-3}$ cm^{-1} (at line center), will this system oscillate? Justify your answer.

(c) What is the orientation of the optical electric field for minimum loss in the cavity (i.e., does $E = E_0 \mathbf{a}_x$ or $E_0 \mathbf{a}_y$ or $E_0 \mathbf{a}_z$?)?

8.14. Suppose the distribution of center frequencies in an inhomogeneously broadened system is approximated by

$$p(f) = \frac{\Delta\nu_D}{2\pi[(f-\nu_0)^2 + (\Delta\nu_D/2)^2]}$$

where $\Delta\nu_D$ is the full width at half maximum of the "pure" inhomogeneously broadened line.

(a) Show that the saturated gain coefficient is given by

$$\gamma(\nu) = \frac{\gamma_0(\nu_0)}{\left(1 + \frac{I_\nu}{I_s}\right)^{1/2}} \cdot \frac{1 + \left(\frac{\Delta\nu_h}{\Delta\nu_D}\right)\left(1 + \frac{I_\nu}{I_s}\right)^{1/2}}{\left\{\left(\frac{\nu - \nu_0}{\Delta\nu_D/2}\right)^2 + \left[1 + \left(\frac{\Delta\nu_h}{\Delta\nu_D}\right)\left(1 + \frac{I_\nu}{I_s}\right)^{1/2}\right]^2\right\}}$$

In this expression $\Delta\nu_h$ is the homogeneous line width, $\gamma_0(\nu_0)$ is the small-signal gain of the "pure" inhomogeneous transition evaluated at line center, I_s is the frequency-independent saturation intensity, and I_ν is the intensity of the wave stimulating the atoms.

(b) Show that one recovers the homogeneous saturation law if $\Delta\nu_h/\Delta\nu_D$ and/or $I_0/I_s \gg 1$.

(c) Identify the circumstances under which one recovers the "pure" inhomogeneous saturation behavior.

(d) Plot the saturated gain coefficient as a function of I/I_s for $\nu = \nu_0$ and $\Delta\nu_h/\Delta\nu_D = 1/2$. Show also the graph of the two limiting forms of the saturation—the extreme homogeneous limit and the inhomogeneous limit.

The problem involves considerable arithmetic so the following is given to alleviate some of the pain.

Define $f - \nu_0 = x$; $\delta = \nu_0 - \nu$; and therefore $(\nu - f)^2 = (x + \delta)^2$;
$a = \Delta\nu_D/2$; $b = \Delta\nu_H/2$ with $\Delta\nu_H = \Delta\nu_h\left(1 + \frac{I_\nu}{I_s}\right)^{1/2}$

A partial fraction expansion will be needed:

$$\frac{1}{(x^2 + a^2)[(x + \delta)^2 + b^2]} = \frac{Ax + B}{x^2 + a^2} + \frac{C(x + \delta) + D}{(x + \delta)^2 + b^2}$$

where
$$B = \frac{\delta^2 + b^2 - a^2}{\delta^4 + 2(b^2 + a^2)\delta^2 + (b^2 - a^2)^2}$$

$$D = \frac{\delta^2 - b^2 + a^2}{\delta^4 + 2(b^2 + a^2)\delta^2 + (b^2 - a^2)^2}$$

The terms involving A and C vanish after integration. Finally, note that

$$\delta^4 + 2(b^2 + a^2)\delta^2 + (b^2 - a^2)^2 = [\delta^2 + (b-a)^2][\delta^2 + (b+a)^2]$$

8.15. A model for one laser being optically pumped by another is shown in the diagram below. Compute the pump intensity (in W/cm²) for the system to reach threshold at $\lambda_{21} = 535$ nm. The following information may be useful: $A_{20} = 5 \times 10^7$ sec^{-1}; $A_{21} = 1 \times 10^8$ sec^{-1}; $\tau_1 = 20$ ns; $M = 138$ amu; $M_p = 1.67252 \times 10^{-27}$ kg; $T = 300$ K; $N_0 + N_1 + N_2 = 10^{14}$ cm^{-3}. The stimulated emission cross section at 351 nm is 10^{14} cm², and the $2 \to 1$ transition is homogeneously broadened with $\Delta\nu = 1$ GHz. Since this is a threshold calculation, one can neglect stimulated emission on the $2 \to 1$ line. You may neglect the attenuation of the pump as it propagates across the laser cavity. Assume steady state.

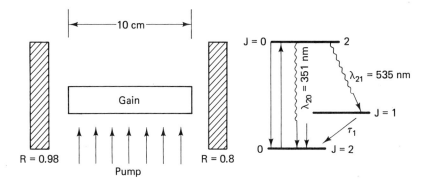

(a) What is the stimulated emission coefficient on the $2 \to 1$ transition?
(b) What gain coefficient on the $2 \to 1$ transition is required to reach threshold?
(c) What pump intensity (in W/cm²) is required to reach threshold for oscillation at λ_{21}?

8.16. The laser shown below utilizes a transition that peaks at $\lambda = 0.55$ nm. The medium has an index of refraction of 1.3 with an inhomogeneous line shape that can be *approximated* by the graph shown.
(a) What is the value of the line shape function at $\nu = \nu_0$?
(b) What is the stimulated emission cross section at line center?
(c) What is the photon lifetime of the *passive* cavity?
(d) What is the orientation of the optical electric field?
(e) If the population of state 1 were 10^{12} cm^{-3}, what must be the density in state 2 to reach threshold?

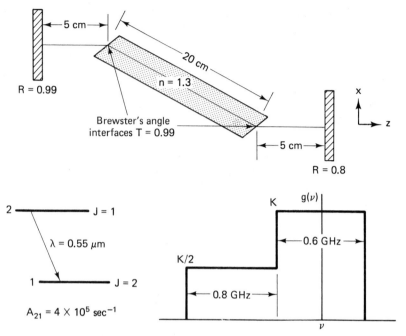

8.17. The energy level diagram shown below illustrates a situation which can occur in gas lasers. The upper state of the $3 \rightarrow 1$ transition is pumped at a rate R_3, and if the lifetime ratio τ_3/τ_1 is favorable, one can obtain gain on the $3 \rightarrow 1$ transition. However, state 2 can also be excited (at a rate R_2), and if sufficient feedback is provided, oscillation can take place on the $2 \rightarrow 1$ transition, which raises the population of state 1 and lowers the gain at λ_{31}. Analyze this case by answering the following two questions. Assume steady state, assume all degeneracies equal to 1, and neglect spontaneous decay from $2 \rightarrow 1$. Define $\tau_3^{-1} = \tau_{30}^{-1} + \tau_{31}^{-1}$.

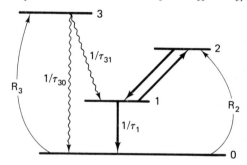

(a) Assume $R_2 = 0$ and thus no oscillation on the $2 \rightarrow 1$ transition. Find an expression of the small-signal gain coefficient on $3 \rightarrow 1$ in terms

of the lifetimes, the pumping rates, and the stimulated emission cross section.
(b) Assume that $R_2 \neq 0$ and that there is a strong saturating signal at the $2 \rightarrow 1$ wavelength such that $N_1 = N_2$. (Neglect spontaneous decay from $2 \rightarrow 1$). Find a new expression for the small-signal gain coefficient at λ_{31}.

8.18. A homogeneously broadened optical amplifier with a small-signal gain of 13 dB is irradiated with a wave with an intensity of 5W/cm². The output intensity is 30 W/cm². What is the saturation intensity? If the saturation intensity were 20 W/cm², what is the maximum power (per unit of area) extractable from this amplifier?

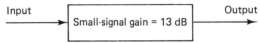

8.19. The saturated gain of a laser oscillator is approximated by

$$G_s^2 = \frac{1}{R_1 R_2} \left[1 - N_2^s c \sigma_{SE} \, h\nu \, \frac{(1 - R_1 R_2)}{P_{\text{out.}}} \right]$$

where N_2^s is the saturated value of the upper state density. This leads to the spectral distribution of power according to

$$S(\omega) = E \cdot E^* = \frac{E_0^2}{(1 - G_s \sqrt{R_1 R_2})^2 + 4 G_s \sqrt{R_1 R_2} \sin^2 \frac{\omega}{c} d}$$

Evaluate the line width $\Delta \nu$ of the laser oscillating in the vicinity of $\omega d/c = q\pi$ for the following conditions: $\nu = 5 \times 10^{14}$; $R_1 R_2 = 0.9$; $d = 50$ cm; $\sigma = 10^{-12}$ cm²; $N_2^s = 10^{12}$ cm^{-3}; $P_0 = 1$ mW.

8.20. Consider an optical amplifier/attenuator that has a Lorentzian line shape with $\Delta \nu_h = 2$ GHz and a "gain" at line center that can be between -30 dB and $+30$ dB. Plot the FWHM of the spontaneous radiation emerging from the end of the cell (normalized to the line width of the transition, i.e., $\Delta \nu_{\text{spont.}}/\Delta \nu_h$).

8.21. The following questions refer to the laser system depicted in the diagram below.
(a) Find the photon lifetime of the passive cavity, that is, with $\gamma_0(\nu) = 0$.
(b) What is the passive cavity Q?
(c) What is the free spectral range of the cavity?
(d) What is the minimum gain coefficient, $\gamma_0(\nu_0)$, necessary to sustain oscillation in this cavity?
(e) If the gain coefficient $\gamma_0(\nu_0)$ were 2×10^{-2} cm^{-1} (at line center) and

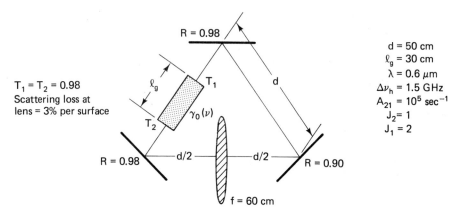

the line shape of the transition was approximated by a Lorentzian with $\Delta \nu_h = 1.5$ GHz, how many $\text{TEM}_{0,0,q}$ modes are above threshold?
- **(f)** What is the stimulated emission cross section (at line center)?
- **(g)** What is the absorption cross section?
- **(h)** The characteristic beam parameter z_0 is related to the dimensions of the cavity and the focal length of the lens by

$$z_0 = \frac{\pi w_0^2}{\lambda} = (3df)^{1/2}\left(1 - \frac{3d}{4f}\right)^{1/2}$$

where $d = 50$ cm and $f = 60$ cm.
- **(1)** Where is $z = 0$ in the cavity?
- **(2)** Find a formula for the resonant frequency of the $\text{TEM}_{m,p,q}$ mode.
- **(3)** What is the *difference* in resonant frequencies (in MHz) of the $\text{TEM}_{0,0,q}$ and $\text{TEM}_{1,0,q}$ modes?

8.22. The fluorescence from a certain system can be approximated by the sketch shown below. What is the value of $g(\nu_0)$?

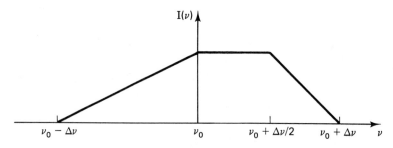

8.23. To make the analysis of the laser system shown in Problem 8.8, assume the following: equal pump rates to states 2 and 1; zero lifetime for state 1; lifetime of state 2 is 100 ns; stimulated emission cross section is

1.3×10^{-17} cm^2 (at line center); no depletion of state 0; and CW operation.
(a) How much pump power in W/cm^3 is required to establish a gain coefficient of 0.05 cm^{-1}?
(b) What is the maximum power (per unit of volume) that can be extracted by stimulated emission?

REFERENCES AND SUGGESTED READINGS

1. A. E. Siegman, *Introduction to Lasers and Masers* (New York: McGraw-Hill Book Company, 1971).
2. A. Yariv, *Introduction to Optical Electronics*, 2nd ed. (New York: Holt, Rinehart and Winston, 1971).
3. A. Maitland and M. H. Dunn, *Laser Physics* (Amsterdam: North-Holland Publishing Company, 1969).
4. A. Yariv, *Quantum Electronics*, 2nd ed. (New York: John Wiley & Sons, Inc., 1975).
5. W. V. Smith and P. P. Sorokin, *The Laser* (New York: McGraw-Hill Book Company, 1966).
6. W. R. Bennett, Jr., "Gaseous Optical Masers," Appl. Opt., Suppl. Opt. Masers, 24–61, 1962.
7. W. R. Bennett, Jr., "Inversion Mechanisms in Gas Lasers," Appl. Opt., Suppl. 2 Chem. Lasers, 3–33, 1965.
8. W. E. Lamb, Jr., "Theory of an Optical Maser," Phys. Rev. *134A*, 1429, 1964.
9. W. R. Bennett, Jr., "Hole-burning Effects in a He-Ne Optical Laser," Phys. Rev. *126*, 580, 1962.
10. A. L. Bloom, "Gas Lasers," Proc. IEEE *54*, 1262, 1966.
11. W. S. C. Chang, *Principles of Quantum Electronics* (Reading, Mass.: Addison-Wesley Publishing Company, Inc., 1969). This book has an extensive bibliography.
12. A. L. Schawlow and C. H. Townes, "Infrared and Optical Masers," Phys. Rev. *112*, 1940, Dec. 1968.
13. E. I. Gordon, "Optical Maser Oscillators and Noise," Bell Syst. Tech. J. *43*, 507–539, 1964.
14. A. Yariv, *Quantum Electronics* (New York: John Wiley & Sons, Inc., 1975).
15. K. A. Jones, *Introduction to Optical Electronics* (New York: Harper & Row, Publishers, Inc., 1987).
16. W. Demtroder, *Laser Spectroscopy*, 2nd Printing (New York: Springer-Verlag, 1982).
17. R. L. Barger and R. L. Hall, "Pressure Shift and Broadening of Methane Line at 3.39 μm Studied by Laser Saturated Molecular Absorption," Phys. Rev. Lett. *22*, 4, 1969.
18. C. J. Borde, J. L. Hall, C. V. Kunasz, and D. G. Hummer, "Saturated Absorption Line Shape: Calculation of the Transit-Time Broadening by a Perturbation Approach," Phys. Rev. A, *14*, 236–262, 1976.
19. J. B. Anderson, J. Maya, M. W. Grassman, R. Laqueskenko, and J. F. Waymouth,

"Monte Carlo Treatment of Resonant Radiation Imprisonment in Fluorescent Lamps," Phys. Rev. A, 31, 2968, 1985.
20. T. Paoli, "Saturation Behavior of the Spontaneous Emission from Double-Heterostructure Junction Lasers Operating High Above Threshold," IEEE J. Quant. Electron. *QE-9*, 267, 1973.
21. V. P. Chebotayev, "Super High Resolution Spectroscopy," in *The Laser Handbook*, Vol. 5, Article 3, Eds. M. Bass and M. L. Stitch (New York: North Holland Publishing Company, 1985). NOTE: The frequency of helium-neon laser at 3.39 μm has been measured to be $\nu(3.39\ \mu m\ line) = 88.3761816029$ THz $\pm$ 1.2kHz.

9

General Characteristics of Lasers

9.1 INTRODUCTION—QUANTUM EFFICIENCY

Chapter 8 discussed general issues applicable to all atomic systems as they approached oscillation and then discussed the saturated condition. However, there remain many important features about all lasers that have yet to be addressed.

For instance, for a given system and state of excitation, can one predict the power out of the laser? How stable is the frequency, and can it be predicted with the ease implied in Chapter 8 (i.e., $\nu_{osc} = qc/2d$)? We have also implicitly assumed steady-state or continuous-wave (CW) oscillation, whereas some lasers are pulsed, for convenience or because of necessity.

This chapter covers these issues, attempting to be as general as possible. However, sometimes, specific laser systems will be chosen to exemplify the particular point.

A critical issue that determines the importance of any laser is the output power. Although the following statement may appear to be obvious, it is amazing how often it is overlooked:

The output power/energy is less than the input excitation.

Having agreed on the obvious, can we put a limit on the fraction of the power input that can appear as an output? An examination of Fig. 9.1 shows that a general upper limit can be placed on any laser. From the discussion in Sec. 8.3, we should now recognize that an ideal laser system is one in which:

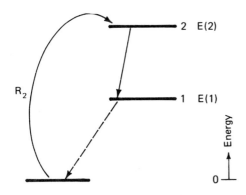

Figure 9.1 Ideal laser system.

1. Only the upper state is pumped.
2. The lower state decays instantaneously—thus $N_1 = 0$.
3. The stimulated emission rate from $2 \rightarrow 1$ overwhelms all other decay processes of state 2.

Thus the following statements are true. Each pumping event costs the external power supply $R_2 E(2)$ watts per unit of volume; stimulated emission can, at most, extract R_2 photons per second per unit of volume with the energy per photon being $E(2) - E(1)$; hence, the maximum efficiency of this laser is

$$\eta_{qe} = \frac{E(2) - E(1)}{E(2)} \qquad (9.1.1a)$$

A more complete analysis involves identifying the fraction of the input excitation used for exciting the upper state η_{pump}; and the fraction of (9.1.1a) that can be extracted by stimulated emission and coupled to the outside world (η_{xtn}). Thus the overall efficiency can be expressed as

$$\eta_T = \eta_{pump} \eta_{qe} \eta_{xtn} \qquad (9.1.1b)$$

Thus any practical system will not meet the quantum efficiency limit although some have come close because of the "kindness of nature" in providing a selective pumping mechanism, and there is a good engineering design of the cavity to optimize the extraction of the photons. The latter topic is the subject of the next section, with the pumping issue being addressed in Chapters 11 to 13.

9.2 OUTPUT POWER (ENERGY)—THE COUPLING PROBLEM

Of course, the efficiency predicted by (9.1.1a) presumes that all of the pumping power is extracted by stimulated emission and all of that is available to the outside world—a most optimistic viewpoint. Hence, our next task is to relate the output power (or energy in the case of pulsed excitation) to the small-signal gain coefficient and the saturation intensity (or energy) of the system. In all of the following parts of this section we assume the model of Sec. 8.3 for the atomic physics part of the problem and consider γ_0 and I_s as being known and thus the pumping efficiency question is bypassed. We also assume homogeneous broadening and oscillation close to line center so that $\overline{g}(\nu) = 1$. (The case of inhomogeneous broadening and $\nu \neq \nu_0$ is reserved for more advanced textbooks.) Our goal is to compute the output intensity for various types of laser configurations.

9.2.1 Traveling Wave Ring Laser

The simplest case that yields an exact analytic solution is that of a unidirectional traveling wave ring laser shown in Fig. 9.2. It is presumed that there is some device incorporated inside the cavity that has an anisotropic attenuation coefficient, and as a result oscillation is constrained to be in *one* direction, such as the counterclockwise one shown. As will become apparent, the one direction is a key simplification used to minimize the mathematical complexity. It is also assumed that there is no intrinsic loss inside the gain medium, but there can still be internal losses inside the cavity because of imperfect reflectivities of the turning mirrors, losses in the "diode" that favors the counterclockwise oscillation, and other parasitic losses. All of these losses are assumed to be independent of intensity.

Fig. 9.2(b) is intended to be a guide for the mathematics and sketches the intensity of the wave as it propagates around the loop. One assumes an intensity I_1 at $z = 0$, just inside the gain cell. It obviously gets bigger going the distance l_g to the output of the gain cell where it is I_2. Then it gets progressively smaller because of the mirror reflectivities and the attenuation due to the "diode." In order to have a steady-state or CW laser, the product of all the survival factors for the photons in the remainder of the cavity times I_2 must reproduce the assumed initial intensity I_1. In other words, the solution must be self-consistent.

Before we jump into the mathematics, it is appropriate to talk our way through the approach to steady state in a manner similar to that used in Sec. 8.2. Suppose the initial value of the intensity $I_1^{(0)}$ is very small, typical of that expected from spontaneous emission from the upper state. Then the full small-signal gain can be obtained from the active medium. Since the net round-trip

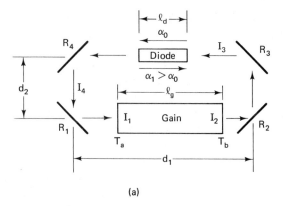

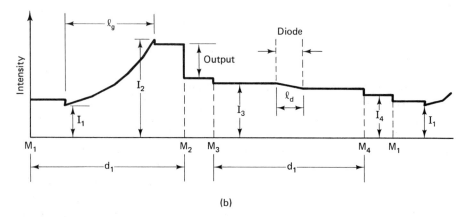

Figure 9.2 Unidirectional traveling wave laser. (a) The geometry. (b) A self-consistent variation of the intensity inside the laser cavity.

gain must be bigger than 1 to even hope for a laser, the amplitude after one round-trip, $I_1^{(1)}$, must be bigger than the initial value $I_1^{(0)}$. (The superscript indicates the round-trip number.)

Now we follow the intensity $I_1^{(1)}$ around the loop to obtain $I_1^{(2)}$ and then $I_1^{(3)}$ and so on, each round-trip taking a time of $(d_1 + d_2 + d_1 + d_2)/c'$ sec. Eventually, $I_1^{(n)}$ is *not* small compared to the saturation intensity and then we can no longer use the simple, intensity-independent, small-signal net gain expression $G_{\text{net}} = (\Pi R_i)(\Pi T_j)[\exp(-\alpha_0 l_d)][\exp(\gamma_0 l_g)]$. How many round-trips does it take for the intensity to become comparable to the saturation intensity enabling this device to be classified as a laser?

An exact answer is not possible here, but a rough "Kentucky windage" estimation follows. If we express all of the loss factors as a lumped attenuation coefficient prorated over the gain length by

Sec. 9.2   Output Power (Energy)–The Coupling Problem

$$\left(\prod_i R_i\right)\left(\prod_j T_j\right) e^{-\alpha_d l_d} \triangleq e^{-\alpha_T l_g} \tag{9.2.1a}$$

or

$$\alpha_T l_g = \alpha_d l_d + \ln\left[\frac{1}{\left(\prod_i R_i\right)\left(\prod_j T_j\right)}\right] \tag{9.2.1b}$$

then the net small-signal gain per round-trip is

$$G_0(\text{net}) = e^{(\gamma_0 - \alpha_T)l_g} \tag{9.2.1c}$$

A reasonable estimate for the number of round-trips needed to obtain oscillation is $(G_0)^n I_1^{(0)} \sim I_s$, an estimation that is consistent with the viewpoint that the small-signal gain is applicable when the intensity is less than I_s. However, it is an answer that depends on the initial intensity provided by the spontaneous emission. (Hence, one can speed up the process by "seeding" the cavity, say with a wave from another laser. This is sometimes referred to as injection locking—more properly it should be called injection "seeding.") Many researchers use the criterion that the exponent in (9.2.1c) should be ~20–30, or in other words, the laser intensity becomes about $\exp(20 \sim 30) = 5 \times 10^8$ to 10^{13} times the spontaneous power in that mode. In any case, we can estimate the number of round-trips and thus the time by applying this criterion and (9.2.1d) below:

$$n \sim \frac{20}{(\gamma_0 - \alpha_T)l_g} \tag{9.2.1d}$$

and thus the laser requires a finite time to establish a more or less steady-state oscillation condition

$$\tau \sim n\tau_{RT} = \frac{20(d_1 + d_2 + d_3 + d_4)}{[(\gamma_0 - \alpha_T)l_g]c} \tag{9.2.1e}$$

This expression indicates a fundamental limitation for pulsed lasers: the excitation may establish gain, but if it does not exist for a period long compared to (9.2.1e), then the intensity does not build up to a significant value nor is the population inversion depleted by stimulated emission.

Now let us return to our primary task and assume that all transients have died out and demand self-consistency. In steady state, intensity in the gain cell is amplified according to (8.3.12) [with $\bar{g}(\nu) = 1$]

$$\frac{dI_{ccw}}{dz} = \frac{\gamma_0 I_{ccw}}{1 + I_{ccw}/I_s} \tag{9.2.2}$$

and thus I_2 is related to I_1 by the following [see the mathematics used in going from (8.3.12) to (8.3.14)]:

$$\ln\left(\frac{I_2}{I_1}\right) + \frac{1}{I_s}(I_2 - I_1) = \gamma_0 l_g \qquad (9.2.3)$$

Now we follow the intensity I_2 around the loop and back to the entrance of the gain cell to find:

$$I_1 = \left[\left(\prod_i R_i\right)\left(\prod_j T_j\right) e^{-\alpha_d l_g}\right] I_2 \qquad (9.2.4)$$

If we substitute (9.2.4) into (9.2.3), we obtain a simple equation for the intensity I_2 at the end of the gain cell.

$$\ln\left[\frac{1}{\left(\prod_i R_i\right)\left(\prod_j T_j\right) e^{-\alpha_d l_d}}\right] + \frac{I_2}{I_s}\left[1 - \left(\prod_i R_i\right)\left(\prod_j T_j\right) e^{-\alpha_d l_d}\right] = \gamma_0 l_g \qquad (9.2.5a)$$

This can be simplified in appearance by using the definition of the loss coefficient prorated over the length of the gain cell given by:

$$\frac{I_2}{I_s} = \frac{(\gamma_0 - \alpha_T)l_g}{1 - e^{-\alpha_T l_g}} \qquad (9.2.5b)$$

where

$$\alpha_T l_g \stackrel{\Delta}{=} \alpha_d l_d + \ln\left[\frac{1}{\left(\prod_i R_i\right)\left(\prod_j T_j\right)}\right] \qquad (9.2.1a)$$

The output intensity (through M_2) is found from (9.2.5b) by

$$I_{\text{out.}} = T_b T_2 I_2$$

or
$$\qquad (9.2.6)$$

$$I_{\text{out.}} = \left[\frac{T_b T_2 (\gamma_0 - \alpha_T) l_g}{1 - e^{-\alpha_T l_g}}\right] I_s$$

9.2.2 Optimum Coupling

A problem of interest is to determine the value of the transmission coefficient of M_2 that yields the maximum output intensity. If T_2 is too large, the losses exceed the gain and no lasing is possible. If T_2 is too small, the laser may be oscillating "brightly" inside the cavity but little is coupled out. Clearly there is an optimum coupling for maximum output. For those of masochistic tendencies, the exact answer requires differentiating (9.2.6) with respect to $T_2 =$

Sec. 9.2 Output Power (Energy)–The Coupling Problem

$1 - R_2$, setting the expression equal to zero, and solving for T_2—a most unpleasant task to do exactly. However, with some judicious approximations, this task can be simplified considerably, yielding reasonably accurate answers:

1. Low loss or the high Q approximation. If the integrated loss, $\alpha_T l_g$, is small, we have a high Q cavity and the exponential term in the denominator of (9.2.6) can be approximated by two terms of a Taylor series.

$$1 - e^{-\alpha_T l_g} \simeq 1 - (1 - \alpha_T l_g) = \alpha_T l_g \quad (9.2.7a)$$

and (9.2.6) becomes

$$I_{\text{out.}} = \left[T_b T_2 \left(\frac{\gamma_0 l_g}{\alpha_T l_g} - 1 \right) \right] I_s \quad (9.2.7b)$$

Now we break up the losses into internal ones and the external coupling loss at M_2 caused by its transmission, expand $ln[1/(1 - T_2)]$ in a Taylor series, and abbreviate freely.

$$\alpha_T l_g = \underbrace{\alpha_d l_d + ln\left[\frac{1}{R_1 R_3 R_4 T_a T_b}\right]}_{\alpha_{\text{int.}} l_g} + \underbrace{ln\left[\frac{1}{R_2 = 1 - T_2}\right]}_{\alpha_{\text{ext.}} l_g} \quad (9.2.7c)$$

$$\simeq \quad L \quad + \quad T_2$$

If we further abbreviate $\gamma_0 l_g$ by g_0, then (9.2.7b) becomes

$$I_{\text{out.}} = \left[T_2 \left(\frac{g_0}{L + T_2} - 1 \right) \right] T_b I_s \quad (9.2.8)$$

Now one can differentiate to find a maximum without a lot of fuss.

$$\frac{dI_{\text{out.}}}{dT_2} = T_b I_s \left[\frac{g_0 L}{(L + T_2)^2} - 1 \right] = 0 \quad \text{for max.}$$

or

$$T_2(\text{opt.}) = -L + (g_0 L)^{1/2} \quad (9.2.9)$$

It is instructive to reinsert (9.2.9) into (9.2.8) to find an expression for the output intensity of the laser with optimum coupling.

$$I_{\text{out.}}(\text{max.}) = T_b I_s \{g_0^{1/2} - L^{1/2}\}^2 \quad (9.2.10a)$$

Again we note that this only makes sense provided the integrated gain g_0 is greater than the losses L. If we go one step further in our approximations and assume $g_0^{1/2} \gg L^{1/2}$ (i.e., high gain compared to internal losses) then the output becomes

$$I_{\text{out.}}(\text{max.}) \cong T_b I_s g_0 = T_b [\gamma_0 I_s l_g] \quad (9.2.10b)$$

One should recognize the term in the brackets as the limiting additive intensity one can extract from an amplifier by stimulated emission; compare with (8.3.16). Although the simplified analysis suffers from the approximations, it is reasonably accurate and is usually used for the first guess—even for a standing wave laser covered in Sec. 9.2.3.

2. High gain and high loss. Some lasers have very high gains and high internal losses; a semiconductor laser is a prime example. Although such lasers are seldom made in the form of a ring geometry analyzed here, it is appropriate to consider that limit because the conclusions are valid even for a simple laser with two mirrors.

For $\alpha l_g > 2 \sim 3$, we can neglect $\exp(-\alpha_T l_g)$ in the denominator of (9.2.6) and again break up the losses into internal plus coupling terms. However, we cannot use the Taylor series expansion for $\ln[1/(1 - T_2)]$ of (9.2.7c). Equation (9.2.6) becomes (after setting $T_b = 1$ to avoid clutter):

$$I_{\text{out.}} = I_s T_2 \left[(\gamma_0 l_g - \alpha_{\text{int.}} l_g) - \ln\left(\frac{1}{1 - T_2}\right) \right] \quad (9.2.11)$$

After differentiating to find an optimum, we obtain a transcendental equation for T_2:

$$\frac{T_2}{1 - T_2} + \ln\frac{1}{1 - T_2} = (\gamma_0 l_g - \alpha_{\text{int.}} l_g) = \text{net integrated gain} \quad (9.2.12)$$

A solution for this coupling is shown in Fig. 9.3. Note that the optimum transmission is quite high for a system with high integrated gain.

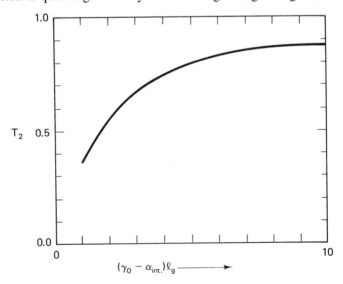

Figure 9.3 Optimum coupling for a high gain–high loss laser.

Sec. 9.2 Output Power (Energy)–The Coupling Problem

A semiconductor laser has such high integrated gain for injected currents above threshold. Consequently, many are made with an antireflection coating on the output facet. Of course, one could not use zero reflection for then there would be no feedback. (More to the point is the fact that it is extremely hard to obtain perfect antireflection coatings for the modes generated by a semiconductor laser.)

9.2.3 Standing Wave Lasers

Most lasers are of the simple variety with just two mirrors for feedback in the manner shown in Fig. 9.4(a), with Fig. 9.4(b) showing the forward (I^+) and reverse (I^-) waves inside the cavity, and Fig. 9.4(c) illustrating a zeroth-order guess as to the spatial variation of those intensities.

One starts a wave $I^+(z = M_1)$, just to the right of M_1 that propagates to the

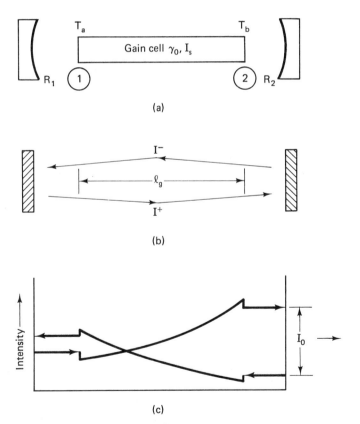

Figure 9.4 A simple standing wave laser. (a) The cavity showing the mirror reflectivities and the transmission coefficients between the mirrors and the gain cell. (b) The counterpropagating intensities. (c) The spatial variation of the waves inside the cavity.

entrance of the gain cell where its amplitude drops because of the imperfect transmission of the gap (windows, scattering, or absorption in that space). Once the forward-going wave is inside the gain cell, it is amplified because of the population inversion until that wave exits at the other end with a corresponding decrease at that window. The wave propagates with a transmission coefficient T_b to the mirror M_2 where only a fraction $R_2 I^+(z = M_2)$ starts the return path and is now identified as I^-. The wave propagating in the negative direction undergoes a similar set of attenuations and amplifications as did I^+ until it is reflected by M_1 and that amplitude must *reproduce* the assumed initial value of I^+. The object of the following is to translate this word story into a simplified mathematical model.

Within the gain medium, the rate of change of the forward and reverse intensities with z is given by

$$\frac{1}{I^+}\frac{dI^+}{dz} = \frac{\gamma_0}{1 + (I^+ + I^-)/I_s} = -\frac{1}{I^-}\frac{dI^-}{dz} \quad (9.2.13)$$

where the minus sign on the right-hand side of (9.2.13) reflects the fact that I^- increases as z decreases. Much to our dismay, we see that the amplification of the forward wave I^+ depends on both I^+ and I^-. Now each atom has *one* package of energy to give; it can give it to the forward wave or the reverse but not to both. This unpleasant mathematical complication has to be faced and makes the exact analysis of this simple laser much more complicated than that of the traveling wave ring laser of the previous section. First it is appropriate to search for an approximate but revealing approach.

9.2.3(a) The high Q approximate solution.

A key simplification results from assuming a high Q passive cavity, because then the losses are small, and as a result the variation of the intensities within the cavity in order to have a self-consistent solution is also small. Thus it is reasonable to neglect the variation of the intensities with z and set the saturated gain coefficient equal to the losses, which are to be prorated on a per unit of length of the active medium (even though most of those losses appear elsewhere inside the cavity).

$$\frac{dI^+}{dz} = \frac{\gamma_0 I^+}{1 + (I^+ + I^-)/I_s} - (\alpha_{\text{int.}} + \alpha_{\text{ext.}})I^+ \simeq 0 \quad (9.2.14)$$

There are several points to be noted about this equation. As noted previously, we have assumed oscillation close to line center so that the frequency dependence of the saturation and the gain coefficient can be ignored $[\bar{g}(\nu) = 1]$. Second, both the forward and reverse waves saturate the gain of the forward one; similar considerations apply to the variation of the reverse wave.* Third, the

---

*Actually there is interference between the fields associated with I^+ and I^-, which leads to a spatial variation of the saturation. However, this is a complication best left for more advanced treatments.

Sec. 9.2 Output Power (Energy)–The Coupling Problem

total loss coefficient is composed of two parts: (1) that which is internal to the laser medium itself or due to internal parts of the laser cavity such as imperfect window transmission, scattering from misaligned or rough surfaces, or anything that represents a useless conversion of photons to heat and (2) the external losses that represent the coupling to the outside world by the intentional choice of a lower reflectivity for mirror M_2; see (9.2.7c).

In the spirit of this high Q approximation, we can assume that the forward and reverse waves are equal in amplitude, and thus the saturating term in the denominator of (9.2.14) is just twice one of them. Solving that equation for the mean circulating intensity yields

$$I = \frac{I_s}{2}\left[\frac{\gamma_0 \cdot 2l_g}{\alpha_T \cdot 2l_g} - 1\right] \tag{9.2.15a}$$

and the output through M_2 is

$$I_{out.} = \frac{T_b I_s}{2}\left[T_2\left(\frac{\gamma_0 \cdot 2l_g}{\alpha_T \cdot 2l_g} - 1\right)\right] = \frac{T_b I_s}{2}\left[T_2\left(\frac{g_0}{L+T} - 1\right)\right] \tag{9.2.15b}$$

Equation (9.2.15b) is identical to (9.2.8) for the ring laser, aside from the extra factor of 2. Even those differences can be explained quite easily:

> The internal loss terms express the integrated gain or loss over a round-trip $2l_g$. The external factor of 2 appears because stimulated emission goes into two directions for the simple lasers but only into one for the ring laser. This points out one of the advantages of the unidirectional geometry.

All of the conclusion from Sec. 9.2.2 regarding optimum coupling apply here and thus will not be repeated.

9.2.3(b) The "exact" analysis (after Rigrod[8]).

If the output mirror has a relatively high transmission coefficient, the approximation of setting the derivative $\partial/\partial z = 0$ is not valid, and we must face up to the fact that the gain coefficient is saturated by both waves and suffer through the mathematical complications. We return to (9.2.13) and normalize the intensities to the saturation value and use a set of mnemonic symbols—$\mathbf{f}$(forward) $= I^+/I_s$ and $\mathbf{r}$(reverse) $= I^-/I_s$. Thus (9.2.13) becomes

$$\frac{1}{\mathbf{f}}\frac{d\mathbf{f}}{dz} = \frac{\gamma_0}{1+\mathbf{f}+\mathbf{r}} = -\frac{1}{\mathbf{r}}\frac{d\mathbf{r}}{dz} \tag{9.2.16}$$

where the minus sign acknowledges the fact that the reverse wave is amplified as it propagates in the negative z direction. Combing the first and second equalities yields a very important relationship between $\mathbf{f}$ and $\mathbf{r}$.

$$\frac{1}{\mathbf{f}}\frac{d\mathbf{f}}{dz} = -\frac{1}{\mathbf{r}}\frac{d\mathbf{r}}{dz}; \quad \frac{d}{dz}[ln(\mathbf{f} \cdot \mathbf{r})] = 0$$

or (9.2.17)

$$\mathbf{f} \cdot \mathbf{r} = k^2 \quad \text{(a constant independent of } z\text{)}$$

Thus even though both waves are functions of z, their product is an (unknown) constant to be evaluated from the boundary conditions at the mirrors. From Fig. 9.4(c), we note that

$$\mathbf{f}(1) = R'_1 \mathbf{r}(1) \quad (9.2.18a)$$

$$\mathbf{r}(2) = R'_2 \mathbf{f}(2) \quad (9.2.18b)$$

where the number in parentheses refers to the planes of Fig. 9.4 for which the boundary conditions apply, and the primes on the reflectivities are used to indicate the *fraction* of the intensity returning from each mirror. For instance, if the one-way transmission coefficient between M_1 and the gain cell were T_a then the quantity R'_1 would be $T_a^2 R_1$. These boundary conditions can be combined to evaluate the constant k^2:

$$k^2 = \mathbf{f}(1)\mathbf{r}(1) = \mathbf{r}(2)\mathbf{f}(2) \quad (9.2.19a)$$

$$= \frac{\mathbf{f}^2(1)}{R'_1} = R'_2 \mathbf{f}^2(2) \quad (9.2.19b)$$

$$= R'_1 \mathbf{r}^2(1) = \frac{\mathbf{r}^2(2)}{R'_2} \quad (9.2.19c)$$

The combination of (9.2.19c) and (9.2.19b) yields the noteworthy fact that

$$\frac{\mathbf{f}(2)}{\mathbf{f}(1)} = \frac{1}{\sqrt{R'_1 R'_2}} = \frac{\mathbf{r}(1)}{\mathbf{r}(2)} \quad (9.2.20)$$

Equation (9.2.20) merely states that the net one-way amplification factor of each of the waves, $\mathbf{f}(2)/\mathbf{f}(1)$ or $\mathbf{r}(1)/\mathbf{r}(2)$, must be equal to the mean fraction of the photons returning from the mirrors.

To integrate (9.2.16) in terms of simple and uncomplicated functions, the loss coefficient internal to the laser medium, α_0, is set equal to zero. Substitute $\mathbf{r} = k^2/\mathbf{f}$ for the left-hand side of (9.2.16) (or $\mathbf{f} = k^2/\mathbf{r}$ on the right) and integrate between the planes (1) and (2) to obtain

$$ln\left[\frac{\mathbf{f}(2)}{\mathbf{f}(1)}\right] + [\mathbf{f}(2) - \mathbf{f}(1)] + k^2\left[\frac{1}{\mathbf{f}(1)} - \frac{1}{\mathbf{f}(2)}\right] = \gamma_0 l_g \quad (9.2.21a)$$

$$ln\left[\frac{\mathbf{r}(2)}{\mathbf{r}(1)}\right] + [\mathbf{r}(2) - \mathbf{r}(1)] + k^2\left[\frac{1}{\mathbf{r}(1)} - \frac{1}{\mathbf{r}(2)}\right] = -\gamma_0 l_g \quad (9.2.21b)$$

Now we are primarily interested in $\mathbf{f}(2)$, the forward wave I^+/I_s, so (9.2.19b) is used to eliminate k^2 and (9.2.20) is used to express $\mathbf{f}(1)$ in terms of $\mathbf{f}(2)$. The solution for $\mathbf{f}(2)$ is

Sec. 9.2 Output Power (Energy)–The Coupling Problem

$$\frac{I^+}{I_s} = \mathbf{f}(2) = \frac{\gamma_0 l_g - \frac{1}{2}\ln\frac{1}{R_1'R_2'}}{(1 - \sqrt{R_1'R_2'})(1 + \sqrt{R_2'/R_1'})} \qquad (9.2.22)$$

Now it is just a matter of propagating this wave through the gap between the active medium and M_2 (call it a transmission coefficient T_b) where only a fraction $1 - R_2$ is available for the output. Eliminating the primes on the effective reflectivities yields an expression for the output intensity in terms of the gain coefficient, saturation intensity, and cavity parameters.

$$\frac{I_{\text{out.}}}{I_s} = \frac{T_b T_2 \left[\gamma_0 l_g - \frac{1}{2}\ln\left(\frac{1}{T_a^2 T_b^2 R_1 R_2}\right)\right]}{(1 - \sqrt{T_a^2 T_b^2 R_1 R_2})(1 + \sqrt{T_b^2 R_2/T_a^2 R_1})} \qquad (9.2.23)$$

It is not a pleasant task to differentiate this equation to find the coupling that yields maximum output power; rather, the result is shown graphically in Fig. 9.5 and compared to the high Q approximate analysis given earlier in this section. It should be clear from this figure that the approximate analysis is fairly accurate overall but especially good at predicting the optimum coupling.

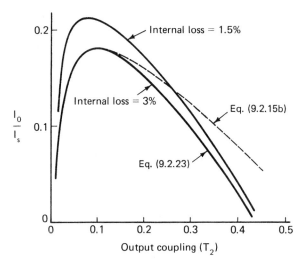

Figure 9.5 Output power as a function of coupling.

Example

Chose a simple laser with

$$T_a = 0.995; \quad R_1 = 0.99$$

$$T_b = 0.995; \quad R_2 \text{ variable from 0.99 to 0.5}; \quad T_2 = 1 - R_2$$

$$\gamma_0 l_g = 0.30 \text{ (i.e., the integrated gain is equal to 30\% per pass)}$$

The total losses are:

$$\alpha_T 2l_g = \ln\left[\frac{1}{T_a^2 T_b^2 R_1}\right] + \ln\frac{1}{R_2}$$

Thus the parameters for use in (9.2.15b) are

$$L = \ln\frac{1}{T_a^2 T_b^2 R_1}; \quad T = \ln\frac{1}{R_2} \sim T_2$$

Also demonstrated in Fig. 9.5 is the sensitivity of the power to the internal nonsaturable losses inside the laser cavity. Even though most of the total loss per round-trip is due to the coupling through M_2 (i.e., T_2), the change of 1.5% in the internal losses makes an appreciable change in the maximum. Although the example considers the effect of window losses or mirror imperfections, it should be clear that any nonsaturable loss placed anywhere would have a similar detrimental effect.

9.2.4 Extraction Efficiency of a Pulse Excited Amplifier

The previous parts of this section have been concerned with a laser operating in a CW mode, and thus the attention was focused on the performance of the oscillator. However, many times the output power of the oscillator is not of primary concern, but rather how well it can be controlled in terms of its beam quality, frequency, and amplitude stability. To obtain these latter goals, a master-oscillator-power-amplifier (MOPA) scheme is used to achieve an overall system optimization. In such circumstances, the efficiency of the amplifier is far more important than that of the oscillator, and, if the application involves extreme peak powers, such amplifiers are usually pulse excited. It is this case that is to be addressed here.

As is apparent from Sec. 8.3, the ideal optical amplifier would have excitation to the upper state only, so that the rate equations 8.3.3a and 8.3.3b become

$$\frac{dN_2}{dt} = R_2(t) - \frac{\sigma I(t)}{h\nu}\left[N_2 - \frac{g_2}{g_1}N_1\right] - \frac{N_2}{\tau_{21}} - \frac{N_2}{\tau_{20}} \quad (9.2.24a)$$

$$\frac{dN_1}{dt} = (0) + \frac{\sigma I(t)}{h\nu}\left[N_2 - \frac{g_2}{g_1}N_1\right] + \frac{N_2}{\tau_{21}} - \frac{N_1}{\tau_1} \quad (9.2.24b)$$

where the (0) in (9.2.24b) indicates our optimistic view of an "ideal" excitation scheme. If we further hope for an excitation rate and extraction by stimulated emission which is fast compared to the normal decay rates of states 2 or 1, then we can simplify even further:

$$\frac{dN_2}{dt} = R_2(t) - \sigma\Gamma_p(t)[N_2 - pN_1] \quad (9.2.24c)$$

$$\frac{dN_1}{dt} = +\sigma\Gamma_p(t)[N_2 - pN_1] \quad (9.2.24d)$$

Sec. 9.2 Output Power (Energy)–The Coupling Problem

where
$$\Gamma_p = I/h\nu = \text{photon flux}$$
and
$$p = g_2/g_1$$

If we add (9.2.24c) and (9.2.24d) and integrate between 0 and t, we obtain

$$N_2(t) + N_1(t) = N_0(t) = \int_0^t R_2(t)\,dt \tag{9.2.24e}$$

Equation (9.2.24e) states that the total number density of atoms promoted to states 2 and 1 must be equal to the integral of the pumping rate $R_2(t)$, which we now assume to be fast compared to the extraction of the energy by stimulated emission. This is equivalent to assuming that the excitation is over by the time the photon flux interrogates the medium. Thus the density N_0 expresses the number of atoms promoted to the upper state by the pump, and it also represents the *sum* of both populations.

Our mathematical goal is to predict the extraction efficiency defined by

$$\eta_{xtn} = \frac{\int_0^t \frac{d\Gamma_p}{dz}\,dt}{\int_0^t R_2(t)\,dt} \tag{9.2.25a}$$

and:

$$\frac{d\Gamma_p}{dz} = \sigma(N_2 - pN_1)\Gamma_p - \alpha_0\Gamma_p \tag{9.2.25b}$$

where $d\Gamma_p/dz$ is the usual amplification of the photon flux by the population inversion, and thus we represent its use by stimulated emission. The quantity α_0 represents losses *internal* to the laser medium.

All variables in (9.2.25) are unknown functions of time, and thus, at first glance, the prospects of a general conclusion might appear to be dim. However, by combining (9.2.24e) with (9.2.24c), one can cast the differential equation for $N_2(t)$ into a form that can be integrated implicitly.

$$\frac{dN_2}{dt} = -\sigma\Gamma_p(t)\{N_2(t) - p[N_0 - N_2(t)]\}$$

or

$$\frac{dN_2}{dt} + (1+p)\sigma\Gamma_p(t)N_2 = p\sigma\Gamma_p(t)N_0 \tag{9.2.26}$$

Multiply both sides by an integrating factor:

$$\exp\left[(1+p)\sigma\int_0^t \Gamma_p(t')\,dt'\right]$$

which enables the left-hand side of (9.2.26) to be expressed as a perfect differential.

$$\frac{d}{dt}\left\{N_2(t)\exp\left[(1+p)\sigma\int_0^t \Gamma_p(t')\,dt'\right]\right\}$$

$$= \left\{\exp\left[(1+p)\sigma\int_0^t \Gamma_p(t')\,dt'\right]\right\}\{p\sigma\Gamma_p(t)N_0\} \quad (9.2.27a)$$

Aside from a trivial integrating factor of $1/(1+p)$ the right-hand side is also a perfect differential, thanks to the assumption of N_0 being independent of time. After including the arbitrary constant of integration, we obtain

$$N_2(t)e^{(1+p)\sigma\int_0^t \Gamma_p(t')\,dt'} = \frac{pN_0}{1+p}e^{(1+p)\sigma\int_0^t \Gamma_p(t')\,dt'} + K \quad (9.2.27b)$$

At $t = 0$, the start of the photon flux, $N_2 = N_0$ by assumption, the integrals are also zero, and thus the constant of integration can be evaluated to be $K = N_0/(1+p)$. Hence the implicit solution for $N_2(t)$, $N_1(t)$, and the net population inversion becomes

$$N_2(t) = \frac{N_0}{1+p}[p + e^{-(1+p)\sigma\theta_p(t)}] \quad (9.2.28a)$$

$$N_1(t) = \frac{N_0}{1+p}[1 - e^{-(1+p)\sigma\theta_p(t)}] \quad (9.2.28b)$$

$$N_2(t) - pN_1(t) = N_0 e^{-(1+p)\sigma\theta_p(t)} \quad (9.2.28c)$$

where

$$\theta_p(t) = \int_0^t \Gamma_p(t')\,dt'$$

$$= \text{the photon fluence (i.e., per unit area)} \quad (9.2.28d)$$

Amazingly, one can predict the time dependence of the population inversion for any pulse shape whatsoever provided it is fast enough.

Now we turn to the extraction efficiency question (9.2.25a), whose denominator is merely N_0 by (9.2.24e). The numerator is considerably more involved because (9.2.25b) and (9.2.25c) must be combined.

$$\eta_{xtn} = \frac{1}{N_0}\int_0^t \{\sigma\Gamma_p(t')[N_0 e^{-(1+p)\sigma\int_0^{t'} \Gamma_p(t'')\,dt''}] - \alpha_0\Gamma_p(t')\}\,dt'$$

$$= \frac{1}{1+p}e^{-(1+p)\sigma\theta_p}\Big|_0^{\theta_p} - \alpha_0\theta_p\Big|_0^{\theta_p} \quad (9.2.29a)$$

or

$$\eta_{xtn} = \frac{1}{1+p}\left[(1 - e^{-(1+p)\sigma\theta_p}) - \frac{(1+p)\alpha_0\sigma}{N_0\sigma}\theta_p\right] \quad (9.2.29b)$$

Now $N_0\sigma = \gamma_0$ is the initial small-signal gain coefficient of the inverted population, α_0 is the inherent loss coefficient of the active medium (i.e., due to scattering, impurity atoms, etc.), and hence $g_s = \gamma_0 - \alpha_0$ is the *net* small-signal gain coefficient of the amplifier. Let us define m as the ratio γ_0/α_0 and rewrite (9.2.29b):

$$\eta_{xtn} = \frac{1}{1+p}\left[1 - e^{-(1+p)\sigma\theta_p} - \frac{(1+p)\sigma\theta_p}{m}\right] \quad (9.2.29c)$$

There is an optimum fluence (i.e., number of photons/area) which maximizes (9.2.29c) given by

$$\theta_p(\text{optimum}) = \frac{\ln m}{(1+p)\sigma} \quad (9.2.30)$$

and the maximum extraction efficiency becomes

$$\eta_{xtn} = \frac{1}{1+p}\left[1 - \frac{(1+\ln m)}{m}\right] \quad (9.2.31)$$

This equation points out some of the inherent difficulties in building efficient pulsed amplifiers and the deleterious effect of internal residual losses, α_0. For instance, let $p = g_2/g_1 = 1$, which is typical of transitions between highly degenerate states, and let $m = \gamma_0/\alpha_0 = 5$; then the maximum fraction of energy used to form the upper state is the product of the quantum efficiency (9.1.1) times (9.2.31) evaluated under this condition. For these numbers $\eta_{xtn} = 0.239$ and if the quantum efficiency is only 50%, then the overall efficiency can be no more than $\sim 12\%$. NOTE: We have yet to address the pumping efficiency question—that is the topic of Chapters 10 to 12—but one can anticipate a significant penalty there also.

9.3 TRANSIENT EFFECTS

Until now we have implicitly assumed a steady-state situation with the intensity of the laser building up from spontaneous emission until the intensity reaches a value where the stimulated emission rate saturates the gain at the loss. Our main emphasis has been on the final equilibrium situation; we have ignored the transient effects in the approach to this final value. This section remedies this deficiency by analyzing practical cases that have dramatic consequences.

9.3.1 Q Switching, Q Spoiling, or Giant Pulse Behavior

The idea of Q switching is a takeoff of some well-known techniques used in low-frequency electronics for the generation of and switching of high power. For instance, consider a radar set operating at 10 to 50 MW of peak radiated power. It is a nontrivial job of modulator design to devise a switch to handle this power level, and it is mind-boggling to consider a "portable" power supply capable of delivering this amount of continuous power.

However, the energy contained in a typical radar pulse—say of duration 0.1 μs—is quite ordinary: $W = (10 \text{ to } 50 \times 10^6) \text{ W} \times 0.1 \times 10^{-6} \text{ sec} = 1$ to 5 joules. The fact that the energy is so minuscule provides a clue to techniques used for its generation. Fig. 9.6 illustrates a simple technique for generating high peak pulsed power.

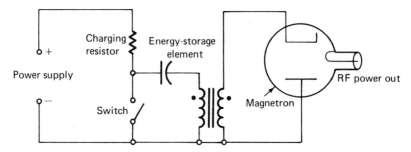

Figure 9.6 Elementary radar transmitter. (Replace the magnetron by a spark plug and it is part of an electronic ignition system.)

1. One extracts energy from the primary source, the power supply, at a reasonable rate limited by the charging resistor to be P_{max} (power supply) $= V^2/R$—call this the *pumping cycle*.
2. This energy is stored by the capacitor, which is prevented from discharging by the open circuit of the switch.
3. Once the switch is closed, the only thing limiting the capacitor discharge current is the impedance of the load—call this the power cycle.

It should be clear that the peak power delivered to the load can be many times the peak instantaneous power extracted from the source.

These same ideas, utilized in the "giant pulse" operation of a laser, use the fact that energy can be stored for future use by creating a population inversion. Obviously, spontaneous emission out of the upper state represents a drain on the stored energy, much as would a leakage path on the capacitor.

However, spontaneous emission causes another difficulty—it is amplified by the population inversion, and if the round-trip gain exceeds 1, will build up to a steady-state value whose intensity is limited by the rate at which energy can be pumped into the system [see (8.3.18)]. To avoid this drain on the population

Sec. 9.3   Transient Effects   249

inversion, the laser is *prevented* from oscillating by making the loss per pass very high while pumping the system. If amplified spontaneous emission can be prevented from saturating the active medium with a single-pass gain length, then considerable energy can be stored in the population difference, $N_2 - N_1$.* This stored energy can be extracted by suddenly *lowering* the loss. Under these circumstances, the gain greatly exceeds the loss and the intensity rapidly builds up from spontaneous noise, reaching a level where further growth is impossible (i.e., when the gain per pass equals the loss per pass).

It is important to remember that *transient* phenomena are being discussed here. This intensity at which further growth is impossible represents the energy stored in the initial population inversion; it is *not* the intensity given by the CW oscillation condition. As we will see, the peak intensity can be many times the CW level.

Let us follow this sequence through with an educated guess as to the behavior before getting buried in the mathematics. As shown in Fig. 9.7, we assume a simple laser geometry with some sort of a closed shutter to spoil the cavity Q and thus prevent oscillation from taking place and using the population inversion. Under such circumstances, one can continually pump energy into the population inversion until some sort of equilibrium is reached between the pump

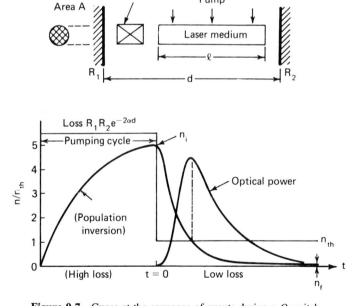

Figure 9.7 Guess at the sequence of events during a Q switch.

*One tends to discount this stored energy until numbers are quoted. For instance, one can store ~50 joules per liter-atmosphere in the population difference of the CO_2 laser transition at 10.6 μm—a figure comparable to the best capacitor available.

and the spontaneous decay processes of the system. This initial population inversion may be many times that required for CW oscillation in the absence of the shutter.

Let us identify $t = 0$ as that point when the shutter is opened. At that instant we have a system that is far above threshold, and thus the spontaneous emission along the axis of the cavity is greatly amplified and, owing to the feedback provided by the cavity, builds up from its low value to one sufficiently strong to start depleting the population inversion. A few numbers will convince one that this occurs on a very short time scale. For instance, assume that the net single-pass gain is 5 [i.e., $(R_1 R_2)^{1/2} \exp(\gamma_0 l) = 5$]. Then in five round-trips (10 single passes), the photon flux would be amplified by $5^{10} \sim 10^7$. As a consequence of this rapid time scale, we neglect any pumping that might occur after $t = 0$.

In view of this large increase in photon flux, we realize that the population inversion will become depleted as the photon number increases. Consequently, we must keep track of the number of excited states as well as the number of photons. (Such a dual bookkeeping situation can lead to a painful headache without careful attention to detail.)

Let us now consider the task of formulating those ideas into a mathematical format so as to make numerical predictions about the amplitude of the intensity produced by this Q-switching operation. The geometry for our analysis is shown in Fig. 9.7 where some of the details indicate that many of the lasers that are candidates for Q switching are solid-state ones and have an index of refraction that is significantly different from 1, and that the electro-optic shutter has still another index. The problem is exceedingly difficult to handle exactly because three spatial dimensions as well as time are involved. We can get rid of at least two of the transverse spatial ones by assigning an area A to them, but we are still left with two dimensions, z and t, and thus we make our first approximation: we treat the cavity as a circuit element and ignore any nonuniformities in the population densities or the photon density with distance inside the simple cavity shown in Fig. 9.8. Let us first find an equation for the time evolution of the total number of photons inside the cavity during the power cycle (i.e., when the switch has a high transmission).

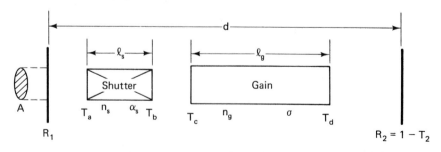

Figure 9.8 The geometry of a Q-switched laser.

Sec. 9.3 Transient Effects

Let N_p be the number of photons inside the cavity. In one round-trip, this number will increase to $\{R_1 R_2 (T_a T_b T_c T_d)^2 \cdot \exp[2\gamma l_g - 2\alpha l_s]\} N_p$. The *change* in the number of photons divided by the time for a round-trip is thus an excellent approximation for the rate of change of N_p with time.

$$\frac{dN_p}{dt} = \frac{[R_1 R_2 (T_a T_b T_c T_d)^2 e^{-2\alpha_s l_s} e^{2\gamma l_g} - 1] N_p}{\dfrac{2(l_{\text{air}} + n_s l_s + n_g l_g)}{c}}$$

or

$$\frac{dN_p}{dt} = \frac{\{\exp[2[\langle\gamma\rangle - \langle\alpha_T\rangle]d] - 1\} N_p}{\tau_{RT}} \qquad (9.3.1)$$

where

$\langle\gamma\rangle = \gamma l_g / d$

= gain coefficient, prorated over the length of the cavity (9.3.2a)

$\langle\alpha_T\rangle = \underbrace{\alpha_s l_s / d + (1/2d)\ln\{1/[R_1(T_a T_b T_c T_d)^2]\}}_{\alpha_{\text{int.}}} + \underbrace{(1/2d)\ln(1/R_2)}_{\alpha_{\text{ext.}}}$ (9.3.2b)

$\langle\alpha_T\rangle$ = loss coefficient, prorated over the length of the cavity (9.3.2c)

$\tau_{RT} = 2(n_s l_s + n_g l_g + l_{\text{air}})/c$ = round-trip time (9.3.2d)

$d = l_s + l_g + l_{\text{air}}$

If the integrated gain minus the loss is small enough, then a Taylor series expansion for the exponential term is in order: $\exp 2(\langle\gamma\rangle - \langle\alpha_T\rangle)d \approx 1 + 2(\langle\gamma\rangle - \langle\alpha_T\rangle)d$ and (9.3.1) becomes

$$\frac{dN_p}{dt} = \frac{2[\langle\gamma\rangle - \langle\alpha_T\rangle]d}{\tau_{RT}} N_p \qquad (9.3.3a)$$

Note that (9.3.3a) is in agreement with our intuition: if the gain coefficient is greater than the loss coefficient then the photon number increases with time; if the reverse is true then it decreases. Note also that this equation predicts the correct value for a steady-state (or CW) laser with the gain coefficient equaling the loss value (even though we are not considering such a case) for $\partial/\partial t = 0$.

It is convenient to manipulate the factors in (9.3.3a) to minimize the number of parameters that need to be specified for the phenomenon. Let us first factor the quantity $\langle\alpha_T\rangle$ out of that equation so that only the ratio of the gain to loss appears in the parentheses. The multiplicative factor left over can be related to the photon lifetime for the *passive cavity*:

$$\frac{1}{\tau_p} = \frac{[1 - \exp(-2\langle\alpha_T\rangle d)]}{\tau_{RT}} \approx \frac{2\langle\alpha_T\rangle d}{\tau_{RT}} \qquad (9.3.4)$$

where the Taylor series expansion for the exponential is in the same spirit as was used for the corresponding term to obtain (9.3.3a) and which can now be cast into a more transparent form:

$$\frac{dN_p}{dt} = \left\{ \frac{2\langle\alpha_T\rangle d}{T_{RT}} \left[\frac{\langle\gamma\rangle}{\langle\alpha_T\rangle} - 1 \right] \right\} N_p = \frac{1}{\tau_p} \left[\frac{\langle\gamma\rangle}{\langle\alpha_T\rangle} - 1 \right] N_p \qquad (9.3.3b)$$

The ratio $\langle\gamma\rangle/\langle\alpha_T\rangle$ is also the ratio of the total population inversion $n = [N_2 - N_1]Al_g$ to the threshold value for oscillation.

$$\frac{dN_p}{dt} = \frac{1}{\tau_p} \left[\frac{n}{n_{\text{th}}} - 1 \right] N_p \qquad (9.3.5)$$

where

$$n = [N_2 - N_1]Al_g$$
= total number for the inverted population in the cavity $\qquad (9.3.6a)$

$$\langle\alpha_T\rangle d = \sigma[N_2 - N_1]_{\text{th}} l_g \qquad (9.3.6b)$$

As the photon number builds up inside the cavity, the inverted population must decrease in a synchronous fashion as shown below. We presume that the build-up takes place on a time scale much faster than any pumping or decay process so that we need to consider only the stimulated emission and absorption terms in the rate equations for N_2 and N_1:

$$\frac{dN_2}{dt} = -\frac{\sigma I}{h\nu}[N_2 - N_1] \qquad (9.3.7a)$$

$$\frac{dN_1}{dt} = +\frac{\sigma I}{h\nu}[N_2 - N_1] \qquad (9.3.7b)$$

We subtract (9.3.7b) from (9.3.7a), multiply both sides by the volume of the active media in contact with the photons, and sprinkle a few factors of one in the form of d/d and $\langle v\rangle/\langle v\rangle$, the effective velocity of propagation within the cavity.

$$\frac{d}{dt}[(N_2 - N_1)Al_g] = -2\left[\sigma(N_2 - N_1)\frac{l_g}{d} \right]\left[\frac{IAd}{h\nu\langle v\rangle} \right]\cdot\langle v\rangle \qquad (9.3.8a)$$

where

$$\langle v\rangle = c\{[l_{\text{air}} + n_g l_g + n_s l_s]/d\}$$
= mean velocity of propagation through the cavity = d/τ_{RT}

The second factor in (9.3.8a), $\{I A d/(h\nu\langle v\rangle)\}$, is equal to the number of photons inside the cavity N_p and thus that equation can be simplified to

$$\frac{dn}{dt} = -2\langle\gamma\rangle\left(\frac{2d}{\tau_{RT}}\right)N_p \qquad (9.3.8b)$$

Our previous work with the photons, (9.3.3b), taught us that the characteristic time scale for changes in N_p was τ_p; hence, we multiply the numerator and denominator of (9.3.8a) by $\langle \alpha_T \rangle$ and identify $1/\tau_p$ with the factor $2\langle \alpha_T \rangle d/\tau_{RT}$. Equation (9.3.8b) becomes

$$\frac{dn}{dt} = -2 \frac{\langle \gamma \rangle}{\langle \alpha_T \rangle} \frac{N_p}{\tau_p} \qquad (9.3.9a)$$

This can be further simplified by recognizing that the ratio of the gain and loss coefficients is the ratio of the inversion number to the threshold value in the active medium.

$$\frac{dn}{dt} = -2 \frac{n}{n_{\text{th}}} \frac{N_p}{\tau_p} \qquad (9.3.9b)$$

This is the last basic formula to be "derived," with quotes being used to emphasize that (9.3.9b) could have been obtained by a small investment in logic and the arithmetic and gyrations between (9.3.3) and here could have been avoided. Every time a photon is added to the cavity, the inversion $N_2 - N_1$ decreases by 2 (for equal degeneracies), and thus (9.3.9b) is just two times the increasing part of (9.3.5). One last bit of manipulation is in order: to define a time scale normalized to the photon lifetime of the passive cavity by $T = t/\tau_p$. Thus our basic equations become

$$\frac{dN_p}{dT} = \left(\frac{n}{n_{\text{th}}} - 1 \right) N_p \qquad (9.3.10a)$$

$$\frac{dn}{dT} = -2 \frac{n}{n_{\text{th}}} N_p \qquad (9.3.10b)$$

Here we run into an obstacle to our progress with the mathematics. These equations are nonlinear and coupled and cannot be solved for N_p and n in terms of elementary functions of time. A computer is needed for this so the issue will be postponed. However, we can find an elementary solution for the photon number N_p in terms of the inversion n. Indeed, if one pays close attention to the physical picture presented in Fig. 9.7 and the physical "walk through" the Q-switch pulse of the prior pages, one can obtain a very good answer for the peak of the power pulse, its energy, and a reasonable estimation for its FWHM without resorting to anything more complicated than a calculator. We start by dividing (9.3.10a) by (9.3.10b), which eliminates time from the equation:

$$\frac{dN_p}{dn} = \frac{1}{2} \left(\frac{n_{\text{th}}}{n} - 1 \right) \qquad (9.3.11)$$

Now we multiply both sides by dn and integrate the left-hand side from the initial value of the photon number $N_p(i)$(which is negligible compared to what it will be) to the photon number, $N_p(\text{max.})$, at the peak of the power pulse, which is the first physical parameter to be determined. But how are we to know when we have reached the peak of the pulse? This is where the walk-through associated with Fig. 9.7 is so important: note that photon number reaches a peak when the inversion crosses the threshold value. Of course, the differential equation (9.3.10) states the same thing: $dN_p/dt = 0$ when $n = n_{\text{th}}$. Thus the limits of integration of the right-hand side are from the initial value of the inversion, n_i, to the threshold value, n_{th}.

$$\int_{N_p(i)\sim 0}^{N_p(\text{max.})} dN_p = \frac{1}{2}\int_{n_i}^{n_{\text{th}}} \left(\frac{n_{\text{th}}}{n} - 1\right) dn$$

$$N_p(\text{max.}) = \frac{n_i - n_{\text{th}}}{2} - \frac{n_{\text{th}}}{2}\ln\left(\frac{n_i}{n_{\text{th}}}\right) \qquad (9.3.12)$$

The photons are being lost to *all* of the various loss mechanisms inside the cavity, but only that part of the loss representing the coupling through M_2 represents power available to the outside. Now $h\nu \cdot N_p$ represents the optical energy stored in the cavity, and the fractional loss of that energy per round-trip as a result of the coupling divided by the time for a round-trip is equal to the power emerging from M_2.

$$P(\text{max.}) = h\nu N_p \cdot \frac{\text{fractional loss per round-trip}}{\text{time for a round-trip}}$$

$$= h\nu N_p \cdot \left[\frac{\langle\alpha_{\text{ext.}}\rangle \cdot 2d}{\tau_{RT}}\right] \qquad (9.3.13a)$$

We would like the factors $(2\langle\alpha_T\rangle d/\tau_{RT})$ to appear together so that they could be replaced by the photon lifetime relationship, $1/\tau_p$. This is accomplished by multiplying by 1 in the form of $\langle\alpha_T\rangle/\langle\alpha_T\rangle$.

$$P(\text{max.}) = \frac{\langle\alpha_{\text{ext.}}\rangle}{\langle\alpha_T\rangle} \cdot \frac{h\nu N_p(\text{max.})}{\tau_p} \qquad (9.3.13b)$$

Let us return to (9.3.11) and integrate over the complete time interval of the pulse so as to estimate the fraction of the inversion which is converted to photons. The photon number N_p starts at a very low value, which we again approximate as zero, reaches a maximum, and returns to this low value at the final stage. Thus the limits of integration on N_p are from zero to zero. However, the limits on n corresponding to those "different" values of zero are n_i and n_f where the latter one is the final inversion number.

$$\int_0^0 dN_p = \frac{1}{2}\int_{n_i}^{n_f} \left(\frac{n_{\text{th}}}{n} - 1\right) dn$$

or

$$\frac{n_i - n_f}{2} - \frac{n_{th}}{2} \ln\left(\frac{n_i}{n_{th}}\right) = 0$$

Rearranging we get

$$\frac{n_f}{n_i} = \exp\left[-\left(\frac{n_i - n_f}{n_{th}}\right)\right] \quad (9.3.14)$$

This is an equation which determines n_f implicitly in terms of the initial value n_i and the threshold value n_{th}. Although it does not have a simple analytic solution, it is fairly simple to solve numerically (or graphically) and the result is present in Fig. 9.9 in a special manner. The horizontal axis is chosen to satisfy our intuition that the development of the Q-switched pulse is strongly influenced by the amount by which the initial inversion exceeds the threshold value at the beginning. The choice of the vertical or y axis is motivated by the practical question: how much of the initial inversion is converted to photons? The fraction of the initial inversion converted to photons is

$$\eta_{xtn} = \frac{n_i - n_f}{n_i} \quad (9.3.15)$$

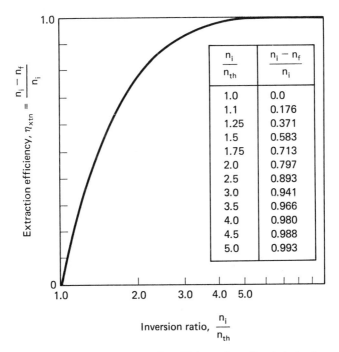

Figure 9.9 The energy extraction efficiency for a Q-switched pulse.

Knowing this fraction, the total energy generated in the form of photons is given by this efficiency times the maximum available. Let N_p be the total number of photons that can be created at the expense of the population inversion.

$$n_i = [N_{2i} - N_{1i}] A l_g \qquad (9.3.16a)$$

$$n_f = [N_{2f} - N_{1f}] A l_g \qquad (9.3.16b)$$

but

$$[N_{2i} - N_{2f}] A l_g = N_p = [N_{1f} - N_{1i}] A l_g$$

If we subtract (9.3.16b) from (9.3.16a), solve for N_p, and multiply by $h\nu$, we obtain the total energy converted into photons:

$$W_p = h\nu \frac{n_i - n_f}{2} = \eta_{xtn} \frac{h\nu n_i}{2} \qquad (9.3.16c)$$

The fraction of this energy in the output is the product of (9.3.16c) and the ratio of the coupling loss to the total loss.

$$W_{out} = \frac{\langle \alpha_{ext.} \rangle}{\langle \alpha_T \rangle} \eta_{xtn} \frac{n_i h\nu}{2} \qquad (9.3.16d)$$

The one remaining issue to be addressed is to estimate the pulse width. For an exact answer, one needs to integrate (9.3.10a) and (9.3.10b) with respect to time, but a very reasonable estimate can be obtained by dividing the energy, (9.3.16b), by the maximum power given by (9.3.13b):

$$P_o(\max.) \Delta t \simeq W_{out} \qquad (9.3.17)$$

A Numerical Example

Let us apply these ideas to the laser system shown below where we presume a ruby laser ($\lambda = 6943$ Å) to be pumped to four times threshold. For simplicity, we assume equal degeneracy of the lasing states even though that is not true for ruby. (See Chapter 10 for a discussion about the details of the ruby laser.) We also assume a residual attenuation of 0.1 cm^{-1} by the switch even when it is in its high transmission state.

We start by computing the threshold gain coefficient and from that quantity the threshold inversion density or number of inverted atoms. We start with (0.1) of this book; that is, round-trip gain must be greater than 1 for an oscillator and equal 1 for threshold.

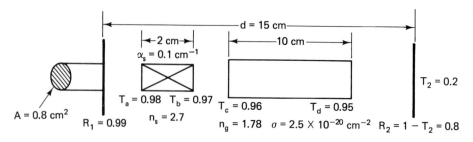

Sec. 9.3 Transient Effects

$$R_1 R_2 (T_a T_b T_c T_d)^2 e^{-2\alpha_s l_s} e^{2\gamma_{th} l_g} = 1$$

or

$$\gamma_{th} = \underbrace{\frac{1}{2l_g} \ln\left[\frac{1}{R_1(T_a T_b T_c T_d)^2}\right]}_{\text{internal cavity losses}} + \underbrace{\alpha_s \frac{l_g}{l_g}}_{\text{switch loss}} + \underbrace{\frac{1}{2l_g} \ln \frac{1}{R_2}}_{\text{external coupling loss}}$$

$$\gamma_{th} = 1.48^{-2} \text{ cm}^{-1} + 2^{-2} \text{ cm}^{-1} + 1.12^{-2} \text{ cm}^{-1}$$
$$= 4.59^{-2} \text{ cm}^{-1}$$

Thus the threshold inversion is found to be

$$(N_2 - N_1)_{th} = \frac{\gamma_{th}}{\sigma} = 1.83 \times 10^{18} \text{ cm}^{-3}$$

The total number of inverted atoms in the cavity at threshold is

$$n_{th} = (N_2 - N_1)_{th} \cdot A \cdot l_g = 1.47 \times 10^{19} \text{ atoms}$$

and because of the specification of being pumped to four times threshold, we also know the initial inversion.

$$n_i = 4 n_{th} = 5.87 \times 10^{19} \text{ atoms}$$

The other parameter of interest is the photon lifetime of the passive cavity:

$$\tau_{RT} = \frac{2(l_{air} + n_s l_s + n_g l_g)}{c} = 1.75 \text{ ns}$$

$$\frac{1}{\tau_p} = \frac{1 - R_1 R_2 (T_a T_b T_c T_d)^2 e^{-2\alpha_s l_s}}{\tau_{RT}}; \quad \tau_p = 2.91 \text{ ns}$$

Thus we first compute the maximum photon number from (9.3.12):

$$N_p(\text{max.}) = \frac{n_i - n_{th}}{2} - \frac{n_{th}}{2} \ln\left(\frac{n_i}{n_{th}}\right) = 1.18 \times 10^{19} \text{ photons}$$

Then we compute the output power at the maximum of the pulse by (9.3.13b):

$$P_0(\text{max.}) = \left(\frac{\alpha_{\text{ext.}}}{\alpha_T}\right) \cdot h\nu \cdot \frac{N_p}{\tau_p} = 282 \text{ MW}$$

which is a rather significant power. The output energy in the Q-switched pulse can be found from (9.3.16d) and from Fig. 9.8, which shows that $\eta_{xtn} = 0.98$:

$$W_{\text{out}} = \left(\frac{\alpha_{\text{ext.}}}{\alpha_T}\right) \cdot \eta_{xtn} \frac{n_i h\nu}{2} = 1.98 \text{ joules}$$

Now we can estimate the pulse width by (9.3.17):

$$\Delta t \approx \frac{W_{\text{out}}}{P_0(\text{max.})} = 7.1 \text{ ns}$$

Fig. 9.10 shows the numerical integration of the basic equations and for $n_i/n_{th} = 4$, the

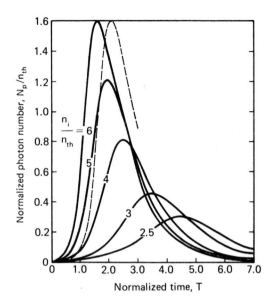

Figure 9.10 Time evolution of Q-switched pulse for different inversions. The initial photon number was taken to be 10^{-3} times the inversion ratio for solid curves, whereas a factor of 10 less was used for the dashed curve.

normalized time for the FWHM of the pulse is $\sim 3.8 - 1.6 = 2.2$ units of photon lifetime or 6.4 ns.

Note that it was not necessary to resort to the definition of the mean gain and loss coefficients to solve this example. Those were defined to simplify the theoretical development and avoid endless multiplicative factors of l_g/d but are not essential parameters. Note also that the exact expression for the photon lifetime was used rather than the approximate expression given by (9.3.4). The slight inconsistency is well within the inaccuracy of treating the laser as a circuit element.

The basic equations of Q switching, (9.3.10a) and (9.3.10b), can be integrated by numerical techniques, given that there are a few photons within the cavity mode to start the stimulated emission process. An example is shown in Fig. 9.10. In this calculation, the initial photon number (divided by the threshold inversion) was taken to be 10^{-3} of the inversion ratio n_i/n_{th}. Although the factor 10^{-3} is arbitrary, the form follows from considerations discussed in the previous chapter (8.8.2).

Note that as the inversion ratio is increased, the time required to develop a more intense pulse decreases. This is due to two causes: (1) with a higher inversion the gain is higher and thus the rate of growth is faster; and (2) a higher

Sec. 9.3 Transient Effects

inversion ratio implies more atoms in the upper state, which will contribute more spontaneous power to start the oscillation.

The solid curves were all computed using the assumption that $N_p/n_{th} = 10^{-3} \times (n_i/n_{th})$, whereas the dashed curve used a factor of 10 less for the initial starting point. It is obvious, then, that the presence or absence of the initial photon number corresponds to a shift in the time axis. One can readily estimate this time shift by the following line of reasoning.

If the initial photon number is decreased by a factor of 10, the media must use a time interval ΔT to amplify the smaller number back to its original value. At these very small photon numbers, we can neglect depletion of the inversion and thus N_p grows according to

$$N_p(\Delta T) = N_p(0) \exp\left[+\left(\frac{n_i}{n_{th}} - 1\right)\Delta T\right] \quad (9.3.18)$$

Thus, for $n_i/n_{th} = 6$, the normalized time interval ΔT to compensate for the arbitrary reduction of a factor of 10 in the initial photon number is 0.461—precisely the shift shown in Fig. 9.10.

One cannot increase the initial inversion too high and obtain arbitrary high peak powers. This is due to two causes:

1. As the initial inversion is increased, the spontaneous emission rate increases proportionately, which is then amplified by the remaining section of the gain medium. Since the single-pass gain is very high because of the high n_i, the amplified spontaneous emission may be sufficient to saturate the gain section, thereby limiting n_i (and also wasting energy). Indeed, it is this phenomenon that *limits* the amount of energy that can be stored in the population inversion in any laser.
2. The amplified spontaneous emission may be sufficient to damage the switch or at least change its characteristics. This last can be used to advantage when the switch is composed of other atoms that absorb at the laser frequency. Then the spontaneous emission "bleaches" the switch by saturating the absorbing cell (i.e., equalizes the population).

At the very least, the switch must have enough loss or hold-off capability so that the round-trip gain is less than 1. For a given switch this limits the initial inversion.

A variety of methods have been used for Q switching, ranging from the brute force technique of spinning one of the mirrors, to sophisticated means using electro-optics switches. One can also use a bleachable absorber that presents a very high attenuation to small intensities but much less as the intensity grows. The physics of this type of absorber will be covered in Sec. 9.3.3.

9.3.2 Gain Switching/Relaxation Oscillations

Another transient case of interest occurs when the population inversion from pumping builds up faster than the photon density in the cavity. This happens in a laser system with a very low probability for spontaneous emission, and thus there are few photons available to start the oscillation process.

For instance, consider the very practical (and spectacular) pulsed TEA* CO_2 laser. A short electrical pulse, on the order of 50 to 100 nsec, is used to excite the gas mixture of helium, nitrogen, and CO_2. As we shall see in Chapter 10, much of the population inversion in this laser is a result of the efficient energy transfer between nitrogen atoms (in the first vibrational level) and the CO_2 molecules. Consequently, the upper state can be created at a rate comparable to the N_2/CO_2 gas kinetic collision rate, which, at the high pressures involved, is on the order of 10^8 to 10^9 sec^{-1}.

In contrast, the spontaneous emission rate from the upper state of CO_2 is very slow, being on the order of 0.3 sec^{-1}, a rate entirely ignorable as compared to other relaxation processes deactivating the upper laser level. (Even these other processes are slow compared to the energy transfer rate between N_2 and CO_2.) As a consequence of this low spontaneous rate, it takes many, many round-trips before the photon density can build up to a level sufficient to affect the population inversion in a significant way.

This "lag" between the build-up in photons and the inversion can also be seen in semiconductor lasers when the injected current is changed (or modulated). (See Chapter 11 for details.)

Thus gain switching is the same as Q switching, with the added complication of following the time dependence of the pumping process. Inasmuch as we avoided the time-dependent problem in the previous section (with a mathematical trick), let us now face up to the issue here.

The laser cavity to be considered has the configuration shown in Fig. 9.11. As will be seen, we will be forced to a numerical solution of the differential equations, but we will first attempt to obtain approximate analytic solutions.

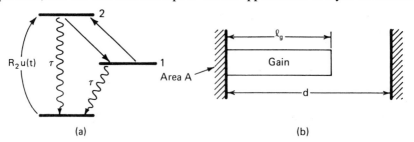

Figure 9.11 Simple example of a gain-switched laser.

---

*The acronym TEA stands for transverse excitation of atomspheric mixture. It was used first in conjunction with the CO_2 system but has been successfully applied to other lasers also.

Sec. 9.3 Transient Effects

Let us assume that the pump is suddenly applied at $t = 0$ and held constant thereafter. To make the arithmetic simple, the model assumes that both states, 2 and 1, decay back to the ground state with the same time constant τ and that only state 2 is pumped. Thus the population densities obey the following differential equations:

$$\frac{dN_2}{dt} = R_2 u(t) - \frac{N_2}{\tau} - (N_2 - N_1)\frac{\sigma I}{h\nu} \qquad (9.3.19a)$$

$$\frac{dN_1}{dt} = -\frac{N_1}{\tau} + (N_2 - N_1)\frac{\sigma I}{h\nu} \qquad (9.3.19b)$$

where $u(t)$ = step function (=0 for $t < 0$; =1 for $t > 0$).

We manipulate these equations in a manner similar to that used for Q switching; compare with (9.3.7) to (9.3.9). Subtract (9.3.19b) from (9.3.19a), multiply by the volume of the active medium, $l_g A$, and recognize that $(IAd/c'h\nu)$ is the number of photons in the cavity, N_p, and that $(N_2 - N_1) l_g A$ represents the inverted population n.

$$\frac{dn}{dt} = (R_2 l_g A) u(t) - \frac{n}{\tau} - 2c' \frac{\gamma l_g}{d} N_p \qquad (9.3.20a)$$

with $\gamma = (N_2 - N_1)\sigma$.

The number of photons within the cavity is treated in exactly the same manner as for Q switching (and has the same approximations and limitations).

$$\frac{dN_p}{dt} = \left(\frac{n}{n_{th}} - 1\right) N_p \qquad (9.3.10) \longrightarrow (9.3.20b)$$

Time is again normalized to that of the photon lifetime of the passive cavity, the inversion density by the threshold value, and the photon number by the CW or steady-state value (to be determined).

$$\frac{d}{dT}\left(\frac{N_p}{N_{p0}}\right) = \left(\frac{n}{n_{th}} - 1\right)\left(\frac{N_p}{N_{p0}}\right) \qquad (9.3.20c)$$

$$\frac{d}{dT}\left(\frac{n}{n_{th}}\right) = \left(\frac{R_2 l A \tau_p}{n_{th}}\right) u(t) - \left(\frac{n}{n_{th}}\right)\frac{\tau_p}{\tau} - 2\left(\frac{n}{n_{th}}\right)\left(\frac{N_{p0}}{n_{th}}\right)\left(\frac{N_p}{N_{p0}}\right) \qquad (9.3.20d)$$

The last two equations for N_p/N_{p0} and n/n_{th} can only be solved by numerical means, but one can obtain limiting forms of the solutions without rushing to the computer.

First, note that if stimulated emission is prevented, say by destroying the Q of the cavity, then the inversion can be found by analytic techniques.

$$n = R_2 \tau \left[1 - \exp\left(-\frac{\tau_p}{\tau} T\right)\right] \qquad (9.3.21a)$$

Now it does not make sense to consider pumping rates, $R_2 A l_g \tau$, less than that required to reach a threshold inversion n_{th}, and therefore there is a minimum pumping rate to be tolerated, which is found by setting $n = n_{th}$ and $N_p = 0$.

$$n_{th} = R_{20} A l_g \tau \qquad (9.3.21b)$$

Second, we note that (9.3.20b) and (9.3.20c) are simple algebraic equations for a steady-state laser after all the transients have died out. For then, $n/n_{th} = 1$, $N_p = N_{p0}$, and all time derivatives are zero

$$\frac{N_{p0}}{n_{th}} = \frac{1}{2} \frac{\tau_p}{\tau} (m - 1) \qquad (9.3.21c)$$

where (9.3.21b) has been used instead of n_{th} and $m = R_2/R_{20}$ is the ratio of the actual pumping rate to threshold value.

A numerical solution to the coupled differential equations (9.3.20c) and (9.3.20d) would yield a large initial "pulse" (similar to that of the previous section) followed by a damped oscillatory approach to a steady-state CW laser.

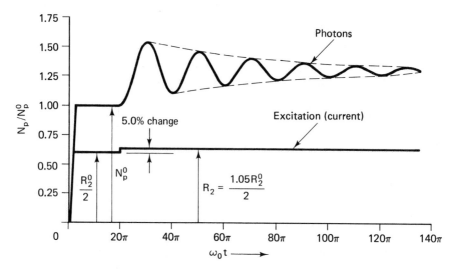

Figure 9.12 The response of a semiconductor diode to a small change in excitation according to the model of Fig. 9.11. The change in excitation is assumed to be a 5% increase in a current which is 1.2 times the threshold.

This damped oscillatory behavior is characteristic of the response of many different types of lasers—gas, solid state, or semiconductors—and is a result of the interchange of energy between the photons and the inversion driven by the pumping. This behavior can be analyzed by finding the "small" change in photon number in response to a "small" change in the pumping rate.

Sec. 9.3 Transient Effects

To be specific, consider the following sequence of events: Imagine a (semiconductor) laser being excited with a pump rate (i.e., current) which is a multiplicitive factor m times threshold (i.e., $m = R_2/R_{20}$); let the pumping rate be changed slightly—say it was increased; what is the corresponding change in the photon number N_p and the inversion population? One can guess the final answer quickly. The photon number will increase but the inversion will resettle at threshold; however, both can go through some wonderous gyrations in attaining a steady state. As the photons increase, the inversion decreases by (9.3.20d), which then causes a decrease in the photons by (9.3.20c), which in turn increases the inversion and so on. . . . It is time for the mathematics!
Let

$$\frac{R_2}{R_{20}} = 1 + \delta r(t) \qquad (9.3.22a)$$

$$\frac{N_p}{N_{p0}} = 1 + \delta p(t) \qquad (9.3.22b)$$

$$\frac{n}{n_{th}} = 1 + \delta n(t) \qquad (9.3.22c)$$

where $\delta r(t)$ is the slight change in pumping and δp and δn are the corresponding small changes in photons and inversion, respectively. We linearize (9.3.21) by using the extreme smallness of products of $\delta n \cdot \delta p$ when (9.3.22) is inserted into (9.3.20). After neglecting such product terms, removing the normalization on time using (9.3.21b) for n_{th}, and inserting $m = R_2/R_{20}$, we obtain

$$\frac{d}{dt}(\delta p) = \frac{\delta n}{\tau_p} \qquad (9.3.23a)$$

$$\frac{d}{dt}(\delta n) = \frac{m\delta r}{\tau} - m\frac{\delta n}{\tau} - (m-1)\frac{\delta p}{\tau} \qquad (9.3.23b)$$

One can combine (9.3.23a) and (9.3.23b) to obtain a single equation for the change in photon number in terms of the change in pumping rate.

$$\frac{d^2}{dt^2}(\delta p) + \frac{m}{\tau}\frac{d}{dt}(\delta p) + \frac{(m-1)}{\tau\tau_p}(\delta p) = \frac{m\delta r}{\tau\tau_p} \qquad (9.3.24)$$

Suppose $\delta r(t)$ was a small step function change, that is, $\delta r(t) = \Delta r u(t)$; what is the time evolution of the photon change δp in response to this change? After a dull, painful exercise in Laplace transforms (or classical differential equations), one obtains (approximately)

$$\delta p \simeq \left(\frac{m}{m-1}\right)\Delta r [1 - e^{-t/\tau_d}\cos \omega_0 t] \qquad (9.3.25a)$$

where

$$\frac{1}{\tau_d} = \frac{m}{2\tau} = \text{damping constant} \qquad (9.3.25b)$$

$$\omega_0 = \frac{m}{2\tau} \left[\frac{4(m-1)}{m^2} \frac{\tau}{\tau_p} - 1 \right]^{1/2} \qquad (9.3.25c)$$

A set of numerical values, typical of a GaAs laser, makes this arithmetic come alive. Assume a "long" GaAs laser with $d = 100$ μm with an index of refraction of 3.6 and facet reflectivity of $R = 0.320$. The lifetime of an injected carrier is $\sim 1 \times 10^{-9}$ sec $= \tau$, but the photon lifetime $(2nd/c)/(1 - R^2)$ is only $= 2.67 \times 10^{-12}$ sec. Thus the ratio $\tau/\tau_p = 374$. For a pumping rate (read "current") $m = 1.2$ times threshold, we obtain the following numerical values for the parameters in (9.3.25a):

damping time constant: $\tau_d = 1.67$ ns

oscillatory rate (radians/sec): $\omega_0 = 8.6 \times 10^9$ sec^{-1}

oscillatory frequency (Hz): $f_0 = 1.37$ GHz

The step response of such a laser to a $\Delta r = 0.05$ is shown in Fig. 9.12. Although one can debate whether the model of Fig. 9.10 is appropriate for a semiconductor laser, the predicted phenomenon is close to what is observed. Since there is a natural "ringing frequency," one would expect a resonance in its frequency response. This is also observed and discussed in detail in Sec. 11.7.2.

Of course, the analysis presented here is valid only for the contrived problem of this section. However, the physical basis for the pulse and/or oscillatory behavior is the same as in more complex and realistic models. The point is that on a transient basis, the population inversion can exceed the threshold value. In Q switching, one adds a controllable loss, whereas in gain switching, the inversion builds up faster than the photons, which have to start from the spontaneous emission into the cavity mode.

9.3.3 Pulse Propagation in Amplifiers

Sec. 8.3 considered the saturation of the gain of an optical amplifier with a CW input signal. Here, we wish to examine this same amplifier when driven by a short pulse at the input terminals. The situation is as depicted in Fig. 9.13. Suppose that we have created a population inversion uniformly throughout an amplifier l_g units long. Into this amplifier, we inject a pulse specified by an arbitrary function of time $f(t)$. Obviously, it takes a finite time nl_g/c for this input pulse to travel the length of the amplifier; consequently, there is an inherent delay between the two signals. Our main goal is to predict the output, $\theta(t)$, given the small-signal gain and the input. If the signal is small enough and stays small

Sec. 9.3 Transient Effects

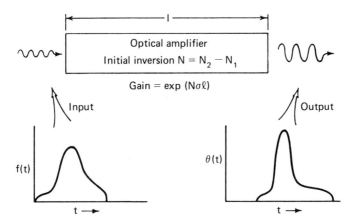

Figure 9.13 Pulse propagation in an optical amplifier.

enough so as to make a negligible perturbation on the population inversion N, then the prediction of the output is trivial. For $n = 1$,

$$\theta(t) = \exp(+N\sigma l) f\left(t - \frac{l_g}{c}\right) \tag{9.3.26}$$

But if total *energy* in this output minus the energy injected by the input pulse is comparable to the energy stored in the population inversion, we are in trouble. Let us first convince ourselves of the fact that there is a problem before attempting to solve it. (Some persons get very nervous when the solutions outnumber the problems!)

Suppose that we had an amplifier with a *small*-signal gain of 10^3 (i.e., 30 dB), at $\lambda = 1.0$ μm, a length of 100 cm, with a stimulated emission cross section of 10^{-18} cm². Then the population inversion is found to be $N_2 - N_1 = 6.908 \times 10^{18}$ cm⁻³. This represents a certain amount of stored energy per cross-sectional area of the amplifier.

$$\frac{W_{\text{stored}}}{\text{area}} = \frac{1}{2} h\nu (N_2 - N_1) l = 68.93 \text{ joules/(cm}^2 \text{ of cross section)}$$

If we injected a pulse, short enough to neglect pumping within its duration, containing 1 joule/cm² energy, then the simple equation (9.3.26) would predict that the output would contain 10^3 joules/cm², obviously incorrect since only 68.9 joules/cm² are available.

Thus the formal limitation to (9.3.26) is that the following inequality be obeyed:

$$\exp(N\sigma l) \int_0^t f(t)\, dt - \int_0^t f(t)\, dt \ll \frac{Nh\nu}{2} \tag{9.3.27}$$

But there is always a fraction of the input wave that obeys the inequality expressed by (9.3.27). Thus we can expect pulse distortion with the leading edge being amplified more than the trailing edge. At any point within the amplifier, the instantaneous amplification factor depends on past history of the wave passing that point.

If we attempt to deal with a universal time, t', with a common time origin for all points in space, we must recognize that the intensity varies in time as well as with distance:

$$\frac{\partial I}{\partial z} + \frac{1}{c}\frac{\partial I}{\partial t'} = N\sigma I \qquad (9.3.28)$$

However, if we deal with local time* and keep track of the propagation delay of each part of the input, then the two-dimensional (z, t') character of the problem disappears and we need only consider the change of I with respect to z.

$$\frac{\partial I}{\partial z} = N\sigma I \quad \text{or} \quad \frac{\partial \Gamma_p}{\partial z} = N\sigma \Gamma_p \qquad (9.3.29)$$

where $\Gamma_p = I/h\nu$, the photon flux.

The neglect of any pumping or decay process that might occur during the passage of a short pulse leads to the following equation for the population inversion:

$$\frac{\partial N}{\partial t} = -2N\sigma \frac{I}{h\nu} \quad \text{or} \quad \frac{\partial N}{\partial t} = -2N\sigma \Gamma_p \qquad (9.3.30)$$

where the factor of 2 accounts for the fact that the inversion $N_2 - N_1$ decreases by 2 for every photon added. By combining (9.3.29) and (9.3.30), we obtain

$$2\frac{\partial \Gamma_p}{\partial z} + \frac{\partial N}{\partial t} = 0 \qquad (9.3.31)$$

To solve (9.3.31), let us *define* a new function P such that[†]

$$\frac{\partial P}{\partial z} = -N\sigma \qquad (9.3.32a)$$

$$\frac{\partial P}{\partial t} = 2\sigma \Gamma_p \qquad (9.3.32b)$$

Then (9.3.31) is satisfied identically and we turn our attention to (9.3.30).

*The local time measures the time after the arrival of the initial part of the pulse.

[†] Please note that *partial* differentiation is involved. This method of solution was first published by E. O. Schultz-DuBois.[11]

Sec. 9.3   Transient Effects

$$-\frac{1}{\sigma}\frac{\partial}{\partial t}\left(\frac{\partial P}{\partial z}\right) = -2\sigma\left(-\frac{1}{\sigma}\frac{\partial P}{\partial z}\right)\left(\frac{1}{2\sigma}\frac{\partial P}{\partial t}\right)$$

or

$$\frac{\partial/\partial t(\partial P/\partial z)}{\partial P/\partial z} = -\frac{\partial P}{\partial t}$$

(9.3.33a)

Integrating with respect to time and recognizing that the "constant" of integration can be an arbitrary function of (only) z leads to

$$ln\left(\frac{\partial P}{\partial z}\right) = -P + ln\left(\frac{\partial F}{\partial z}\right) \qquad (9.3.33b)$$

where the function $ln[\partial F(z)/\partial z]$ is that arbitrary function. Rewriting makes the equation a little simpler in appearance.

$$e^P \frac{\partial P}{\partial z} = \frac{\partial F}{\partial z} \qquad (9.3.33c)$$

Now we integrate with respect to z and recognize that the integration "constant" may now be an arbitrary function of time.

$$\int e^P \frac{\partial P}{\partial z} dz = \int \frac{\partial F}{\partial z} dz + K(t)$$

$$P = ln[F(z) + K(t)] \qquad (9.3.34)$$

So much for the mathematical chicanery; let us now turn to evaluating these arbitrary functions $F(z)$ and $K(t)$. Our boundary conditions are that population inversion is a given constant for $t < 0$, and the photon flux Γ_p is specified at $z = 0$ by $\Gamma_0 f(t)$. Thus, from the definition of P, (9.3.32b), we have

$$\Gamma_0 f(t) = \frac{1}{2\sigma}\frac{\partial P}{\partial t}\bigg|_{z=0} = \frac{1}{2\sigma}\frac{\partial K/\partial t}{F(0) + K(t)}$$

which has a solution

$$ln\left[\frac{K(t) + F(0)}{K(0) + F(0)}\right] = \frac{2\sigma}{h\nu}\left[h\nu\Gamma_0 \int_0^t f(t)dt\right] \qquad (9.3.35a)$$

$$= \frac{w_{in}(t)}{w_s} \qquad (9.3.35b)$$

The term in the square brackets of (9.3.35a) is the time-dependent energy (per unit of area) of the optical input, and thus the parameter $h\nu/2\sigma$ is the saturation energy of this amplifier.

$$\boxed{\text{Saturation energy} = w_s = \frac{h\nu}{2\sigma}} \quad (9.3.36)$$

The function $F(z)$ can be related to the initial population inversion N_0 by combining (9.3.34) with (9.3.32a).

$$N_0 = -\frac{1}{\sigma}\frac{\partial P}{\partial z}\bigg|_{t=0} = -\frac{1}{\sigma}\frac{\partial F/\partial z}{F(z) + K(0)}$$

The solution for $F(z)$ is

$$\frac{F(z) + K(0)}{F(0) + K(0)} = \exp(-N\sigma z) \triangleq G_0^{-1}(z) \quad (9.3.37)$$

In (9.3.37), we recognized the familiar collection of terms, $\exp(-N_0 \sigma z)$, as being the inverse of the small-signal gain. Only a few more manipulations remain to obtain the desired result—the output flux at $z = l_g$. Add (9.3.35) to (9.3.37) and solve for $F(z)$ and $K(t)$:

$$F(z) + K(t) = [F(0) + K(0)]\left\{\exp\left[\frac{w_{in}(t)}{w_s}\right] + G_0^{-1}(z) - 1\right\} \quad (9.3.38a)$$

Hence, the function $P(z, t)$ is given by (9.3.34) to be

$$P(z, t) = \ln[K(0) + F(0)] + \ln\left\{\exp\left[\frac{w_{in}(t)}{w_s}\right] + G_0^{-1}(z) - 1\right\} \quad (9.3.38b)$$

The photon flux at $z = l_g$ can be found from (9.3.32b), noting (with a sigh of relief) that the differentiation relieves us of the necessity of evaluating the constants $K(0)$ and $F(0)$.

$$\Gamma_p(l, t) = +\frac{1}{2\sigma}\frac{\partial P}{\partial t}\bigg|_{z=l_g} = \Gamma_0 f(t)\frac{\exp[w_{in}(t)/w_s]}{\exp[w_{in}(t)/w_s] + G_0^{-1} - 1} \quad (9.3.39)$$

Equation (9.3.39) is the main result of this section. Note that if the input energy is very small compared to the saturation value, then all exponential terms are ~ 1, and (9.3.39) degenerates to the simple form given by (9.3.26).

$$\Gamma_p(l, t) = \theta(t) = G_0 \Gamma_0 f(t) \quad (9.3.26)$$

In other words, the output is a faithful reproduction of the input but amplified by G_0.

When the input energy is comparable to w_s, the output wave shape is severely distorted. For example, let the input be given by a "smooth pulse" of the form

$$\Gamma_p(0, t) = \Gamma_0 \sin^2\left(\frac{\pi t}{2T}\right), \quad 0 < t < 2T \quad (9.3.40)$$

Sec. 9.3   Transient Effects

where T is the pulse width (FWHM). The energy contained in this pulse is given by

$$w_{\text{in}}(t) = w_0 \left[\frac{t}{2T} - \frac{1}{2\pi} \sin\left(\frac{\pi t}{T}\right) \right]$$

with w_0 being the energy in the entire pulse.

Fig. 9.14 shows the output flux for various values of the ratio w_0/w_s for a 6-dB amplifier ($G_0 = 4$). Note that for very small inputs, the output is just 4 times larger than the input. However, if the input energy is equal to that stored in the amplifier, the pulse shape is severely distorted. One cannot extract any more energy than is stored in the amplifier. One can quickly verify this fact by comparing the area under the curve marked $w_0/w_s = 1$ to the area under the input of Fig. 9.14. If we do not fuss too much about accuracy, then the peak output is 2 units high, with a FWHM of about T yielding $2T$ units of energy, one of which was supplied by the input.

This energy accounting can be performed more precisely in the case of a "square" input pulse of the form $\Gamma_0 [u(t) - u(t - T)]$. Then the output energy is given by

$$w_{\text{out}}(t) = h\nu \int_0^t \Gamma(l_g, t)\, dt$$

$$= h\nu \Gamma_0 T \int_0^t \frac{\exp(w_0 t/w_s T)}{\exp(w_0 t/w_s T) + G_0^{-1} - 1} \frac{dt}{T} \quad (9.3.41a)$$

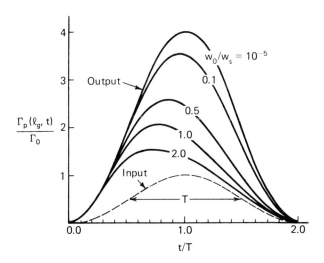

Figure 9.14 Saturation of a 6-dB amplifier by an input pulse of the form $\Gamma_0 \sin^2(\pi t/2T)$.

where

$$w_{in}(t) = h\nu\Gamma_0 t = w_0 t/T; \quad w_0 = \text{total energy in the pulse}$$

By letting $x = \exp[(w_0 t)/(w_s T)]$, this function can be integrated easily:

$$w_{out}(t) = w_s \ln G_0 + w_s \ln\left[\exp\left(\frac{w_0 t}{w_s T}\right) + G_0^{-1} - 1\right] \quad (9.3.41b)$$

Now $G_0 = \exp(N_2 - N_1)\sigma l_g$, $w_s = h\nu/2\sigma$, and, if $(w_0 t/w_s T)$ is large enough so as to neglect 1 in the last term, then (9.3.41b) simplifies to the obvious:

$$w_{out}(t) = \frac{h\nu(N_2 - N_1)l}{2} + h\nu\Gamma_0 t \quad (9.3.41c)$$

The first term is the energy that *was* stored in the population inversion; after that is extracted, the output is just equal to the input *without* amplification.

It should be noted that the theory presented in this section also applies to a saturable absorber. In that case $N(z, t = 0) = (N_2 - N_1)_0$ will be a negative number and the small-signal "gain" will be less than 1 (i.e., it is an attenuator). Without resolving the problem, we should be able to guess at the answer as to pulse distortion.

For the attenuator problem, we would guess a response similar to Fig. 9.14, but with time running backwards. The leading edge will be absorbed heavily, but succeeding packets of photons will eventually bleach the system, allowing unity transmission (see Problem 9.27).

9.4 MULTIMODE EFFECTS—MODE LOCKING

In the previous sections on Q switching, gain switching, and pulse distortion in amplifiers, we have treated the gain medium as if it were homogeneously broadened. However, we know that some transitions are inhomogeneously broadened, and thus some lasers oscillate at more than one frequency within a given transition. This fact can be utilized to generate extremely short pulses by a technique called "mode locking." Although this scheme does not generate the extreme energy per pulse as does Q switching, it does lead to very short pulses occurring at a repetitive rate.

Consider a laser oscillating on an inhomogeneous transition with many modes above threshold and lasing, such as that shown in Fig. 9.15. One can readily appreciate that there can be a large number of modes oscillating in such a system—just by considering the numerics. For instance, consider the He:Ne transition at 0.6328 μm, which has a Doppler width of 1500 MHz. (This sounds big, but $\Delta\lambda$ is small, 0.02 Å.) If the length of the optical cavity was 150 cm, 15 or more modes could oscillate simultaneously.

Sec. 9.4 Multimode Effects—Mode Locking

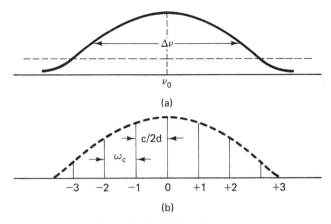

Figure 9.15 Multimode laser.

A general equation describing the classical electromagnetic field at a particular point in space is

$$e_T(t) = \sum_{-(N-1)/2}^{+(N-1)/2} E_n e^{j[(\omega_0 + n\omega_c)t + \phi_n]} \quad (9.4.1a)$$

The factor E_n indicates the amplitude of the nth mode, which is oscillating at the (angular) frequency $(\omega_0 + n\omega_c)$. This phase term ϕ_n is very important. It is tempting to set it equal to zero, but that would not be justified at this stage. There is nothing in the physical system to indicate that mode 0 and mode 1 should have the same common time origin. Quite to the contrary, modes 0 and 1 can be perfectly coherent waves described by $E_0 \sin \omega_0 t$ and $E_1 \cos(\omega_1 t - 26.8°)$, respectively. Obviously, the number 26.8° was picked at random to illustrate that the phase ϕ could be anything. If the phase and amplitude of that mode are constant in time, then these particular waves represent perfectly monochromatic coherent oscillators.

It is very easy to prove that the total electric field *repeats* itself every $T = 2d/c$ sec, irrespective of the values of the phases.

$$\begin{aligned} e(t + T) &= \sum E_n \exp\left\{j\left[(\omega_0 + n\omega_c)\left(t + \frac{2d}{c}\right) + \phi_n\right]\right\} \\ &= \sum E_n \exp\left\{j\left[(\omega_0 + n\omega_c)\left(t + \frac{2\pi}{\omega_c}\right) + \phi_n\right]\right\} \\ &= \sum E_n \exp\left\{j\left[(\omega_0 + n\omega_c)t + \underbrace{\frac{2\pi n\omega_c}{\omega_c}}_{n2x} + \underbrace{\frac{2\pi\omega_0}{\omega_c}}_{q2x} + \phi_n\right]\right\} \\ &= e(t) \end{aligned} \quad (9.4.1b)$$

For many lasers, the phases ϕ_n and the amplitudes E_n fluctuate slowly with time (compared to ν^{-1}) in a completely uncorrelated fashion and thus the periodicity of $\mathbf{e}(t)$ is destroyed unless a method is found to establish a definite phase relationship between the modes. Let us examine the consequences of an orderly relationship before discussing methods to achieve it.

Consider Fig. 9.16, which is a phasor diagram of the modes shown in the

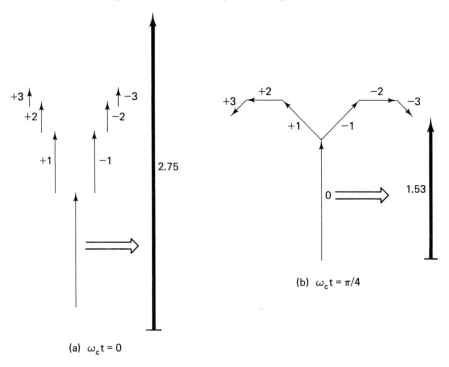

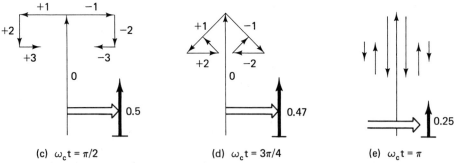

Figure 9.16 Phasor addition of the modes shown in the spectrum of Fig. 9.15. The mode amplitudes were chosen according to proportional relation: $1:0.5:0.25:0.125$.

Sec. 9.4 Multimode Effects—Mode Locking

frequency representation in Fig. 9.15. Because there are slightly different frequencies in this spectrum, the phases of the various modes change with respect to our reference, which is the mode at line center. Each part of Fig. 9.16 is a graphic or phasor addition of the fields of each of the modes, as expressed by (9.4.2), for various fractions of a round-trip time. The time interval from $\omega t = \pi$ to 2π is the mirror image of that from 0 to π, so only the latter sequence is shown. Because the central mode is chosen to be the reference, the phasors for $\nu > \nu_0$ rotate counterclockwise; those with $\nu < \nu_0$ rotate in the opposite sense. The ratio of peak *field* (at $\omega_c t = 0$) to that at the minimum (at $\omega_c t = \pi$) is $2.75/0.25 = 11.1$, and thus the ratio of intensities is the square of the field ratio or $11^2 = 121$. Obviously, we have a laser generating a sequence of pulses spaced a round-trip time apart, with a peak value many times the minimum value. The mathematical analysis that follows attempts to define this pulse in terms of the envelope of the spectral distribution of modes, the number of modes, and the CW power of the laser.

9.4.1 Simplified Mathematical Analysis

Let us "lock" each mode to a common time origin and assign each phase, ϕ_n, to be zero. (Once we have the results, we can also imagine methods of "teaching" each mode where $t = 0$ is.) To make the mathematics simple, we assign equal amplitude to each mode (i.e., $E_n = E_0$):

$$\frac{e(t)}{E_0} = \sum_{-(N-1)/2}^{+(N-1)/2} e^{j(\omega_0 + n\omega_c)t} = e^{j\omega_0 t} \sum x^n \qquad x = e^{j\omega_c t} \qquad (9.4.2)$$

If

$$S = x^{-k} + x^{-k+1} + \cdots + x^{k-1} + x^k$$

then

$$xS = x^{-k+1} + \cdots + x^k + x^{k+1}$$

Hence,

$$(1 - x)S = x^{-k} - x^{k+1}$$

or

$$S = \frac{x^{k+1} - x^{-k}}{x - 1} = \frac{x^{1/2}}{x^{1/2}} \frac{x^{k+(1/2)} - x^{-[k+(1/2)]}}{x^{1/2} - x^{-1/2}}$$

$$k = \frac{N - 1}{2}$$

Therefore our total electric field is given by

$$e(t) = E_0 e^{j\omega_0 t} \frac{\sin(N\omega_c t/2)}{\sin(\omega_c t/2)} \quad (9.4.3a)$$

The intensity of the laser is given by

$$I(t) = \frac{e(t)e^*(t)}{2\eta_0} = \frac{E_0^2}{2\eta_0}\left[\frac{\sin(N\omega_c t/2)}{\sin(\omega_c t/2)}\right]^2 \quad (9.4.3b)$$

When $\omega_c t/2$ is an integral multiple of π, the intensity is N^2 times the intensity per mode. But, of course, we have assumed N equal amplitude modes—hence the average power of the laser is N times the power per mode. Consequently, we obtain the (approximate) relation that the peak power is N times the average.

$$P_{\text{peak}} = N \times P_{\text{average}} \quad (9.4.3c)$$

The numerator of (9.4.3b) is a rapidly oscillating function, going to zero when $\omega_c t/2 = \pi/N$ while the denominator slowly increases to 1 at $\omega_c t/2 = \pi/2$. Hence, the instantaneous power would appear as shown in the dashed curve of Fig. 9.17.

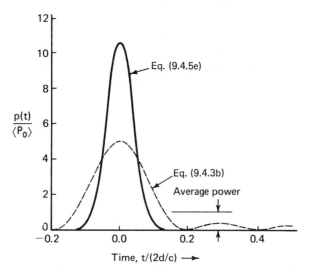

Figure 9.17 Comparison of the two theories of the mode-locked pulse. The two expressions were evaluated for the same average power and for $\Delta\omega/\omega_c = 5 = N$.

We can estimate the pulse width by the relation

$$(P_{\text{peak}})\tau_p \simeq (P_{\text{average}}) \cdot \tau_{RT}$$

or $\quad (9.4.3d)$

$$\tau_p \approx \frac{2d}{c} \div N$$

Sec. 9.4   Multimode Effects—Mode Locking

But the number N of oscillating modes (which are hopefully locked together) is approximately the line width $\Delta\nu$ divided by the cavity mode spacing $c/2d$:

$$N \sim \Delta\nu/(c/2d)$$

or   (9.4.3e)

$$\tau_p \sim \frac{2d}{c} \div \frac{\Delta\nu}{c/2d} \sim \frac{1}{\Delta\nu}$$

This can be a very short pulse. Even the "narrow" line width of the He:Ne laser, $\Delta\nu = 1500$ MHz, would produce a pulse of ~ 0.6 ns. Such a pulse would occupy only 7 inches of free space. The capabilities of solid-state lasers, such as ruby and Nd-glass, stagger the imagination. Consider the Nd-glass system with $\Delta\nu = 3 \times 10^{12}$ Hz ($\Delta\lambda \sim 100$ Å) around $\lambda = 1.06$ μm. This indicates a capability of 0.3 ps with a peak power 10^4 times the average power. The pulse would only occupy *0.3 mm* of space. Thus there are an enormous number of photons packed into a very small portion of space. One can readily appreciate the fact that the measurement of these ultrashort pulses is a difficult problem.

9.4.2 Gaussian Amplitude Distribution

One can obtain a better representation of the mode-locked pulse if one is willing to make a few approximations. Our first one is to assign a Gaussian envelope to the mode intensities.

$$I_n = I_0 \exp\left\{-4(ln\ 2)\left[\left(\frac{n\omega_c}{\Delta\omega}\right)^2\right]\right\} \quad (9.4.4a)$$

where I_0 is the mode intensity at line center, $\Delta\omega$ is bandwidth occupied by the modes that are greater than one half of the amplitude of the central mode; and, as before, n is the mode number, starting with zero at line center and terminating at $(N - 1)/2$, where the small-signal gain coefficient equals the loss coefficient. It must be emphasized that (9.4.4a) is merely a convenient mathematical representation of the mode amplitudes; in particular, one should not identify $\Delta\omega$ with the Doppler width of the inhomogeneously broadened transition.*

Given this representation, the electric field associated with each mode is

$$E_n = E_0 \exp\left[-2(ln\ 2)\left(\frac{n\omega_c}{\Delta\omega}\right)^2\right] \quad (9.4.4b)$$

where $E_0^2/2\eta_0$ is the intensity of the mode at line center.

Thus the total electric field is

---
*See Problem (9.30) for a development of the actual profile, a comparison with (9.4.4a), and a relationship between $\Delta\omega$ and $\Delta\omega_D$.

$$e(t) = E_0 e^{j\omega_0 t} \sum_{-(N-1)/2}^{+(N-1)/2} (1) \exp\left[-2(\ln 2)\left(\frac{n\omega_c}{\Delta\omega}\right)^2\right] \exp(jn\omega_c t) \quad (9.4.4c)$$

If the laser is well above threshold (i.e., $[\gamma_0(\nu_0)/\alpha]^2 > 1$), the central modes are much larger than those on the wings, where the gain coefficient is close to the loss. Hence, we add an insignificant amount of power if we arbitrarily extend the summation limits to $\pm\infty$. We also "smear" each mode into a continuum of sinusoids spread over the interval, ω_c, which is symmetrically placed around the original mode. Let

$$n\omega_c = x$$

Then

$$(n + \tfrac{1}{2})\omega_c - (n - \tfrac{1}{2})\omega_c = \Delta x \to dx$$

Then we replace the factor 1 in (9.4.4c) by dx/ω_c and we obtain an integral representation for the field rather than a summation.

$$e(t) = \frac{E_0 e^{j\omega_0 t}}{\omega_c} \int_{-\infty}^{+\infty} \exp\left[-\frac{2(\ln 2)x^2}{(\Delta\omega)^2}\right] \exp(+jxt)\, dx \quad (9.4.5a)$$

Now we complete the square of the expression in the exponent of the integrand and compensate for this outrageous bit of mathematics by multiplying by the inverse factor outside the integral.

$$\frac{e(t)}{E_0 e^{j\omega_0 t}} = \frac{1}{\omega_c} \exp\left\{-\left(\frac{\Delta\omega t}{2(2\ln 2)^{1/2}}\right)^2\right\}$$

$$\times \int_{-\infty}^{+\infty} \exp\left\{-\left[\left(\frac{(2\ln 2)^{1/2} x}{\Delta\omega}\right)^2 - jxt - \left(\frac{\Delta\omega t}{2(2\ln 2)^{1/2}}\right)^2\right]\right\} dx$$

$$(9.4.5b)$$

$$= \frac{1}{\omega_c} \exp\left\{-\left(\frac{\Delta\omega t}{2(2\ln 2)^{1/2}}\right)^2\right\}$$

$$\times \int_{-\infty}^{+\infty} \exp\left\{-\left[\frac{(2\ln 2)^{1/2} x}{\Delta\omega} - j\frac{\Delta\omega t}{2(2\ln 2)^{1/2}}\right]^2\right\} dx$$

By making the obvious substitution of u for the terms in brackets and inserting the appropriate integrating factors, we have a tabulated integral.

$$\frac{e(t)}{E_0 e^{j\omega_0 t}} = \frac{\Delta\omega}{(2\ln 2)^{1/2}\omega_c} \exp\left[-\left(\frac{\Delta\omega t}{2(2\ln 2)^{1/2}}\right)^2\right]\left(\int_{-\infty}^{+\infty} e^{-u^2}\, du = (\pi)^{1/2}\right)$$

$$= \left(\frac{\pi}{2\ln 2}\right)^{1/2} \frac{\Delta\omega}{\omega_c} \exp\left[-\left(\frac{\Delta\omega t}{2(2\ln 2)^{1/2}}\right)^2\right] \quad (9.4.5c)$$

The power is the integral of the Poynting vector $\left(\dfrac{1}{2\eta_0}\right) \int e(t) \cdot e^*(t)\, dA$ over

Sec. 9.4 Multimode Effects—Mode Locking

the cross section of the beam and is the quantity that would be displayed with a fast detector and an oscilloscope.

$$\frac{p(t)}{P_0} = \frac{\pi}{2 \ln 2} \left(\frac{\Delta\omega}{\omega_c}\right)^2 \exp\left[-\left(\frac{\Delta\omega t}{2(\ln 2)^{1/2}}\right)^2\right] \quad (9.4.5d)$$

P_0 = power in mode at line center

This equation is *not* periodic, a fact that is a result of our replacement of the summation by an integral. But if we keep this obvious limitation in mind, (9.4.45d) is a much better description of the mode-locked output than was obtained previously in (9.4.3b). We can obtain the pulse width directly from (9.4.5d) by finding the time interval $\Delta t_{1/2}$ over which the power is greater than one half of its peak value;

$$\Delta t_{1/2} = \frac{4 \ln 2}{\Delta\omega} = \frac{4 \ln 2}{2\pi} \left(\frac{\omega_c}{\Delta\omega}\right) \frac{2d}{c}$$

or (9.4.6)

$$\Delta t_{1/2} = 4 \ln 2/(2\pi\Delta\nu) = 0.44/\Delta\nu$$

The last form was obtained by multiplying the first by 1 in the form of ω_c/ω_c and letting $\omega_c = 2\pi(c/2d)$. This is the shortest pulse width possible given that the energy for this pulse was derived from a finite portion of the spectrum $\Delta\omega$. As such, it is sometimes called a "band-limited" or a Fourier-transform limited pulse.* You cannot do better!

Equation (9.4.5d) can be placed into a more convenient form by relating the power in the central mode to the total power in all modes, a quantity that is easily measured. We can add the individual powers of each mode [i.e., $\langle P \rangle = \Sigma P_n$ from (9.4.4a) or compute the average of (9.4.5d) with an averaging time of $2d/c$, the pulse repetition period.

$$\langle P \rangle = \frac{1}{2d/c} \int_{-d/c}^{+d/c} p(t) \, dt$$

$$\frac{\langle P \rangle}{P_0} = \frac{\omega_c}{2\pi} \int_{-\pi/\omega_c}^{+\pi/\omega_c} \left(\frac{\pi}{2 \ln 2}\right) \left(\frac{\Delta\omega}{\omega_c}\right)^2 \exp\left[-\left(\frac{\Delta\omega t}{2(\ln 2)^{1/2}}\right)^2\right] dt \quad (9.4.7a)$$

Now the exponential term is sharply peaked around $t = 0$ and is essentially zero at the times indicated by the limits of integration. With little error, then, the limits can be extended to $\pm\infty$, and we obtain a tabulated integral.

---

*Equation (9.4.5d) appears to describe a pulse different from that allowed by the uncertainty relations. This is a result of following the convention of using half-power points to define a bandwidth and a pulse width, whereas the uncertainty relation deals with the root of the mean-square deviation from the central value.

$$\frac{\langle P \rangle}{P_0} = \frac{1}{2}\left(\frac{\pi}{\ln 2}\right)^{1/2}\left(\frac{\Delta\omega}{\omega_c}\right) \tag{9.4.7b}$$

Thus the formula for the instantaneous power, (9.4.5d), can be rewritten in terms of the average power and the pulse width $\Delta t_{1/2}$, (9.4.6).

$$p(t) = \left(\frac{\pi}{\ln 2}\right)^{1/2}\left(\frac{\Delta\omega}{\omega_c}\right)\langle P \rangle \exp\left[-4(\ln 2)\left(\frac{t}{\Delta t_{1/2}}\right)^2\right] \tag{9.4.5e}$$

This function is plotted as the solid curve in Fig. 9.17. This figure shows that the simplified procedure (9.4.2) → (9.4.3e) is only a fair representation of the pulse predicted by the more complicated analysis. But both theories predict an extremely short pulse, taxing the time response of the detectors and writing speeds of our best oscilloscopes.

9.4.3 Spatial Variation of the Field

Let us return to the simplified analysis of Sec. 9.4.1 and examine the field as a function of time *and* space, assuming that the cavity consists of highly reflecting mirrors (Fig. 9.18).

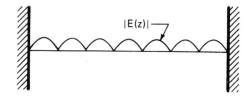

Figure 9.18 Standing waves in a cavity.

$$e(z, t) = \sum_n E_n \sin\left[(\omega_0 + n\omega_c)t\right] \sin k_n z \tag{9.4.8a}$$

with

$$k_n = \frac{\omega_0 + n\omega_c}{c}$$

Thus

$$e(z, t) = \sum_n E_n \sin\left[(q + n)\frac{2\pi c}{2d}t\right] \sin\left[(q + n)\frac{2\pi z}{2d}\right] \tag{9.4.8b}$$

Using the trigonometric formula

$$\sin\theta \sin\phi = \tfrac{1}{2}[\cos(\theta - \phi) - \cos(\theta + \phi)]$$

$$e(z, t) = \sum_n \frac{E_n}{2}\left\{\cos\left[(q + n)\frac{\pi}{d}\left(t - \frac{z}{c}\right)\right] \cos\left[(q + n)\frac{\pi}{d}\left(t + \frac{z}{c}\right)\right]\right\} \tag{9.4.9}$$

Sec. 9.4 Multimode Effects—Mode Locking

The first term in (9.4.9) represents the pulse traveling to the right, whereas the second is a pulse moving to the left. Hence, we could interpret this mode locking as a tightly bunched packet of photons bouncing back and forth between the mirror, a very simple and useful picture.

This viewpoint suggests methods of mode locking and "teaching" the modes to oscillate with a common phase. Imagine a fast shutter in the cavity at one mirror (Fig. 9.19), which is opened for a short period of time to allow a photon packet to pass through and return from the mirror. If the packet arrives at the proper time, it experiences zero loss, but the photons that arrive at the wrong time experience a very high loss.

Although we have implied a shutter driven by some external mechanism, that is not a necessity. If one can find a medium that can be changed by the initial portion of the wave, the photons can operate the "toggle switch" on the shutter!

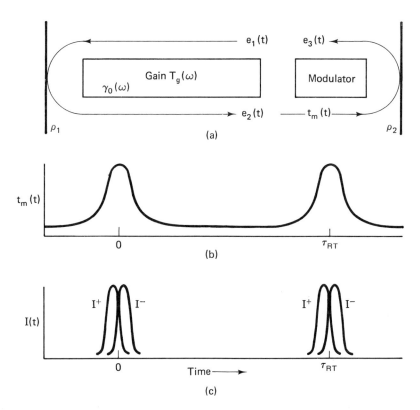

Figure 9.19 Mode locking of a laser. (a) Geometry showing the external modulator and the fields inside the laser cavity. (b) The transmission coefficient of the modulator. (c) The intensities arriving at the modulator.

If one were to insert an attenuating medium inside the laser cavity such that the initial part of the pulse would "bleach" it, the remainder of the pulse could pass with little attenuation. After passage of the main body of the pulse, the equalized population of the bleached attenuator relaxes back to the normal state. Then it is hit again by the pulse on the next round-trip and the process repeats itself. If the relaxation rate is faster than the repetition frequency, then the laser can mode-lock by virtue of the saturable absorber.

The extra medium inserted into the laser cavity could be a gain cell rather than a bleachable attenuator, and, if this is so, the laser medium can serve for both functions—to provide gain and to provide an intensity-dependent medium. A moment's consideration will indicate that the net use of the pumping will be higher if the laser is model-locked, which is the reason that some lasers lock spontaneously. However, the conditions distinguishing a free-running and a spontaneously mode-locked laser are very tenuous, and thus most mode-locked lasers are driven by an external modulator controlling either the amplitude of its transmission or its phase. In the next subsection, such a system will be considered, and it will be found that even a homogeneously broadened laser can be mode-locked. The modulator generates "sidebands" that are automatically locked to the central mode (or carrier in communications terminology) and thus give rise to short pulses.

9.4.4 Active Mode Locking in a Homogeneously Broadened Laser

In principle, a laser operating on a homogeneously broadened transition should oscillate on a cavity mode with the highest gain-to-loss ratio to the demise of the other cavity modes. Based on previous discussion on mode locking in a multimode (inhomogeneous) laser, one would not expect similar phenomena in a homogeneously broadened system—since there are no other modes! However, there is no law against inserting the time-varying modulator with either an amplitude or phase modulation, as shown in Fig. 9.19, into *any* laser cavity (the atoms have no vote on our actions). Our task is to describe the response of the total system, gain plus modulator, to the driven transfer function of the shutter.

Fig. 9.19 suggests that such a ploy would be successful. If a pulse (however it gets started) arrives at the modulator when there is a high transmission coefficient, it will pass through it with little absorption; if it arrives at the wrong time—when there is a lot of absorption—then that pulse does not survive. Thus we search for a self-consistent temporal (or pulse) solution for the fields in the laser cavity rather than a CW spectral representation. In spite of the differences in approaches, the two viewpoints lead to a common answer.

One assumes an electric *field* (not intensity) entering the gain cell from the right given by Gaussian envelope function;

Sec. 9.4 Multimode Effects—Mode Locking

$$e(t) = A\exp\left[-2(\ln 2)\left(\frac{t}{\tau_p}\right)^2\right]e^{j\omega_0 t} = Ae^{-at^2}e^{j\omega_0 t} \quad (9.4.10a)$$

with

$$\omega_0 = 2\pi c/\lambda_0 \quad \text{and} \quad a = 2(\ln 2)/\tau_p^2 \quad (9.4.10b)$$

The parameter a is just a convenient abbreviation for the factors involved with the pulse width. Follow that pulse through the gain cell and back to the entrance of the modulator whose time-varying (one-way) transmission coefficient is given by

$$t_m(t) = \exp\left[-\delta^2 \sin^2\left(\frac{\omega_m t}{2}\right)\right] \quad (9.4.11)$$

where ω_m is the (radian) modulation frequency that will turn out to be very close to the cavity mode spacing. As we will see, the choice of (9.4.11) for the transmission coefficient is for mathematical convenience with the quantity δ merely a measure of the on-to-off *field* transmission coefficient of the modulator.

The logic path for obtaining the self-consistent solution is as follows:

1. Start with $e_1(t)$ given by (9.4.10a). Find its spectral representation $E_1(\omega)$ by use of the Fourier transform.
2. Propagate $E_1(\omega)$ through the gain cell, to the mirror M_1 with a field reflection coefficient ρ_1 and back again to the exit to find $E_2(\omega)$. This requires us to find the complex transfer function, (amplitude and phase) of the gain cell.
3. Go back to the time domain, via the inverse Fourier transform, to find $e_2(t)$, the field incident on the modulator.
4. Multiply $e_2(t)$ by the two-way transmission coefficient of the modulator, the square of (9.4.11), and the field reflection coefficient of M_2, ρ_2, to find $e_3(t)$, the field incident on the gain cell.
5. Compare $e_3(t)$ to $e_1(t)$. The field $e_3(t)$ must be the same as $e_1(t)$, arriving at the proper time (in the modulator period) and having the same shape in order to have obtained a self-consistent solution.

These logic paths use the following mathematics:

1. Spectral representation of $e_1(t)$. We use the Fourier transform defined by

$$E_1(\omega) = \int_{-\infty}^{+\infty} e_1(t) e^{-j\omega t}\, dt \quad (9.4.12a)$$

to find

$$E_1(\omega) = A\left[\left(\frac{\pi}{a}\right)^{1/2} e^{-[(\omega-\omega_0)^2/4a]}\right] \quad (9.4.12b)$$

2a. Transfer function of the gain cell. This is a homogeneously broadened transition, by assumption, and thus even the saturated line shape is a Lorentzian. However, we must use a *complex* gain coefficient to account for the amplitude *and phase* of the field. It is easy to show (see Chapter 14) that this complex gain coefficient is given by

$$\frac{\gamma(\omega)}{|\gamma(\omega_0)|} = \frac{1}{1 + j\left(\dfrac{\omega - \omega_0}{\Delta\omega/2}\right)} \quad (9.4.13a)$$

where $|\gamma(\omega_0)|$ is the usual *intensity* gain coefficient at line center and $\Delta\omega$ is the FWHM of that gain coefficient. A Taylor series expansion is used for (9.4.13a):

$$\frac{\gamma(\omega)}{|\gamma(\omega_0)|} = 1 - j\left(\frac{\omega - \omega_0}{\Delta\omega/2}\right) - \left(\frac{\omega - \omega_0}{\Delta\omega/2}\right)^2 \quad (9.4.13b)$$

Thus the single-pass transfer function through the gain cell is

$$T_g(\omega) =$$
$$\exp\left\{-jkl + \frac{|\gamma(\omega_0)l_g|}{2}\left[1 - j\left(\frac{\omega - \omega_0}{\Delta\omega/2}\right) - \left(\frac{\omega - \omega_0}{\Delta\omega/2}\right)^2\right]\right\} \quad (9.4.14a)$$

where $|\gamma(\omega_0)|l_g/2$ must be used since we are dealing with fields rather than intensities.

2b. The field $E_2(\omega)$. The field $E_1(\omega)$ propagates to the left impinging on M_1 and thus is reduced by a factor ρ_1 and returns through the gain cell to the starting point of our analysis.

$$E_2(\omega) = \rho_1 T_g^2(\omega) E_1(\omega) \quad (9.4.14b)$$

$$E_2(\omega) = \rho_1 A [\pi/a]^{1/2} \{\exp |\gamma(\omega_0)l_g|\}$$
$$\cdot \left\{\exp\left[-\frac{(\omega - \omega_0)^2}{4a} - |\gamma(\omega_0)l_g|\left(\frac{\omega - \omega_0}{\Delta\omega/2}\right)^2\right]\right\}$$
$$\cdot \left\{\exp\left[-j(k \cdot 2d) + |\gamma(\omega_0)l_g|\left(\frac{\omega - \omega_0}{\Delta\omega/2}\right)\right]\right\} \quad (9.4.14c)$$

The total round-trip linear phase delay of the passive cavity, $k2d$, is used in (9.4.14c) to avoid repeating this calculation for each individual element or space. The imaginary terms in the exponent of (9.4.14c) correspond to the group delay of the pulse for a round-trip. If we abbreviate $|\gamma(\omega_0)|l_g$ as g_0; rewrite k as

$$k = \frac{\omega}{c'} \equiv \frac{\omega_0}{c'} + \frac{(\omega - \omega_0)}{c'} \quad (9.4.15a)$$

Sec. 9.4 Multimode Effects—Mode Locking

and equate the imaginary terms of (9.4.14c) to a Taylor series expansion of the phase delay, we obtain the group velocity or group delay:

$$\frac{\omega_0}{c'} 2d + \left(\frac{\omega - \omega_0}{c'}\right) 2d + g_0 \left(\frac{\omega - \omega_0}{\Delta\omega/2}\right) \triangleq \left[\beta_0 + \frac{\partial \beta}{\partial \omega}(\omega - \omega_0)\right] 2d \quad (9.4.15b)$$

or

$$\frac{\partial \beta}{\partial \omega} \triangleq \frac{1}{v_g} = \frac{1}{c'} + \frac{g_0}{2d} \frac{1}{(\Delta\omega/2)} \quad (9.4.15c)$$

Thus

$$T_{RT} = \frac{2d}{v_g} = \frac{2d}{c'} + \frac{g_0}{(\Delta\omega/2)} \quad (9.4.15d)$$

with

$$g_0 = |\gamma(\omega_0)| \, l_g = \text{single-pass integrated gain} \quad (9.4.15e)$$

Thus the modulation frequency ω_m must be chosen correctly to allow passage of the pulse when the transmission coefficient is a maximum (at $\omega_m t/2 = m\pi$) in (9.4.11); this idea yields our first "design" equation:

$$\omega_m T_{RT} = m\pi$$

or

$$f_m = m \left[\frac{1}{\frac{2d}{c'} + \frac{g_0}{\Delta\omega/2}}\right] = \left[\frac{m\Delta\nu_c}{\left(1 + \frac{g_0 \Delta\nu_c}{\pi \Delta\nu}\right)}\right] \quad (9.4.16)$$

Inasmuch as the correction factor in the denominator is small compared to one, (9.4.16) states that the modulation frequency should be slightly less than a multiple of the cavity intermode spacing $\Delta\nu_c = c'/2d$.

3. Find $e_2(t)$. Go back to the time domain by use of the inverse Fourier transform:

$$e_2(t) = \frac{1}{2\pi} \int_{-\infty}^{+\infty} E_2(\omega) e^{j\omega t} \, d\omega \quad (9.4.17a)$$

and assume that the modulation frequency is chosen to obey (9.4.16) so that the phase factors of (9.4.14c) add to a multiple of 2π. Patience with the arithmetic yields

$$e_2(t) = \frac{\rho_1 A}{2\sqrt{qa}} (e^{g_0})(e^{-(t^2/4q)})(e^{j\omega_0 t}) \quad (9.4.17b)$$

where

$$q = \frac{1}{4a} + \frac{g_0}{(\Delta\omega/2)^2} \quad (9.4.17c)$$

4. Find $e_3(t)$. One multiplies (9.4.17b) by the square of the transmission coefficient of the modulator (9.4.11) (because of a double pass) and by the field reflection coefficient of M_2 to find $e_3(t)$. Assuming that the pulse arrives near $\omega_m t/2 = m\pi$, and expanding $\sin x \sim x$, we obtain

$$e_3(t) = \rho_1 \rho_2 \frac{A}{2} \frac{1}{\sqrt{qa}} e^{j\omega_0 t} \exp\left\{-\left[\frac{1}{4q} + 2\delta^2\left(\frac{\omega_m}{2}\right)^2\right]t^2\right\} \quad (9.4.18)$$

where time t now refers to one round-trip later than was chosen for (9.4.10a).

5. Force self-consistency. We compare (9.4.18) to (9.4.10a) to obtain

$$A e^{j\omega_0 t} \Leftrightarrow \rho_1 \rho_2 e^{g_0} \frac{A}{2} \frac{1}{\sqrt{qa}} e^{j\omega_0 t} \quad (9.4.19a)$$

and

$$at^2 \triangleq 2 \ln 2 \left(\frac{t}{\tau_p}\right)^2 \Leftrightarrow \left[\frac{1}{4q} + 2\delta^2 \left(\frac{\omega_m}{2}\right)^2\right] t^2 \quad (9.4.19b)$$

where (9.4.10b) has been used to recover the definition of a. The solution for the FWHM of the pulse, τ_p, from (9.4.19b) is

$$\tau_p = \frac{\sqrt{2 \ln 2}}{\pi} \left(\frac{2g_0}{\delta^2}\right)^{1/4} \cdot \frac{1}{(f_m \Delta\nu)^{1/2}} \quad (9.4.20)$$

where it is assumed that the parameter $4g_0 a/(\Delta\omega/2) = (8 \ln 2)g_0/(\Delta\omega\tau_p/2)^2$ is small; f_m is the modulation frequency given by (9.4.16), $\Delta\nu$ is the homogeneous line width, g_0 is the single-pass integrated gain, and δ is a measure of the on-to-off transmission of the modulator.

The other half of the self-consistency demand (9.4.19a), forces the integrated net gain, averaged over a round-trip time, to be equal to 1; that is,

$$\rho_1 \rho_2 e^{g_0} \frac{1}{\sqrt{qa}} = 1 \quad (9.4.19a)$$

Now, by combining (9.4.10a), the definition of a, with (9.4.17c) for q, we have

$$qa = a\left[\frac{1}{4a} + \frac{g_0}{(\Delta\omega/2)^2}\right] = \frac{1}{4}\left[1 + \frac{(8 \ln 2)g_0}{(\Delta\omega\tau_p/2)^2}\right]$$

Sec. 9.4 Multimode Effects—Mode Locking

If we square both sides of (9.4.19a), use this relationship for qa, and take the natural log of each side, we obtain a somewhat familiar expression:

$$2g_0 \triangleq |\gamma(\omega_0)l| = \ln\frac{1}{R_1R_2} + \ln\left[1 + \frac{(8\ln 2)g_0}{(\Delta\omega\tau_p/2)^2}\right] \quad (9.4.21)$$

The first term on the right-hand side of (9.4.21) is merely the conventional round-trip cavity loss of a simple laser. The second term is the time-averaged loss introduced by the modulator. Some of the field passes through this modulator when $t_m < 1$, and thus there are photons absorbed, with the fraction lost per round-trip equal to this last term. Recall that we assumed a Taylor series expansion for the gain coefficient [cf. (9.4.13b)], which tacitly assumed that the bandwidth of the pulse was much less than the available bandwidth of the gain. This amounts to assuming $\Delta\omega\tau_p/2 \gg 1$; thus the time-averaged loss in (9.4.21) is not large.

Thus we can "mode-lock" a homogeneously broadened laser by generating the modes with the amplitude modulator. Although the pulse width (9.4.20) is not simply related to the amplifying bandwidth as was found in Sec. 9.4.2 and Sec. 9.4.3, a few numbers will convince us that the pulse is quite short.

Let $\Delta\nu = 2.5$ GHz, $\Delta\omega = 1.57 \times 10^{10}$ rad/sec
$d = 50$ cm, $l_g = 40$ cm, $c' = c = 3 \times 10^{10}$ cm/sec
$\therefore \Delta\nu_c = c/2d = 300$ MHz
$g_0 = 0.12$ (i.e., $|\gamma(\omega_0)| = 0.3\%$/cm)
$\delta = 1.0$ [assumes $t_m(\text{on}) = 1.0$ and $t_m(\text{off}) = e^{-1} = 0.368$]
$\tau_{RT} = 2d/c + g/(\Delta\omega/2) =$
  $3.33 \times 10^{-9} + (1.53 \times 10^{-11}) = 3.35 \times 10^{-9}$ sec
$f_m = 298.6$ MHz [for $m = 1$, see (9.4.16)]
$\therefore \tau_p = 0.3$ ns

This pulse requires a radian bandwidth of $\sim 1/(2\tau_p) = 1.6 \times 10^9$ sec^{-1}, which is about 1/10 of that assumed to be available. Such a pulse occupies only 4 inches of space. For the above example, the modulator introduces a time-averaged round-trip loss of 11%.

One can also use a phase modulator and obtain short pulses. We again presume a "pulse" solution for the fields within the cavity, which arrives at the proper time of the phase excursion of the modulator. We assume that the excess phase shift introduced by the modulator can be expressed as

$$t_m(t) = \exp\left[-j2\delta_\phi \cos \omega_m t\right] \quad (9.4.22a)$$

The proper time is near $t = 0$ so we expand the cosine into a Taylor series and use the following form for the transmission coefficient of the modulator:

$$t_m(t) \simeq \exp\left\{\mp j\left[2\delta + \frac{\delta\omega_m^2 t^2}{2}\right]\right\} \quad (9.4.22b)$$

This modulator impresses a time-varying phase on any field propagating through it, and thus we anticipate that the pulse may have "chirp"; that is, the frequency may be a function of time. Therefore our assumed "pulse-like" solution should reflect this possibility.

$$e_1(t) = Ae^{-at^2} \exp\left[j\left(\omega_0 + \frac{\Delta\omega t}{T}\right)t\right] \quad (9.4.23a)$$

where a is defined by 9.4.10b and the parameter $\delta\omega/T = b$ is a "chirp" parameter expressing the assumed linear variation of radian frequency with time

$$\omega(t) = \omega_0 + (\delta\omega/T)t = \omega_0 + bt \quad (9.4.23b)$$

The same logic path is followed as was used for the amplitude modulator, but because of the chirp one encounters more involved arithmetic. A formal method to avoid repeating many of the manipulations is to replace a in the prior analysis by $a - jb$ for the present one. Thus, for instance, $E_1(\omega)$ is given by

$$E_1(\omega) = A\left[\left(\frac{\pi}{a-jb}\right)^{1/2} e^{-\frac{(\omega-\omega_0)^2}{4(a-jb)}}\right] \quad (9.4.24)$$

This field is propagated through the gain cell and back to and through the modulator in a round-trip time as discussed previously. For the pulse arriving at the extremities of the phase modulator, the self-consistent requirement leads to

$$b = \pm a = \pi^2 f_m \,\Delta\nu \left(\frac{\delta_\phi}{2g_0}\right)^{1/2} \quad (9.4.25)$$

and the pulse width is given by

$$\tau_p = \frac{\sqrt{2\ln 2}}{\pi}\left(\frac{2g_0}{\delta_\phi}\right)^{1/4}\left(\frac{1}{f_m \Delta\nu}\right)^{1/2} \quad (9.4.26)$$

Equation (9.4.26) is nearly the same as (9.4.19) except that δ appears rather than δ^2. The two possible solutions expressed by (9.4.25) represent the arrival at the two extremes of the phase modulator.

9.5 DISPERSION EFFECTS

9.5.1 Material Dispersion

In all of the discussions to date, we have implicitly treated the index of refraction as a constant, independent of wavelength. We have made this approximation to avoid complicating issues and obscuring the physical understanding.

Sec. 9.5 Dispersion Effects

Now it is time to "pay our dues" to the real world. The index of refraction is wavelength dependent, and that fact does change some of our prior conclusions.

In semiconductor lasers, the gain-bandwidth is so large ($\Delta\lambda = 1$ to 30 nm) and the cavities are so small ($d = 10$ to 100 μm) that it is imperative to account for the wavelength dependence of n. As will be seen in the following development, the spacing between modes in such a laser depends on n and $dn/d\lambda$.

Our previous statement that there must be an integral number of half wavelengths (in the material) is still valid, but we must display the wavelength dependence of n explicitly. Thus, for two adjacent modes, we have

$$q\lambda_q = 2n(\lambda_q)d \qquad (9.5.1a)$$

$$(q-1)\lambda_{q-1} = 2n(\lambda_{q-1})d \qquad (9.5.1b)$$

In our "zeroth"-order approximation to date, we have assumed $n(\lambda)$ to be evaluated at some convenient nearby wavelength λ_0. Now we expand $n(\lambda)$ in a Taylor series about that same wavelength.

$$n(\lambda) = n(\lambda_0) + \frac{dn}{d\lambda}(\lambda - \lambda_0) \qquad (9.5.2)$$

Substitute this into (9.5.1) and subtract the two equations

$$(q-1)\lambda_{q-1} = 2d\left[n(\lambda_0) + (\lambda_{q-1} - \lambda_0)\frac{dn}{d\lambda}\right]$$

$$q\lambda_q = 2d\left[n(\lambda_0) + (\lambda_q - \lambda_0)\frac{dn}{d\lambda}\right]$$

$$\overline{-\lambda_{q-1} + q(\lambda_{q-1} - \lambda_q) = 2d(\lambda_{q-1} - \lambda_q)\frac{dn}{d\lambda}}$$

Solving for the difference, $\lambda_{q-1} - \lambda_q = \Delta\lambda$ and dropping the distinction between λ_{q-1} and λ_0 leads to

$$\Delta\lambda = \frac{\lambda_0}{q - 2d(dn/d\lambda)} \qquad (9.5.3)$$

If we define an effective index of refraction such that our simple resonance formula applies (i.e., $q\lambda = 2n_{\text{eff}}d$), (9.5.3) provides us with a formula for n_{eff}, namely, let

$$q\lambda \triangleq 2n_{\text{eff}}d, \qquad q = \frac{2n_{\text{eff}}d}{\lambda}$$

then

$$\Delta\lambda = \frac{\lambda^2}{2d[n_0(\lambda) - \lambda(dn/d\lambda)]} \qquad (9.5.4)$$

If the linear dispersion term were zero, then $\Delta\lambda = \lambda^2/2nd$; hence $n_{\text{eff.}}$ is given by

$$n_{\text{eff.}} = n(\lambda_0) - \lambda \frac{dn}{d\lambda} \qquad (9.5.5)$$

Fig. 9.20 shows raw experimental data of the spontaneous emission intensity emitted by a semiconductor sample 0.043 cm long.* This sample is driven at a level that is below threshold, and thus the Fabry-Perot modes are visible but are not overwhelmingly intense. In the region between 648.5 and 649.0 nm, there are $4\frac{3}{4}$ modes for an average $\Delta\lambda$ of 1.053 Å. Using a mean wavelength of $(648.5 \times 649.0)^{1/2}$ and (9.5.4) gives an effective index of 4.65. Performing the same type of calculation in the region between 649.5 and 650.5 nm yields an effective index of 4.18. Inasmuch as the normal index of refraction for this type of a semiconductor (GaAs) is 3.5, one can see that $dn/d\lambda$ plays an important role.

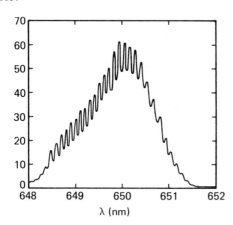

Figure 9.20 Spontaneous emission spectra from a $GaAs_{1-x}P_x$ sample driven below threshold (Sample No. 373–2).

9.5.2 Anomalous Dispersion

Of course, in a dense solid material such as GaAs, one can expect changes of the order calculated in the example given previously. But there is even a higher-order effect due to the presence of gain, which is measurable even in a dilute system such as a gas.† This higher-order effect arises because of the frequency dependence of the gain coefficient.

To appreciate this last statement, one should recognize that the gain coefficient, $\gamma(\nu)$, is twice the real part of a complex propagation coefficient of an electromagnetic wave, and the imaginary part is the phase shift (per unit of length) of that wave. It is a fundamental fact of life that the real and imaginary

---

*These data were supplied by N. Holonyak, Jr.

†For a gas system the material dispersion (i.e., $dn/d\lambda$) is negligible since n is so close to 1. For instance, air at STP has $n = 1.000281$.

Sec. 9.5 Dispersion Effects

parts of *any* transfer function of *any* physical system obeying causality* are related by a complicated set of integrals, called the Kramers-Kronig relationships. Since the gain of a laser is a rather violent function of frequency in the vicinity of the line center, one should expect extraordinary changes in the phase shift in this same band. In Sec. 9.4.4 we used this fact to develop the transfer function of a homogeneously broadened gain cell; compare with (9.4.13).

Recall that the complex gain coefficient was given by

$$\frac{\gamma(\omega)}{|\gamma(\omega_0)|} = \frac{1}{1 + j\left(\frac{\omega - \omega_0}{\Delta\omega/2}\right)} \simeq 1 - j\frac{\omega - \omega_0}{(\Delta\omega/2)} - \left(\frac{\omega - \omega_0}{\Delta\omega/2}\right)^2 \quad (9.4.13b)$$

This can (and will) be derived in Chapter 15 by assuming an active electron bound to the core with a "spring" yielding the resonant frequency ω_0, and it can be derived from the Kramers-Kronig formulas of Appendix II. For now, let us acknowledge that the real part of (9.4.13) agrees with our traditional Lorentzian line shape, and thus it is representative of any bell-shaped curve.

Since the field propagates as $\exp[(\gamma z)/2 - jkz]$, the imaginary part of (9.4.13) can be interpreted as a change in the phase of the wave.

$$\phi = \left[k + \frac{\omega - \omega_0}{\Delta\omega} \cdot |\gamma(\omega_0)|\right] z \quad (9.5.6)$$

Now we invoke the condition for resonance: in a cavity of length d with a gain length l_g:

$$\frac{\omega}{c'} 2d + \frac{(\omega - \omega_0)}{\Delta\omega} |\gamma(\omega_0)| 2l_g = q \cdot 2\pi \quad (9.5.7a)$$

or

$$\nu\left(\frac{2d}{c'}\right) = q + \left(\frac{\nu_0 - \nu}{\Delta\nu}\right) \frac{|\gamma(\omega_0) l_g|}{\pi} \quad (9.5.7b)$$

A graphic solution shows the physics of the phenomenon much better than analytic chicanery. In Fig. 9.21, we consider a Lorentzian-like gain profile in (a), and then a plot of the right-hand side of (9.5.7b) for various values of q in Fig. 9.21(b). The left-hand side of (9.5.7b) is a straight line of slope $2d/c$ and the intersections are the solutions to the equation.

If $\gamma_0(\nu) = 0$, the right-hand side of (9.5.7b) would be horizontal lines differing by integers -1, 0, and $+1$ in the vertical direction, and the intersections would be spaced $c'/2d$ apart. Because of the dispersion of the gain of the active medium, the modes are *pulled* toward the line center. It is left as a problem to show that the modes are separated by

---

* Almost every physical system obeys causality. In this simplest form, causality requires that the output follow the input in time sequence (i.e., the output may not precede the input), a most reasonable requirement.

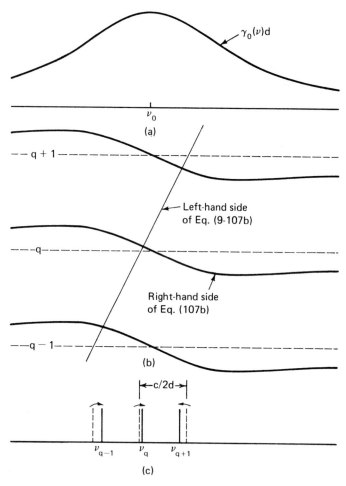

Figure 9.21 Mode pulling in a gas laser: (a) the gain; (b) the variation of $n(\nu) - 1$; (c) the oscillating spectra of the laser.

$$\nu_q - \nu_{q-1} \doteq \frac{c'}{2d}\left[1 - \frac{\gamma_0(\nu)}{\pi}\frac{l_g}{\Delta\nu}\frac{c/2d}{}\right] \quad (9.5.8)$$

If the laser is now allowed to oscillate, each cavity mode will drill a "hole" in the gain profile, which can be considered a narrow-band "absorption." This, in turn, creates a change in refractivity in the opposite sense as gain leading to modes "pushing" away from each other. As one can readily appreciate, this can become quite complicated.

PROBLEMS

9.1. The laser transition in neon at 6328 Å is Doppler broadened by the thermal motion of the atoms at a gas temperature of $\sim 300°C$ (assume pure Ne^{20}). (Volume = 1 cm^3.)
  (a) What is the full width at half maximum of the transition? (Ans. 1.81 GHz.)
  (b) How many blackbody modes couple to this transition [use $\Delta\nu$ from (a)]? (Ans. 3.79×10^8 modes.)
  (c) Suppose that a He:Ne laser tube is placed inside a cavity 1 m long. How many $TEM_{0,0,q}$ modes are within $\pm\Delta\nu/2$ of the line center? (Ans. 12.)
  (d) What is the relative probability of a group of neon atoms radiating into one of the $TEM_{0,0,q}$ modes as opposed to all of the blackbody modes? (Ans. $p = 3.2 \times 10^{-8}$.)
  (e) The energy of the upper state, $3s_2$, is 166,658.484 cm^{-1} above the ground state. Express the upper- and lower-state energies in eV.
  (f) What is the quantum efficiency of this laser?
  (g) Find the stimulated emission cross section given that $A_{21} = 6.56 \times 10^7$ sec^{-1}.
  (h) If the density of excited neon atoms in state 1 ($J_1 = 2$) is 10^{10} cm^{-3}, how many excited atoms in state 2 ($J_2 = 1$) are required to establish a small-signal gain coefficient of 5%/m?

9.2. Consider the atomic system shown below, where the A coefficients are given.

$A_{21} = 5 \times 10^7$
$A_{20} = 2 \times 10^7$
$A_{10} = 10^8$

  (a) What are the wavelengths of the various transitions? Express all transitions in terms of units in common use (eV, Hz, Å, nm, cm^{-1}).
  (b) What is the lifetime of state 2? (Ans. 14.3 ns.)
  (c) What is the branching ratio of the $2 \to 1$ transition? (Ans. 0.71.)
  (d) Suppose that 10^{14} atoms/cm^3 are excited from state 0 to state 2 at $t = 0$ by some external mechanism. Describe the time evolution of the various populations in state 1 and state 2.
  (e) Suppose that this external agent is stong enough to keep a steady-state population of 10^{14} atoms/cm^3 in state 2.
    (1) How much power is required? (Ans. 3.84 kW/cm^3.)

(2) What is the steady-state population of state 1? (Ans. 5×10^{13} cm^{-3}.)

(3) What power is radiated spontaneously in the $2 \rightarrow 1$ transition? (Ans. 1.84 kW/cm^3.)

(4) What is the quantum efficiency of the $2 \rightarrow 1$ transition? (Ans. 67.7%.)

9.3. An argon-ion laser at $\lambda_0 = 0.5145$ μm generates a TEM$_{0,0}$ Hermite-Gaussian beam with the laser cavity shown below.

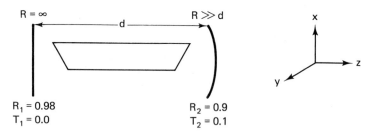

(a) Specify the polarization of the optical electric field.

(b) If the minimum spot size is 1 mm and the *output* power is 8 W, what is the peak electric field inside the laser cavity incident on the spherical mirror?

9.4. Suppose we had a laser system with a stimulated emission cross section of 10^{-16} cm^2 and an initial inversion of 10^{14} cm^{-3} filling a cavity 50 cm long. The cavity uses mirrors with reflectivities $R_2 = 0.9$ ($T_2 = 0.1$) and $R_1 = 0.98$ ($T_1 = 0$) and contains a nonsaturable loss of 2% per round-trip. Assume homogeneous broadening.

(a) By what factor is the system above threshold (i.e., what is the ratio of $\Delta N/\Delta N_t$)?

(b) Estimate how long it would take for the laser intensity to increase (from noise) by a factor of 10^5. (Ignore saturation.)

(c) If the wavelength is 1 μm and the lifetime of the upper state is 1.0 μs, find the output intensity of this laser. (Assume $\tau_1 \sim 0$ and $dI/dz \sim 0$ within the laser cavity.)

9.5. The cavity shown in Fig. 9.4 with $R_1 = 0.95$ ($T_1 = 0$), $R_2 = 0.8$ ($T_2 = 0.2$), $l_g = 10$ cm, and $d = 15$ cm is used for a laser at $\lambda_0 = 7200$ Å. The stimulated emission cross section is 10^{-18} cm^2 and the saturation intensity is 20 kW/cm^2. (You may assume saturation according to the homogeneous law.)

(a) What is the photon lifetime of the passive cavity?

(b) What is the inversion density $[N_2 - (g_2/g_1)N_1]$ necessary to reach threshold for CW oscillation?

(c) Suppose that the cavity mode is "seeded" with a running wave of 10^8 photons/cm^2 (at $\lambda = 7200$ Å at $t = 0$); then an inversion of 2×10^{17}

cm^{-3} is created instantaneously (by an external pump). *Estimate* the time interval required for the internal intensity of the laser to reach half of the saturation value. (NOTE: An approximate solution is desired here, but justify your approach with some physical reasoning.)

(d) Is the actual time interval longer or shorter than your estimate?

9.6. An optical amplifier of length $l_g = 10$ cm uses a homogeneously broadened transition centered at $\lambda_0 = 760$ nm, with a stimulated emission cross section of 2×10^{-20} cm^2, an upper-state lifetime of 1.54 ms, and a negligible lower-state lifetime. There is some residual loss of $\alpha = 0.01$ cm^{-1}, which is independent of intensity whereas the gain coefficient saturates according to the homogeneous law.

(a) Find the inversion necessary to obtain a net small-signal amplification of 6 dB.

(b) What is the value of the saturation intensity (in W/cm^2)?

(c) At what value of the input intensity will the net gain of this amplifier be 3 dB (i.e., $I_{out}/I_{in} = 2$)?

9.7. The following questions refer to the ring laser shown below. Assume that the active medium is an atomic gas (copper) with the energy levels as shown and that the transition is homogeneously broadened with a Lorentzian line width of 3.5 GHz. (State 1 has a long lifetime, and thus the copper vapor laser is not a CW system. However, that fact does not affect the following.)

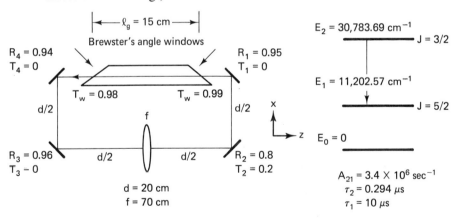

(a) Find the wavelength in nanometers. (Ans. 510.7 nm.)
(b) What is the line width (FWHM) in angstrom units? (Ans. 0.0304 Å.)
(c) What are the degeneracies of the states? (Ans. $g_2 = 4$, $g_1 = 6$.)
(d) What is the stimulated emission cross section (in cm^2)? (Ans. 6.42×10^{-14} cm^2.)
(e) What is the quantum efficiency of this laser? (Ans. 63.6%.)
(f) What is the minimum required inversion density to reach threshold for CW oscillation? (Ans. 4.23×10^{11} cm^{-3}.)

(g) Because of Brewster's angle windows the laser should be linearly polarized. Specify the direction of the optical electric field if the plane of the page is the xz plane and the output is along z. (Ans. $\mathbf{E} = E_0 \mathbf{a}_x$.)

(h) Is the cavity stable? Demonstrate your ability to analyze such cavities by
   (1) Showing an equivalent lens waveguide starting at M_1 and proceeding counterclockwise around the cavity.
   (2) Identifying a unit cell with the lens being the first element.
   (3) Finding the $ABCD$ matrix for this unit cell in terms of d and f.
   (4) Justifying your answer for the stability.

(i) Suppose the active medium acted as a *negative* gas lens (which it does). *Indicate* the place and order of matrix multiplication for the unit cell defined previously. (Do not expand, just indicate the operation.)

9.8. The following questions refer to the laser system shown below. Assume that only the clockwise wave is oscillating and that the transition is homogeneously broadened with $\Delta \nu_h = 2$ GHz and $A_{21} = 5 \times 10^5$ sec^{-1}.

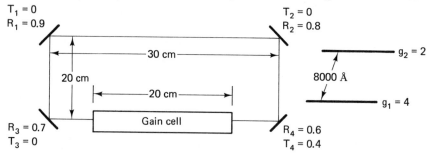

(a) What is the stimulated emission cross section? (Ans. 4.05×10^{-14} cm^2.)

(b) What is the population inversion necessary to reach threshold for oscillation in this cavity? (Ans. 1.48×10^{12} cm^{-3}.)

(c) If the lifetime of the upper state (due to all spontaneous and collisional causes) is 0.9 μs and that of the lower state is 10 ns, what is the saturation intensity? (Ans. 6.8 W/cm^2.)

(d) Express the line width in Å and in cm^{-1}. (Ans. 0.043 Å and 0.067 cm^{-1}.)

(e) Assume that an external pumping agent is sufficient to create a small-signal gain coefficient of 0.1 cm^{-1} (which is higher than the threshold requirement). What is the output intensity? (Ans. 3.13 W/cm^2.)

9.9. The following questions refer to a ring laser similar to that of Problem 9.8. The cavity specifications are as follows: $R_1 = 0.96$, $R_2 = 0.8$, $R_3 = 0.97$, $R_4 = 0.98$, $T_1 = T_3 = T_4 = 0$, $T_2 = 0.2$. The wavelength of the transition is 760 nm, which originates at a state at $E_2 = 3.2$ eV; the stimulated emission cross section is 2×10^{-20} cm^2; the upper-state lifetime is 1.54 ms and is pumped directly from the ground state at a rate R_{02}; and the lower-state lifetime is negligible (i.e., $N_1 = 0$ always). For the purpose of this problem, assume only the counterclockwise wave is oscillating.

(a) Find the upper-state density required to reach threshold, assuming steady-state operation.
(b) Estimate the pumping power (per unit volume) R_{02} required to reach threshold. (NOTE: One can neglect stimulated emission for this part of the calculation.)
(c) Compute the output intensity (W/area) of the laser if the pumping rate was 1.5 times that required to reach threshold.

9.10. Consider the ring laser shown in problem 9.8 with $R_1 = 0.9$, $R_2 = 0.7$, $R_3 = 0.6$, $R_4 = 0.95$, and $T_1 = T_2 = T_4 = 0$, $T_3 = 0.2$. Assume that the system is excited so that the small-signal gain coefficient is three times the threshold value in the absence of stimulated emission, that there is homogeneous broadening, and that laser oscillation is restricted to the counterclockwise wave near line center. Compute the *output* intensity if $I_s = 5$ W/cm^2 using the exact analysis and compare that with the high Q approximation method in which $dI/dz \sim 0$.

9.11. Consider an optically pumped laser system depicted in the diagram below. A strong optical pump tuned to line center of the $0 \rightarrow 2$ transition is incident on the sample from the side and promotes atoms from state 0 to 2. The atoms in state 2 can decay back to 0 by the indicated spontaneous emission ($A_{21} = 10^5$ sec^{-1}) and/or quenching processes ($k_{20} = 5 \times 10^6$ sec^{-1}), or to state 1 by spontaneous emission ($A_{20} = 10^6$ sec^{-1}) and/or stimulated emission. The spacing $E_1 - E_0$ is much larger than kT, and thus one can neglect any initial population in state 1. Atoms in state 1 find their way back to 0 at a rate $\tau_1^{-1} = 10^7$ sec^{-1}. In order to make the problem simple, assume that the density of state 2 (or 1) is always much less than that of state 0, and thus one need not worry about the conservation of atoms (or a rate equation for state 0), that is, $N_0 =$ constant. Assume steady state and a homogeneous line shape for all transitions with $\Delta \nu_h = 10$ GHz.

(a) What is the *absorption* cross section for the pump wave at the frequency ν_{20}?
(b) What is the stimulated emission cross section for the $2 \rightarrow 1$ transition?
(c) What is the lifetime of state 2?
(d) If the optical pump intensity is weak, the $2 \rightarrow 1$ transition does not lase and stimulated emission at ν_{21} can be neglected. Under these circumstances, what is the ratio of the population densities N_2/N_1?

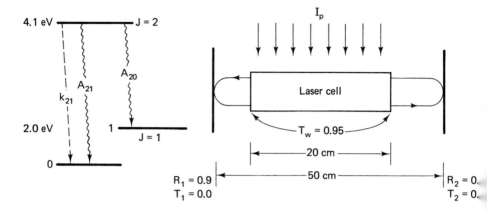

(e) What is the minimum inversion density, $N_2 - (g_2/g_1)N_1$, required to achieve oscillation on the $2 \to 1$ transition?

(f) Formulate the steady-state rate equation for states 2 and 1 in terms of σ_p, I_p, σ_{21}, and I_{21} and the appropriate atomic relaxation rates. Use these equations to determine the intensity of the pump to reach threshold on the $2 \to 1$ transition.

(g) If the pump is 10 times the value required to reach threshold at ν_{21}, then the stimulated emission rate on the $2 \to 1$ transition overwhelms all other loss processes for state 2 and pumping processes for state 1. Use this fact to neglect the appropriate terms in the rate equations and thus predict the *output* intensity of this laser. (The approximations are intended to simplify matters, but if they confuse the issue, do it the long way—it is not that much more difficult.)

(h) If the pump intensity is very large, then the approximation that the ground state density does not change is not valid. The atoms are pumped to state 2, immediately stimulated to make a transition to 1, and then slowly relax back to 0. Using this scenario, predict the pump intensity at which the ground state density is depleted by 30%.

9.12. A certain laser medium is described by the following specifications: center of the transition = 2 eV; line width of the homogeneously broadened transition = 1.5 GHz; stimulated emission cross section at line center = 10^{-16} cm^2; lifetime of the upper state τ_2 = 1 msec; the branching ratio of the upper state = 0.8; lifetime of lower state τ_1 = 10 μs. Assume equal degeneracies of the states. The cavity is similar to that shown in Fig. 9.4 with $R_1 = 0.98$ ($T_1 = 0$), $R_2 = 0.9$ ($T_2 = 0.1$); a gain length l_g = 30 cm; and a cavity d = 50 cm.

(a) What is the threshold population inversion to achieve oscillation?

(b) If the system is pumped to 1.5 times its threshold value, explain how the laser selects the frequency (or frequencies) of oscillation.

(c) Estimate the intensity (in W/cm²) generated by this laser when pumped to 1.5 times threshold. Use the assumption that $dI/dz \sim 0$.
(d) Use the Rigrod analysis of Sec. 9.2.3(b) to compute the output intensity and compare with (c).

9.13. Read and report on the paper "Short Pulse Amplification in the Presence of Absorption" by M. Tilleman and J. Jacob, Appl. Phys. Lett. *50*(3), 121, 1987. Your report should include a discussion of
(a) The formulation and meaning of the various terms in Eq. 1 to 3 of the paper.
(b) The derivation of Eq. 4.
(c) If the optical bandwidth is 0.3 nm, they restrict their attention to pulses longer than 1 ps. Why?
(d) What is the asymptotic value of the energy from an amplifier in terms of the small-signal integrated gain to loss ratio γ_0/α?
(e) Recompute the information required to construct Figs. 2 and 3 for $\gamma_0/\alpha = 10$.

9.14. A homogeneously broadened ring laser with a small-signal gain coefficient of 0.2 cm⁻¹ and a saturation intensity of 10 mW/cm² oscillates in the cavity shown in Fig. 9.2(a) with $R_1 = R_3 = R_4 = 0.98$, $T_1 = T_3 = T_4 = 0$, and $T_2 = 1 - R_2$. The loss coefficient of the "diode,"* which constrains oscillation to the counterclockwise direction, is 0.5 cm⁻¹.
(a) Assume that the oscillation frequency is close to the center of the transition and that the transmission of M_2 is variable. Plot the output power as a function of the transmission coefficient T_2 using the various theories of Sec. 9.2; that is,
  (1) Use the self-consistent theory of Sec. 9.2.1 (use a solid curve).
  (2) Assume that the losses are prorated over the gain length and that $dI/dz = 0$ within the gain cell as was done in the simplified analyses of Sec. 9.2.3(a) (use a dashed curve).
(b) If one uses (9.2.9) to predict the optimum coupling, what percentage error in the output intensity results?

9.15. Assume that oscillation in the ring laser shown below is constrained to the counterclockwise direction by some unidirectional optical device. The gain coefficient of the active medium is 4.66%/cm, and there is a 2% scattering loss at the entrance and exit of the cell.
(a) What is the minimum value of the reflectivity R_3 that permits oscillation?
(b) Solve the problem self-consistently by demanding that the input to the

---

*For information on such "diodes" see "Design and Performance of a Broad-Band Optical Diode to Enforce One-Direction Traveling Wave Operation of a Ring Laser" by T. F. Johnston, Jr., and W. Proffitt, IEEE J. Quant. Elect. *QE-16*, 483–489, 1980.

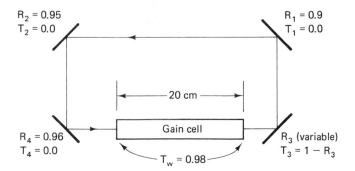

gain cell be the feedback from the output. Plot the output intensity (normalized to the saturation value) as a function of the transmission coefficient of M_3.

(c) Solve for the output intensity by making the high Q approximation in which the saturated gain coefficient is set equal to the losses prorated over the gain length. Compare with the answer of (b).

9.16. Redo the development of Sec. 9.3.1 on Q switching for the case $g_2 \neq g_1$; that is, show that (9.2.29b), (9.3.1), and (9.3.5) are modified to

$$\frac{dn}{dt} = -\left(1 + \frac{g_2}{g_1}\right)\frac{n}{n_{\text{th}}}N_p \qquad (9.3.10\text{b})$$

$$N_p(\max.) = \frac{n_i - n_{\text{th}}}{(1 + g_2/g_1)} - \frac{n_{\text{th}}}{(1 + g_2/g_1)}\ln\left(\frac{n_i}{n_{\text{th}}}\right) \qquad (9.3.12)$$

$$W_p = \frac{n_i h\nu}{(1 + g_2/g_1)}\eta_{\text{xtn}} \qquad (9.3.16\text{c})$$

where $n = [N_2 - (g_2/g_1)N_1]Al_g$.

All other equations and Fig. 9.8 remain the same.

9.17. Consider a ruby laser to be Q switched as shown in Fig. 9.7. The ruby rod has a cross-sectional area of 1 cm^2, is 15 cm in length, and is placed in a cavity 20 cm long. The Cr^{3+} doping density is 1.58×10^{19} atoms/cm^3 and the stimulated emission cross section is $\sigma = 1.27 \times 10^{-20}$ cm^2. The pumping agent creates an initial population of 10^{19} atoms/cm^3 in the upper laser state with negligible population in the pumped states (i.e., $N_3 \sim 0$). Assume that the pump band is centered at $\lambda = 4500$ Å; every atom pumped to state 3 relaxes to state 2; lifetime of state 2 is 3 ms; $g_1 = g_2$; and the mirror specifications are $R_1 = 0.95$, $T_1 = 0$, $R_2 = 0.7$, $T_2 = 1 - R_2$.

(a) How much pump power is required to maintain the population in state 2 at 10^{19}/cm^3?

(b) How much power is radiated spontaneously before the Q switch is operated?

(c) What is the peak output power of the Q-switched pulse?
(d) How much energy is contained in the pulse?
(e) Estimate the pulse width by using (9.3.17) and the predictions of Fig. 9.10.
(f) Plot the peak power and energy contained in the Q-switched pulse as the transmission coefficient, T_2, is varied from 0 to 1.0.

9.18. Consider a ruby laser to be Q switched as shown in Fig. 9.7. The rod is 10 cm long and 1 cm^2 in area and is placed in a cavity 30 cm long. The doping density is 1.58×10^{19} atoms/cm^3 and the absorption cross section at 6943-Å is 1.27×10^{-20} cm^2. The pumping agent creates an initial population of 10^{19} atoms/cm^3 in the upper state N_2 with negligible population in pumped states ($N_3 \sim 0$). Assume that the pumping occurs by virtue of the absorption at $\lambda = 4500$ Å and that 90% of the atoms pumped to state 3 relax to state 2. The radiative lifetime of state 2 is 3 ms. The index of refraction of the ruby rod is 1.78 and that of the electro-optic switch is 2.35. The switch is 8 cm long. The mirror characteristics are $R_1 = 0.95$, $T_1 = 0$; $R_2 = 0.7$, and $T_2 = 0.3$. There is a 3% scattering loss at each air-to-ruby interface and a 5% scattering loss at each interface of the switch, and the open switch absorbs 2% of radiation passing through it. The degeneracies of the states are $g_2 = 2$, $g_1 = 4$.
(a) How much spontaneous power is radiated by the rod before the switch is opened?
(b) How much pump power is required to maintain the population of state 2 at 1×10^{19} cm^{-3}?
(c) What is the peak output power of the Q-switched pulse?
(d) How much energy is contained in the output pulse?
(e) Estimate the pulse width using (9.3.12).

9.19. A laser at $\lambda = 1.0$ μm, which is to be Q switched, has an initial population inversion at 10^{18} atoms, a factor of 3 above the threshold value required for CW oscillation in the high Q cavity. There are three types of photon losses in the cavity: coupling through one mirror to the outside world at a rate of 5×10^7 sec^{-1}; loss in the electro-optic switch at a rate of 10^7 sec^{-1}; and useless losses at the air-rod-switch interfaces at a rate of 2×10^7 sec^{-1}. The inversion after the Q-switched pulse is over 5.9×10^{16} atoms.
(a) Find the peak power out of this laser.
(b) Find the output energy in the Q-switched pulse.
(c) Estimate the pulse width.

9.20. Let us reconsider Problem 9.17 and allow a 4% loss per pass at each air-ruby interface ($T = 0.96$). Let the Q-switch element be 2 cm long, have an index of refraction of 1.5, and have a 2% residual loss per pass when open. Find the peak power in the output, the output energy, and the pulse width from Fig. 9.10 as well as (9.3.12).

9.21. The system shown in the diagram below is to be used to generate a Q-switched pulse in the cavity shown there. If the system is pumped to four times the threshold value, find the output energy of the pulse. (CAUTION: This system is very different from ruby or the usual models chosen for Q-switching analysis. The lower state decays very fast, even faster than a round-trip transit time.)

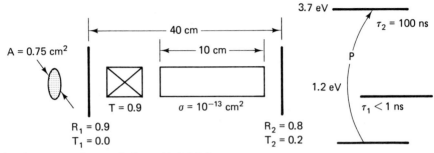

9.22. Derive (9.2.30) from (9.2.29c).

9.23. Derive (9.2.28) from (9.2.27b) and the boundary conditions.

9.24. Return to the rate equations (9.2.24a) and (9.2.24b) and make the ideal CW laser approximation of τ_1 and $d/dt = 0$.
  (a) Show that the extraction efficiency can be written as

$$\eta_{xtn} = \frac{\Gamma}{\Gamma + \Gamma_s} - \frac{1}{m}\frac{\Gamma}{\Gamma_s}$$

where

$$\Gamma_s = \frac{I_s}{h\nu} = \text{saturation flux} = \frac{1}{\sigma \tau_2}$$

  (b) For what value of Γ is this a maximum?
  (c) What is the maximum extraction efficiency?
  (d) Evaluate for $m = 5$.

9.25. Find the exact solution to (9.3.24) and evaluate the error in the approximate solution (9.3.25). Use the same numerical parameters as were given in the text.

9.26. An amplifier with a small-signal gain of 10 dB (i.e., $G_0 = 10$) is irradiated by a square pulse of 100-ns duration containing 50 mJ/cm² energy at the input. The saturation energy of the amplifier is 100 mJ/cm².
  (a) Sketch the output intensity as a function of time.
  (b) Identify the time at which the output intensity is half of its peak value.
  (c) What is the maximum energy (per cm²) that can be extracted from this amplifier?

9.27. This problem predicts the results of an experiment in which one irradiates an *absorption* cell with an optical pulse approximated by

$$I(t) = I_0 \sin^2\left(\frac{\pi t}{T}\right) \quad (T = 2 \text{ ns})$$

The optical frequency is tuned to the center of the transition, which has an absorption cross section of 10^{-14} cm^2. The "small"-signal transmission through the cell is -30 dB. Graph the output intensity as a function of time for three values of energy contained in the pulse (a) $w = 2$ μJ/cm^2; (b) $w = 20$ μJ/cm^2; (c) $w = 200$ μJ/cm^2. ($\lambda_0 = 5889$Å).

9.28. Suppose that we had a homogeneously broadened optical amplifier at 1.315 μm that is 1 m long and has a stimulated emission cross section of 10^{-20} cm^2 (at the line center), a line width of 5 GHz, and an inversion of 2×10^{18} cm^{-3}. This amplifier is irradiated with a "square" pulse that is 10 nsec wide and is detuned from the line center by 2.5 GHz.
(a) What is the small-signal gain G of this amplifier?
(b) Describe, with words and sketches, the output wave shape as the energy (per cm^2) of the pulse is varied. Estimate the energy per cm^2 when appreciable pulse distortion occurs—say the trailing edge is 50% of the leading edge.
(c) What is the maximum energy per cm^2 that can be extracted from this amplifier?
(d) Does the answer to (c) change if the input is tuned to the line center? Explain.

9.29. A homogeneously broadened laser cell has the following specifications: a gain coefficient of $\frac{1}{2}$%/cm (at the line center); a saturation intensity of 25 W/cm^2, a FWHM of 2 GHz, and length of 80 cm. The simple optical cavity of Sec. 9.2 is 100 cm long, with mirrors $R_1 = 0.98$ and R_2 variable; all other losses are negligible.
(a) Assume oscillation near the line center and plot the output intensity as a function of the coupling $(1 - R_2)$. Show two graphs, one using the simple theory and the other using the more exact computation.
(b) What is the coupling to achieve maximum power, and what is that power? Assume a spot size $w = 1$ cm.
(c) If one tunes the mirror spacing by mounting R_1 on a translator, one can move the cavity mode across the gain profile. If this is done with the simple cavity shown in Fig. 9.2 with $R_2 = 0.7$, the output does *not* vary significantly. Explain why it does *not*.

9.30. Consider an inhomogeneously broadened (Doppler) laser transition whose small-signal gain coefficient, at the line center, exceeds the loss (prorated over the length) by some specified factor K; that is,

$$\gamma_0(\nu) = \gamma_0(\nu_0) \exp\left[-(4\ \ln 2)(\nu - \nu_0)^2/\Delta\nu_D^2\right]$$

where $\gamma_0(\nu_0)/\alpha = K$.

(a) Derive an expression for the bandwidth over which oscillation is possible.
(b) If the length of this laser is d units, how many $\text{TEM}_{0,0}$ modes can oscillate? Assume that the central mode is at the line center.
(c) If we again prorate the losses over the gain length and assume $dI/dz \cong 0$ and an *inhomogeneous* saturation law [cf (8.6.11)], find the intensity of each mode in terms of $\gamma_0(\nu)$ and α.
(d) Derive an explicit formula for I_q/I_s in terms of the gain coefficient at the line center $\gamma_0(\nu_0)$, the loss α, and the frequency separation from the line center.
(e) This expression is almost a Gaussian. Derive an expression that relates the FWHM of the mode intensities to the Doppler width and the gain-to-loss ratio, $\gamma_0(\nu_0)/\alpha$.

9.31. Suppose that the amplitudes of the modes of a laser had a Lorentzian-like distribution of power such that

$$P = P_0 \sum_{-\infty}^{+\infty} \frac{(\Delta\omega/2)^2}{(n\omega_c)^2 + (\Delta\omega/2)^2}$$

where P_0 is the average power of the mode at the line center and $\Delta\omega$ is the FWHM of the oscillating spectrum. Compute the following quantities (as a function of $\Delta\omega$, and the power P_0).
(a) The total average power produced by this laser. (Ans. $\langle P \rangle = [P_0(\pi/4)(\Delta\omega/\omega_c)]$).
(b) The peak power of the mode-locked pulse $[(4/\pi)(\Delta\omega/\omega_c)\langle P \rangle]$.
(c) The pulse width (FWHM) of the mode-locked pulse ($\Delta t = 2 \ln 2/\Delta\omega$).

9.32. Suppose that the power in the modes of a laser had a $[(\sin x)/x]^2$ distribution.

$$P(\omega) = P_0 \sum_n \frac{\sin^2(n\pi\omega_c/\Delta\omega)}{(n\pi\omega_c/\Delta\omega)^2}$$

where

$$-N < n < +N \text{ and } \frac{\Delta\omega}{\omega_c} = N,\ \omega_c = 2\pi\frac{c}{2d}$$

where P_0 is the average power of the mode *at the line center* and $\Delta\omega$ the half width of the spectrum of oscillation. Find the following (as a function of average power and $\Delta\omega$).
(a) The peak power.

(b) The pulse width (FWHM) of the mode-locked pulse.
(c) Sketch the time dependence of the mode-locked pulse, labeling the peak power, the pulse width, and the pulse repetition frequency.
(NOTE: It is necessary to make various approximations to obtain an analytic answer to this problem. Do not hesitate to do so, but explain why the approximations are justified or not.)

9.33. Suppose the spectral distribution of power of a mode-locked laser is given by

$$P(\omega) = P_0 \sum_{n=-\infty}^{+\infty} \text{sech}^2\left(\frac{n\omega_c}{\Delta\omega}\right)$$

where n is the mode number measured from line center ω_0; P_0 is the power in the central mode; and $\Delta\omega$ is a characteristic width of the spectra. Assume $\Delta\omega/\omega_c = 10$; $\nu_c = \omega_c/2\pi = 100$ MHz; $P_0 = 10$ mW.
(a) What is the average power?
(b) What is the peak power of the mode-locked pulse?
(c) What is the width (full width at half maximum) of the mode-locked pulse?
The approximate relations $P_{\text{peak}} = N\langle P\rangle$, $\Delta t = (2d/c)/N$ can be used as a check, but an exact answer is required here.

9.34. The frequency spacing between modes of a low gain laser, such as the 6328-Å laser, is given by $\nu_{0,0,q+1} - \nu_{0,0,q} = c/2d$. For high gain lasers, the dispersion caused by the inverted population modifies this formula; see (9.5.7). Consider the He:Ne laser transition at 3.39 μm ($3s_2 - 3p_4$) with a small-signal gain coefficient of 30 dB/m; line width of ~300 MHz, and $A_{21} = 2.87 \times 10^6$ sec^{-1} with $g_2/g_1 = 3/5$ (see Table 10.3).
(a) Evaluate the stimulated emission cross section.
(b) Express 30 dB/m in terms of a fraction per centimeter.
(c) What must be the inversion, $N_2 - (g_2/g_1)N_1$ to obtain the 30 dB/m gain coefficient?
(d) Derive and evaluate the mode spacing (9.5.8) for the two modes "close" to ν_0.

REFERENCES AND SUGGESTED READINGS

1. A. G. Fox and T. Li, "Resonant Modes in a Maser Interferometer," Bell. Syst. Tech. J. 40, 453–488, Mar. 1961.
2. H. Kogelnik and T. Li, "Laser Beams and Resonators," Appl. Opt. 5, 1550–1567, Oct. 1966.
3. A. E. Siegman, "Unstable Optical Resonator for Laser Applications," Proc. IEEE 53, 277–287, Mar. 1965.
4. A. E. Siegman, *Introduction to Lasers and Masers* (New York: McGraw-Hill Book Company, 1971).

5. W. R. Bennett, Jr., "Gaseous Optical Masers," Appl. Opt., Suppl. Opt. Masers, 24–61, 1962.
6. W. R. Bennett, Jr., "Inversion Mechanisms in Gas Lasers," Appl. Opt., Suppl. 2 Chem. Lasers, 3–33, 1965.
7. W. S. C. Chang, *Principles of Quantum Electronics* (Reading, Mass.: Addison-Wesley Publishing Company, Inc., 1969). This book has an extensive bibliography.
8. W. W. Rigrod, "Saturation Effects in High-Gain Lasers," J. Appl. Phys. *36*, 2487, 1965.
9. W. G. Wagner and B. A. Lengyel, "Evolution of the Giant Pulse in a Laser," J. Appl. Phys. *34*, 2042, 1963.
10. R. W. Hellworth, "Theory of the Pulsation of Fluorescent Light from Ruby," Phys. Rev. Lett. *6*, 9, 1961.
11. E. O. Schultz-DuBois, "Pulse Sharpening and Gain Saturation in Traveling-Wave Masers," Bell Syst. Tech. J. *43*, 625, 1964.
12. "FM and AM Modelocking of the Homogeneous Laser Part 1, Theory; Part II, Experiment," IEEE J. Quant. Electron. *QE-6*, 694, 1970. See also Siegman and Kuezenga, Appl. Phys. Lett. *14*, 181, 1969.
13. T. F. Johnston and W. Proffitt, "Design and Performance of a Broad-band Optical Diode to Enforce One-direction Traveling Wave Operation of a Ring Laser," IEEE J. Quant. Electron. *QE-16*, 483, 1980.
14. D. L. Huestis, see the 5 volume set, *Applied and Atomic Collision Physics*, Eds. Massey, McDaniel, and Beterson (New York: Academic Press, Inc., 1982). Parts of the article by D. L. Huestis (Vol. 3, Chapter 1, p 1, Eds. McDaniel and Nighan) entitled, *Introduction and Overview*, were the basis for sec. 9.2.4.
15. A. E. Siegman, *Lasers*, (Mill Valley, Calif.: University Science Books, 1986), Chaps. 5–6.
16. A. Yariv, *Quantum Electronics* (New York: John Wiley & Sons, Inc., 1975).
17. Peter W. Smith, "Mode Locking of Lasers," Proc. of IEEE, *58*, 1342–1356, 1970. This contains over 150 additional references.

10

Laser Excitation

10.1 INTRODUCTION

This chapter introduces the "facts of life" of real lasers and discusses the physical processes that can lead to a population inversion. The pumping process is the most crucial issue facing laser research. It is well and good to talk about such esoteric topics as cavity modes, energy levels, and stimulated emission, but without adequate pumping, the system will not lase.

Fortunately, it is not as difficult as it might appear at first glance. Some people state emphatically that "something in everything will lase if hit hard enough." The fact that ordinary Jello (the dessert) will lase gives considerable substance to the statement (and also becomes the first edible laser[1]). Thus it is not news for another laser to be found, but it is news to understand the excitation route of any laser.

Almost any energy source can be used, even another laser. The following is a partial listing of the pumping agents that have been used successfully:

1. Optical
   (a) Incoherent—a flash lamp
   (b) Coherent—another laser
2. Electrons
   (a) A swarm—a gas discharge (DC, RF, or pulsed)
   (b) An energetic electron beam (>500 kV)
3. A thermal oven

4. A chemical reaction
   (a) A chemical burn—a flame
   (b) A rapid burn—an explosion
5. Heavy particles
   (a) Ion beams
   (b) Fission products from a reactor-like environment
6. Ionizing radiation
   (a) A nuclear bomb (the ultimate in a one-shot experiment, i.e., singular!)
   (b) An x-ray source

Although considerable prejudice toward the gas laser has been used in compiling this list, many of the same techniques can be used for other types, such as the semiconductor laser. In that case, the most attractive means is to inject the electron-hole pairs into the appropriate cavity for recombination; but semiconductors have been pumped with electron beams, other lasers (gas, solid state, or semiconductor), and with incoherent sources. Semiconductor lasers have become so important that they will be covered in a separate chapter (see Chapter 11).

In view of this wide diversity of techniques and materials, we will have to restrict our attention to a few examples chosen to illustrate the principles involved, rather than to make the reader an expert in laser excitation. We restrict our attention to the first four categories, because most efficient lasers are excited in this manner.

Let us make a final general observation: we start talking about *real* systems, and real states in real systems have spectroscopic names. As interesting as the quantum rules and regulations are for assigning these names, we will not cover these issues here. Rather, we assume that you have some familiarity with spectroscopic notation* and get on with the business at hand.

10.2 THREE- AND FOUR-LEVEL LASERS

Most lasers are classified as having either three or four energy levels that are of significance in pumping and/or lasing processes. This, of course, is a gross approximation; a practical laser has N levels, where N can be a frighteningly large number! Even though such a classification is an overly simplistic approximation, some general ideas can be obtained about the ease or difficulty in pumping such a laser.

---

* Some of the elementary ideas and formulas associated with atomic and molecular notation are given in Chapter 15. The section on vibrational-rotational transitions should be read before attempting to read the sections beyond Sec. 10.5.4.

Sec. 10.2 Three- and Four-Level Lasers

Consider Fig. 10.1, where three-level and four-level systems are shown. A system is classified as a three-level laser if the lower state, 1, is heavily populated by natural causes. In Fig. 10.1(a), state 1 is the ground state of the active atom and thus is heavily populated unless a very unusual set of circumstances prevails.

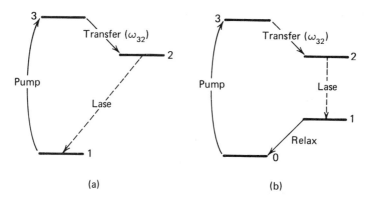

Figure 10.1 Three- and four- level laser systems: (a) three-level laser; (b) four-level laser.

If the energy difference, $E_1 - E_0$, in Fig. 10.1(b) is large compared to the characteristic energy of the environment, then the Boltzmann factor, $\exp[-(E_1 - E_0)/kT]$, is much less than 1 and state 1 is naturally empty.

Whether a state is naturally populated or not controls how hard we must work to create a population inversion. Let us do some elementary arithmetic on the three-level system to appreciate the distinction.

Suppose that the active media had a density of N active atoms; therefore, $N_1(t = 0) = N$. As one pumps the $1 \rightarrow 3$ transition, the density of state 1 decreases, since atoms must be conserved:

$$N_1 + N_2 + N_3 = N \tag{10.2.1}$$

If the relaxation $3 \rightarrow 2$ is very fast, all the atoms pumped to 3 are converted to state 2, and thus the population N_3 is small and can be neglected. If the pump is strong enough, we can first *equalize* the population between (2) and (1) and then proceed to create an inversion. But note that we must work fairly hard to obtain zero gain (i.e., $N_2 = N_1 = N/2$).

If, as we hope, state 2 can only decay by radiative processes, we must supply this *critical fluorescence* power by our pump.

$$\frac{P_f}{V} = \frac{h\nu_{21}}{\tau_{sp.}} \frac{N}{2} \tag{10.2.2}$$

Since N can be a large number, this fluorescence power is large also, and more is required to obtain a laser. Fortunately, once this point given by (10.2.2) is reached everything goes into the creation of an inversion.

The four-level system requires far less power to create an inversion because the lower-state population is minimal.

10.3 OPTICAL PUMPING

10.3.1 Overview

Probably the most straightforward technique for creating a population inversion is to use another photon source as a pump. Historically, this approach to lasers grew out of earlier successes in the microwave portion of the spectrum (i.e., maser) and is still widely used. We use a coherent or an incoherent source of radiation at the frequency $1 \rightarrow 3$ (or $0 \rightarrow 3$) in Fig. 10.1 to create the population inversion. The examples given below illustrate how this is accomplished.

10.3.2 Ruby Lasers

The ruby laser was the first laser[2] and operated in the visible portion of the spectrum at $\lambda_0 = 6943$ Å. The chemical composition of ruby consists of a crystal of sapphire (Al_2O_3) in which a small amount of the aluminum is replaced by chromium by adding Cr_2O_3 to the melt in the growth process. The pure host crystal possesses a rhombohedral unit cell shown in Fig. 10.2(a) where the axis of symmetry is the so-called c axis. Because of the arrangement of the atoms, the crystal is uniaxial with different indices of refraction depending on whether the electric field, E, is perpendicular to c (an ordinary wave with $n_0 = 1.763$) or parallel to c (an extraordinary wave with $n_e = 1.755$).

The chromium atoms are active in the lattice as triply ionized ones, Cr^{3+}, and give rise to the energy-level diagram shown in Fig. 10.2(b). The states participating in the laser transition are labeled by 2 and 1 to correspond to our previous notation. Note that the upper state is split into two levels, denoted by $2\bar{A}$ and $\bar{E}$, which are separated by 29 cm^{-1}. Transitions originating from the $2\bar{A}$ are denoted by R_2; those from $\bar{E}$ are called R_1. The lifetime of the E states because of spontaneous emission is 3 ms, which is very long by atomic process standards, and thus these levels are sometimes classified as *metastable*. Because of the anistropy, the spontaneous emission (and thus stimulated emission) depends on the orientation of the optical electric field with respect to the axis (more about this later). The degeneracy of each of the upper states is 2, whereas that of the ground state is 4; hence one needs to have the population in the $\bar{E}$ level to be half of that in the $^4A_2$ state in order to obtain optical transparency, $N_2 - (g_2/g_1)N_1 = 0$, on the R_1 transition. Note that the $2\bar{A}$ level will have an almost equal population to that of the $\bar{E}$ state, and thus roughly half of the doped

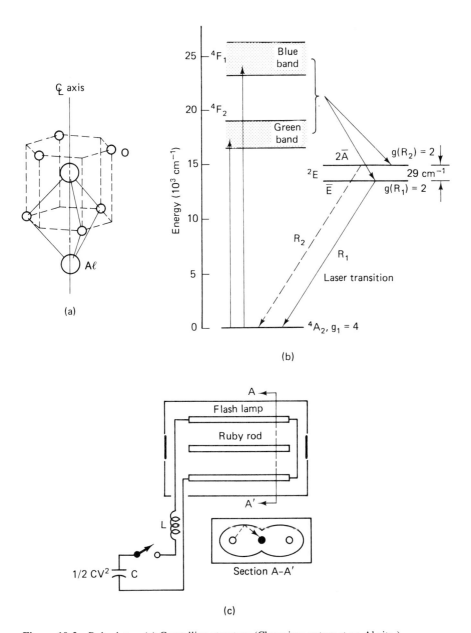

Figure 10.2 Ruby laser. (a) Crystalline structure (Chromium enters at an Al site.). (b) Energy-level diagram for Cr^{3+} in Al_2O_3. (c) Typical pumping scheme using a dual elliptical cavity.

chromium atom must be promoted to the ^2E manifold to obtain transparency on the R_1 line. Further pumping yields gain and oscillation if proper feedback is supplied.

One of the very fortunate and desirable features of ruby is the presence of very strong absorption bands corresponding to the $^4A_2 \rightarrow {}^4F_2, {}^4F_1$ transitions shown. These bands are nearly 1000 Å wide and are located in the green (18,000 cm^{-1}) and violet (25,000 cm^{-1}) portions of the spectrum. Because of their width and strength of optical attenuation they can absorb a significant fraction of light emitted by an incoherent flash lamp that radiates more or less as a blackbody source at an elevated temperature of about 7000° K to 9000° K. Fortunately the lifetimes of the pumped states, $^4F_2$ and $^4F_1$, are very short, with most of the atoms returning to the ^2E levels rather than back to the ground state. The mean quantum efficiency of the pumping process for the two bands is about 70%; that is, 70% of the atoms promoted to the ^4F levels appear in the ^2E levels. A summary of data for a typical ruby rod is given in Table 10.1, and some is presented graphically in Fig. 10.3.

TABLE 10.1 PHYSICAL DATA ON A TYPICAL RUBY SYSTEM*

Item	Value
Cr_2O_3 doping (% by weight)	0.05
Cr^{3+} concentration	1.58×10^{19} cm^{-3}
Output wavelength (at 25° C)	R_1: 14403 cm$^{-1} \leftrightarrow$ 6943 Å
	R_2: 14432 cm$^{-1} \leftrightarrow$ 6929 Å
Spectral line width (300° K)	11 cm^{-1} or 5.3 Å
Quantum efficiency (of pumping)	0.7
Absorption cross section of R_1 laser line	1.22×10^{-20} cm^2
Stimulated emission cross section	2.5×10^{-20} cm^2
Residual scatter losses in crystal at R_1	~0.001 cm^{-1}
Major pump bands	
Blue (404 nm)	$\alpha_\parallel = 2.8$ cm^{-1} $\alpha_\perp = 3.2$ cm^{-1}
Green (554 nm)	$\alpha_\parallel = 2.8$ cm^{-1} $\alpha_\perp = 1.4$ cm^{-1}
Refractive indices	
Ordinary ray ($E \perp c$)	1.763
Extraordinary ray ($E \parallel c$)	1.755

*Data from Koechner.[24]

Note that the absorption cross section for the normal (i.e., unpumped) rod shown in Fig. 10.3 is very strong for the green and violet bands—much stronger than for the laser transition. Indeed the absorption at 6943 Å is hardly "on the map" of Fig. 10.3(a); the details of that absorption are shown with a greatly expanded scale in Fig. 10.3(b). It is important to realize that the data presented in Fig. 10.3 were taken with a low-level signal, small enough to neglect any significant depletion of the ground state population. When the rod is pumped, the $^4A_2$ state density decreases, as does the absorption on all of the bands originating there, hopefully converting what was absorption at 6943 Å to gain.

Sec. 10.3 Optical Pumping

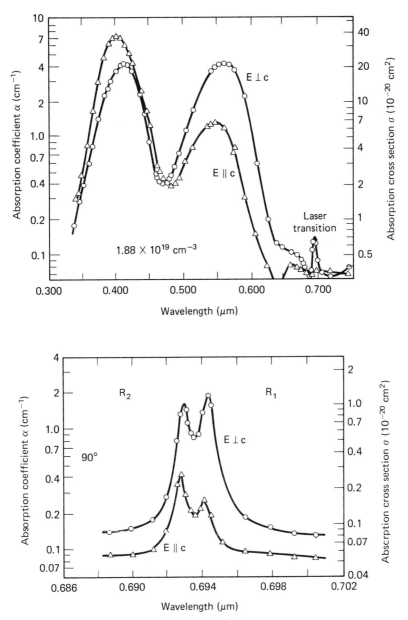

Figure 10.3 Absorption coefficient and cross section for ruby with $Cr^{3+} = 1.58 \times 10^{19}$ cm^{-3}.

The stimulated emission cross section at the R_1 and R_2 lines can be found directly from the data shown in Fig. 10.3(b) with due attention paid to the degeneracies shown in Fig. 10.2(b). The stimulated emission cross section is related to the absorption cross section by

$$\sigma_{SE} = \left[\frac{g_1}{g_2}\right] \sigma_{abs.} \qquad (7.7.3)$$

Hence, $\sigma_{SE} = 2.5 \times 10^{-20}$ cm^2 for the R_1 line with **E** perpendicular to the c axis.

Obviously, ruby is a three-level laser with all the attendant difficulties. The pump-to-laser route is as follows:

1. The optical radiation from the flash lamps in the wavelength region between 400 and 600 nm is absorbed by the atoms in the ground state ($^4A_2$), promoting them to the $^4F_2$ state. This same radiation attempts to stimulate these atoms to return to the ground level. In competition with this stimulated return is a fast relaxation of level 3 to the upper laser level.
2. Because of this fast relaxation and very slow spontaneous emission from the upper laser level back to the ground, a considerable number of atoms can accumulate in the ^2E state. If the $1 \to 3 \to 2$ process is fast compared to ($\tau_{sp.}^{-1}$), an inversion of population can be achieved on the $2 \to 1$ transition.
3. We have a laser if proper feedback is provided.

There are several questions that could (or should) be asked concerning this scenario:

1. Where did the energy go in the fast relaxation between levels 3 and 2? Answer: Most, if not all, goes into heating the crystal lattice.
2. Why is the pump radiation not absorbed at the R_1 and R_2 transitions? Answer: It is, but observe the magnitude of the absorption cross section in the vicinity of 694.3 nm: $\sigma \leq 1.2 \times 10^{-20}$ cm^2 [see Fig. 10.3(b)]. This is 20 times lower than the absorption cross section at 550 nm.
3. Why is the pump radiation not amplified on the R_1 and R_2 wavelengths? Answer: It is, but the amplification path is only the diameter of the rod, the stimulated cross section is small, and the pump radiation is comparatively weak compared to the much more intense laser radiation (when it starts).

A mathematical analysis can be quite involved even for such a "clean" optical pumping scheme. First, we need to determine the spectral distribution of the flash lamp. Obviously, if most of its radiation is concentrated outside the absorption bands, few atoms will be promoted to state 3 and thus to 2. On the other hand, if the power is concentrated in a spectral region where the absorption is too high, most of the pumping occurs on the outer radius of the rod. Thus matching the spectral characteristic of the lamp to the active medium to optimize the pumping is not a simple engineering job.

Typical emission characteristics are shown in Fig. 10.4 for flash lamps filled with different gases and excited in a variety of ways. According to Ref. 4,

Sec. 10.3   Optical Pumping

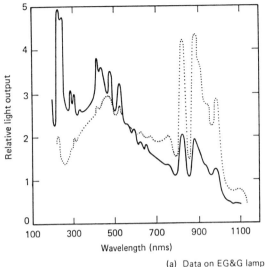

Spectral data curves show the effect of operating the FX-38C-3 (4 mm I.D., 3 inch arc length, xenon filled—400 torr) flash tube at 50 joules input at low and high voltage to enhance either the IR or the UV output.

Solid line curve shows data at 50 joules input or 1400 VDC, 51 μF, 0 inductance. Dotted line curve shows data at 50 joules input or 700 VDC, 205 μF, 0 inductance. Shifts in spectral data are primarily due to changes in current density J, A/cm². Higher current density J shifts spectral output toward the shorter wavelengths or blue region, UV. Conversely lower current density J shifts spectral output toward the longer wavelength or red region, IR.

Data was taken with EG&G 580/585 Spectroradiometer system. Spectral resolution 10 nm in the 200-700 nm region; and 20 nm in the 700-1200 nm region.

(a) Data on EG&G lamp

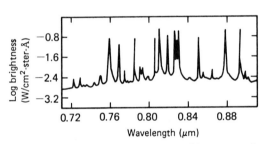

(b) Brightness of a CW-pumped krypton arc lamp

Figure 10.4 Spectral characteristic of typical lamps: (a) data on EG & G lamp (Courtesy of EG & G Electro-Optics.); (b) brightness of a CW-pumped krypton arc lamp. (From Koechner et al.[4])

55% of the electrical energy into the lamp of Fig. 10.4(b) appears as radiant energy* in the wavelength interval 300 to 1000 nm; a similar percentage applies for the lamp of Fig. 10.4(a). This figure illustrates that the spectral characteristics depend on the fill gas—xenon or krypton,—and as Fig. 10.4(a) shows, the current density in the lamp.

Given the above information, we could in principle do a fair job of modeling the ruby laser. However, it is quite tedious and requires a computer to do an accurate job. Some of the computation difficulties and some "obvious" physical limits to the pumping sequence are given below.

*This efficiency is far better than that of the best visible laser.

1. One must conserve atoms, that is,

$$N_1 + N_2 + N_3 = N_0 \tag{10.3.1}$$

or the sum of the atom densities in $^4A_2$ (state 1), the sum of those in $\bar{E} + 2\bar{A}$ (state 2), and the sum of those in $^4F_2 + {}^4F_1$ (state 3) must equal the original doping density N_0. Fortunately the lifetime of ^4F levels is very short, and hence their density can be neglected.

2. The time scale for the interchange of population between the $\bar{E}$ and $2\bar{A}$ states is short ($\sim$1 ns) but not instantaneous. For pulses short compared to this time, only the population in a particular state—say the E level—can be extracted by stimulated emission by the R_1 radiation. For pulses longer than this time scale, the ^2E manifold can be considered as one level, with, however, a stimulated emission cross section of the R_1 line (2.5×10^{-20} cm^2).

3. Because of the short time for interchange of populations between $2\bar{A}$ and $\bar{E}$ one can consider those states in local thermodynamic equilibrium at the lattice temperature, the ratio of populations being given by the Boltzmann factor.

$$\frac{N(2\bar{A})}{N(\bar{E})} = \frac{2}{2} \cdot \exp\left[-\frac{\Delta E}{kT}\right] \tag{10.3.2a}$$

$$= 0.87 = K \tag{10.3.2b}$$

for $\Delta E = 29$ cm^{-1} and $kT = 209$ cm^{-1} (300 K). Thus if we name the sum $\bar{E} + 2\bar{A}$ as state 2, there are

$$N(2\bar{A}) = \frac{K}{1+K} \cdot N_2 \tag{10.3.2c}$$

$$N(\bar{E}) = \frac{1}{1+K} \cdot N_2 \tag{10.3.2d}$$

atoms in the respective levels of the ^2E manifold. In order to obtain optical transparency at 6943 Å, we need $N(\bar{E}) = 1/2 N(^4A_2) = 1/2 N_1$. Thus

$$\frac{N_2}{1+K} - \frac{1}{2}N_1 = 0 \quad \text{(optical transparency)} \tag{10.3.3a}$$

$$N_2 + N_1 = N_0 \quad \text{(conservation of atoms)} \tag{10.3.3b}$$

or

$$N_2 = \frac{1+K}{3+K}N_0 = 0.483 N_0 \tag{10.3.3c}$$

$$N_1 = \frac{2+K}{3+K}N_0 = 0.517 N_0 \tag{10.3.3d}$$

Sec. 10.3   Optical Pumping

Note those values are just about the same as one would obtain by assuming equal degeneracies between N_2 and N_1 and ignoring the splitting in the upper state as was done in the section for Q switching.

4. There is considerable spontaneous fluorescence emitted even at the condition of zero optical gain.

$$P_{\text{spont.}} = h\nu \cdot \frac{N_{2c}}{\tau_{\text{sp.}}} = \frac{1 + K}{3 + K} \cdot \frac{h\nu N_0}{\tau_{\text{sp.}}} = 727 \text{ W/cm}^3 \quad (10.3.3e)$$

for $N_0 = 1.58 \times 10^{+19}$ cm^{-3} and $\tau_{\text{sp.}} = 3$ ms. (This is sometimes called the critical fluorescence power.) Any further pumping yields gain.

5. If we assume a quantum efficiency of the pumping process of 0.7, then we must absorb:

$$P_{\text{ab.}} = \frac{P_{\text{sp.}}}{\eta} = 1.04 \text{ kW/cm}^3 \quad (10.3.4)$$

from the flash lamp in order to just reach the condition of optical transparency.

6. If we assume a conversion efficiency of electrical energy stored in the capacitor to emitted radiation of the flash lamp into all wavelengths of 55%; with 20% of that radiated energy being absorbed in the two bands; and a time scale for the pumping of 3 ms (it cannot be much longer—otherwise the atoms decay as fast as they are produced), then we require a stored energy in the capacitor ($\frac{1}{2} CV^2$) of

$$W = \frac{1.04 \times 10^3 \text{ W/cm}^3 \times (3 \times 10^{-3} \text{ sec.})}{0.55 \times 0.2} = 28.3 \frac{\text{joules}}{\text{cm}^3 \text{ of rod}} \quad (10.3.5)$$

Obviously, more energy is required to obtain a laser.

7. A better computation is frustrated by the fact that the pump radiation is propagating across the chord of a cylindrical rod and the number of atoms absorbing that radiation is changing with time. It is safe to say that the calculation is tedious.

10.3.3 Neodymium Lasers

Another common solid-state material is made by doping a rare earth, neodymium, into a variety of host materials, with the most common ones being amorphous glass and crystalline yttrium aluminum garnet ($Y_3Al_5O_{12}$), or YAG. In each case, the active atoms participate as if they were triply ionized, Nd^{3+}, with energy levels and broadening of the states dependent on the host lattice. (The crystal is, of course, charge neutral, the three electrons being bonded to the neighboring atoms of the host.) The dopant (1% to 2%) goes into the amorphous glass at random sites, and thus each Nd^{3+} ion "sees" a slightly different environment. On the other hand, the Nd^{3+} substitutes for Y^{3+} in the cubic crystal of YAG, and thus each of these dopant atoms sees more or less identical environments. It should come as no surprise then that the glass laser transition is inhomogeneously broadened (but not by the Doppler phenomenon) with a comparatively wide line, whereas the YAG transition line widths are much smaller (although still inhomogeneously broadened by lattice vibrations). While the YAG and the glass laser resemble each other, both lasing at $\lambda_0 = 1.06$ μm, they are sufficiently different to warrant separate discussions.

10.3.4 Neodymium-YAG Lasers

Pure YAG, $Y_3Al_5O_{12}$, is an optically isotropic crystal with a cubic structure characteristic of garnets. Because of the difference in size of the Nd^{3+} which is substituted for Y^{3+} (about 3%), one is limited in the amount of neodymium that can be included to about 1%; otherwise, the crystal becomes severely strained. Because the active atoms are in a well-defined environment, the energy levels are well defined and narrow; they are shown in Fig. 10.5(a) with a greater detail in Fig. 10.5(b) where the dominant laser transition at $\lambda_0 = 1.064$ is shown as the transition between the upper state of the $^4F_{3/2}$ (at 11507 cm^{-1}) and the lower state $^4I_{11/2}$ at 2110 cm^{-1}. In Fig. 10.5(c) the energy levels are shown presuming that the host lattice was cooled to liquid nitrogen temperatures. Specific data on the various transitions, arranged in order of wavelength, are given in Tables 10.2 and 10.3. The last column of Table 10.3 is "saved" as a problem at the end of this chapter. The other transitions will lase given a suitable wavelength selective cavity.

The typical pumping route, such as from a flash lamp, is also indicated in Fig. 10.5(a) with approximately 10% of the absorbed energy going directly to the upper laser manifold $^4F_{3/2}$ and the remainder going to the higher states. For the most part, the ions promoted to 4, 5, and so on return to the $^4F_{3/2}$ level for participation in the laser action. The lower laser level, $^4I_{11/2}$, decays with a lifetime of ~30 ns, but that radiation is strongly absorbed by the host lattice and thus that energy shows up as heat. Fortunately, YAG has a high thermal conductivity and this unwanted energy can be removed by conduction. The system can be operated either CW or pulsed, mode locked or Q switched (or combinations thereof).

The fact that the states are well defined is a double-edged sword. The system is obviously a four-level laser, and because of the narrow line widths the stimulated emission cross section is large and the pumping threshold is low. However, the absorption bands are also narrow, and thus one does not make

Sec. 10.3 Optical Pumping

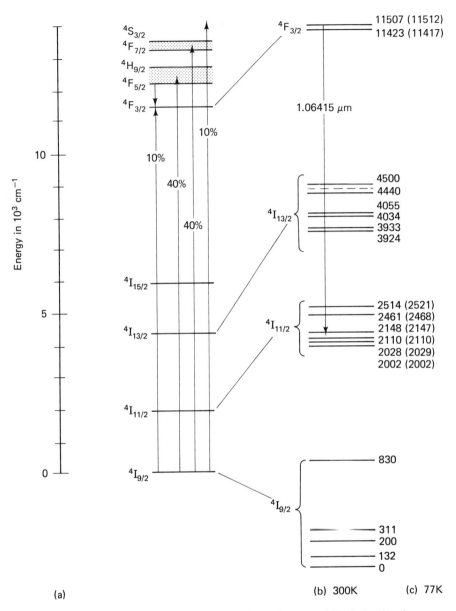

Figure 10.5 Energy level for neodymium in YAG. (a) Structure of YAG showing the pumping routes with the percentages referring to a pump with a broad spectral output. (b) Details of the manifold at 300° K showing the dominant transition. (c) Energy levels at 77° K. (Data from Kaminiski.[25] See also Koechner.[24])

TABLE 10.2 CHARACTERISTICS OF A TYPICAL NEODYMIUM-YAG LASER ROD

Bulk Parameters	Value
Chemical formula	$Nd: Y_3Al_5O_{12}$
Weight % (Nd)	0.725
Atomic % (Nd)	1.0
Density of Nd atoms	$1.38 \times 10^{+20}\,cm^{-3}$
Index of refraction	1.81633 (1.06 μm, 1% doping)
Scattering losses	$0.002\,cm^{-1}$
Dominant transition	(11,507 → 2111 cm^{-1}) 1.064 μm
Fluorescent lifetime of $^4F_{3/2}$	255 μs
Fluorescent efficiency of $^4F_{3/2}$	99.5%
Stimulated emission cross section	$2.7 - 8.8 \times 10^{-19}\,cm^2$ (doping and temperature dependent)
Lower state lifetime ($^4I_{11/2} \rightarrow\, ^4I_{9/2}$)	~30 ns

From Koechner.[24]

TABLE 10.3 DETAILED DATA ON $^4F_{3/2} \rightarrow\, ^4I_{13/2},\, ^4I_{11/2},\, ^4I_{9/2}$ TRANSITIONS

Transition	Wavelength (μm)		Branching Ratio	Line Width (FWHM in cm^{-1})	σ ($10^{-19}\,cm^2$)
$^4F_{3/2} \rightarrow\, ^4I_{9/2}$	0.86-0.95 μm		0.3		
$^4F_{3/2} \rightarrow\, ^4I_{11/2}$	1.0521	C	0.0383	4.5	
	1.0549		0.0023	4.5	
	1.0615	B	0.0799	3.6	
	1.06415	A	0.1275	5.0	3.0
	1.0644	A'	0.0533	4.2	*
	1.0682		0.0340	6.5	
	1.0737		0.0657	4.6	
	1.0779		0.0463	7.0	
	1.1055		0.0145	11.0	
	1.1119		0.0297	10.2	
	1.1158		0.0356	10.6	
	1.1225		0.0328	9.9	
		$\Sigma =$	0.56		
$^4F_{3/2} \rightarrow\, ^4I_{13/2}$	1.3-1.4		0.14		
$^4F_{3/2} \rightarrow\, ^4I_{15/2}$	1.74-2.13		~0.01		

* Because of their proximity the A and A' transitions contribute to each other. Thus the effective stimulated emission coefficient at 1.06415 μm increases to $3.3 \times 10^{19}\,cm^2$ when the shift, line width, and Boltzmann factor are included in the calculation.
From Koechner.[24]

Sec. 10.3   Optical Pumping   319

good use of the radiation emitted by an efficient Xe flash lamp. For that reason, one attempts to use other gases in the pumping lamp, such as krypton, which matches the pumping bands.

Recently, there has been a significant breakthrough for low power Nd-YAG by using the output of semiconductor lasers to pump the $^4F_{3/2}$ level directly. Inasmuch as the efficiency of semiconductor lasers to convert electrical energy to photons is quite high ($>40\%$), this enables one to pump the YAG laser with simple power supplies (i.e., flashlight batteries as a last resort). By using frequency doubling techniques, the output wavelength can be converted to the "green" ($\lambda_0 = 1.064/2 = 532$ nm) with a diffraction-limited $\text{TEM}_{0,0}$ mode as the output. The entire system can be contained in a "cigar box" or smaller. Although this system concept is limited to low power (tens of milliwatts), this can be the front end of a master oscillator power amplifier (MOPA) chain that generates very large powers with radiation characteristics (mode pattern and wavelength) determined by the initial state.

10.3.5 Neodymium-Glass Lasers

Glass, being a technology that has been studied since ancient times, can be made very uniform, doped with a variety of donor substances, polished to fantastic accuracy, and cast into large ingots. The details for Nd^{3+} in a typical glass host are shown in Fig. 10.6 with the energy-level diagram shown in (a), a typical spontaneous emission profile for the dominant laser transition in various types of glass in (b), and the absorption of ED-2 glass shown in (c).

Some of the more important physical details of the fluorescence from the $^4F_{3/2}$ manifold (necessary to solve the problems) are given in Table 10.4. Because of the amorphous character of glass, the absorption bands are much broader than those of YAG, which is good. However, this same characteristic leads to wide fluorescent lines and thus to a smaller stimulated emission cross section that, to some, may appear undesirable.

However, gain is not the only criterion for the "goodness" of the medium. If one is dealing with an optical amplifier, for instance, (which is the main use of glass), then energy storage in the population inversion is of primary concern. Because of the high doping densities permitted in glass, with each active atom storing $h\nu/2$ joules of energy, the glass amplifiers have a very high figure of merit. Furthermore, the wide band width over which gain is significant permits the amplification of very short pulses, with tens of picoseconds being common.

One last comment should be made about glass: it is a very poor thermal conductor—about a factor of 40 less than that of YAG. It is difficult to remove the waste heat, and consequently the repetition rate of glass lasers is severely limited.

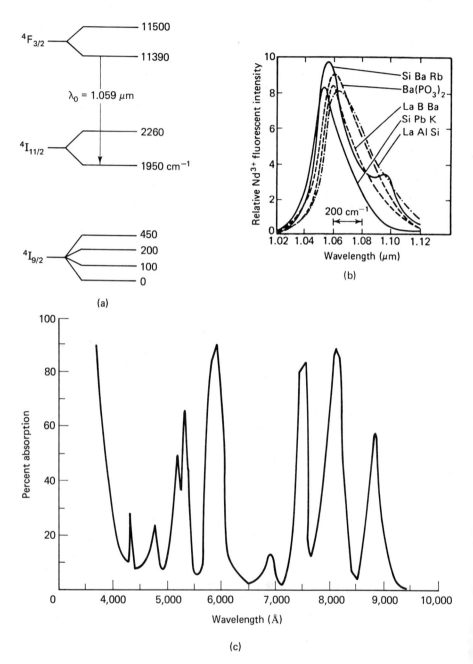

Figure 10.6 Neodymium-glass laser. (a) Diagram of the manifold showing the laser levels. (b) Spontaneous emission profile of Nd^{3+} in various glasses. (c) Absorption of a 6.3 mm slab of ED-2 glass. (From Koechner.[24])

Sec. 10.4   Dye Lasers   321

TABLE 10.4   PROPERTIES OF COMMON NEODYMIUM-GLASS SYSTEMS

Glass Designation	ED-2	LG-630	LCG-11	LGH-5	EV-2
Type	Silicate	Silicate	Silicate	Phosphate	
$[Nd^{3+}](10^{20}$ cm$^{-3)}$	2.8	2.8	3.8	3.17	3.1
Index	1.555	1.509	1.546	1.531	1.504
$\lambda(\mu m)$	1.0623	1.0623	1.0624	1.0560	1.054
$A_T(^4F_{3/2}$ in 10^3 sec$^{-1})$	3.33	1.56	1.74	3.45	1.77
$\Delta\lambda$(FWHM in Å)	260	220	290	186	170
$\sigma(10^{-20}$ cm$^2)$	3.03	2.1	2.0	3.9	4.7
Lifetime ($^4I_{11/2}$)*	~11 ns	(50-100 ns)			

*Data from Koechner[24] and from Owens-Illinois. See also Brown[26] (page 68), which contains data on many more glasses.

10.4 DYE LASERS

Whereas the active atoms in the previous three examples could be readily identified, it is most difficult to assign the fluorescence emitted by a dye to any one or even a pair of atoms associated with the molecular chain of, say, a common dye such as rhodamine B dissolved in ethylene glycol. A reading of the chemical formula of this dye, $C_{28}H_{31}ClN_2O_3$, and a glance at the structure diagram, shown in Fig. 10.7, convince us that it is not a simple matter to make this identification. Nevertheless, such a system lases quite spectacularly, and in spite of the apparent complexity of its structure, it behooves us to obtain a feeling for such a laser, even if based on an imperfect picture.

The medium usually consists of a dye dissolved in a solvent such as water, alcohol, or ethylene glycol (antifreeze) at 10^{-4} M concentration. The pumping agent is usually the light from a flash lamp or another laser. More about the pump later. A first-order approximation to the energy-level diagram for such a system is shown in Fig. 10.8. In constructing this figure, we have assumed an energy-level diagram commonly found in a diatomic molecule, with vibrational and rotational states superimposed on an electronic energy level, which, in turn, depends on some coordinate in the molecule.

Whether such a model is even approximately correct must be judged based on its agreement with experimental detail. It does. Indeed, to put the matter in proper perspective, the model was constructed to explain the absorption and fluorescence spectrum shown in Fig. 10.9. Note that in this figure, the singlet absorption and fluorescence are nearly mirror images of each other (when proper attention is paid to the scale factors!).

This is consistent with the singlet absorption starting at the short wavelengths corresponding to transitions from low-lying levels in S_0 to high vibrational levels of S_1 of Fig. 10.8. Once the system is in these high vibrational states of S_1, it relaxes very quickly to the lowest state of this manifold, and then

Dye	Structure	Solvent	Wavelength
Acridine red	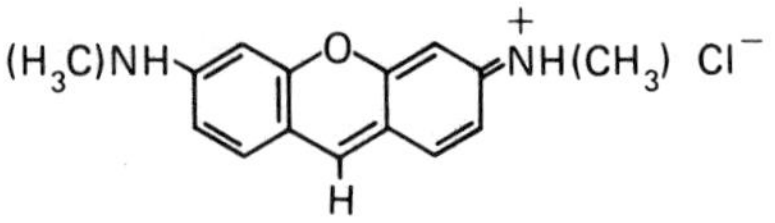 (H₃C)NH–[xanthene]–NH(CH₃)⁺ Cl⁻	EtOH	Red 600–630 nm
Puronin B	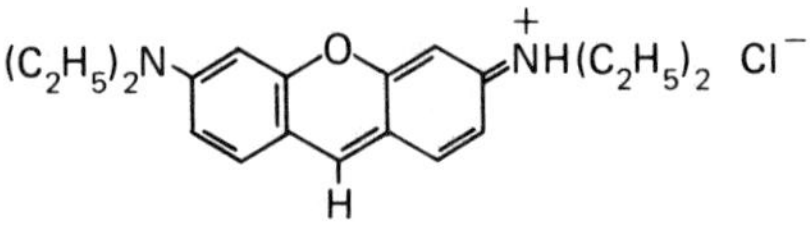 (C₂H₅)₂N–[xanthene]–NH(C₂H₅)₂⁺ Cl⁻	MeOH, H₂O	Yellow
Rhodamine 6G	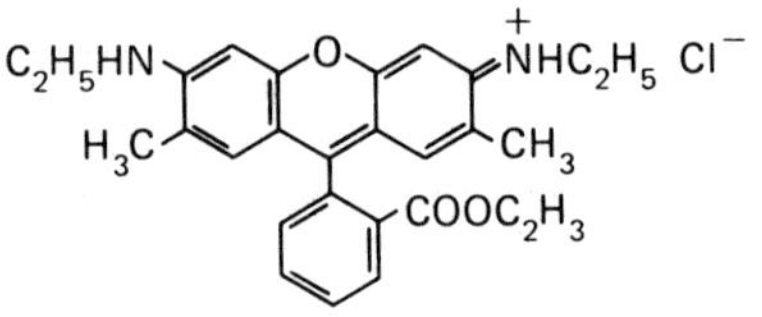 C₂H₅HN–[xanthene with H₃C, CH₃, COOC₂H₅]–NHC₂H₅⁺ Cl⁻	EtOH, MeOH, H₂O, DMSO, Polymethylmethacrylate	Yellow 570–610 nm
Rhodamine B	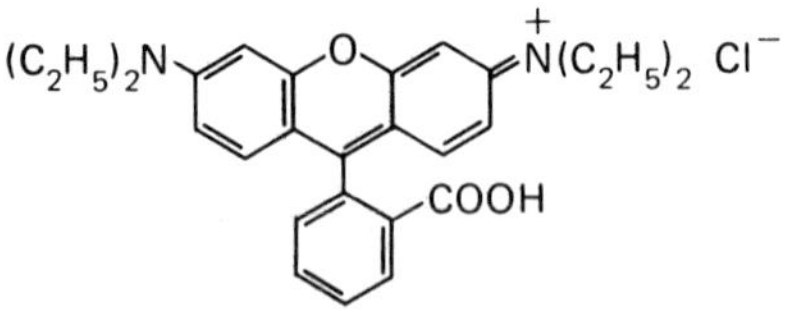 (C₂H₅)₂N–[xanthene with COOH]–N(C₂H₅)₂⁺ Cl⁻	EtOH, MeOH, Polymethylmethacrylate	Red 605–635 nm
Na-fluorescein	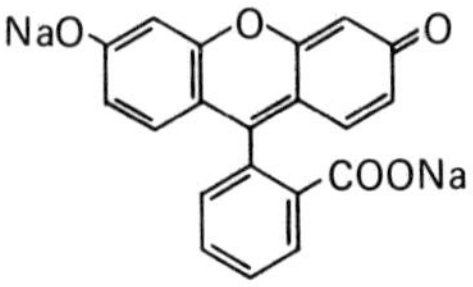 NaO–[xanthene with COONa]=O	EtOH, H₂O	Green 530–560 nm
2,7-Dichlorofluorescein	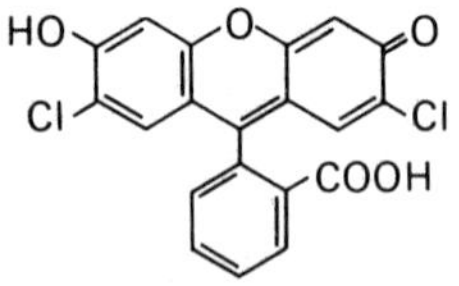 HO–[xanthene with Cl, Cl, COOH]=O	EtOH	Green 530–560 nm
7-Hydroxycoumarin	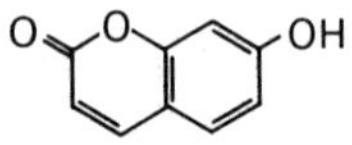 [coumarin with OH]	H₂O (pH ~ 9)	Blue 450–470 nm
4-Methylembelliferone	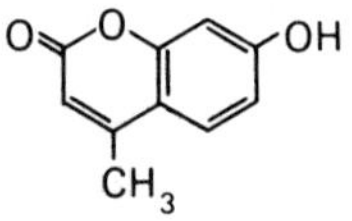 [coumarin with OH, CH₃]	H₂O (pH ~ 9)	Blue 450–470 nm
Esculin	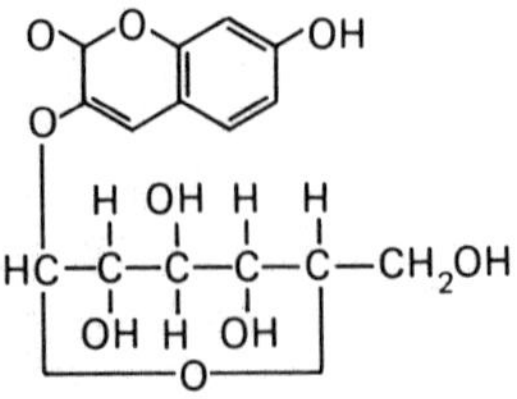 [coumarin with OH and glucoside: HC–C–C–C–CH₂OH with OH H OH]	H₂O (pH ~ 9)	Blue 450–470 nm

Figure 10.7 Molecular structure, laser wavelength, and solvents for some laser dyes. (Data from Snavely.[8])

Sec. 10.4   Dye Lasers   323

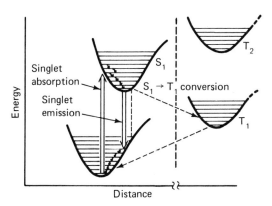

Figure 10.8 Energy-level diagram typical of a dye. (Data from Bass et al.[9])

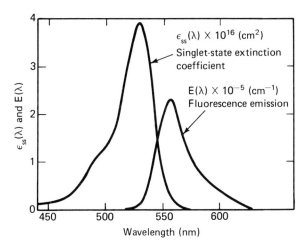

Figure 10.9 Singlet-state absorption and fluorescence spectra of rhodamine 6G obtained from measurements with a 10^{-4} M ethanol solution of the dye. (Data from Snavely.[8])

reradiates a photon—fluorescence—in making a transition back to the high vibrational states of S_0. The fact that the fluorescence and absorption are mirror images is convincing evidence that S_1 state is displaced along the coordinate axis with respect to the S_0 state and is consistent with the Franck-Condon principle.*

Thus the pumping sequence is to pump the ground-state molecules into the S_1 state fast enough to overcome the thermal population of the high vibration states of S_0. Whereas the radiative lifetime of the upper laser level of ruby or neodymium was hundreds of microseconds, here the lifetime is submicroseconds—typically 10 ns. Consequently, the pumping must be intense and *fast*.

*The Franck-Condon principle is discussed in Chapter 15. In its simplest terms, it states that transitions must be vertical, or the relative coordinates of a molecule cannot change in absorbing or emitting a photon or in any electronic process.

This last requirement is necessary because of the presence of the triplet system in the dye—a system that is normally empty. Unfortunately, some of the molecules in the S_1 state are converted to a triplet configuration—by interaction with the solvent or by a hundred other processes—and accumulate in the state T_1. Now absorptive transitions can take place between T_1 and T_2. By conforming to Murphy's law,* the T_1 and T_2 absorptions coincide with part of the emission profile of S_1 to S_0.

Thus what started out to be a four-level system has many of the undesirable features of the three-level one. We can enhance the reconversion of that triplet back to the S_0 state by adjusting the dye chemistry. For instance, the lifetime of T_1 can be as long as 10^{-3} sec or as short as 10^{-7} sec in a solution in which oxygen is dissolved. But even this short lifetime is long compared to the radiative lifetime of S_1. As a consequence, a dye laser tends to be self-terminating, owing to the buildup of the triplet T_1 population. Thus most flash-lamp pumped dye lasers are pulsed with pulse widths on the order of 100 to 1000 ns.

But, of course, the striking beauty of a dye laser is its tunability over a majority of its fluorescence spectrum. Indeed, the dye laser has provided the answer to the dream of having a tunable oscillator in the visible portion of the spectrum. This device is every bit as useful as the corresponding tunable oscillator is for the low-frequency domain (maybe even more).

The pulsed dye laser tends to be of the high-power, short-pulse variety. The former characteristic is caused by the very intense fast pumping and the latter by self-quenching by the triplet buildup. However, there is a way to avoid this last difficulty—flow the dye out of the cavity—to prevent the triplets from building up too high of a level.

This simple line of reasoning has lead to a CW dye laser that has the form shown in Fig. 10.10. Here the dye solution flows across the cavity at a position

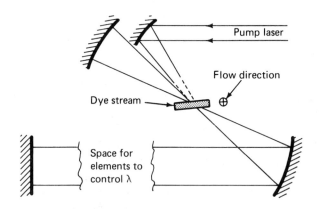

Figure 10.10 Typical configuration for a CW dye laser.

*"If anything can go wrong, it will."

where the spot size of the $TEM_{0,0}$ dye laser mode is a minimum so as to enhance the stimulated emission rate. Furthermore, the continuous stream of dye solution is oriented at Brewster's angle to enhance the Q of the dye-laser cavity.

On the other hand, the angle of the incoming pump—usually another laser, such as an argon-ion or krypton-ion laser—is adjusted to give maximum gain along the region where the pump and dye laser beams overlap.

The used dye is collected and recirculated. Now, there are literally minutes available for the triplet T_1 to be reconverted back to S_0. The more pertinent time interval is the one corresponding to the passage of the stream past the focus of the pump beam. If this is a small spot size, a nominal flow replenishes the dye before the triplets can quench the laser action.

10.5 GASEOUS-DISCHARGE LASERS

10.5.1 Overview

Shortly after the demonstration of the ruby laser, Javan, Bennett, and Herriott[10] demonstrated oscillation in an RF excited gas mixture of helium and neon at a wavelength of 1.1523 μm. Since that time, literally hundreds of gases have been excited and made to lase at literally thousands of different wavelengths. In fact, Ref. 11 lists 159 different transitions for neon alone, with wavelengths ranging from the UV at 267.9 nm to 132.8 μm at the ultramicrowave portion of the spectrum. (One now asks the question of where [in λ] and with what efficiency a gas will lase, rather than if it will lase.) Obviously, we have to restrict our attention to a few of the more common ones, namely, the helium-neon laser, the argon ion laser, and the CO_2 laser, all being excited by a simple discharge.

Gas lasers have many advantages over other types of lasers, primarily because the active medium—a gas—is very homogeneous even when mistreated in a violent fashion. As we will see later, they are relatively efficient. Because of the homogeneity, most gas lasers produce a nearly perfect Gaussian beam mode (with proper optics), and some can be scaled to large volumes and extreme powers.

However, these very desirable characteristics have disadvantages. Gas lasers are big devices—huge compared to a semiconductor laser, for instance. They require voltages and currents for excitation that are not normally compatible with most modern solid-state electronic devices. Nevertheless, the gas laser, in particular the CO_2 laser, has a commanding lead in CW power and efficiency. In many respects, the gas laser is the simplest to understand and analyze.

10.5.2 Helium-Neon Laser

The energy-level diagram for the helium-neon system is shown in Fig. 10.11 together with some common laser transitions shown as solid lines. For instance, the very common "red" laser line at 6328 Å is a transition from the $3s_2$

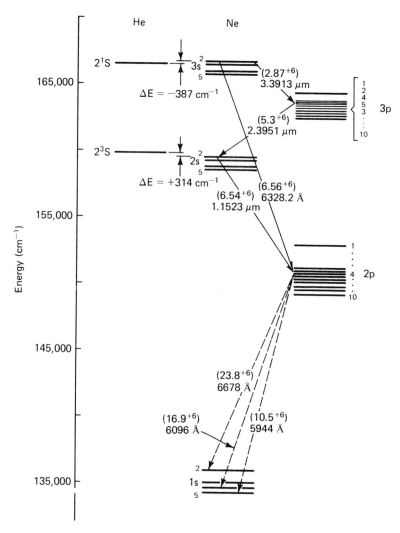

Figure 10.11 Energy-level diagram for the helium-neon laser. The solid line represents the common laser line; the dashed lines are spontaneous. The numbers in parentheses are the A coefficients.

Sec. 10.5   Gaseous-Discharge Lasers   327

to $2p_4$ state.* The numbers in parentheses are the A coefficients. Also shown in this figure, as dashed lines, are the spontaneous decay routes from the lower laser level, $2p_4$, down to the $1s$ states.

These latter transitions, $2p_4 \rightarrow 1s_{2,4,5}$, and many others originating at the rest of the $2p$ states, give neon signs their characteristic red color. Note that the A coefficients (the numbers in parentheses in Fig. 10.11) for these spontaneous transitions are quite large compared with that of the laser transition. This illustrates an important point referred to in Chapter 8, where a generalized pumping scheme of a laser was discussed. We do *not* look for strong spontaneous lines for a laser; rather, it is the weak lines, which terminate on the upper states of the strong ones, that are prime candidates for a laser.

To illustrate this point further, consider the data given in Table 10.5 for the two states of interest in the 6328-Å transition. Here we list the many transitions out of the two states and their relative intensities as listed in the *Handbook of Chemistry and Physics* (CRC Press). *All* the $3s \rightarrow 2p_{1-8}$ transitions have lased, and none of the $2p_4 \rightarrow 1s_{2-5}$ have oscillated (in CW operation). This is in spite of the fact that the A coefficient is in the numerator of the laser gain equation:

$$\gamma_0(\nu) = A_{21} \frac{\lambda^2}{8\pi} g(\nu) \left(N_2 - \frac{g_2}{g_1} N_1 \right) \qquad (10.5.1)$$

However, as we will see, it is difficult to obtain an inversion on the $2p \rightarrow 1s$ transitions; hence, they do not lase. But the rapid spontaneous decay of the $2p$ states empties the lower level of the $3s \rightarrow 2p$ laser transitions.

A typical helium-neon laser is a glass tube 10 to 100 cm long, 2 to 8 mm in diameter, and filled with helium and neon gas in a 5:1 to 20:1 ratio to a pressure prescription of (12).

$$p \cdot d \sim 0.36 \text{ torr-cm} \qquad d = \text{diameter of bore, cm}$$

Typical currents through the discharge tube range from 5 to 100 mA for CW operation.

Such a prescription satisfies the scaling laws of a positive column discharge, a subject to be discussed in greater detail in Chapter 12. For now, we need only the experimental facts of life about low-pressure discharges.

1. The electron temperature[†] is directly related to the ratio of the electric field to number density of gas atoms (i.e., $T_e \propto E/N$ or E/p).

---

*The spectroscopic notation for neon is particularly confusing but widely used. Paschen notation, as used in Fig. 10.11, was an attempt to fit the neon spectra to a hydrogen-like theory. It did not work! Nevertheless, the notation is still with us. As far as we are concerned, the numbers and letters are names of the states, nothing more.

[†]The word "temperature" implies a Maxwellian distribution of electron velocities. In fact, it is not! But it is a reasonable approximation to use for the initial understanding.

TABLE 10.5 DATA ASSOCIATED WITH THE VARIOUS STATES OF NEON

Transition	J_{upper}	J_{lower}	λ(Å)	$A(10^6 \text{ sec}^{-1})$	Relative Intensity	
$3s_2 \to 2p_1$	1	0	7304.9	0.48	30	
$3s_2 \to 2p_2$	1	1	6401.1	0.6 (est.)	100	
$3s_2 \to 2p_3$	1	0	6351.9	0.7	100	
$3s_2 \to 2p_4$	1	2	6328.2	6.56	300	(common "red" laser)
$3s_2 \to 2p_5$	1	1	6293.8	1.35	100	
$3s_2 \to 2p_6$	1	2	6118.0	1.28	100	
$3s_2 \to 2p_7$	1	1	6046.1	0.68	50	
$3s_2 \to 2p_8$	1	2	5939.3	0.56	50	
$3s_2 \to 2p_9$	1	3	¦5882.5¦	Forbid $\Delta J = 2$	Not observed	
$3s_2 \to 2p_{10}$	1	1	5433.6	0.59	250	
$3s_2 \to \Sigma 2p$	1	—	Red-orange	12.8	—	
$3s_2 \to 3p_4$	1	2	33913	2.87	—	
$3s_2 \to \Sigma 3p$	1	—	IR	5.24	—	
$2p_4 \to 1s_2$	2	1	6678.3	23.8	500	
$2p_4 \to 1s_3$	2	0	¦6234.5¦	Forbid $\Delta J = 2$	Not observed	
$2p_4 \to 1s_4$	2	1	6096.2	16.9	300	
$2p_4 \to 1s_5$	2	2	5944.8	10.5	500	
$2p_4 \to \Sigma 1s$			Red-orange	51.2		

Other transitions	ΣA		λ	$A(\times 10^6)$	
$2p_1 \to \Sigma 1s$	87.9	$2s_2 - 2p_1$	1.5231 μm	0.802	
$2p_2 \to \Sigma 1s$	116.6	$2p_2$	1.1767 μm	4.089	
$2p_3 \to \Sigma 1s$	61.7	$2p_3$	1.1602 μm	0.801	(first gas laser)
$2p_4 \to \Sigma 1s$	51.7	$2p_4$	1.1523 μm	6.537	
$2p_5 \to \Sigma 1s$	53.3	$2p_5$	1.1409 μm	2.301	
$2p_6 \to \Sigma 1s$	53.6	$2p_6$	1.0844 μm	7.543	
$2p_7 \to \Sigma 1s$	49.3	$2p_7$	1.0621 μm	0.816	
$2p_8 \to \Sigma 1s$	41.2	$2p_8$	1.0295 μm	0.726	
$2p_9 \to \Sigma 1s$	43.3	$2p_9$	Forbidden	—	
$2p_{10} \to \Sigma 1s$	33.6	$2p_{10}$	0.8895 μm	1.708	

2. To a reasonable approximation, E/N is a function only of the pressure times diameter product.

3. The electron temperature (or E/N) is either constant or decreases slightly with increasing current.

Thus keeping the *pd* product constant implies a discharge with more or less the same voltage drop per unit length and same electron temperature. Typical values for T_e are $\sim 80{,}000°$K and $E/p = 28$ V/cm/torr at $pd = 0.36$ torr-cm.

The pumping sequence of the red laser is as follows: helium, being the majority gas present, is excited by the energetic electrons in the high-energy tail of the Maxwellian distribution. This is represented by the following chemical equation:

$$e(\text{K.E.}) + \text{He}(1^1S) \longrightarrow \text{He}(2^1S) + e(\text{K.E.} - 20.6 \text{ eV}) \quad (10.5.2)$$

Sec. 10.5 Gaseous-Discharge Lasers

where the potential energy of the He(2^1S) state relative to the ground state is 20.6 eV.

Now this particular state is metastable; that is, it cannot decay back to the ground state by emitting a photon, as such a transition would require $J = 0 \rightarrow J = 0$ step. Consequently, such a state will live for a long time—many seconds in fact—unless something else collides with the excited helium atom. That something else may be another electron, which could excite it further up the helium ladder, or the electron could reverse the arrow of (10.5.2). The excited helium state could also diffuse to the wall to be deactivated there (and warm up the universe). None of these possibilities help the laser a bit; in fact, they are deleterious to its operation.

But a neutral neon atom can be that something else that collides with that excited helium atom. The potential energy of the He state is transferred to the neon according to

$$\text{He}(2^1S) + \text{Ne}(\text{ground}) + 387 \text{ cm}^{-1} \longrightarrow \text{He}(1^1S) + \text{Ne}(3s_2) \quad (10.5.3)$$

where the excess energy, 387 cm^{-1}, is provided by the kinetic energy of the colliding atoms.

Thus we have a *selective* mechanism for pumping the upper laser level. Furthermore, since the lower $2p_4$ state has a much shorter radiative lifetime than the $3s_2$ level, it contributes significantly to the creation of the population inversion.

Unfortunately, the $2p_4$ level can also be excited by the discharge independently of the $3s_2 \rightarrow 2p_4$ route, and this fact ultimately dooms the He:Ne laser system to low power. For instance, the helium 2^3S state is excited at an even greater rate than is the 2^1S state. This lower helium metastable state transfers its energy to the $2s$ manifold in neon, which, in turn, radiates to $2p$ level. Indeed, this is the excitation route for the laser transition at 1.1523 μm ($2s_2 - 2p_4$)—the first gas laser. If the 6328-Å transition is desired, the $2^3S \rightarrow 2s_2 \rightarrow 2p_4$ route is an unavoidable pumping of the lower laser level. Alternatively, the $2^1S \rightarrow 3s_2 - 2p_4$ is bad for the 1.1523-μm laser. This is an example where one laser transition tends to fill the lower state of another.

A classic example of the competition for the upper state is provided by the 6328-Å and 3.39-μm transitions. The latter, being of longer wavelength and thus having a smaller Doppler width, has a much higher stimulated emission cross section than does the visible transition. Thus, unless special precautions are taken to reduce the feedback at 3.39 μm, the $3s_2 \rightarrow 3p_4$ transition will lase and deplete the population inversion of the $3s_2 \rightarrow 2p_4$.

There is another route for pumping the $2p$ levels: electron impact excitation from the neon ground state or from the neon $1s$ manifold. After all, the neon sign depends on the $2p \rightarrow 1s$ transitions for its characteristic color, and it does indeed work! Therefore the $2p$ states will be excited by a discharge. As we will see shortly, this process is the ultimate culprit for the termination of the laser action at high current.

To model this common 6328-Å laser is not a simple task, because there are at least four $3s$, ten $3p$, four $2s$, ten $2p$, and four $1s$ neon levels and the two helium metastable states to be related to the voltage and current through the discharge. This would require the simultaneous solution of 34 rate equations for the atomic populations and one rate equation for the electron and ion density. Obviously, that is not feasible here.

But, with some judicious approximations, one can gain an understanding of the gross behavior of the laser as the current through the discharge is varied. This is done in the following example.

Example: Pumping of the helium-neon 6328-Å laser

It is obvious that we must reduce the number of variables and cross couplings to even remotely handle the situation by analytic techniques. The model chosen is shown in Fig. 10.12 (a). Our approximations are

1. Ignore the rate equation for the electrons and ions and assume that the electron density is directly proportional to the current, in agreement with experiment.[12]
2. The excitation of the helium metastable proceeds by electron impact with helium in the ground state. Once formed, M can transfer its energy to neon at a rate r_t (per helium metastable atom per neon atom), diffuse to the wall and be deactivated, or be ionized by a second collision with another electron. We ignore the fact that the neon state, N_2, can transfer its potential energy back to helium and also ignore the deactivation of M by superelastic collisions with electrons.
3. The only pumping mechanism allowed for the upper laser level is the transfer indicated; N_2 decays by spontaneous emission to the [1] manifold and subsequently to the [0] state (the $1s$ levels).
4. There are two primary pumping routes for the lower laser level; electron impact excitation from the ground state and from the neon $1s$ manifold N_0. We ignore the pumping of N_1 by spontaneous decay from N_2 as being small compared to the electron pumping rate. The lower state decays by spontaneous decay back to N_0.
5. The neon metastable states are excited by electron impact with ground-state neon atoms and by spontaneous decay from N_2. One can balance the spontaneous decay into N_0 against the electron deexcitation to N_1 from N_0 and the direct electron excitation of N_1. The net deexcitation routes are diffusion to the walls and ionization by a second electron collision.

With these assumptions, the 35 rate equations simplify to four:

$$\frac{d[M]}{dt} = n_e \langle \sigma_M v_e \rangle [\text{He}] - \frac{[M]}{\tau_M} - r_t [M][\text{Ne}] - n_e \langle \sigma_i v_e \rangle [M] \quad (10.5.4)$$

$$\frac{d[N_2]}{dt} = r_t [M][\text{Ne}] - A_2 [N_2] \quad (10.5.5)$$

Sec. 10.5   Gaseous-Discharge Lasers

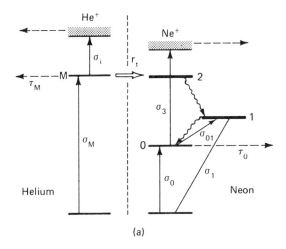

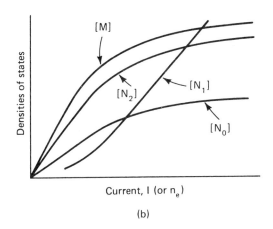

Figure 10.12 (a) Simplified model of the helium-neon laser. The solid lines represent electron excitation routes, the dashed lines represent diffusion to the walls of the tube, and the wavy lines represent radioactive decay; and (b) the variation of the states with current.

$$\frac{d[N_1]}{dt} = n_e\{\langle\sigma_1 v_e\rangle[\text{Ne}] + \langle\sigma_{01} v_e\rangle[N_0]\} - A_1[N_1] \tag{10.5.6}$$

$$\frac{d[N_0]}{dt} = n_e\langle\sigma_0 v_e\rangle[\text{Ne}] + A_1[N_1] - n_e\langle\sigma_{01} v_e\rangle[N_0] + A_2[N_2] \\ - \frac{[N_0]}{\tau_0} - n_e\langle\sigma_3 v_e\rangle[N_0] \tag{10.5.7}$$

where all quantities in square brackets are densities and the symbol $\langle \sigma v_e \rangle$ indicates that the cross section for the process must be multiplied by the electron velocity and averaged over the electron distribution function (see Chapter 12).

We set all time derivatives equal to zero and set about the unpleasant task of solving these coupled equations. The first two are simple:

$$[M] = \frac{n_e \langle \sigma_m v_e \rangle [\text{He}]}{\tau_M^{-1} + r_t[\text{Ne}] + n_e \langle \sigma_i v_e \rangle} \tag{10.5.8}$$

$$[N_2] = \frac{r_t[\text{Ne}][M]}{A_2} = \frac{1}{A_2} \frac{n_e \langle \sigma_m v_e \rangle r_t[\text{He}][\text{Ne}]}{\tau_M^{-1} + r_t[\text{Ne}] + n_e \langle \sigma_i v_i \rangle} \tag{10.5.9}$$

By combining (10.5.6) with (10.5.7) we first find the density of the neon metastable state $[N_0]$:

$$[N_0] = \frac{n_e \langle \sigma_0 v_e \rangle + \langle \sigma_1 v_e \rangle [\text{Ne}]}{\tau_0^{-1} + n_e \langle \sigma_3 v_e \rangle} + \frac{A_2[N_2]}{\tau_0^{-1} + n_e \langle \sigma_3 v_e \rangle} \tag{10.5.10}$$

where $[N_2]$ is given by (10.5.9). Then we return to (10.5.6) to relate $[N_1]$ to all the prior terms.

$$[N_1] = n_e \langle \sigma_1 v_e \rangle [\text{Ne}] + n_e \langle \sigma_{01} v_e \rangle [N_0] \tag{10.5.11}$$

To evaluate these equations is to experience an arithmetic nightmare! The trend should be clear if one mentally substitutes current I for the electron density in the above and sketches the variation of the states with this externally controlled variable.

This is done in Fig. 10.12(b). In examining this sketch, one should follow the order of the previous four equations. For instance, (10.5.8) indicates that $[M]$ should increase linearly with n_e until ionization collisions become more frequent than the deactivation rates of diffusion to the walls or transfer to neon. Equation (10.5.9) indicates that $[N_2]$ has more or less the same functional dependence on current as does $[M]$.

Equation (10.5.10) is complicated unbearably by the last term, but since $[N_2]$ saturates (i.e., becomes independent of current) in the same manner as does the first term, the variation of $[N_0]$ is similar to $[M]$.

The two terms of (10.5.11) play a crucial role in establishing a limitation to the helium-neon laser and relegate it to the low-power domain. Both terms state that the lower laser level varies linearly with current. Thus, since $[N_2]$ saturates at high currents, $[N_1]$ is bound to catch up and exceed* $[N_2]$—the inversion is destroyed and the laser ceases operation!

Although the helium-neon laser is doomed to small power, it has practically cornered the market for alignment and other low-power applications. Even its paltry 1 to 10 mW is very intense compared to incoherent light sources. In numbers produced and sold, it is far in front of whichever gas laser is in second place!

*A more complicated model would show that eventually the ratio of N_2/N_1 would be g_2/g_1 $\exp(-\Delta E/kT_e)$—a normal population equilibrated at the electron temperature.

10.5.3 Ion Lasers

The atomic levels associated with a laser can be the excited states of an ion, and many elements have been used in this manner.* Two of the most common ion lasers use the heavy rare gases krypton and argon for the donor atoms.

The energy-level diagram for the Ar^+ laser is shown in Fig. 10.13. As is shown in this figure, one can obtain nearly 50 different laser transitions in the excited neutral argon atom at wavelengths ranging from 0.7 to 30 μm, and "visible" transitions in argon ion with the common ones shown there.

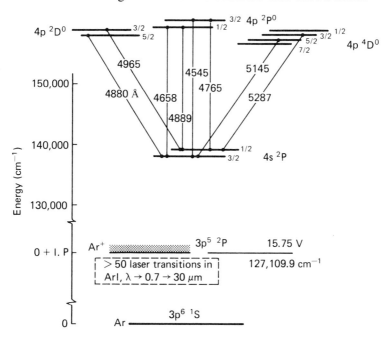

Figure 10.13 Energy-level diagram for the argon-ion laser.

Such a diagram allows us to review the selection rules for atomic transition (see Chapter 15). The small numbers to the right of each energy level are J values of each state, and the superscript preceding the capital letter is the multiplicity $(2S + 1)$ of the group of states. Note that all transitions obey $\Delta J = \pm 1, 0$ and $\Delta L = \pm 1, 0$. As indicated in Chapter 15, the $\Delta S = 0$ rule is usually the first to be broken, and the strong 5145-Å line is an example.

---
*We have laser lines in the excited ions of Cd, Se, Hg, Ge, Ag, Au, Pb, and so on. Unfortunately, many of these elements require extreme temperatures to obtain a reasonable vapor pressure. (For instance, the vapor pressure of gold at 900°C is only 10^{-11} torr.)

It is also important to note that the energy scale of Fig. 10.13 has *two* breaks; one between the ground state of neutral argon and the first ionization at 15.75 V and the other between this value and the lower laser level. To reach the upper laser level of the 4880-Å line at 158,731.20 cm^{-1}, we must provide an additional 127,109.9 cm^{-1} to first create an ion, or a total of 35.43 V.

There is a reason for the emphasis on numbers here. First, the maximum efficiency of this laser—the quantum efficiency—is low, 7.2%. Furthermore, the upper state of the laser is higher than the highest first ionization potential of any gas. Thus we cannot hope for a selective excitation transfer mechanism as found in the helium-neon laser; rather, we must pump the upper level from the ion ground state or by direct excitation from the neutral. In view of the fact that the high-energy tail of the electron distribution drops as $\exp(-\epsilon/kT)$, there are many more electrons capable of first ionizing the atom (15.75 V) and then another electron exciting the ion (19.68 V) than have the energy to excite the $4^2pD_{5/2}^0$ state directly from the neutral argon ground state.

One consequence of this two-step excitation process is that an argon-ion laser discharge is very intense, with amperes being carried by a small-bore discharge, which, in turn, requires water cooling to survive such treatment. Indeed, the survival of the wall material is the major limitation to the generation of intense visible CW radiation.

Typical efficiencies of the laser are far worse than the quantum efficiency noted above, with 5-W output (in all lines) being typical from a laser run from a power supply requiring ~12 kW from a three-phase 460-V, 60-Hz plug. This is an overall efficiency of 0.03%! Thus there is considerable laser engineering in the not-so-trivial problem of building the power supply.

In spite of the poor efficiency, the argon (and krypton) ion lasers are very common, a major use being the optical pump for the tunable dye lasers. In that application, one can expect a conversion efficiency of 10% to 30% between the pump and the dye laser output.

10.5.4 CO$_2$ Lasers

The CO$_2$ laser is one of the most spectacular, most efficient, and because of these characteristics, most useful and most studied lasers discovered to date. Whereas most gas laser efficiencies are measured in tenths of a percent, the CO$_2$ laser is measured in tens. A "small" laser can produce tens of watts and, by using a master oscillator power amplifier (MOPA) chain, one can achieve terawatts of pulsed power;* or one can obtain hundreds of kilowatts in a CW mode. The latter two devices are not small. Applications of this laser have been numerous and diverse; for instance, one uses the CO$_2$ in trimming operations in industrial

---

*At that power level, the laser has to be pulsed. After all, the total installed power-generating capacity in the United States in 1975 was only 525 million kilowatts, or 0.525 terawatt.

manufacture, pattern cutting, communications, welding and cutting of steel (many inches thick), weaponry, and laser fusion, to list only a few applications.

Oscillation can be obtained on more than 200 vibrational-rotational (VR) transitions in the 8 to 18 μm wavelength range. Many of the desirable characteristics of the CO_2 laser can be attributed to the kindness of nature in providing us with a linear symmetric molecule that has a characteristic vibrational energy in a near-perfect match with the characteristic vibrational energy of nitrogen (N_2).

The pertinent energy levels of the CO_2 and N_2 molecules are shown in Fig. 10.14, where some classic artifacts are added to aid the interpretation.

One usually describes the states by the number of quanta in the characteristic vibration of the classic linear molecule shown at the top. Thus the 10°0 state has one quanta in the characteristic ν_1 or symmetric stretching mode, the 02°0 state has two quanta in the bending mode, and the 00°1 state has a single quanta in the asymmetric stretch.* As shown in the diagram, lasing takes place between the *odd* rotational levels of the 001 state and the *even* rotational states of either the 100 or 020 states.[13, 14]

Note that the first vibrational level of N_2 coincides very closely with the upper laser level (001). Indeed, most of the lower vibrational levels of N_2 from $v = 1$ to 8 are spaced to have an excellent match with the (000) to (001) separation of CO_2. Recall that N_2 is a homonuclear molecule, and thus radiative transitions between vibrational levels are strictly forbidden (see Chapter 15). In other words, the vibrational levels of N_2 are metastable. Thus, if one can vibrationally excite these N_2 molecules, they can transfer this trapped energy selectively to the upper laser level.

Finally, note that the excited-state energy levels of helium could not be placed on this diagram without adding 68 additional pages pasted side by side in a foldout fashion. The first excited state of helium occurs at 159,850 cm^{-1}, whereas the upper laser (001) level is only at 2349 cm^{-1}. This contrast explains in part why the CO_2 laser is so much more efficient than the helium-neon laser. One needs only to supply $\sim \frac{1}{2}$ eV to excite the (001) state of CO_2, whereas ≥ 19.8 eV is required to excite the upper state of the helium-neon laser.

An elementary pumping sequence for an electrical discharge in a CO_2-N_2-He mixture (for a typical 1:2:3 pressure ratio) is as follows:

1. The electrical power is transferred to the electrons (as is the case in *all* discharges) by the electric field.

2. The electrons transfer this power by collisions to the neutral gas atoms. This power is apportioned to the gas in three different categories:

---

*Unfortunately, this simple one-to-one identification with the classic vibration is only an approximation. However, it suffices for the initial introduction to this laser. In the same spirit, we ignore the meaning of the superscript on v_2.

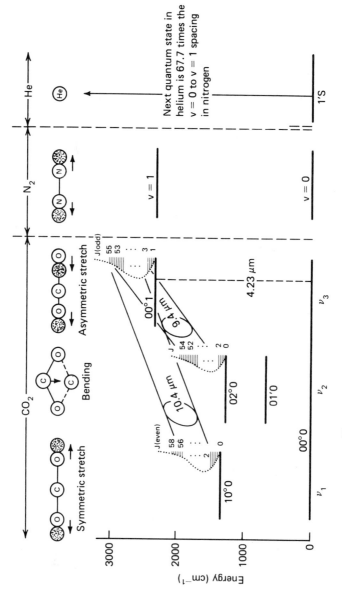

Figure 10.14 Energy-level diagram of the CO_2-N_2-He laser.

Sec. 10.5 Gaseous-Discharge Lasers   337

a. *Gas heating:* This is caused by the "elastic" collisions of the very light electrons with the more massive neutral atoms. Although these collisions are mostly elastic, there are many such collisions, and some energy is expended in raising the kinetic temperature of the gas.

b. *Vibration excitation:* This is an inelastic collision process represented by the following chemical equations:

(1) For the upper state:

$$e + N_2(v = 0) \longrightarrow N_2(v = n \leq 8) + (e - KE) \quad (10.5.12)$$

followed by

$$N_2(v = n) + CO_2(000)$$
$$\longrightarrow N_2(v = n - 1) + CO_2(001) \quad (10.5.13)$$

or

$$e + CO_2(000) \longrightarrow CO_2(001) + (e - KE) \quad (10.5.14)$$

(2) For the lower state:

$$e + CO_2(000) \longrightarrow CO_2(010)$$
$$\text{or } CO_2(020) \quad (10.5.15)$$
$$\text{or } CO_2(100) + (e - KE)$$

c. *Electronic excitation and ionization:* Although ionization is essential to maintain an active discharge, the fraction of the electrical power used to do so is usually insignificant in discharges in molecular gases.

3. Theory and experiment show that 60% of the electrical power can be funneled into pumping the upper laser level (see Chapter 12).

Now the quantum efficiency of the 9.4-μm laser is 45%, and with 60% entering the upper state, we could expect a "wall plug" efficiency of ~27%. Such numbers have been approached.

To understand why this is achieved in a CO_2-N_2-He laser (but not in most other molecular lasers) requires us to examine the details of the foregoing reactions. This will be done shortly. Before doing so, let us return to the energy-level diagram (Fig. 10.14) to discuss some of the details of the laser transitions.

Table 10.6 lists some of the pertinent data for CO_2. The fundamental "frequencies" indicate the energy levels of the various states involved, ignoring rotation. Thus the $J = 0 \rightarrow J = 0$ of the $00°1 - 10°0$ transition would appear at 960.8 cm^{-1} (were it not for the fact that such a transition is forbidden). The position of the rotationless $J = 0$ state of $00°1$ is 2349.3 cm^{-1}, which is the sum of the corresponding frequencies of $10°0 \rightarrow 00°0$ and $00°1 \rightarrow 10°0$,

TABLE 10.6 DATA ASSOCIATED WITH THE CO_2 LASER

Fundamental frequencies

$01^10 \rightarrow 00°0 = 667.3$ cm^{-1} $\quad 00°1 \rightarrow 00°0 = 2349.3$ cm^{-1}
$02°0 \rightarrow 00°0 = 1285.5$ cm^{-1} $\quad 00°1 \rightarrow 10°0 = 960.8$ cm^{-1}
$10°0 \rightarrow 00°0 = 1388.3$ cm^{-1} $\quad 00°1 \rightarrow 02°0 = 1063.6$ cm^{-1}

Rotational data

$B(00°1) = 387140.44 \times 10^{-6}$ cm^{-1}
$D(00°1) = 13.252 \times 10^{-8}$ cm^{-1}
$B(10°0) - B(00°1) = 3047.389 \times 10^{-6}$ cm^{-1}
$D(10°0) - D(00°1) = -1.8366 \times 10^{-8}$ cm^{-1}
$B(02°0) - B(00°1) = 3340.757 \times 10^{-6}$ cm^{-1}
$D(02°0) - D(00°1) = 2.3816 \times 10^{-8}$ cm^{-1}

Transition probabilities (Einstein A coefficients)

	$00°1 \rightarrow 00°0$* (4.2-μm band)	$00°1 \rightarrow 10°0$ (10.4-μm band)	$10°0 \rightarrow 02°0$ (9.4-μm band)
P branch	2.1×10^2 (sec^{-1})	0.34	0.20
R branch	2.0×10^2	0.33	0.19
N_2 data:	$\omega_e = 2359.1$ cm^{-1} $B_e = 2.010$ cm^{-1}	$\omega_e x_e = 14.456$ cm^{-1}, $\alpha_e = 0.0187$ cm^{-1}	$\omega_e y_e = 0.00751$ cm^{-1} $r_e = 1.094$ Å

*Resonance trapping of this radiation has a profound effect on the effective radiative lifetime. For instance, the P branch has an effective transition probability of 10.1 sec^{-1} rather than 210 sec^{-1}, as listed.

Data from Cheo.[13]

$1388.3 + 960.8 = 2349.1$, or the sum of $(02°0 \rightarrow 00°0) + (00°1 \rightarrow 02°0)$, $1285.5 + 1063.6 = 2349.1$. The fact that these sums do not exactly match the specified energy of the $00°1$ state indicates a small error in one or more of the measurements.

Note the value of the Einstein A coefficients—or the inverse, the radiative lifetime for the laser transitions. This lifetime is 3 to 5 seconds—or 3 to 5 heartbeats! Even the resonance transition $00°1$ to $00°0$ has a paltry 10/sec rate. These numbers illustrate a very important point about the CO_2 laser (and others operating on VR transitions):

> *Spontaneous* radiation is not a significant loss process for the laser states and is usually ignored in a kinetic rate equation.

The energy exchange between colliding gas molecules is overwhelmingly more significant than is spontaneous emission.

There are two consequences that might seem paradoxical at first glance, given the wonders and efficiency of the laser:

1. In a pulsed CO_2 laser, it takes a significant amount of time to generate

Sec. 10.5   Gaseous-Discharge Lasers

enough photons in the proper cavity mode to start the laser oscillating. Indeed, the gain can build up faster than the photons and leads to a *gain-switched* pulse (see Sec. 9.3.2).

2. The stimulated emission cross section is small, typically less than 10^{-18} cm^2.

Consequently, there must be an appreciable fraction of the molecules excited to obtain a reasonable gain.

The last item explains the high-power, high-energy characteristics of the CO_2 laser. If we excite a large percentage of the fill gas to the proper state, each molecule can contribute one photon by the stimulated emission process and we obtain many, many photons. Let us illustrate this point by an example.

Example

Consider a gas fill of $1:2:3$ mixture of CO_2-N_2-He at 10 atm and assume that the excitation enters the 001 level by the N_2 route (10.5.12) and (10.5.13). Let us assume a pulsed discharge that is to create a vibrational "temperature" in N_2 of 0.5 eV. If every vibrational state in N_2 from $v = 1$ to $v = 8$ is capable of creating a 10.6-μm CO_2 photon, how much energy is involved?

Solution

Partial pressure of $N_2 = 2/(1 + 2 + 3) \times 10 = 3.33$ atm. Thus the concentration of N_2 is $3.33 \times 2.69 \times 10^{19} = 8.96 \times 10^{19}$ cm^{-3}. The number of the nitrogen molecules in vibrational states $v = 1$ to 8 is given by the probabilities that the energies, $hc\omega_e(v + \frac{1}{2})$, can occur, divided by the sum of the probabilities for *all* vibrational states, $v = 0$ to $v = \infty$, times the concentration of nitrogen molecules.

$$N_2(v = 1 \to 8) = [N_2] \frac{\sum_{1}^{8} \exp[-hc\omega_e(v + \frac{1}{2})/kT]}{\sum_{0}^{\infty} \exp[-hc\omega_e(v + \frac{1}{2})/kT]} \quad (10.5.16)$$

where the anharmonic terms for the vibrational energy levels have been ignored. For $\omega_e = 2359.1$ cm^{-1} (see Table 10.6) and $kT/e = 0.5$ eV, the ratio of the two series is 0.553, or there are 4.48×10^{19} cm^{-3} molecules capable of producing a 10.6-μm photon (0.1169 eV/photon). Thus, in principle, we should be able to extract 0.838 J/cm^3 or 838 J/liter. We do not do quite this well, but the calculation is close. Such numbers indicate the energy-storage capability of the CO_2-N_2 laser system.

As mentioned earlier, the CO_2 laser can oscillate on any of the P or R branches shown in Fig. 10.14. In a simple laser cavity without any wavelength discrimination, the P branch of the 10.4-μm band will dominate. This is because the factors in the laser-gain equation

$$\gamma(\nu) = A_{21} \frac{\lambda^2}{8\pi} g(\nu) \left(N_2 - \frac{g_2}{g_1} N_1 \right)$$

all favor this collection of lines: the A coefficient is bigger for the $00°1 \to 10°0$ group of transitions (see Table 10.6), the wavelength is larger, and the P branch is always favored over the R branch (see Chapter 15 for an analogous situation in CO). Thus the P-branch oscillation on the 10.4-μm band competes for and wins the photons associated with the $001 \to 100$ (or 020) inversion at the expense of the gain on the R branch of the 10.4-μm band and both branches of the 9.4-μm band.

With sufficient frequency discrimination in the cavity, such as that provided by a grating, the P (4), P (6), . . . , P (56) and R (4), . . . , R (54) of the $00°1 \to 10°0$ transition lase as well as the P (4) to P (60) and R (4) to R (56) of the $00°1 \to 02°0$ band.* To compute the wavelengths of these transitions is a straightforward but dull arithmetic task (which is best left as a problem). Incidentally, some of these wavelengths have been measured by beating the laser against a high harmonic of a microwave-crystal–controlled source, and this accounts for the extreme precision of the data presented in Table 10.6.

Let us now return to the excitation process. The figures presented for the energy storage are so much chicanery if we cannot establish a vibrational temperature with electron collisions. We can! As mentioned earlier, nature kindly provides us with a large cross section for electron excitation of the vibrational levels of N_2. Figure 10.15(a) shows the total excitation cross section of $v = 1$ to 8 states, with Fig. 10.15(b) showing only the excitation of the $v = 1$ state as measured by G. Schulz.

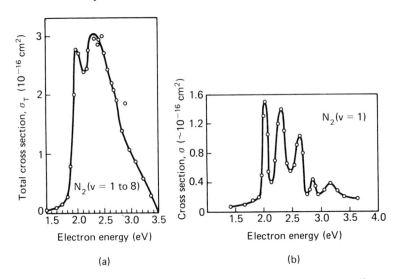

Figure 10.15 Vibrational excitation cross section. (Data from Schultz.[15])

*Remember that only even J values appear in the rotational structure of the 100 and 020 states, and only odd J values appear in the 001 state.

Sec. 10.6 Chemical Lasers

There is considerable physics buried beneath such simple-looking curves. First note that the cross section is virtually zero below about 1.5 eV. It only takes an electron of energy $G(1) - G(0) \simeq 0.3$ eV to excite the $v = 1$ state. Why is the apparent threshold so high? The reason is that the excitation proceeds via a complex route. The electron is first attached to the N_2 molecule, makes a few swings around it, and then flies off, leaving the N_2 in the vibrational excited state. The chemical equations describing the process are

$$e(E) + N_2 \longrightarrow N_2^- | \longrightarrow N_2(v) + e(E - G(v))$$

Thus the upward shift in the energy scale is caused by the shift of N_2^- with respect to N_2.

We studiously avoided any and all discussion of the role of helium except to note that none of its excited states are even on the map of the energy-level diagram of Fig. 10.14. Yet, more often than not, it is the majority gas. Let us now indicate its role.

Helium has many purposes in the CO_2 laser, some rather mundane and others quite subtle. A rather mundane role is to help transport the useless and deleterious heat generated by the discharge to the walls of the container for cooling. A more subtle role is to relax the lower laser level. However, its primary purpose is to control the E/N (or E/p) and thus the electron "temperature" of the discharge, which, in turn, controls the relative division of the electrical power to the processes noted previously. We postpone discussion of this most important role until Chapter 12.

The CO_2 laser is probably the most extensively studied and analyzed system in existence. It certainly has a commanding lead in performance.

10.6 CHEMICAL LASERS

10.6.1 Background

If one is used to thinking in terms of energy in common electrical circuits, or even optical circuits, terms such as kilojoules, megajoules, and kilowatts are impressive, especially so when applied to lasers. However, the energy released in common chemical reactions exceeds these large units by huge factors.

For instance, consider the case of a small automobile powered by a 100-hp engine (1 hp = 746 W), which will transport four adults 30 miles to their destination in about 30 minutes using a chemical reaction of 1 gallon of gasoline with air. If the engine were operating at rated power, this requires a minimum expenditure of energy of 134 megajoules (MJ), assuming, of course, a 100% efficient engine. Obviously, the last assumption is in error by a large margin, and hence 1 gallon of gasoline is capable of delivering considerably more energy than 134 MJ.

For another example, consider the burn—a rather violent burn—of TNT, commonly called an explosion. One pound (~0.45 kg) of high explosive (HE) yields about 4 MJ.

Thus it is a natural question to ask whether some of this enormous chemical energy can be released in the form of coherent photons, rather than all going into a splitting of the eardrums and a destruction of the local environment. The answer is yes.

10.6.2 HF Chemical Lasers

Consider a mixture of common molecular hydrogen H_2 and the rather nasty gas F_2. *If* the temperature is low enough and *if* the mixture is not perturbed by some external excitation, the two gases can coexist indefinitely. However, if a "spark" or some other excitation occurs, the mixture will rapidly "burn" and produce HF.

$$H_2 + F_2 + \text{perturbation} \longrightarrow \cdots \longrightarrow 2HF \quad (10.6.1)$$

There is considerable chemistry represented by the dots in (10.6.1), which we will not address here. Rather, we restrict our attention to the possibility that the perturbation in (10.6.1) generates some free fluorine (or hydrogen) atoms. Then the following chemical reactions are possible*:

$$F + H_2 \longrightarrow HF(v \leq 3) + H, \quad \Delta H = -31.7 \text{ kcal/mole} \quad (10.6.2a)$$

$$H + F_2 \longrightarrow HF(v \leq 9) + F, \quad \Delta H = -97.9 \text{ kcal/mole} \quad (10.6.2b)$$

This is an example of a chain reaction with the H atom from (10.6.2a) being used in the reaction (10.6.2b); its product, in turn, yields the F atom required in (10.6.2a). Both reactions are exothermic, and thus the reaction products share the chemical energy released. Some of this excess energy increases the temperature of the gas, which, in turn, increases the rate of the reaction. Obviously, the chain reaction consumes the donor molecules, and thus it will slow down when the densities of H_2 and F_2 are depleted. These are problems in chemical kinetics that will be left to textbooks devoted to that topic.

Our main attention should be focused on the fact that many of the HF molecules are formed in a high vibrational state—$v = 3$ for the "cold" reaction (10.6.2a) and $v = 9$ for the "hot" reaction (10.6.2b). This is ideal from a laser point of view—a process leading to a selective pumping of the upper laser level.

However, nature is seldom as generous as is implied by the preceding paragraph. Some of the HF is apparently formed in the vibrational states lower than the upper limits indicated in the reactions. Furthermore, the molecules will

---

*It is important to become acquainted with different ways of specifying energy, such as joules, eV, and cm^{-1}, and (10.6.2a) and (10.6.2b) introduce a chemical unit, kilocalories (kcal) per mole. The conversion factor is 1 eV = 23.062 kcal/mole.

Sec. 10.7 Excimer Lasers: General Considerations   343

accumulate in lower vibrational states, thereby destroying an initial population inversion. This problem can be alleviated to some degree by fast flow, and very high power CW chemical lasers have been reported.

The chemical reaction can be initiated by various means: UV light may be used to dissociate some of the molecules, an energetic electron beam may be injected into the mixture, or a simple electrical discharge between two electrodes can be employed. These techniques usually lead to a pulsed laser, but some have produced enormous energies—1.7 kJ in a 60 ns pulse.[16]

10.7 EXCIMER LASERS: GENERAL CONSIDERATIONS

The rare gases (helium, neon, argon, krypton, and xenon) have always been considered models for chemical *inertness*. For the most part, these gases do not form compounds with anything else, and hence they normally exist as simple monatomic gases. The crude and simple explanation usually given is that these gases have "closed" or "filled" shells for the orbiting electrons and thus do not participate in a chemical reaction.

However, it was recognized that if the rare gas was excited, this simple line of reasoning would no longer apply. For then, one of the electrons would be in the next "shell" and thus an excited rare gas (Rg*) would appear as the next element over in the periodic chart. Thus excited neon (Ne*) might act as Na in a chemical reaction, Ar* as potassium, Kr* as rubidiun, and Xe* as cesium. Since an alkali-halogen reaction such as

$$K + F_2 \longrightarrow (K^+F^-) + F \qquad (10.7.1)$$

proceeds with vigor, one might also expect a reaction such as

$$Ar^* + F_2 \longrightarrow (Ar^+F^-)^* + F \qquad (10.7.2)$$

to do likewise. (Note that the "salts," such as NaCl, can be considered as bound by ionic attraction, and thus we expect the state formed by (10.7.2) to be similar.) Here, however, we would expect the rare gas–halide "salt" to be in an excited state whose energy is related to the initial excited state of the rare gas. If the rare gas were not excited, no reaction with the halogen donor should take place because of the filled shells. They would merely repel each other at small nuclear spacing. This simple line of reasoning led to the invention of the *r*are *g*as–*h*alide (RgH) lasers, which are part of a class of lasers called "excimers"—a system that only exists for a measurable time in an excited state.

10.7.1 Formation of the Excimer State

As a specific example, consider Fig. 10.16, which shows the energy level diagram associated with interaction of an argon ion, an electron, and a fluorine atom. At very large inner nuclear spacing, one could have an argon ion plus a

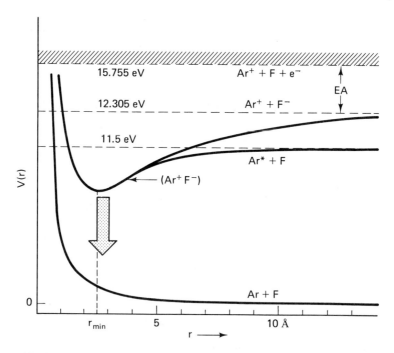

Figure 10.16 Energy-level diagram associated with the formation of the (Ar^+F^-) excimer.

free electron and a neutral fluorine atom with a total potential energy of the interacton equal to the ionization value of argon (15.755 eV). Now let us get rid of the electron by attaching it to fluorine making a negative ion. The electron affinity of F is about 3.45 eV; hence the asymptotic limit of potential energy between Ar^+ and F^- at large nuclear spacing is $15.755 - 3.45 = 12.305$ eV. These two ions attract each other according to Coulomb's law, and thus the potential energy curve decreases as the internuclear spacing decreases. At very small distances, $1 \rightarrow 2$ Å, other repulsive short-range forces come into play so as to prevent two atoms being at the same place at the same time, and thus there is a potential minimum at about 2.3 Å (from experiment).

Let us estimate the potential energy depression between $r = \infty$ and $r = 2.3$ Å due to this Coulomb attraction between Ar^+ and F^-. The electrostatic energy is simply

$$\Delta V = \frac{e}{4\pi\epsilon_0 r_{min}} = 6.54 \text{ eV} \tag{10.7.3}$$

Thus the potential depression due to the Coulomb attraction is 6.54 eV, and the total energy is now $12.305 - 6.54 = 5.77$ eV. Now we have to add some of the repulsive interaction to prevent $r \rightarrow 0$, about which we know little except that it exists, but 5.77 eV is a reasonable estimate for the potential energy of the lowest ionically bound state of $(Ar^+F^-)^*$. (The actual value is ~6.7 eV.)

If neither of the participants, Ar and F, is excited, the ground state of the

Sec. 10.7 Excimer Lasers: General Considerations

complex consists merely of a repulsive potential preventing the two atoms from getting too close together and the actual energy at $r_{min} = 2.2$ Å amounts to ~0.29 eV. Any argon and fluorine atoms appearing at this spacing will immediately fly apart in a time estimated by the following simple calculation.

They only have to move 1 Å before the repulsive forces are negligible and in that distance accelerate to a kinetic velocity corresponding to 0.29 eV. Thus setting $\frac{1}{2}Mv^2 = $ e (0.29), and using M as the sum of the two masses (Ar = 40 amu and F = 19 amu), we obtain a velocity of 9.7×10^4 cm/sec. If they start at zero velocity and end with 9.7×10^4 cm/sec, pick an average velocity of 4.85×10^4 cm/sec for their speed. Thus the lifetime of the lower state of ArF at $r_{min} = 2.2$ Å is only $(10^{-8}$ cm$) \div 4.8 \times 10^4$ cm/sec $= 2 \times 10^{-13}$ sec, which is quite an adequate approximation to zero! It is so short that there is no reason to make a better calculation.

The radiative lifetime of the ionically bound excimer state at $r = 2.2$ Å is short by normal electronic standards but huge compared to the lower state lifetime. Thus we have one of the best approximations to an *ideal laser* system known at the present time. The main question is whether we can *produce* this ionically bound upper state in a practical sense. Obviously, the previous kinetic sequence requires copious quantities of Ar$^+$ and F$^-$ (hence it is called the ionic channel) and something to stabilize the complex at $r_{min} = 2.2$ Å. Let us postpone the details of the production process for a moment so as to describe another channel for the formation of this ionically bound excimer.

This second process is very similar to the "burning" of an alkali metal in a halogen gas as described by (10.7.1). Since (10.7.2) implies the "alkali-like" atom is an excited state of the rare gas, when it and a neutral halogen donor, F_2, NF_3, and so on, get close together at $r \sim 3$ to 4 Å, the excited electron of the rare gas finds that it is energetically favorable to jump to the halogen atom creating a negative ion and leaving behind a positive rare gas ion. These two attract and follow the same interaction curves discussed previously. This extraction of a halogen from its donor by the "alkali-like" excited state of a rare gas is aptly named "*the harpooning reaction.*" It occurs with nearly unit efficiency; that is, every collision of Rg* with these donors yields (Ar$^+$F$^-$)*. The extra atoms of the fluorine donor and the (Ar$^+$F$^-$)* complex share the excess energy of the reaction and stabilize the excimer in the upper state. The excited rare gas precursor is usually a metastable state, and thus this last process is called the "metastable channel."

The choice of the ArF system was motivated by the fact that the photon obtained by the radiation from the excimer state to ground at $(6.7 - 0.29 = 6.41$ eV) is one of the shortest wavelengths available ($\lambda = 193$ nm). However, other rare gases and other halogens can be used to generate a variety of wavelengths. Table 10.7 illustrates the matrix formed by considering combinations between the various rare gases and various halogens. Note that the wavelengths become shorter as one uses lighter rare gases, which have higher ionization

TABLE 10.7 RARE GAS–HALIDE WAVELENGTHS (nm)*

Halogen	EA		Rare Gas			
			Neon	Argon	Krypton	Xenon
		IP(eV)	21.56	15.755	13.996	12.127
		M(eV)	16.6	11.55	9.92	8.31
			nm	nm	nm	nm
Fluorine	(3.45)		108	**193**	**249**	**351**
Chlorine	(3.61)		—	175	**222**	**308**
Bromine	(3.36)			161	**206**	**282**
Iodine	(3.06)				185	253

* IP = ionization potential; M = metastable level; EA = electron affinity. Wavelengths in boldface type refer to the peak of the laser; those in lightface type refer to the fluorescence assignable to an excimer (see text) and have not yet lased.
Data from Brau.[32]

potential and metastable levels, and heavier halogens, which have lower electron affinities. This follows from the scenario described for ArF (and allows the student to follow it and predict the wavelengths).

All the combinations in Table 10.7 follow the same general scenario as described for ArF. There is a slightly amusing wrinkle for XeF in that the ground state is actually bound by about 1200 cm^{-1}, and thus there are a few stable vibrational states in it. (This proves once again that the molecules cannot read our textbooks!) However, since the binding is so small (only ~ 6 kT), it is easily dissociated and does not prevent lasing.

Table 10.8 presents some of the pertinent data for the common rare gas–halogen lasers and was taken from the work by Brau.[32]

TABLE 10.8 DATA ON RARE GAS–HALIDE LASER SYSTEMS*†

Excimer	r (Å)	ω_e (cm^{-1})	σ (10^{-16} cm^2)	τ (ns)	λ (nm)
XeBr	3.1	120	2.2	12–17.5	282
XeCl	2.9	194	4.5	11	308
XeF	2.4	309	5.3	12–18.8	351
KrCl	2.8	210	—	—	222
KrF	2.3	310	2.5	6.7–9	249
ArCl	2.7	(280)‡	—	—	175
ArF	2.2	(430)	2.9	4.2	193

* r_e = minimum of the lowest ionically bound excimer state; ω_e = vibration constant representative; σ = stimulated emission cross section; τ = radiative lifetime; λ = dominant laser wavelength.
† These lasers hold a commanding lead as far as efficiency in the production of UV and near-UV power.
‡ Values in parentheses are estimates.
Data from Brau.[32]

10.7.2 Excitation of the Rare Gas–Halogen Excimer Lasers

It should be clear that any excitation scheme requires the production of copious quantities of rare gas ions, or excited states, or both. Before describing various techniques for producing them, let us estimate a lower bound for the pumping power required to establish a typical gain coefficient of 10%/cm. Using a typical stimulated cross section from Table 10.8 of $\sim 4 \times 10^{-16}$ cm^2, one thus requires the upper state density of 2.5×10^{14} cm^3, presuming that the lower state is empty. Now a typical fluorescence lifetime is 10 ns, yielding a photon energy of $h\nu/e = 5$ eV; hence the spontaneous power is

$$P_{\text{spont.}} = \left(\frac{h\nu}{e}\right) \cdot \left(\frac{N_2 e}{\tau_{\text{sp.}}}\right) = 20 \text{ kW/cm}^3$$

This is not an insignificant power loading, and we have not included any other losses associated with the formation of the excimer; one loses 6.54 eV via the ion channel (10.7.3). Thus it is safe to estimate that one needs at least 100 kW/cm^3 to excite such a laser. The power supply electronics to provide this excitation is a tough electrical engineering problem, and removal of the waste heat is a good mechanical one. In any case, such lasers are pulsed excited.

There are two primary methods: nearly relativistic electron beam excitation (hundreds of kVs, hundreds of amp/cm^2) and a discharge excitation with fast radar-type modulator circuits. Sometimes the two are combined whereby the E-beam current only sustains the discharge and the main power is supplied by the modulator circuit. It is not appropriate to discuss the electronic circuitry here, other than to say that it is a significant problem. However, the physical mechanisms for achieving the excited state are considerably different. Thus we discuss the physics of the excitation separately.

10.7.2.1 E-beam (or nuclear) excitation. The geometry of this type of excitation is shown in Fig. 10.17 where it is presumed that an energetic, high-current electron beam is injected through the foil separating the high vacuum where the E-beam is formed and the high-pressure gas where its energy is deposited. The key parameter in such a deposition is the so-called W value, which is the necessary energy deposited in a neutral gas to create 1 electron-ion pair, which in the case of argon is about 26 eV for electron energies greater than 1 keV. Thus, for instance, a 500 keV electron will generate some $5 \times 10^{+5}/26 \sim 19{,}000$ secondary electron-ion pairs. Inasmuch as the ionization potential of argon is 12.755 eV, not 26 eV, the energetic beam electron will also produce some excited states that decay to the metastable level. For argon, which is representative of all rare gases, only one third of an excited state is produced for each electron-ion pair. Consequently, most of the excitation from an E-beam proceeds through the ion channel.

The rare gas ions, of course, are exactly the items necessary for the formation of the excimer. The electrons are rapidly attached to a halogen by a

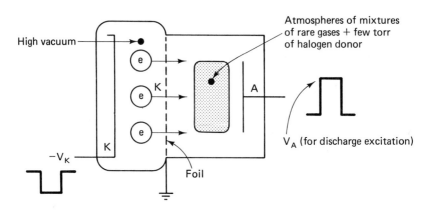

Figure 10.17 Excitation of the rare gas–halide laser.

dissociative attachment process typified by

$$e + NF_3 \longrightarrow F^- + \text{Products (NF}_2\text{?)} \qquad (10.7.4)$$

Such processes occur very fast with rate coefficients k_a between 10^{-9} and 10^{-7} cm^3/sec. Thus, if one had a background NF$_3$ pressure of 3 torr, then its density is ([NF$_3$] = 3 torr × 3.54^{16} cm^{-3}/torr = 1.06^{17} cm^{-3}), and the electrons have a lifetime for conversion into a negative ion of $k_a[NF_3]^{-1} = 9.4$ ns, assuming $k_a = 10^{-9}$ cm^3/sec; it will be much shorter if the higher rate coefficient is used. The negative and positive ions get together in the presence of a third body, usually another rare gas atom, to form the excimer.

$$Ar^+ + F^- + Ar \longrightarrow (Ar^+F^-)^* + Ar \qquad (10.7.5)$$

The third body is necessary so that it as well as the complex can share the excess energy of the reaction, which is ~6.54 eV as computed by (10.7.3). If the pumping is intense enough, enough excimer states will be formed to yield enough gain for the spontaneous emission to build up to a coherent amplitude so as to stimulate the complex down to the ground state (after which it dissociates in about 0.1 psec).

Now there are many competing channels for this energy, so many that one has to consult the current literature and monographs for a more complete model. However, there is one basic point to keep in mind: the energy flow to the excimer state with E-beam excitation is from above—starting at the ion level and terminating in the excimer state.

10.7.2.2 Discharge pumping. In any gas discharge, the power enters the system through the electron gas with the electric field moving the typical electron *up* the energy ladder. (This is discussed in detail in Chapter 12 for the CO$_2$ laser, but the physics is the same for any discharge.) When the electron energy is low, it can only make "elastic" collisions with the majority rare gas. The small amount of energy lost in such a collision, a "ping-pong ball bouncing off a battleship," merely warms the gas. The high energy tail of the electron distribution can make *inelastic collisions,* creating an excited state such as

$$e(E) + Ar \longrightarrow Ar^* + e(E - E_x) \qquad (10.7.6)$$

This is the primary means by which the electron loses energy. (In a fluorescent lamp, nearly 65% of $V \cdot I$ goes into such a process.) The remainder of the power appears as gas heating with a small amount, typically $\leq 2\%$, going into ionization. Thus the ion channel is not significant for the formation of the excimer state.

However, the "harpooning" reaction described by (10.7.2) using the excited state of (10.7.6) can lead to the excimer.

$$Ar^* + F_2 \longrightarrow (Ar^+F^-)^* + F \qquad (10.7.2)$$

Such reactions occur very fast and with nearly unity efficiency. Thus, in principle, the discharge pumping could be very efficient.

Unfortunately, nature conspires against us and generates a number of obstacles. We obviously need electrons for (10.7.6) to take place. However, the halogen attaches the electrons (10.7.4), and now the negative ion does no good whatsoever, other than causing trouble in that a detachment reaction can occur.

$$e + F^- \longrightarrow F + 2e \qquad (10.7.7)$$

which leads to a negative dynamic resistance and an unstable discharge. Worse yet, (10.7.6) is only efficient at high electron temperatures, which implies high electric fields (see Chapter 12), and this is also the regimen for (10.7.7).

Thus compromises have to be made, and this reduces the promise of high efficiency dramatically. One can use a low-current, high-energy electron beam to sustain the discharge and then use the maximum electric field compatible with discharge stability. (UV preionization also helps.) If too many electrons are present, a *superelastic* process robs the energy from the excimer state.

$$e(E) + (Ar^+F^-)^* \longrightarrow Ar + F + e(E + E_{excimer}) \qquad (10.7.8)$$

In spite of the problems, the discharge-pumped laser works quite well with wall-plug efficiencies on the order of 1% to 2%. Most small commercial tabletop excimer lasers are of this variety and yield pulsed UV energy on the order of 50 to 100 mJ/pulse in a width of ~ 10 ns (power = 10 MW), with a repetition rate of ~ 500 pps (average power ~ 50 W). These lasers are surely the most efficient source of high-intensity UV radiation.

10.8 THE FREE ELECTRON LASER

The free electron laser (FEL) is the newest class of oscillators to generate coherent electromagnetic radiation in the ultramicrowave (mm wavelengths) to the visible portion of the spectrum. Such devices can be analyzed by classic (relativistic) dynamics or by quantum theory with nearly identical results. Thus it could be debated whether it should be classified as a *l*ight *a*mplifier by *s*timulated *e*mission of *r*adiation (LASER), but the rest of the world has accepted the classification, and we do likewise.

A precise mathematical description of a FEL requires considerable investment in the theory of relativity, but some of the basic concepts can be appreciated without doing so. Every student in elementary physics runs into the fact that an accelerated electron radiates. (How else can one introduce Bohr's theory?) If one recognizes the fact that the radiation from a simple dipole antenna is caused by the fact that charges are accelerated back and forth along a stationary path, one obtains the (cosine)^2 dipole pattern found in every elementary text in electromagnetics and illustrated in Fig. 10.18(a).

Most accept the fact of radiation, but few have actually worked out the details or thought about the physical reason for its appearance. A simple, handwaving explanation is that the moving accelerated charge interacts with its *own*

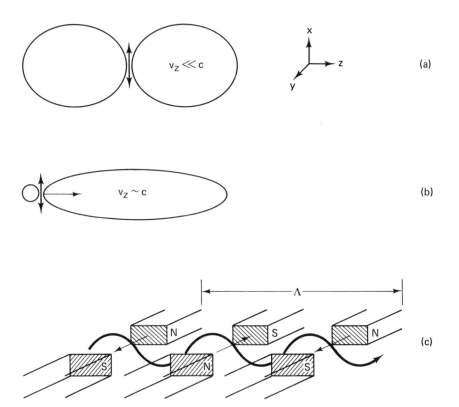

Figure 10.18 The radiation by an accelerated electron. (a) The pattern when the electron's velocity $v_z \ll c$; (b) $v_z \sim c$; and (c) the electron trajectory in a wiggler.

Sec. 10.8   The Free Electron Laser

field, which was established at an earlier time when the electron was at a different position along its path. Because of the finite velocity of light, the near field of a charge does not respond instantaneously to the position of the charge, and thus some of the electron's kinetic energy is lost (i.e., radiated) as a result of interaction.

The symmetrical figure-eight pattern of Fig. 10.18(a) is distorted if the charge is also translating along the z axis with a velocity comparable to c. This is shown in Fig. 10.18(b) and points out that the emission is now primarily in the same direction as the electron velocity. The simple physical reason for this distortion is that the moving and accelerated (in x or y direction) electron now interacts with its own field established from a prior position along z and at an earlier time.

Note that the radiation does *not* go into 4π as would spontaneous emission from a group of randomly polarized excited atoms; rather, it is heavily weighted in the direction of the electron's motion. Now spontaneous radiation is never very important to *any* laser other than to provide the initial seed for subsequent amplification by stimulated emission, and that from an accelerated electron is no exception. Furthermore, the FEL output is not simply the addition of all the radiated powers of each electron. The beam must be bunched into inhomogeneous packets. In fact, if the electrons were regularly spaced along z and the system were infinitely long, there would be no radiation along z! The field radiated by one electron would be cancelled by that emitted by another. Only by virtue of a finite size that limits the number of electrons in the cavity to a finite value, N, and thus permits statistical fluctuations along z (on the order of $N^{1/2}$) is there any spontaneous power whatsoever.

As with any laser, gain comes from stimulated emission, and the classic term of electromagnetics corresponding to it is $\mathbf{E} \cdot \mathbf{i}$, the rate of interchange of energy between the field and the current. This must be *negative* for an amplifier, signifying that the electrons give energy to the field. We have now reached the point requiring the investment in relativity to make further progress. We will not do so. We have defined some basic requirements for an FEL:

1. One must generate a relativistic beam with as much current (electrons) as possible. (Since the interaction scales as $\mathbf{E} \cdot \mathbf{i}$, high currents are an obvious requirement.)
2. One must periodically accelerate these electrons as they translate along the z axis.

There are many schemes for accomplishing this last task, and one is illustrated in Fig. 10.18(c). One uses a periodic array of permanent magnets with alternate polarities such that $\mathbf{B}_w$ is perpendicular to z. (A $\mathbf{B}_z$ field can also be included.) Such a device is aptly named a **wiggler**. Shown also in Fig. 10.18(c) is a typical undulating trajectory of an electron passing through the wiggler, and hence some prefer to call it an undulator.

If the current is high enough, and if the strength of the wiggler is strong enough, and if feedback is provided, one obtains a coherent output at a free space wavelength given (approximately) by

$$\lambda_0 \sim \Lambda/2\gamma^2 \tag{10.8.1}$$

where $(\gamma - 1)\, m_0 c^2$ = the kinetic energy of the electrons; $\gamma = (1 - \beta^2)^{-1/2}$, $\beta = v_z/c$, m_0 = the rest mass of the electron, and Λ = the mechanical period of the wiggler.

The dependence of λ_0 on the square of the relativistic factor* is due to a double Doppler shift—once to move the permanent magnet field into the beam frame of the electrons and once to move the radiation back into the laboratory. This dependence accounts for much of the excitement for FELs. One can control λ_0 by the kinetic energy and by a mechanical dimension Λ. Furthermore, by designing the wiggler to account for the conversion of some of the beam energy to photons, and by recovering or reusing the electron in a racetrack geometry, one can predict quite high efficiencies. Such lasers are not of the tabletop variety.

PROBLEMS

10.1. It is known from experiment that if the c axis of ruby is perpendicular to the optic axis, the laser will oscillate in a linearly polarized mode. Assume the c axis along the x direction with the optical axis along z.
  (a) What is the direction of the electric field ($\mathbf{a}_x$ or $\mathbf{a}_y$)? Why? HINT: Consider Fig. 10.2(b).
  (b) If the difference in the cross sections for $\mathbf{E} \perp c$ and $\mathbf{E} \parallel c$ were due to the different indices of refraction (Table 10.1), what would be the ratio? Which would be bigger? Which is bigger?

10.2. While the time scale for interchange of populations between the **E** and the **2A** states is short—1 ns or so—it is not instantaneous. Discuss how this fact would influence your modeling of a Q-switched ruby laser or a system used for amplification.

10.3. Consider the possibility of pumping a ruby laser rod, 1 cm in diameter, placed at the focal point of a cylindrical parabolic reflector aimed at the sun. The spectra from the sun have the shape of a blackbody distribution with $T = 5000°$K and a total integrated intensity (over all frequencies) of 1.4 kW/m². Estimate the size of the parabola needed to achieve equal

---

*An apology is offered for using the symbol γ for the relativistic factor since it is the same symbol as used for the gain coefficient. However, that convention is too ingrained in all literature to even think of changing it for this small section.

populations in the upper and lower laser levels. Assume the data of Fig. 10.3. Pick a mean absorption cross section of 10^{-20} cm^2 for 3500 Å $< \lambda <$ 4300 Å and 7×10^{-19} cm^2 for 5300 Å $< \lambda <$ 6000 Å. Ignore differences between E $\perp$ and $\parallel$ to c.

10.4. Use the data provided in Table 10.3 for the following questions. Present the answers in tabular form whenever possible.

(a) Identify the upper and lower laser levels for each of the wavelengths listed.

(b) Use the fact that the spontaneous lifetime of the $^4F_{3/2}$ level is 255 μs, the given branching ratios, and the line widths of the various transitions to compute the stimulated emission cross section for the various wavelengths. Assume a Lorentzian line shape for each.

(c) Assume that the atoms in the $^4F_{3/2}$ level are in local thermodynamic equilibrium with the lattice at 300°K. What are the percentages in the two levels of that state? (Ans. 40.1% in R_2 and 59.9% in R_1.)

(d) If there are 5×10^{17} atoms/cm^{-3} excited to the $^4F_{3/2}$ manifold, compute the small-signal gain coefficients for each of the transitions. Assume the concentration of the $^4I_{1\ 1/2}$ to be zero.

(e) At what lattice temperature would the dominant transition be different from that of 1.06415 μm? Suppose the crystal were cooled to 77°K (liquid nitrogen). Ignore any change in the line widths (or that all change in a synchronous fashion) and assume 5×10^{17} atoms/cm^3 were excited to the $^4F_{3/2}$ manifold. Repeat (d). Which transition has the highest gain coefficient?

(f) Find the saturation intensity for each of the transitions (at 300°K).

10.5. Use the data of Fig. 10.2 to compute the wavelengths of the absorption between the $^4I_{9/2}$ and $^4F_{3/2}$ manifolds. Note that some of these transitions match the emission of GaAs or AlGaAs semiconductor lasers.

10.6. Assume a Lorentzian line shape for all of the transitions listed in Table 10.2. Compute the effective stimulated emission cross section at 1.06415 μm including the contribution from the wings of the 1.0644 transition.

10.7. Use the data given in Table 10.5 to compute the fluorescent power emitted by the glass laser systems when pumped to achieve a gain of 1%/cm. Assume a doping of 1 wt%.

10.8. Use the ideas discussed in Sec. 8.7 to estimate the maximum optical energy that can be extracted from a glass rod 10 cm long and 1 cm^2 in area with a doping density of 1 wt% Nd.

10.9. Assume that the glass laser system indicated in Table 10.4 were incorporated in a cavity 40 cm long, pumped to $1.5 \times$ threshold, and mode-locked by one means or another. Without being too complicated, estimate the pulse width.

10.10. Use the data of Table 10.5 to compute the following parameters for the

helium-neon laser system. Assume low-pressure gas with Doppler broadening being dominant ($T = 400°C$).
- **(a)** Stimulated emission cross sections for the following transitions: $\lambda = 3.39$ μm, $\lambda = 1.1523$ μm, and $\lambda = 0.6328$ μm. Name the transitions.
- **(b)** What is the radiative lifetime of the $3s_2$ and $2s_2$ states of neon?
- **(c)** What is the saturation intensity for these transitions? (Assume that $\tau_1 \ll \tau_2$.)
- **(d)** Assume a laser tube 50 cm long with a population difference $[N_2 - (g_2/g_1)N_1]$ equal to 10^{10} cm^{-3} for each of the transitions listed above. What is the maximum intensity that can be extracted from this laser on these three transitions? [This part requires the solution to parts (a) and (c) and the theory from Sec. 9.2.1.]
- **(e)** The laser sequence $3s_2 \rightarrow 3p_4 \rightarrow 2s_2 \rightarrow 2p_4$ is sometimes called the push-pull laser. Why? What wavelengths are generated?
- **(f)** The transition at 6401 Å ($3s_2 \rightarrow 2p_2$) seldom lases. Why?

10.11. The gain on the $3s_2 \rightarrow 3p_4$ transition in the helium-neon laser at 3.39 μm is very large—say, 30 dB in a 1-m-long tube.
- **(a)** Assume a low-pressure gas mixture at 400°K and use the data to compute the population inversion necessary to provide this gain. [Ans. $N_2 - (g_2/g_1)N_1 = 2.04 \times 10^9$ cm^{-3}.]
- **(b)** Suppose that this same population inversion was obtained on the common "red" line at 6328 Å ($3s_2 \rightarrow 2p_4$). What would be the small-signal gain in a 1-m tube? (Ans. 0.452 dB/m.) Why is the gain so different?
- **(c)** Use the data given in Table 10.5 and Fig. 10.11 to compute the saturation intensity for the two laser transitions. Assume a homogeneous line width of 20 MHz and that $\tau_1 \ll \tau_2$. [Ans. $I_s(3.39$ μm$) = 2.5$ mW/cm^2; $I_s(0.6328$ μm$) = 0.17$ W/cm^2.]

10.12. By providing sufficient wavelength discrimination, one can make all of the $3s \rightarrow 2p$ transitions in neon lase. Assume a simple cavity, without prisms, gratings, or etalons, and equal inversion densities on the various transitions. What must be the finesse of the cavity at the various wavelengths compared to that at 6328 Å to ensure lasing? (HINT: Compute the gain-to-loss ratios and compare to that at 6328 Å.)

10.13. Use the NBS tables to construct an energy level diagram for the Cd$^+$(ion) laser in a manner similar to that shown in Fig. 10.10. Include the helium metastable state, 2^3S and 2^1S, and pick your scale so that the upper and lower levels of the 4415 and 3250 Å Cd$^+$ laser are shown.

10.14. A Penning reaction, which is exemplified by

$$\text{He}(2^1\text{S or } 2^3\text{S}) + \text{Cd} \rightarrow \text{He(ground)} + (\text{Cd}^+)^* + e(\text{Kinetic energy})$$

is thought to be responsible for some part of the pumping of the (Cd$^+$)

laser. Use the data found in Problem 10.13 to compute the excess energy of this reaction (which is usually transferred to the free electron).

10.15. Compute the quantum efficiency of the argon-ion laser at 4880 Å.

10.16. Construct a graph showing the amount of energy stored in the vibrational levels ($v = 1$ to 8) of N_2 per liter-atm of gas as a function of temperature between 500 and 1500°C.

10.17. Evaluate the small-signal gain coefficient for the P(22) and R(20) transitions of the CO_2 laser using the A coefficients given in Table 10.4, the rotational constant (pick a mean value for the upper and lower states), a rotational temperature of 500°K, and a ratio of population in the 001 to 100 states of 1.1. Assume a density in the 100 state to be 10^{15} cm^{-3} and a homogeneous line width of 1 GHz.

10.18. The following is a model for an excitation transfer laser. State 3 in gas A is excited by an external source at a rate R (cm^{-3}/sec), which can decay back to the ground state at a rate $A_{30}N_3$ or transfer its excitation to state 2 of gas B at a rate $\omega_{32}N_3$. The fate of state 2 follows the usual routes (i.e., spontaneous and stimulated decay) but can also transfer its energy back to state 3 of gas A at a rate $\omega_{23}N_2$.
(a) Show the rate equations for this laser.
(b) Find the value of R that enables this system to have a gain of 0.01 cm^{-1}. Assume a stimulated emission cross section of $\sigma = 10^{-16}$ cm^2. (Ans. $R = 1.5 \times 10^{22}$ cm^{-3}/sec.)
(c) Assume that this laser is pumped far above threshold and its intensity is such that the populations N_2 and N_1 are equalized (for all practical purposes) by stimulated emission. What is the efficiency of this laser? (Ans. 35%.)

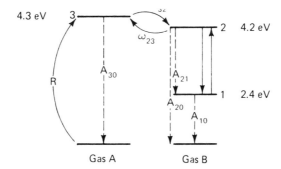

10.19. The argon-ion laser is inhomogeneously (Doppler)-broadened with a width of 5.0 GHz and is contained in a cavity 100 cm long. The average power is 4 W when the small signal gain is three times the threshold value.
(a) How many TEM$_{0,0,q}$ modes are above threshold? (Assume that the central mode is at line center.)

(b) How much power is carried by the central mode? (Assume that the distribution of power among modes is Gaussian; i.e., $I(\nu_q) = I(\nu_{q1} = \nu_0) \exp\{-4 \ln 2[(\nu_q - \nu_0)/\Delta\nu_0]^2\}$.
(c) What is the peak power of the laser when it is mode-locked?
(d) What is the pulse width (FWHM) of the mode-locked output?
(e) If the spot size is 2 mm, what is the peak *density* (i.e., photons/volume) of the mode-locked pulse?

10.20. Consider the hypothetical molecular system shown, where it is assumed that the N molecules are formed in the ($v = 4$) vibrational state at $t = 0$. If one neglects all relaxation processes except stimulated emission for lowering the vibrational quantum number:
(a) What is the maximum energy that can be extracted from this system as optical power?
(b) What is the final distribution of vibrational states after stimulated emission ceases? [This problem can be approximated by a chemical laser in which the upper state is formed by a chemical reaction A + B → AB* ($v = n$). Note that there is no ground-state population initially.] To make the problem easier, neglect threshold requirements and assume that lasering occurs provided that $N_{\text{upper}} > N_{\text{lower}}$. Also ignore the rotational structure, by computing the energy only in the vibrational bands.
(c) Evaluate the optical power assuming 1 atm of atoms in the $v = 4$ state and lasing occurring for 1 μsec.

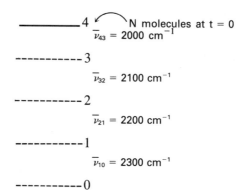

10.21. The helium-neon laser transition is Doppler broadened with a FWHM of 1.5 GHz. Assume the pumping and thus the small-signal gain coefficient are four times the threshold value, and the cavity is 100 cm long.
(a) find the number of $TEM_{0,0}$ modes that can oscillate simultaneously.
(b) Suppose all of the modes were locked together:
 (1) What is the repetition frequency of the pulses?
 (2) Estimate the pulse width.

10.22. The problem is to model the following "Gedanken" experiment. One has an ideal laser medium (i.e., $\tau_1 = 0$ and hence $N_1 = 0$ always) with a stimulated emission coefficient of 10^{-15} cm^2; an upper state lifetime of 50 ns; a branching ratio of 0.6; a length of 15 cm, and an area of 0.6 cm^2; and the wavelength is $\lambda = 5000$ Å. It is placed in the cavity shown below and the excitation rate of state 2, R_2, is varied from zero to three times the threshold for oscillation. There are two detectors shown: one is used to measure the spontaneous emission from the upper state and emerging from the side of the cavity; the other is to measure the laser output along the axis of the cavity. Assume that the output of each detector is a current that is proportional to power.

Construct a graph showing the variation of both signals as a function of a normalized pumping rate $R_2/R_\text{threshold}$. Plot the *total* spontaneous and laser power in watts from this system. (The side-light detector would only capture a fraction of the total spontaneous amount, but that fraction would not be a variable.)

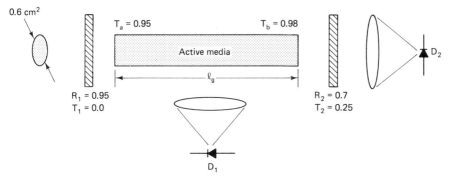

10.23. Suppose the short-range nuclear potential repelling the krypton and fluorine atoms can be expressed as

$$V(r) = V_0 \left(\frac{a}{r}\right)^6 \text{ with } V_0 = 54.7 \text{ eV and } a = 1.5 \text{ Å}$$

The attractive force is due to the coulomb forces. Use the above and the IP of krypton in Table 10.7 to estimate the wavelength of the (KrF)* excimer radiation. The following steps may be helpful:
(a) Find the "equilibrium radius" r_min of the upper state. (It will be close to but not equal to the 2.3 Å specified in Table 10.8)
(b) Find the total energy of the (Kr$^+$F$^-$)* state at r_min.
(c) Find the total energy of the Kr + F at r_min by evaluating $V(r)$ at r_min.
(d) Subtract the answer of (c) from that of (b). This will be the energy of the transition and will be close but not equal to that of Table 10.8.
(e) It helps to draw the interaction potentials and to label r_min on that sketch.

10.24. The following are in reference to an FEL with a beam energy of 75 MeV.
 (a) Evaluate the relativistic factor $\gamma = m/m_0 = [1 - (v/c)^2]^{-1/2}$. (Ans. $\gamma = 147.8$.)
 (b) Evaluate the quantity $\beta = v/c$ for this electron. (Ans. $\beta = 0.999977$.)
 (c) If the periodicity of the magnets in Fig. 10.18 were 3 cm, what would be the wavelength of the laser? (Ans. $\lambda = 0.687$ μm.)

REFERENCES AND SUGGESTED READINGS

1. T. A. Hänsch, M. Pernier, and A. L. Schalow, "Laser Action of Dyes in Gelatin," IEEE J. Quant. Electron. *QE-7*, 47, Jan. 1971.
2. (a) T. H. Maiman, "Stimulated Optical Radiation in Ruby Masers," Nature *187*, 493, 1960.
 (b) T. H. Maiman, "Optical and Microwave-Optical Experiments in Ruby," Phys. Rev. Lett. *4*, 564, 1960.
 (c) T. Maiman, "Stimulated Optical Emission in Fluorescent Solids I. Theoretical Considerations," Phys. Rev. *123*, 1145–1150, Aug. 15, 1961.
 (d) T. H. Maiman, R. H. Hoskins, I. J. D'Haenens, C. K. Asawa, and V. Evtuhov, "Stimulated Optical Emission in Fluorescent Solids. II Spectroscopy and Stimulated Emission in Ruby," Phys. Rev. *123*, 1151–1157, Aug. 15, 1967.
3. D. C. Cronemeyer, "Optical Absorption Characteristics of Pink Ruby," J. Opt. Soc. Am. *56*, 1703, 1966.
4. W. Koechner, L. C. DeBenedictis, E. Matovich, and G. E. Meyer, "Characteristics and Performance of High Power CW Krypton Arc Lamps for Nd:YAG Laser Pumping," IEEE J. Quant. Electron. *QE-8*, 310, 1972.
5. E. Snitzer and C. G. Young, "Glass Lasers," in *Lasers*, Vol. 2, Ed. A. K. Levine (New York: Marcel Dekker, Inc., 1968), p. 199.
6. H. G. Danielmey, "Progress in YAG Lasers," in *Lasers*, Vol. 4, Eds. A. K. Levine and A. J. DeMaria (New York: Marcel Dekker, Inc., 1976), p. 8.
7. T. Kushida, H. M. Marcos, and J. E. Geusic, "Laser Transition Cross-section and Fluorescence Branching Ratio for Nd^{3+} in Yttrium Aluminum Garnet," Phys. Rev. *167*, 1289, 1969.
8. B. B. Snavely, "Flash Lamp Pumped Dye Lasers," Proc. IEEE *57*, 1374, 1969.
9. M. Bass, T. F. Deutsch, and M. J. Weber, "Dye Lasers," in *Lasers*, Vol. 3, Eds. A. K. Levine and A. DeMaria (New York: Marcel Dekker, Inc., 1971), p. 275.
10. A. Javan, W. R. Bennett, Jr., and D. R. Herriott, "Population Inversion and Continuous Optical Maser Oscillation in a Gas Discharge Containing He-Ne Mixtures," Phys. Rev. *6*, 106, 1961.
11. W. R. Bennett, Jr., "Inversion Mechanisms in Gas Lasers," Appl. Opt., Suppl. 2 Chem. Laser, 3–33, 1965.
12. E. F. Labuda and E. I. Gordon, J. Appl. Phys. *35*, 1647, 1964.
13. P. K. Cheo, "CO_2 Lasers," in *Lasers*, Vol. 3, Eds. A. K. Levine and A. DeMaria (New York: Marcel Dekker, Inc., 1971).
14. The even-odd sequence is due to the symmetry of the CO_2 molecule (see G.

Herzberg, *Infrared and Raman Spectra* (Princeton, N.J.: D. Van Nostrand Company, 1966), Chap. 1.
15. G. J. Schultz, "Vibrational Excitation of N_2, CO, and H_2 by Electron Impact," Phys. Rev. *135A*, 988–994, 1964.
16. (a) E. L. Patterson and R. A. Gerber, "Characteristics of a High Energy Hydrogen Fluoride (HF) Laser Initiated by an Intense Electron Beam," IEEE J. Quant. Electron. *QE-11*, 642–647, Aug. 1975.
    (b) G. N. Hays, J. M. Hoffman, and G. C. Tisone, "Phoenix 2: Sandia's 1 kJ HF Laser System," Paper WE 1 Topical Meeting on Inertial Confinement Fusion, San Diego, Calif., Feb. 26–28, 1980.
17. (a) A. Yariv, *Introduction to Optical Elecronics,* 2nd ed. (New York: Holt, Rinehart and Winston, 1971), Chap. 7.
    (b) A. Yariv, *Quantum Electronics*, 2nd ed. (New York: John Wiley & Sons, Inc., 1975), Chap. 10.
18. B. G. Streetman, *Solid State Electronic Devices*, 2nd ed. (Englewood Cliffs, N.J.: Prentice-Hall, Inc., 1980), Chap. 10.
19. W. R. Bennett, Jr., "Gaseous Optical Masers," Applied Opt., Suppl. Opt. Masers, 24–64, 1962.
20. C. K. N. Patel, "High Power Carbon Dioxide Lasers," Sci. Am. *219*, 22–23, 1968.
21. D. R. Herriott, "Applications of Laser Light," Sci. Am. *219*, 141–156, 1968.
22. A. B. Siegman, *An Introduction to Lasers and Masers* (New York: McGraw-Hill Book Company, 1971).
23. A. E. Siegman, *Lasers* (Mill Valley, Calif.: University Science Books, 1986).
24. W. Koechner, *Solid State Engineering* (New York: Springer-Verlag, 1976).
25. A. A. Kaminiski, *Laser Crystals*, in *Springer Series in Optical Sciences* (New York: Springer-Verlag, 1981).
26. H. Brown, *High Power Glass Lasers* (New York: Springer-Verlag, 1976).
27. See "Third Special Issue on Free-electron Lasers," IEEE J. Quant. Electron. *QE-21*, No. 7, 1985. This contains 40 papers on free-electron lasers.
28. Thomas C. Marshall, *Free Electron Lasers* (NewYork: MacMillan Publishing Co., 1985). This has a comprehensive bibliography of 206 references and a very readable introduction. See also IEEE J. Quant. Electron. *QE-23*, No. 9 (1987) for additional 27 articles.
29. (a) J. E. Velazco and D. W. Setser, "Quenching Studies of Xe ($^3P_2$) Metastable Atoms," IEEE J. Quant. Electron. *QE-11*, 708–709, Aug. 1975.
    (b) L. G. Piper, J. E. Velazco, and D. W. Setser, J. Chem. Phys. *59*, 3323, 1973.
    (c) J. E. Velazco and D. W. Setser, Chem. Phys. Lett. *25*, 197, 1974.
30. M. Rokni, J. A. Mangano, J. H. Jacob, and J. C. Hsia, "Rare Gas Fluoride Lasers," IEEE J. Quant. Electron. *QE-14*, 464–481, July 1978.
31. C. A. Brau, "Rare Gas Halogen Excimers," in *Excimer Lasers*, Ed. C. K. Rhodes, Topics in Applied Physics, Vol. 30 (New York: Springer Verlag, 1979).
32. John M. Madey, "Stimulated Emission of Bremsstrahlung in a Periodic Magnetic Field," Journal of Applied Physics, *42*, 1906, 1971.
33. H. Motz, "Application of the Radiation from Fast Electron Beams," Journal of Applied Physics, *22*, 527, 1951.
34. J. A. Pasour, "Free Electron Lasers," IEEE Circuits and Devices Magazine, 55–63, 1987.

35. J. Madey, "Stimulated Emission of Radiation in Periodically Deflected Electron Beam," U.S. Patent 3,822,410, July 2, 1974.
36. J. J. Ewing, "Excimer Lasers," in *The Laser Handbook*, Vol. 3, Article A4, Ed. M. L. Stitch, (New York: North Holland Publishing Company, 1979).
37. C. J. Ultee, "Chemical and Gas Dynamic Lasers," in *The Laser Handbook*, Vol. 3, Article A5, Ed. M. L. Stitch, (New York: North Holland Publishing Company, 1979).

11

Semiconductor Lasers

11.1 INTRODUCTION

11.1.1 Overview

Semiconductor lasers are the last class of optical oscillators to be considered and are the most important for communications and control applications because of their convenience, efficiency, and natural compatibility with the rest of modern electronics. They have the simplest structure imaginable; two pigtail leads to be connected to a power supply; a p-n junction where most of the physics resides; and cavity mirrors formed by cleaving (i.e., breaking!) the crystal. A typical laser is shown in Fig. 11.1 with some artistic artifacts added to identify some of the problems to be addressed in this chapter.

As implied by Fig. 11.1, it is the injected current density which provides the optical gain to a wave whose spatial extent may be much larger than the active region. This is an electromagnetic problem that will be addressed in Chapter 13. Note also the small dimensions indicated in Fig. 11.1 and think about the implications based on our prior discussion of the larger gas or solid-state lasers. In order for the round-trip gain (RTG) to exceed one, the gain coefficient must be quite large. For instance, let the facet reflectivity be 0.3 (typical) and apply the RTG equation to compute $\gamma = 4.82 \times 10^1 \, \text{cm}^{-1}$, a value unheard of in our previous work. The injected current spreads in the plane of the junction so that the effective area is larger than the $10 \times 250 \, \mu\text{m}^2$ of the contact.

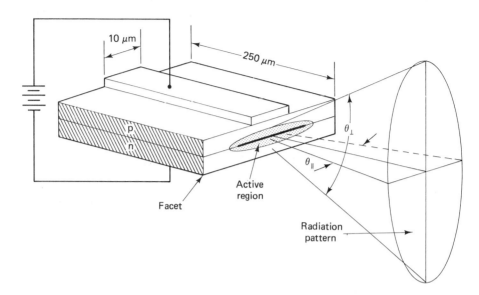

Figure 11.1 A generic *p-n* junction laser. The active region is at the junction of *p* and *n* regions where the current is flowing and is shown as solid black. The mode inside the semiconductor is more or less like an ellipsoid with a spatial extent much larger than the active region and is different perpendicular and parallel to the junction. This results in a radiation pattern with different spreading angles.

Let us arbitrarily pick a factor of 4 increase to account for this spreading. For a drive current of 100 mAmp, the current density is a rather startling 1 kAmp/cm^2. Such a laser might well operate at >50% efficiency. The voltage source has to be some bigger than the band gap ~ 1.43 V, so pick 2.0 V for this estimation. The electrical power used to drive this generic device is 200 mW with 100 mW emerging as optical radiation from a device smaller than this exclamation mark! No wonder there is so much excitement.

Some of their performance characteristics are considerably less than what one can achieve with other types of lasers. For instance, the output radiation is *not* a simple TEM$_{0,0}$ mode as it can be with a gas or solid-state laser; rather, it is usually a field whose dimensions along the plane of the junction are much larger than those perpendicular to it, and in either case, is on the order of a free space wavelength or so. Because of the small dimensions the beam diverges with angles on the order of a few degrees, and, because of the unequal dimensions, the beam spreads unequally in the two directions. The optical bandwidth over which oscillation is possible is enormous by gas or solid-state laser standards, and thus multimode oscillation is the norm unless special precautions are taken. Further, the spectral purity of the oscillating modes tends to be much worse than the corresponding values for the larger gas or solid-state types (but still better than the resolution of most optical instruments).

11.1.2 Populations in Semiconductor Laser

There are many phenomena that a semiconductor laser has in common with the larger ones which have been used in most of our analysis. For instance, both use an *inverted* population; hence our first task is to decide what constitutes such an inversion, or even more basic, what are the "populations"?

Let us establish an answer to this last question before anything else. We are interested in the stimulated recombination of holes and electrons generating a photon according to the following "chemical" equation.

$$e + h \longrightarrow (h\nu) > E_g \qquad (11.1.1)$$

and thus the electron and hole populations are involved. Only direct band-gap semiconductors permit this reaction with high probability for transitions between the lowest allowed states in the respective bands. The situation is shown in Fig. 11.2 where we plot the number of available states, shaded as being occupied, white as being empty, for the two types of semiconductors as a function of the momentum of a carrier. In that figure, we presume a few electron states in the conduction band are occupied and there are a few unoccupied states (i.e., holes) in the valence band. Note that in either case, the electrons will give up an energy roughly equal to the band-gap value in making the transition from a state *up* in the conduction band to one *down* in the valence band. For the most part, it does so by emitting a photon (11.1.1) for a direct material but finds some nonradiative process for the indirect material that competes for the electron population, and thus few photons are emitted. Why?

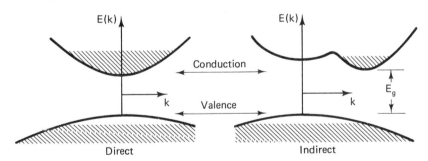

Figure 11.2 $E(k)$ vs. k (momentum) for direct and indirect semiconductors. The shaded areas are intended to imply states filled with an electron, and the white areas are empty.

The reason is contained in the fact that both energy and momentum must be conserved in any reaction. Now conservation of energy is usually the easiest and most transparent. Initially, the electron is at or above the conduction edge, and it makes a transition to a state at or below the valence band edge. Hence, it yields an energy somewhat greater than $E_c - E_v = E_g$ to either a photon or to this other unnamed process.

Conservation of momentum requires a more careful scrutiny of the process since we require the vector sum of the final momentum of all partners to be equal to the initial value. Let us make this accounting for the average electron in the direct band-gap case. As we proceed, the numbers will point out the difficulty for the indirect case:

$$\hbar \mathbf{k}_f + \hbar \mathbf{k}_p = \hbar \mathbf{k}_i \qquad (11.1.2)$$

where $\mathbf{k}_{f,p,i}$ equals the wave vector of the final electron state, the photon, and the initial state. Now $|\hbar \mathbf{k}|$ is the momentum of the electron, and we can approximate it by the effective mass, m_e^*, times its mean kinetic velocity found by equating its kinetic energy *in the band*, $m^* v^2/2$, to $3kT_e/2$ and letting the electron temperature equal the lattice temperature. For $m_e^* = 0.067\ m_0$ (typical of GaAs) and $kT/e = 0.026$ V, we find the random velocity for an electron in the conduction band to be

$$v_e = (3kT/m_e^*)^{1/2} = 2.61 \times 10^5 \text{ m/sec} \qquad (11.1.3a)$$

Thus

$$|m_e^* v_e| = \hbar k_i = 1.59 \times 10^{-26} \text{ kg-m/sec} \qquad (11.1.3b)$$

Consider the momentum of a photon with a wavelength of 8400 Å: $k_p = 2\pi/\lambda_0$, and

$$\hbar k_p = h/\lambda_0 = 7.89 \times 10^{-28} \text{ kg-m/sec} \qquad (11.1.4)$$

Note that the momentum of the photon is very small compared to that of the average electron, and thus the momentum of the hole or final state must be just about equal to the starting value. In other words, *the transition must be vertical* and

$$\Delta \mathbf{k} = \mathbf{k}_i - \mathbf{k}_f \approx 0 \qquad (11.1.5)$$

This is usually referred to as the "**k** conservation" rule.

Usually there are enough scattering events in a direct material to relax the **k** conservation rule somewhat, but the problem becomes much worse for the indirect materials where the electrons at the indirect minima of Fig. 11.2(b) have a **k** vector on the order of π/a (a = lattice constant). For a typical distance—say 5.45 Å(GaP)—the momentum is enormous compared to the previous numbers: $\hbar k = 6.08 \times 10^{-25}$ kg-m/sec. This is 40 times the kinetic momentum of the average electron, which in turn was 20 times that of the photon!

Optical transitions can and do take place in indirect materials—the "green" LED using GaP is an experimental existence proof—but the oscillator strength for radiation is much weaker than that of a direct material. The amount of radiation that a group of electron-hole pairs produces is really a matter of competition between radiative and nonradiative rates. The latter tends to dominate in indirect materials, especially in the presence of impurities or defects. We will concentrate on the direct materials listed in Table 11.1 in which the radiative rates are quite fast.

TABLE 11.1 PROPERTIES OF COMMON SEMICONDUCTING MATERIALS

Material	Band Gap (eV)	ϵ_r	μ_e	μ_h	m_e^*/m_0	Lattice (Å)
			(cm^2/V-sec)			
C(i)	5.47	5.7	1800	1200	0.2	3.5668
GaP(i)	2.26	11.1	1600	100	0.82	5.4512
AlAs(i)	2.16	10.9	180	—	—	5.6605
GaAs(d)	1.43	13.2	8500	400	0.067	5.6533
InP(d)	1.35	12.4	4600	150	0.077	5.8686
Si(i)	1.12	11.9	1500	450	1.1	5.4309
GaSb(d)	0.72	15.7	5000	850	0.042	6.0957
InAs(d)	0.36	14.6	33,000	460	0.023	6.0584
InSb(d)	0.17	17.7	80,000	1250	0.0145	6.4794

There are many similarities between semiconductor lasers and the much larger gas and solid-state ones, but there are distinct differences also. Hence our first job is to reestablish our connection with elementary semiconductor theory such as that used to describe a *p-n* junction, transistor, or photodiode so as to identify an "inverted" population.

11.2 REVIEW OF ELEMENTARY SEMICONDUCTOR THEORY

It is assumed that most of the readers are familiar with many of the issues associated with semiconductors, and many of the elementary ideas that have been used again and again in elementary electronic courses will be skipped. However, there are a few ideas that are crucial to laser understanding, and these will be discussed in detail. The first task is to determine how many electronic states are available to be filled by an electron in the conduction band or emptied in the valence band. We proceed in two steps: first to identify the number available and then to compute the probability that the state is filled (or empty). In a loose sort of way, it is the filled states in the conduction band minus the filled states in the valence band which correspond to $N_2 - N_1$ of our earlier description of lasers.

11.2.1 Density of States

The fact that elementary particles, such as electrons, can be described by a wave function is an idea that has been established again and again in elementary courses in physics, chemistry, and electronics. We need the idea here also. It should also come as no surprise that much of the notation used to describe these electronic wave functions in a crystal is the same as that used in electromagnetic theory. After all, both are wave phenomena and both have quantization forced on the wave functions for the same reasons, namely, boundary conditions.

For a semiconductor of size L_x, L_y, L_z, one invokes the most rudimentary and reasonable type of requirement; namely, the electron (or hole) should be found somewhere inside the volume (and not outside!). We assume that the wave function for an electron varies as

$$\psi(\mathbf{r}) = u(r) \exp(-j\mathbf{k} \cdot \mathbf{r}) \qquad (11.2.1)$$

where $u(r)$ reflects the effect of the lattice through which the electron wave is propagating and is the function that generates the familiar valence and conduction bands. Let us concentrate on the physics of the propagating factor. For a rectangular box of semiconductor with sides L_x, L_y, and L_z, one must have standing waves between the extremes and thus the components of k must be related to these dimensions by

$$k_x = m\pi/L_x; \qquad k_y = p\pi/L_y; \qquad k_z = q\pi/L_z \qquad (11.2.2)$$

$$\text{for } \mathbf{k} = k_x \mathbf{a}_x + k_y \mathbf{a}_y + k_z \mathbf{a}_z \qquad (11.2.3)$$

In order to compute the density of states in the bands we need to count (yes, count!) the number of combinations of quantum numbers m, p, and q that are permitted for various energies with respect to the band edges. The drudgery of this seemingly uninspiring calculation can be minimized by constructing a "mode" diagram similar to that used in Chapter 7 for electromagnetic modes in a blackbody cavity. However, let us proceed independently of that prior development and attempt to describe the quantization requirement (11.2.2) by the "k-space" diagram of Fig. 11.3. The axes are k_x, k_y, k_z, and an allowed value of $\mathbf{k}$ is the vector from the origin to one of the triplets of points labeled by the quantum numbers (m, p, q) with the appropriate multiplicative factors $\pi/L_{x,y,z}$ understood. There is one last item to be noted about the construction of Fig. 11.3: there is exactly one set of quantum numbers (or modes) per elementary volume in k space: $V_k = (\Delta k_x = \pi/L_x)(\Delta k_y = \pi/L_y)(\Delta k_z = \pi/L_z)$. Thus each state occupies a volume $\pi^3/L_x L_y L_z$ in k space, or

Density of modes in k-space = (1 mode)/(volume of each mode)

$$= L_x L_y L_z / \pi^3 \qquad (11.2.4)$$

Now comes the step that saves us the drudgery of counting. We recognize that these points are very close together for a macroscopic piece of semiconductor. For instance, let $L_x = L_y = L_z = 1$ mm $= 10^{-3}$ m. Choose some multiplicative factor—say $(m^2 + p^2 + q^2)^{1/2} = 100$—and multiply it by (π/L_x) to obtain a value of $k \approx 3.14 \times 10^5$ m^{-1} as a test case. Let us translate this value into a more comfortable one by computing the momentum and kinetic energy of the electron with this value of k. The momentum is trivial, $p = \hbar k = 3.31 \times 10^{-29}$ kg-m/sec. If we assume an effective mass $m^* = 0.067\, m_0$, we find an equivalent speed of this electron equal to 543 m/sec—hardly blinding. The kinetic energy is also unimpressive, $E = (\hbar k)^2 / 2m^* = 8.99 \times 10^{-27}$ joules =

Sec. 11.2 Review of Elementary Semiconductor Theory

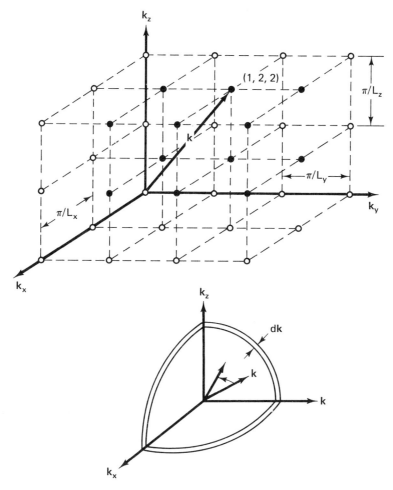

Figure 11.3 Mode diagram for a "large" semiconductor. The top sketch shows each individual allowed value of **k** whereas the insert implies that the points are very close together so that they could be considered as a continuum.

5.6×10^{-8} eV. (This energy is to be added to the conduction band value or subtracted from E_v).

The point to be drawn from these numbers is that the mode numbers must be huge, $(m^2 + p^2 + q^2)^{1/2} \approx 10^5$ or so, before the energy is comparable to the thermal value $kT/e = 0.0259$ eV, or the spacing between them must be very small for ranges of **k** of interest. The smaller scaled version shown as an insert in Fig. 11.3 illustrates the simple computation involved in counting the states. The wave vector can range from along k_z, k_y, or k_x or any combination. For $|\mathbf{k}|$ = constant, this describes one eighth of a sphere and if we allow **k** to expand to $\mathbf{k} + d\mathbf{k}$, this puts a shell thickness dk on this fraction of that sphere. The volume of that shell, $(4\pi k^2 dk/8)$ times the density of points in k-space is the number of states, $N_k dk$ encompassed by allowing k to vary from k to $k + dk$.

$$N_k dk = \{L_x L_y L_z/\pi^3\}\{1/8\}\{4\pi k^2 dk\} \quad (11.2.5)$$

Remember that we are dealing with momentum states of an electron that also has two possible spin orientations. Hence, (11.2.5) must be multiplied by a factor of 2 and the density of states in k-space per unit of volume becomes

$$\rho_k\, dk \cdot V = 2 \cdot N_k dk$$

or

$$\rho_k\, dk = 2N_k dk/L_x L_y L_z = k^2 dk/\pi^2 \quad (11.2.6)$$

Most persons feel more comfortable expressing the density of states in terms of the energy of the carrier (electron or hole) beyond the band edge (i.e., above the conduction band or below the valence band edge). The energy and momentum *in a band* are related by

$$E = (\hbar k)^2/2m^* = (\epsilon - E_c) \text{ or } (E_v - \epsilon) \quad (11.2.7)$$

where ϵ is the total energy and E is an energy measured from the edge of a band. Thus

$$k = (2m^* E/\hbar^2)^{1/2}$$

$$dk = \left(\frac{2m^*}{\hbar^2}\right)^{1/2} \frac{dE}{2E^{1/2}}$$

Substituting these relations into (11.2.6), we obtain the density of states in either the conduction or valence band.

$$\rho_{c,v}(E)dE = \left\{\frac{1}{2\pi^2}\right\}\left\{\frac{2m^*}{\hbar^2}\right\}^{3/2} \{E^{1/2}dE\} \text{ (in m}^{-3}\text{)} \quad (11.2.8)$$

where E is measured from the band edge, that is, up into the conduction band or down into the valence band, and m^* is the effective mass of each of the carriers. This is sketched in Fig. 11.4 for a semiconductor with an effective mass of the electron equal to $0.067\, m_0$ and that of the "heavy or normal" hole* of ~ 4 times that value. (For GaAs, the actual value of the mass of the heavy hole is $0.55\, m_0$. It should be apparent from (11.2.8) and Fig. 11.4 that the curvature of the band is related to the effective mass in that band, and if the actual value of m_h had been used, the curvature of the valence band would be $1/23.5$ of that of the conduction and the lower solid curve of Fig. 11.4 would have appeared as a straight line.)

* In the valence band of III-V semiconductors, there is a light hole band that has a maximum energy identical to that of the normal or heavy hole at $k = 0$. However, because of its smaller mass (more or less that of an electron), the density of states of the light hole is much less than that of the normal hole, and thus most holes at a given energy in that band are "heavy" holes.

Sec. 11.2 Review of Elementary Semiconductor Theory

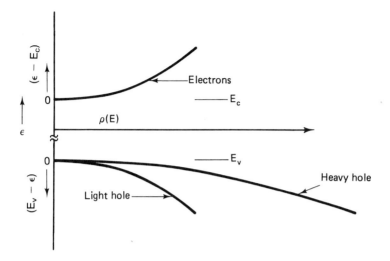

Figure 11.4 Density of states in a semiconductor with $m_{hh}^* = 4 \cdot m_e^*$ and $m_{lh} = m_e^*$.

There are a couple of points to be noted. First, the density of states for holes per unit of energy dE is much greater than that for electrons—a fact that is a consequence of the disparity in effective masses. The second point is that this represents a fairly large number of quantum states to be filled. For instance, suppose we wished to fill every available state in the conduction band from E_c to $E_c + \Delta E$ with the rest being empty. Thus the density of electrons that must be created in this sample, either by injection or by some other means, is found by integrating (11.2.8) from 0 to ΔE. This presumes that every state is filled or emptied with equal probability (an issue that will be addressed more precisely in the next section).

$$N_e = \int_0^{\Delta E} \rho_c(E)dE = \frac{1}{3\pi^2}\left[\frac{2m_e^*\Delta E}{\hbar^2}\right]^{3/2} \text{ (in m}^{-3}) \qquad (11.2.9)$$

Suppose $\Delta E = 0.1$ V $= 1.6 \times 10^{-20}$ joules and $m_e^* = 0.067\, m_0$ using GaAs as an example. Then we would need 2.5×10^{24} electrons/m^3 or 2.5×10^{18} cm^{-3} to fill every state that is less than 0.1 V above the conduction band edge. If we simultaneously created the same number of holes in the valence band (so that the electrons could make an optical transition to an empty state) then we can use (11.2.9) to find the empty energy interval occupied by the holes. Performing the same type of arithmetic we find that the empty interval of the valence band is only $(m_e^*/m_h^*)\Delta E_c = 0.0122$ V. Of course, the above calculations presume a complete filling (or emptying) of one of the bands up to the levels involved, which does not happen at a finite temperature, but it does illustrate the numerical factors involved.

Let us use the example considered above to estimate the distribution of photon energies resulting from the recombination of electrons within 0.1 eV of

E_c with empty states, of the holes, in the valence band that are 0.0122 eV below E_v. Let us arbitrarily pick the center of the recombination spectra to be midway between the allowed extremes: $E_g < h\nu < E_g + \Delta E_c + \Delta E_v$. For $E_g = 1.43$ eV, we obtain

$$h\nu/e = E_g + (0.1 + 0.0122)/2 = 1.486 \text{ V (i.e., } \lambda_0 = 0.834 \text{ } \mu\text{m)}$$

A reasonable estimate for the spectral width is one fourth of the energy spread within the two bands:

$$h\Delta\nu/e = (0.1 + 0.0122)/4 = 0.02805 \text{ V}$$

Now using $\dfrac{\Delta\nu}{\nu} = \dfrac{\Delta E}{E}$ one obtains $\Delta\nu = 6.788$ THz. This is a rough estimation for the bandwidth over which oscillation might be possible; it is enormous by gas or solid-state laser standards.

Of course, the numbers are only approximate because of the assumption of completely filled (or emptied) states. We need to fold in the probability of a state being occupied—a task to which we now turn.

11.3 OCCUPATION PROBABILITY: QUASI-FERMI LEVELS

In the previous section, we merely identified the states and blindly assumed all states filled to a certain level with equal probability. If the lattice temperature were 0°K, this is legitimate, but at a finite temperature, one must fold in the probability of a state being occupied. The Fermi function:

$$f(\epsilon) = \frac{1}{\exp\left[(\epsilon - E_f)/kT\right] + 1} \quad (11.3.1)$$

describes the probability of a state of energy ϵ, if it exists, being occupied and thus $1 - f(\epsilon)$ is the probability that it is empty. Obviously, at $\epsilon = E_f$, the probability is 1/2 (at all temperatures). If the Fermi level is in the forbidden gap, $E_v < E_f < E_c$, and if the difference $(\epsilon - E_f) \gg kT$, then the Fermi function can be approximated by the Boltzmann factor and the probability of an electron state being occupied at an energy ϵ above E_f is given by

$$f(E) \approx \exp - \left[(\epsilon - E_f)/kT\right]$$

Such an approximation is quite often used for conventional semiconductor electronics. It is not valid for the active region of a semiconductor laser because, as we shall see, the Fermi level must be *within a band* and the inequality is not obeyed. Hence we will always use the complete expression given by (11.3.1).

We have another much more serious deviation from semiconductor electronics—we need *electrons and holes* to exist simultaneously at the same place. You should recognize that (11.3.1) presents a problem (before we try to

Sec. 11.3   Occupation Probability: Quasi-Fermi levels

solve it). If the Fermi level moves up to the conduction band, then (11.3.1) indicates that most of the states below E_f are *filled*! Thus the electron cannot make a transition to the valence band—all of the available states are taken. The problem is that (11.3.1) deals with an equilibrium situation whereas a laser uses an *inverted* system that is far from thermodynamic equilibrium. To handle this case, we borrow Shockley's concept of a quasi-Fermi level, F_n, to describe the probability of an electron state being occupied and F_p to describe the statistics of the hole state. At equilibrium, $F_n = F_p = E_f$.

A rigorous justification for this idea is beyond the scope of this book, but a rough justification comes from the following reasoning. The electrons (or holes) interact with each other on a time scale of 10^{-12} sec, which is extremely fast compared to recombination times of 10^{-9} sec. Hence it is reasonable to consider each group of carriers, electrons or holes, to be in equilibrium with each other in a band and described by a quasi-Fermi level pertinent to each. Thus we attach a subscript "n" to $f(\epsilon)$ of (11.3.1) and use F_n instead of E_f to describe the occupation probability of the electron states in the conduction band.

$$f_n(\epsilon) = \frac{1}{\exp[(\epsilon - F_n)/kT] + 1} \quad (11.3.2)$$

Similar considerations apply to the holes. The probability of a state being empty (i.e., a hole) is related to the quasi-Fermi level for the hole F_p by

$$f_p(\epsilon) = 1 - \frac{1}{\exp[(\epsilon - F_p)/kT] + 1}$$
$$= \frac{1}{\exp[(F_p - \epsilon)/kT] + 1} \quad (11.3.3)$$

Now we multiply the quasi-Fermi functions by the density of states to find the density of electrons or holes at an energy ϵ in an interval $d\epsilon$ in the various bands:

$$n_c(\epsilon)d\epsilon = \frac{1}{2\pi^2}\left[\frac{2m_e^*}{\hbar^2}\right]^{3/2} \frac{(\epsilon - E_c)^{1/2}d\epsilon}{\exp[(\epsilon - F_n)/kT] + 1} \quad \epsilon > E_c \quad (11.3.4)$$

$$p_v(\epsilon)d\epsilon = \frac{1}{2\pi^2}\left[\frac{2m_h^*}{\hbar^2}\right]^{3/2} \frac{(E_v - \epsilon)^{1/2}dE}{\exp[(F_p - \epsilon)/kT] + 1} \quad \epsilon < E_v \quad (11.3.5)$$

where the inequalities merely reflect the fact that there are no states to be filled (or emptied) for $E_v < \epsilon < E_c$ (with no doping).

Figure 11.5 is a sketch of the distribution of electrons for the case considered previously for the two different values of temperature; $kT/e = 0$ (0°K) and $kT/e = 0.0259$ V, with $F_n = E_c + 0.1$ eV. At 0°K, all of the states up to F_n in the conduction band are filled with all of the states above F_n being empty.

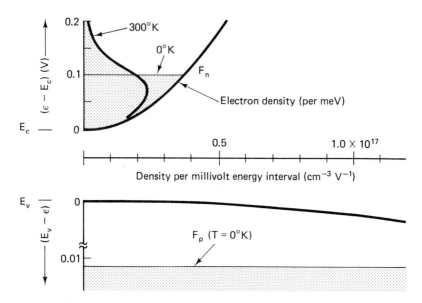

Figure 11.5 Distribution of electron and hole populations at $T = 0°K$ and $T = 300°K$. Note that the energy scales are different for the conduction and valence bands. The sketch is shown for GaAs with $m_e^* = 0.067\, m_0$, $m_{hh} = 0.55\, m_0$, and $kT/e = 0.0259$ V. The presence of light holes has been ignored.

11.4 OPTICAL ABSORPTION AND GAIN IN A SEMICONDUCTOR

We are now in the position to consider situations that lead to gain in a semiconductor, by recognizing that this is just the opposite to those leading to absorption. For instance, consider the "Gedanken" or (thought) experiment shown in Fig. 11.6 where a small-signal variable frequency source irradiates a slab of semiconductor, and one plots the log of $I_{out}(\nu)/I_{in}(\nu)$ as a function of frequency (or wavelength) as data. To make our initial task simple, let us assume $T = 0°K$ and consider an intrinsic material with $F_n = F_p = E_f$ at mid-gap.

If $h\nu_s < E_g$ there are plenty of occupied states in the valence band (all of them!) but no state for the electron to occupy if it is promoted to a higher energy state. Hence there is no absorption. When $h\nu_s > E_g$ true absorption starts. One promotes an electron from a filled state in the valence band to an empty state in the conduction band. Without being complicated and esoteric, we would guess that this absorption coefficient is proportional to the number of states available for absorption provided there is the appropriate number of empty states to receive the electrons. Obviously, as $h\nu$ becomes bigger than E_g, both numbers become larger, varying as the density of states in each band and hence the absorption coefficient also gets larger. The "data" are sketched in Fig. 11.6(c).

Sec. 11.4 Optical Absorption and Gain in a Semiconductor

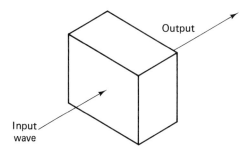

(a) The experiment

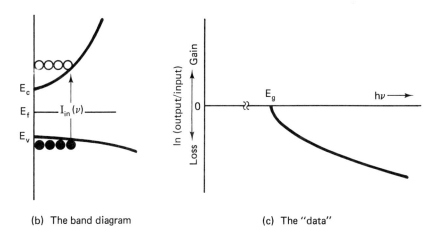

(b) The band diagram

(c) The "data"

Figure 11.6 Optical absorption experiment: (a) indicates a wave propagating through a slab of semiconductor, (b) shows the band diagram of a "normal" semiconductor, and (c) indicates the data.

Now let us force the system away from equilibrium by using a very strong optical pump I_p whose path overlaps the probing wave in the manner shown in Fig. 11.7. We presume that this pump wave is strong enough to maintain a significant number of electrons in the conduction band. Because intraband relaxation to the extremities of each is very fast, the states from E_c to F_n fill and the states from E_v to F_p empty. Because of our assumption of 0°K, all states $E_g < E < F_n$ are filled and all states $E > F_n$ are empty. (Similar comments apply to the valence band.)

The exact position of $F_n - F_p$ relative to the pump photon energy, $h\nu_p$, depends on the intensity I_p, which must maintain the electrons in the conduction band against their "leaking" out downward to the hole states by recombination.

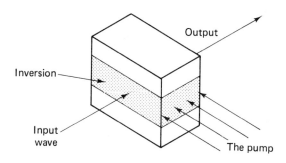

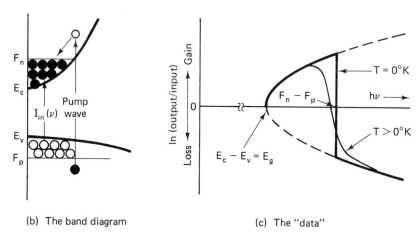

(a) The experiment

(b) The band diagram   (c) The "data"

Figure 11.7 An optically pumped amplifier with the same conditions as were presumed for Fig. 11.6 with the addition of a strong pump creating an inversion by moving the quasi-Fermi levels into the bands.

It is similar to one attempting to fill a bucket with a few holes in it. If the water flow is small, little water accumulates in the bucket. If the water flow from the hose (read the pump intensity) is large enough, the bucket (i.e., conduction band) can fill up to a level such that the "leaks" equal the input. More about the kinetics of these leaks later.

Let us turn our attention to the small-signal, variable-frequency probing wave and talk our way through its interaction with the crystal as $h\nu_s$ is varied. As before, there is no absorption or gain if $h\nu_s < E_g$ because there are no states in the forbidden gap. If $h\nu_s > E_g$ there are states separated by the photon energy but *inverted* from that considered previously. Now we have occupied states *up*, and unoccupied states *down*. The effect on the probing wave is just the opposite to that found for the equilibrium case—*it experiences gain in the same proportion as it had experienced absorption*.

Sec. 11.4 Optical Absorption and Gain in a Semiconductor

Thus the gain coefficient is the mirror image about the horizontal (or frequency) axis of the absorption coefficient as shown in Fig. 11.7. If you recall from Chapter 7 that stimulated emission and absorption are just reverse processes, then this change of absorption to amplification is just a natural result of the inversion with filled states up (i.e., electrons) and empty states (holes) down. This recognition rescues us from the task of performing detailed quantum calculations to obtain the photon-electron-hole interaction rates. The gain does *not* persist for all frequencies. If $h\nu_s > F_n - F_p$ we have a normal situation with filled states down and empty states up and there is absorption. Consequently the necessary condition for amplification in a semiconductor is that a pumping mechanism creates an *"inversion"* expressed by

$$F_n - F_p > h\nu > E_g \tag{11.4.1}$$

This was first pointed out by Bernard and Duraffourg[16] and was used by Dumke[17] in estimating the possibility of amplification in a semiconductor.

A finite temperature $T > 0°K$ rounds off the sharp points and makes the transition from gain to absorption in a smooth fashion as is shown in the development of the next section.

11.4.1 Gain Coefficient in a Semiconductor

Let us consider optical transitions from a small band dE_b centered at E_b that is ΔE_c above E_c to a band dE_a centered at E_a located at ΔE_v below the valence band as is shown in Fig. 11.8.

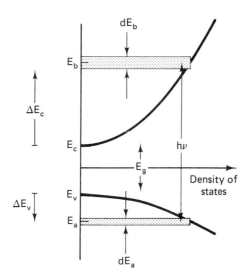

Figure 11.8 Optical transitions in a semiconductor.

The picture presented by Fig. 11.8 is deceptively simple. It appears that one merely adjusts the position of E_b and E_a such that the difference is the photon energy under consideration (either amplification or absorption). Can it be any combination of positions such that $E_b - E_a = h\nu_s$ with E_a, E_b in the respective bands? No, these are optical transitions and k_c, measured in the conduction band, and k_v, measured in the valence band, must be equal before and after the absorption or emission of a photon (for a "pure" material). Thus the first question is: given $h\nu$ where are E_b and E_a with respect to E_c and E_v?

To answer this we return to the relationship between the allowed values of electron momentum, $\hbar k_{c,v}$, and energy in the respective bands:

$$\hbar k_c = [2m_e^*(E_b - E_c)]^{1/2} \qquad (11.4.2a)$$

$$\hbar k_v = [2m_h^*(E_v - E_a)]^{1/2} \qquad (11.4.2b)$$

Since the transitions must conserve momentum, $k_c - k_v = k_{\text{opt.}} \approx 0$, we obtain

$$\Delta E_c = (E_b - E_c) = \frac{m_h^*}{m_e^*}(E_v - E_a) = \frac{m_h^*}{m_e^*}\Delta E_v \qquad (11.4.3)$$

which again points out that most of the "action" takes place in the conduction band because of the heavy mass of the holes.

The photon energy is $h\nu = E_b - E_a$; hence

$$E_g + (E_b - E_c) + (E_v - E_a) = h\nu = E_g + \Delta E_c + \Delta E_v \qquad (11.4.4)$$

After combining (11.4.4) with (11.4.3) we obtain

$$E_b - E_c = \frac{m_h^*}{m_h^* + m_e^*}(h\nu - E_g) \qquad (11.4.5a)$$

and

$$E_v - E_a = \frac{m_e^*}{m_h^* + m_e^*}(h\nu - E_g) \qquad (11.4.5b)$$

If we combine (11.4.5a) and (11.4.5b), we obtain the obvious from Fig. 11.8; that is, $(E_b - E_c) + (E_v - E_a) = h\nu - E_g$.

Now we ask the second question: Are the differential energy spreads shown in Fig. 11.8 equal? The answer is again no. From (11.4.3) we obtain

$$dE_b = \frac{m_h^*}{m_e^*}d(-E_a) \qquad (11.4.6)$$

where the minus sign reflects the fact that hole state energy increases downward in Fig. 11.8.

Not all states in the energy interval dE_b and dE_a can participate in the optical transition. The spin of the state must also be conserved, and this fact divides the total density of states between E_b and $E_b + dE_b$ by a factor of two.

Sec. 11.4 Optical Absorption and Gain in a Semiconductor

The conservation of momentum (i.e., $\hbar k$) puts an additional restriction on the number of states available, for not all states in dE_b can terminate anywhere in dE_a. To examine this issue, we return to (11.2.6) and divide it by 2 to find the density of states in the interval k to $k + dk$.

$$\rho_{c,v}(k)dk = \frac{1}{2\pi^2}k^2 dk \text{ (one spin state)} \qquad (11.4.7)$$

This formula applies to either or both bands, and thus the subscripts (c, v) are not used to identify k. However, most are more comfortable with an energy level diagram, such as Fig. 11.8, rather than a plot of $\rho_c(k)$ vs. k, so we proceed along that line instead. The *reduced density* of states is defined to reflect the number of states at E_b and E_a, within dE_b and dE_a, which *can* participate in the transition at $h\nu$, and which do conserve spin and momentum. We noted that

$$\rho_{\text{red.}}(h\nu)dE = \rho(k)dk = \frac{1}{2\pi^2}k^2 dk \qquad (11.4.8)$$

Hence

$$\rho_{\text{red.}}(h\nu) = \frac{1}{2\pi^2}k^2 \frac{dk}{dE} \text{ (evaluated at } h\nu = E_b - E_a) \qquad (11.4.9)$$

and thus $\rho_{\text{red.}}(h\nu)$ is the density of states (L^{-3}) per energy interval between $h\nu$ and $h\nu + d(h\nu)$ that can participate in the photonic process while conserving momentum. (Whether the participation is by absorption or stimulated emission comes later.)

We add the differential form (11.4.2a) and (11.4.2b) to obtain an expression for dE:

$$dE = dE_b - dE_a = \hbar^2 \left[\frac{1}{m_e^*} + \frac{1}{m_h^*}\right] k\,dk$$

Solving for dk/dE and substituting the result into (11.4.9), one obtains

$$\rho_{\text{red.}}(h\nu) = \frac{1}{2\pi^2 \hbar^2}\frac{m_e^* m_h^*}{m_e^* + m_h^*} k \qquad (11.4.10)$$

Now we invert (11.4.2a) to find k in terms of $E_b - E_c$ and use (11.4.5a) to express that difference in terms of $(h\nu - E_g)$.

$$\hbar k = \left\{\left[\frac{2m_e^* m_h^*}{m_e^* + m_h^*}\right](h\nu - E_g)\right\}^{1/2} \qquad (11.4.11)$$

and finally

$$\rho_{\text{red.}}(h\nu) = \frac{1}{4\pi^2}\left(\frac{2m_r}{\hbar^2}\right)^{3/2}(h\nu - E_g)^{1/2} \qquad (11.4.12a)$$

where

$$m_r = \frac{m_e^* m_h^*}{m_e^* + m_h^*} = \text{the reduced mass} \qquad (11.4.12b)$$

Some prefer to write (11.4.12a) as

$$\rho_{\text{red.}}(h\nu) = \frac{1}{2}\left[\frac{1}{\rho_c(E_b)} + \frac{1}{\rho_v(E_a)}\right]^{-1} \qquad (11.4.12c)$$

which is an obvious problem "saved" for the student.

Having identified where and what states *can* be involved, we can now ask how many transitions (per sec) *do* take place. We proceed in an identical fashion to that used in Chapter 7; the transition *rate* from b → a or from a → b is proportional to

1. An Einstein B coefficient just as in the elementary lasers
2. The number of photons per unit of volume at ν in the interval ν to $\nu + d\nu$, which is just the energy density $\rho(\nu)$ divided by $h\nu$, that is, $P(h\nu) = \rho(\nu)/h\nu$
3. The reduced density of states in that interval given by (11.4.12)
4. The probabilities that the states a and b will be occupied times the probability that b (or a) will be empty and thus "accept" the electron after it has interacted with the photons

The product of items 3 and 4 takes the place of the population inversion and line shape function, $(N_2 - N_1)g(\nu)$, in simple gas lasers. Following the above word statement, we find the number of absorptive transitions $R_{a\to b}$ caused by a wave in the energy interval dE to be

$$R_{a\to b} = [B_{12}] \cdot [P(h\nu)\,d(h\nu)] \cdot [\rho_{\text{red.}}(h\nu)] \cdot \{f_v(E_a)[1 - f_c(E_b)]\} \qquad (11.4.13a)$$
$$|\leftarrow 1 \rightarrow| \, |\leftarrow\!\!-\!\!-2\!\!-\!\!-\rightarrow| \, |\leftarrow\!\!3\!\!\rightarrow| \, |\leftarrow\!\!-\!\!-\!\!-4\!\!-\!\!-\!\!-\rightarrow|$$

and for the stimulated transition rate $R_{b\to a}$:

$$R_{b\to a} = [B_{21}] \cdot [P(h\nu)\,d(h\nu)] \cdot [\rho_{\text{red.}}(h\nu)] \cdot \{f_c(E_b)[1 - f_v(E_a)]\} \qquad (11.4.13b)$$
$$|\leftarrow 1 \rightarrow| \, |\leftarrow\!\!-\!\!-2\!\!-\!\!-\rightarrow| \, |\leftarrow\!\!3\!\!\rightarrow| \, |\leftarrow\!\!-\!\!-\!\!-4\!\!-\!\!-\!\!-\rightarrow|$$

where the numbered factors correspond to the list above.

If we subtract (11.4.13a) from (11.4.13b) to find the *net rate of stimulated transitions*, we find that the products of the Fermi functions cancel, and a relatively simple form for the net stimulated rate results (for $B_{12} = B_{21}$, which will be shown later):

$$R_{b\to a} - R_{a\to b}$$
$$= [B_{21}] \cdot [P(h\nu)\,d(h\nu)] \cdot [\rho_{\text{red.}}(h\nu)] \cdot \{f_c(E_b) - f_v(E_a)\} \qquad (11.4.14)$$

Sec. 11.4 Optical Absorption and Gain in a Semiconductor

Evaluating the B coefficient of (11.4.14) can be a chore involving detailed quantum calculations and thus is best left to more advanced texts. However, the functional form of (11.4.14) and our prior work with the rate equations in Chapters 8 and 9 suggest a direct relationship between the absorption coefficient of an unpumped semiconductor (which can be measured) and the net stimulated emission rate given by the above. The gain coefficient can be expressed in words by

$$\gamma(h\nu) = \frac{1}{I(\nu)} \frac{dI(\nu)}{dz}$$

$$= \frac{\text{power emitted per unit of volume}}{\text{power per unit of area crossing the volume}}$$

The power emitted per unit of volume is just $h\nu$ times the net number of optical transitions as given by (11.4.14). The power per unit of area crossing the volume is $h\nu$ times the number of photons per unit of volume, $P(h\nu)$, times the group velocity,* which is equal to c/n_g. Hence we obtain an expression for the gain coefficient.

$$\gamma(\nu) = B_{21} \frac{n_g}{c} [\rho_{\text{red.}}(h\nu)] \cdot [f_c(E_b) - f_v(E_a)] \quad (11.4.15a)$$

with

$$h\nu = E_b - E_a$$

Equation (11.4.15a) holds for any combination of Fermi functions describing any state of excitation of the semiconductor. In particular it holds for a "normal" or a semiconductor in equilibrium with its environment in which case absorption is the rule rather than gain. If for instance $T = 0°K$, then $f_c(E_b) = 0$ and $f_v(E_a) = 1$ and (11.4.15a) predicts a negative number, that is, absorption. This example teaches us that the factors preceding the braces in (11.4.15a) make up the linear absorption coefficient and thus we can express the gain in terms of it.

$$\gamma(h\nu) = \alpha_0(h\nu)\{f_c(E_b) - f_v(E_a)\} \quad (11.4.15b)$$

where the subscript "0" is used to imply $T = 0°K$. However, even though our logic path used $T = 0°K$, (11.4.15b) holds for all temperatures and for all combinations of the Fermi functions. Its importance lies in the fact that $\alpha(h\nu)$ can, in principle, be measured, and thus we do not have to suffer through the quantum calculations. It also states that the gain coefficient $\gamma(h\nu)$ can be at most between $\pm\alpha(h\nu)$. Finally, recall that the reduced density of states contains the factor $(h\nu - E_g)^{1/2}$ [cf (11.4.12a)], so that in practice one can "hide" most of the factors in (11.4.14) and express the gain coefficient as

$$\gamma(h\nu) = K(h\nu - E_g)^{1/2}[f_c(E_b) - f_v(E_a)] \quad (11.4.15c)$$

where K can be determined from experiment.

* The group index is given by $n_g = n(\lambda) - \lambda(dn/d\lambda)$.

Equation (11.4.15) manages to be both informative and one that saves us the chore of complicated calculations. In principle $\alpha(h\nu)$, as sketched in Fig. 11.7, can be considered as a known quantity or, at least, one that can be measured. If, for instance, one pumps harder, F_n increases, F_p decreases, and the crossover between gain and absorption goes to higher photon energies. Furthermore, it does not take much imagination to recognize that the peak gain coefficient is a strong function of lattice temperature.

While most of our discussion has presumed $T = 0°K$ one can now apply (11.4.15) at a finite temperature by using the appropriate Fermi functions. Note that the last factor $[f(E_b) - f(E_a)]$ must be greater than zero in order to obtain positive gain. It is left for a problem to show that this is implied by the expression

$$h\nu = (E_b - E_a) < (F_n - F_p) \tag{11.4.1}$$

for all temperatures.

11.4.2 Spontaneous Emission Profile

In addition to relating the absorption to the gain coefficient, one can also relate it to the spectral distribution of the spontaneous emission resulting from recombination of the electrons and holes. To obtain this connection, we use an argument identical to that presented by Einstein in his explanation of the blackbody spectrum and covered in Sec. 7.3. Consider a semiconductor contained within a heated cavity, both at a temperature T and both in thermodynamic equilibrium. (Why one would ever do such an experiment is somewhat far-fetched, but in any case the theory should handle the problem.) The semiconductor will absorb some of the blackbody radiation emitted by the atoms in the cavity, and then re-emit its own spontaneous profile because of electron-hole recombination. The key point is that the combination of absorption of the familiar blackbody radiation emitted by the atoms in the cavity wall and the production of new photons by electron-hole recombination must *still* reproduce the original blackbody spectrum. This is an example of the use of the principle of *detailed balancing,* a topic covered in more detail in Appendix I.

Thus we require a detailed balance between the generation of the carriers by the absorption of photons between ν and $\nu + d\nu$ from the ambient blackbody radiation and the radiative recombination rate $R(\nu)d\nu$, which emits photons into this same interval. At thermodynamic equilibrium this balancing must take place in every frequency interval $d\nu$, even though other, possibly more important, processes are taking place simultaneously. The rate of absorption of photons in this frequency interval is the spatial coefficient $\alpha(\nu)$, times the group velocity c/n_g, times the number of photons $P(h\nu)$ at the frequency ν in the interval $d\nu$. At thermodynamic equilibrium, the rate of recombination of electron-hole pairs yielding a photon between ν and $\nu + d\nu$ must be *exactly* equal to the absorption

Sec. 11.4 Optical Absorption and Gain in a Semiconductor

rate of photons in this same interval from the ambient (blackbody) background radiation. Hence

$$R(\nu)\,d\nu = \alpha(\nu)\frac{c}{n_g}\left\{\frac{8\pi n^2 n_g \nu^2}{c^3}\left[\frac{d\nu}{\exp(h\nu/kT) - 1}\right]\right\} \quad (11.4.16)$$

where the last brace is just the blackbody energy density, (7.2.10), divided by $h\nu$ to convert it into photon density.

Because $h\nu/kT \gg 1$ for most of the applications of interest, one has

$$R(\nu) = \alpha(\nu) \cdot \frac{8\pi \nu^2 n^2}{c^2} \cdot \exp\left[-\frac{h\nu}{kT}\right] \quad (11.4.17a)$$

or

$$\alpha(\nu) = \frac{\lambda_0^2}{8\pi n^2} \cdot R(\nu) \cdot \exp\left[+\frac{h\nu}{kT}\right] \quad (11.4.17b)$$

Although a thermodynamic equilibrium environment was used to derive (11.4.17), the relation between R and α holds for any means of excitation of a semiconductor. This result is similar to the relationship between A_{21} and B_{21} that was originally discussed (in Chapter 7) for thermodynamic equilibrium, but, because it is a property of the atom, not the environment, it held for all cases. By the same reasoning, the relationship between the absorption coefficient and the spontaneous emission profile as expressed by (11.4.17) holds for the semiconductor in any case also. Now we have an even more explicit and convenient expression for the gain coefficient in a semiconductor:

$$\gamma(\nu) = \frac{\lambda_0^2}{8\pi n^2} \cdot R(\nu) \cdot \left[\exp\left(\frac{h\nu}{kT}\right)\right]\{f_c(E_b) - f_v(E_a)\} \quad (11.4.18)$$

One should start to recognize familiar factors in the expression for the absorption coefficient (11.4.17) and for the gain coefficient (11.4.18). Note the factor $\lambda_0^2/8\pi n^2$, a familiar friend from elementary laser theory. The next two factors $R(\nu) \exp[h\nu/kT]$ express how fast the carriers recombine and the spectral distribution of the emission. In elementary lasers, the corresponding factors would be $A_{21}(N_2 - N_1)g(\nu)$. For semiconductors, it is customary to write $1/\tau$ for *rate* of recombination of n carriers, and thus the identification with elementary lasers is complete.

Fig. 11.9 illustrates the wavelength dependence of the spontaneous emission from a p-type GaAs sample that was computed from the measured absorption coefficient.

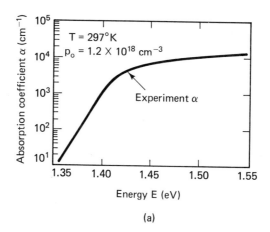

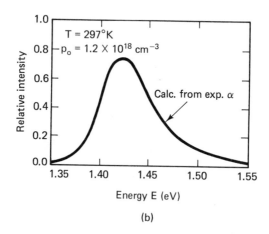

Figure 11.9 (a) Measured absorption coefficient in GaAs doped with an acceptor concentration of 1.2×10^{18} cm^3. (b) Spontaneous emission profile calculated from the absorption. (Data from Casey and Stern.[19])

11.4.3 The Inverted Semiconductor: An Example

It is appropriate to take a break in the theoretical development and compute the density of electrons and holes required to obtain an inverted population at a finite temperature. For this example, we will choose intrinsic GaAs and ignore the light holes. Because the whole question of an inverted population revolves around the position of the Fermi levels, we compute the density of carriers in each band as a function of these levels. This computation will enable us to estimate the injection current required to maintain this population of electron-hole pairs against recombination. The density of electrons in the conduction

Sec. 11.4 Optical Absorption and Gain in a Semiconductor

band is the integral of the density of states times the probability that the state will be filled—the Fermi function.

$$n = \frac{1}{2\pi^2}\left[\frac{2m_e^*}{\hbar^2}\right]^{3/2} \int_{E_c}^{\infty} \frac{(E - E_c)^{1/2} dE}{\exp\left[(E - F_n)/kT\right] + 1} \quad (11.4.19a)$$

The number of holes is a corresponding integral over states in the valence band:

$$p = \frac{1}{2\pi^2}\left[\frac{2m_h^*}{\hbar^2}\right]^{3/2} \int_{-\infty}^{E_v} \frac{(E_v - E)^{1/2} dE}{\exp\left[(F_p - E)/kT\right] + 1} \quad (11.4.19b)$$

If we make the substitution $E = E_c + u$ in (11.4.19a) and $E = E_v - u$ in (11.4.19b) and identify

$$x = u/kT; \quad a = \exp\{[F_p - E_v]/kT\}; \quad b = \exp\{[E_c - F_n]/kT\}$$

Then (11.4.19a) and (11.4.19b) become

$$n = \frac{1}{2\pi^2}\left[\frac{2m_e^* kT}{\hbar^2}\right]^{3/2} \cdot \left[\exp - \left(\frac{E_c - F_n}{kT}\right)\right] \int_0^{\infty} \frac{x^{1/2} dx}{e^x + 1/b} \quad (11.4.19c)$$

$$p = \frac{1}{2\pi^2}\left[\frac{2m_h^* kT}{\hbar^2}\right]^{3/2} \cdot \left[\exp - \left(\frac{F_p - E_v}{kT}\right)\right] \int_0^{\infty} \frac{x^{1/2} dx}{e^x + 1/a} \quad (11.4.19d)$$

Now in conventional semiconductor electronics (transistors, diodes, FETs, etc), the Fermi levels are in the gap and thus factors $(a$ or $b) \gg 1$. Hence the integrals in (11.4.19c) and (11.4.19d) can be evaluated in closed form.

$$I = \int_0^{\infty} \frac{x^{1/2} dx}{e^x + 1/(a, b)} \rightarrow \int_0^{\infty} x^{1/2} e^{-x} dx = \frac{\sqrt{\pi}}{2}$$

$$[\text{for } (a, b) \gg 1] \quad (11.4.20)$$

If the semiconductor were in equilibrium so that $F_n = F_p = E_f$, then the product of (11.4.19c) and (11.4.19d) yields a constant that depends only on the gap, $E_c - E_v = E_g$, the temperature, and the effective masses in the bands, but not on the position of the Fermi level. That constant is, of course, the square of the intrinsic carrier density and the product yields the familiar relation, $n \cdot p = n_i^2$.

A semiconductor laser is not a system in equilibrium, and furthermore, the replacement of the denominators of (11.4.19c) and (11.4.19d) by the exponential is not valid because at least one of the quasi-Fermi levels must be *in* a band for an inversion and thus (a, b) may be less than 1. Thus (11.4.19) is multiplied and divided by $\pi^{1/2}/2$, and the integral evaluated numerically:

$$I_2 = \frac{2}{\sqrt{\pi}} \int_0^{\infty} \frac{x^{1/2} dx}{e^x + 1/(a, b)} \quad (11.4.21)$$

Values for the integral and the electron hole populations in an intrinsic semiconductor are given in Table 11.2. The evaluation of n, p used $m_e^* = 0.067\, m_0$,

TABLE 11.2 CARRIER DENSITIES IN AN INTRINSIC SEMICONDUCTOR AS A FUNCTION OF THE POSITION OF THE QUASI-FERMI LEVELS

(a, b)	$(E_c - F_n)/kT$ $(F_p - E_v)/kT$	I_2	n (cm^{-3})	p (cm^{-3})	Comment
10^3	6.91	1.000	4.36^{14}	1.02^{16}	
10^2	4.61	0.996	4.34^{15}	1.02^{17}	
50	3.91	0.993	8.66^{15}	2.04^{17}	
20	3.00	0.983	2.14^{16}	5.04^{17}	
10	2.30	0.967	4.22^{16}	9.92^{17}	
7.75	2.05	0.957	5.40^{16}	1.27^{18}	Fermi
5	1.61	0.936	8.16^{16}	1.92^{18}	levels
2	0.69	0.860	1.87^{17}	4.41^{18}	in gap ↑
1	0.00	0.765	3.34^{17}	7.84^{18}	—
0.5	−0.69	0.641	5.59^{17}	1.31^{19}	Fermi ↓
0.2	−1.61	0.457	9.96^{17}	2.34^{19}	levels
0.129	−2.05	0.373	1.27^{18}	2.99^{19}	within
0.1	−2.30	0.329	1.43^{18}	3.37^{19}	bands

$m_h^* = 0.55\, m_0$, and $kT/e = 0.0259$ V in (11.4.19). Note that the "sacred" relationship $n \cdot p = n_i^2$ does *not* hold when the Fermi levels are within the bands or when the system is not in equilibrium.

Values for the electron and hole densities for the various positions of the Fermi levels are also given in Table 11.2 in addition to the evaluation of the integral. There are important points to be gained from the study of those numerical values.

1. It only takes about 3.34×10^{17} electron/cm^3 to move the quasi-Fermi level for electrons to the edge of the conduction band, but it takes 23.5 times as many holes to make $F_p = E_v$. Hence it is advantageous to create (or inject) electrons into a region that is heavily doped p type. (However, if a material is so heavily doped, the impurity levels merge with the bands and the gap is no longer a sharp cutoff for allowed states. Furthermore, the k selection rule is also relaxed. We leave those considerations for more advanced texts).

2. Note that for equal electron and hole densities of 1.27×10^{18} cm^{-3}, the quasi-Fermi level for the electrons is inside the conduction band, $(E_c - F_n)/kT = -2.05$, whereas the quasi-Fermi level for the holes is still in the forbidden gap, $(F_p - E_v)/kT = 2.05$. However, the difference $F_n - F_p$ *equals* the band gap.

If one would pump harder and create more electron-hole pairs, one obtains net optical gain.

This critical density can be used to estimate the pumping rate required to reach threshold by some sort of a pumping scheme, say by optical pumping with another laser or flash lamp, E-beam excitation, or carrier injection. The excess

Sec. 11.4 Optical Absorption and Gain in a Semiconductor

electrons (or holes) created by this pumping scheme recombine according to the following kinetics:

$$\frac{dn}{dt} = -\beta \cdot n \cdot p + G \quad (11.4.22)$$

where β is the recombination rate and is equal to 2×10^{-10} cm^3/sec (by experiment) for GaAs and G is the generation (or pumping) rate to be discussed later. To sustain a steady-state electron-hole pair density of $n = p = 1.27 \times 10^{18}$ cm^{-3}, we require a generation (or pumping) rate of

$$G = \beta \cdot n \cdot p = 2 \times 10^{-10} \cdot 1.27 \times 10^{18} \cdot 1.27 \times 10^{18}$$

$$= 3.23 \times 10^{26} \frac{e - h \text{ pairs}}{\text{cm}^3 - \text{sec}} \quad (11.4.23)$$

If these carriers are supplied by current injection and if they recombine in a width $d = 1$ μm, the current density required (in amp/cm^2) is

$$J = ed \cdot \left[\frac{dn}{dt}\right]_{\text{recomb.}} = 1.6 \times 10^{-19} \cdot 1 \times 10^{-4} \cdot 3.23 \times 10^{26}$$

$$= 5.17 \text{ kAmp/cm}^2 \quad (11.4.24)$$

This is a threshold type of calculation based on the numerics of Table 11.2. If the current is bigger than this value, then one can anticipate gain and oscillation with proper feedback. As indicated previously, band-tailing effects from heavy doping introduce modifications, but the numbers are typical.

It should also be obvious from (11.4.24) that there is considerable virtue in reducing the width of the recombination region. For instance, if the width over which recombination occurred were reduced to 100 Å (0.01 μm), then the current would only be 51.7 amp/cm^2. If the plane of the junction were 10 μm wide and 500 μm long (see Fig. 11.11), then the circuit current would only be 2.6 mAmp—small enough for most flashlight batteries. This is one of the major advantages of heterostructures (to be discussed in the next section).

The densities in Table 11.2 are a very strong function of lattice temperature varying as

$$T^{3/2} \exp - [\Delta E/kT]$$

For instance, let $T = 77°K$ (liquid N$_2$ temperatures). Then the densities required to obtain $(F_n - E_c)/kT = -2.05$ and $(F_p - E_v)/kT = +2.05$, that is, $(F_n - F_p)/kT = E_g/kT$), would be reduced to $n = 1.27 \times 10^{18} (77/300)^{3/2} = 1.65 \times 10^{17}$ cm^{-3}; and the threshold current density of (11.4.24) becomes (assuming β is a constant, which it is not) $J = 87.3$ amp/cm^2 (at 77°K). The exponential factor is the controlling one, and thus many researchers report the temperature dependence of the threshold current as

$$J = J_0 \exp\left[\frac{T - T_0}{T_0}\right] \quad (11.4.25)$$

where a good laser device has T_0 of 400°K (i.e., 100°C) or so.

11.5 THE DIODE LASER

Recall that the necessary condition for optical gain as given by (11.4.1) is

$$F_n - F_p > h\nu > E_g \quad (11.4.1)$$

Our task is to realize this condition by the injection of carriers into the region where they recombine and generate photons.

11.5.1 The Homojunction Laser

Figure 11.10 illustrates the geometry employed in the earliest type of semiconductor lasers. It used heavily doped p and n regions on either side of a junction leading to the energy level diagram along the ($y = 0$, z) plane as shown below the sketch of the metallurgical junction. As is the case for every equilibrium p-n junction, the Fermi level is constant throughout the device with no current flowing.

When the junction is biased in the forward direction, as in Fig. 11.10(c), the Fermi level splits because of the injection of minority carriers (electrons into the p region, holes into the n region) and there exists a region near the junction where there is *simultaneously* a high density of electrons and a high density of holes. Because of the much higher mobility of electrons compared to that of holes, most of the injection is by electrons into the p region. These electrons recombine with the majority hole after diffusing a distance, d, which is more or less the diffusion length $d = L_n = (D\tau)^{1/2}$. If we take GaAs as our example, some typical parameters pertinent for heavily doped materials are

$$\mu_n = 600 \text{ cm}^2/\text{V-sec} \qquad D_n = 15.5 \text{ cm}^2/\text{sec}$$
$$\mu_p = 30 \text{ cm}^2/\text{V-sec} \qquad D_p = 0.78 \text{ cm}^2/\text{sec} \quad (11.5.1)$$

where the Einstein relation $D/\mu = kT/e$ has been applied.

The minority carriers, the electrons, recombine with the "sea of holes" made available through doping, according to the following kinetics:

$$\frac{dn_p}{dt} = -(\beta p_p) \cdot n_p = -\frac{n_p}{\tau_r} \quad (11.5.2)$$

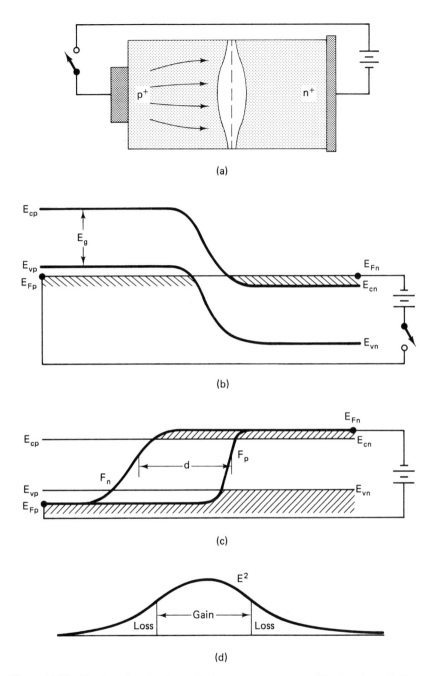

Figure 11.10 The homojunction laser: (a) shows a cross section of the junction with the bowed area being due to current spreading; (b) and (c) show the band diagram in equilibrium and with injected current; (d) illustrates the electromagnetic mode experiencing gain and loss.

The numbers of Sec. 11.4.3 (Table 11.2) indicate that the acceptor doping density of $\approx 9 \times 10^{18}$ cm^{-3} would make the p side degenerate (E_f in the valence band). Thus, for $\beta = 2 \times 10^{-10}$ cm^3/sec, one obtains a recombination lifetime for the electron injected into the p side of

$$\frac{1}{n_p}\frac{dn_p}{dt} = -\frac{1}{\tau_r} = \beta p_p = -1.8 \times 10^9 \text{ sec}^{-1}$$

or

$$\tau_r = 0.56 \text{ ns}$$

The diffusion length and thus the cross-sectional distance d over which gain is possible can be estimated to be

$$d \sim L_n = (D_n \tau_n)^{1/2} = 0.93 \ \mu m$$

One can make a rough estimate of the current density (amp/cm^2) necessary to create this population inversion by a few simple calculations. Let us assume that the quasi-Fermi level for electrons is 0.050 eV above E_c and that we can treat their distribution in energy as if $T = 0°K$ (to make the arithmetic easy). Thus

$$n = \frac{1}{3\pi^2} \cdot \left\{\frac{2m_e^* \Delta E}{h^2}\right\}^{3/2} = 8.79 \times 10^{23} \text{ m}^{-3} = 8.79 \times 10^{17} \text{ cm}^{-3}$$

Now these carriers are going to recombine in a distance of roughly 0.93 μm at a rate of 1.8×10^9/sec (i.e., $1/\tau_n$). Hence the current density required to maintain this nonequilibrium condition is

$$J = \frac{n e d}{\tau_r} = 23.5 \text{ kAmp/cm}^2 \quad (11.5.3)$$

That is a fair amount of current and is typical of homojunction lasers. [It differs from (11.4.24) because of the assumptions, but it is close.] There are a few more issues to be noted about this homojunction laser.

1. The distance d over which carrier inversion is achieved (i.e., $F_n - F_p > h\nu$) is *not* a controlled parameter, and there is very little "design" involved.
2. The distance d is on the order of the wavelength being amplified. Although it is not appropriate now to become involved in the electromagnetics of the laser mode, it is reasonable to recognize that the mode might extend over even a larger distance as shown in Fig. 11.10(d). Shown here is the case where the central part of the electromagnetic mode experiences gain, whereas the edges experience loss. It is a nontrivial task to address this problem and will be discussed in Chapter 13 on advanced electromagnetics.

Sec. 11.5 The Diode Laser

3. Finally, we can guess that the spatial extent of the field in the y direction, which is controlled by the width of the contact (~ 10 μm), is considerably larger than in the x direction, which in turn is controlled by the diffusion length. This fact has a profound effect on the free-space radiation pattern of the laser. As shown in Fig. 11.11, the small width in the direction perpendicular to the junction implies a much larger radiation angle $\theta_\perp$ compared to that measured in the plane of the junction. One does not obtain a nice cylindrically symmetric $TEM_{0,0}$ mode from an injection laser.

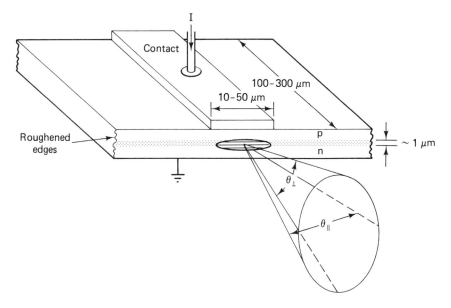

Figure 11.11 The radiation field of a semiconductor laser.

Comments 1 and 2 are significantly modified for heterojunction lasers (covered in the next section), but much of comment 3 applies to all semiconductor lasers. It may be the one instance in which a semiconductor laser is at a disadvantage compared to the larger gas or solid-state laser.

11.5.2 Heterojunction Lasers

There are many problems associated with the simple p-n junction lasers described in the previous section, which can be attributed to the fact that one is using the same material (i.e., GaAs) for both the p and n region. Two of the critical ones are

1. The injected minority carriers are "free" to diffuse where they will—a fact that dilutes the spatial distribution of recombination and thus the gain.

2. There is very little guiding and confinement of the electromagnetic wave being amplified. There is a small bit of wave guiding due to the slight decrease of refractive index on the *n* side (due to the free electrons) and on the *p* side due to the small change in E_g with acceptor doping. However, these changes are very small, and we face the unpleasant fact that the central part of the wave may be amplified with the tails, which extend into the noninverted regions, being attenuated.

Both of these problems can be ameliorated by the use of *heterostructures* to form the active portion of the laser. These are junctions between two dissimilar materials such as GaAs with $Al_xGa_{1-x}As$, with x being the fraction of gallium being replaced by aluminum. It is a fortuitous fact of nature that GaAs and AlAs semiconductors have almost identical lattice constants (see Table 11.1) and thus can be mixed and can be grown on top of each other with little strain involved and a very small density of traps at the interface.* This metallurgical fact is critical to the success of *making* the junction.

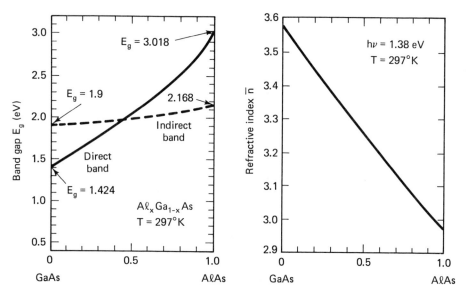

Figure 11.12 Mole fraction AℓAs, x. Dependence of the band gap and index of refraction of $Al_x Ga_{1-x} As$ on the amount of aluminum. (Data from Casey and Panish.[22])

*The GaAs-$Al_xGa_{1-x}As$ system is the best known case involving three different atoms and is the consequence of the nearly equal sizes of Al and Ga. By substituting a bigger and a smaller one from the Col. III and V elements in Table 11.1, other quaternary lattice matches can be achieved such as (GaIn)(AsP) and (GaIn)(AsSb).

Sec. 11.6 Quantum Size Effects

Two other physical factors play a critical role. As the percentage of aluminum is increased (x ↑), the band gap *increases* and the index of refraction goes *down*, and this asynchronous behavior is true for quaternary alloy combinations also. This fact is truly God's gift to the semiconductor laser field, for it greatly alleviates both of the above problems. Fig. 11.12 illustrates the dependence of the band gap and the index of refraction on the mole fraction of Al substituted for gallium.

Fig. 11.13 illustrates a laser using a double heterostructure geometry and is shown biased in the forward direction. Shown also is the variation of the

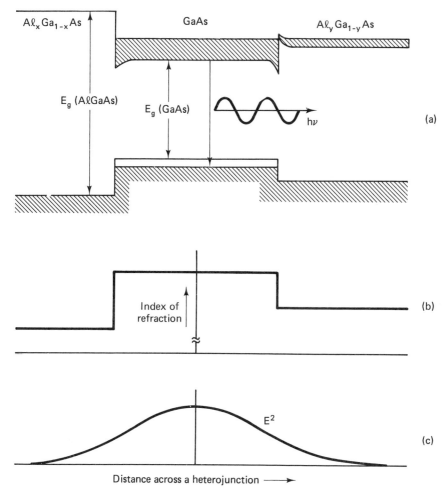

Figure 11.13 The band diagram for a forward-biased heterostructure in (a), the refractive index in (b), and a sketch of the light intensity in the vicinity of the active region in (c).

refraction index and the light intensity in a plane perpendicular to the junction. Note that the electrons injected from the n-type $Al_xGa_{1-x}As$ material are confined to and recombine in lower band-gap p-type GaAs. Furthermore, note that now there is a well-defined dielectric slab waveguide formed similar to that analyzed in Chapter 4. This causes the electromagnetic intensity to be maximized in the same region as the inversion, a condition that also maximizes the stimulated emission rate. In addition, the photon energy of the "tails" of the electromagnetic mode is *less* than the band gap of the confining layers. Hence, most of the built-in absorption problems have also disappeared. All of the above effects have caused a dramatic reduction in threshold currents and have caused a virtual explosion in the application of semiconductor lasers. For low power applications, such as communication and control, there is no other serious competitor.

Before we leave this section, we should point out that the construction of the energy level diagram of Fig. 11.13 involves considerable semiconductor physics. For instance, how much of the band discontinuity $E_g(Al_xGa_{1-x}As) - E_g(GaAs)$ should be apportioned to the conduction band and how much to the valence band? This is determined by the requirement that the vacuum level be *continuous* and *parallel* to the conduction and valence bands. After due allowance for the electron affinities, this implies 65% of ΔE_g being assigned to the conduction band discontinuity and 35% to the valence band.* In order to ensure continuity of the electric displacement flux and to ensure a constant Fermi level in equilibrium, one obtains the "spikes" at the interfaces (similar to the "spikes" found at a Schottky junction). The details of this are best left to texts on semiconductors.

11.6 QUANTUM SIZE EFFECTS

The ability to grow very thin layers ($L \approx 40\text{-}100$ Å) of semiconductor materials has brought about a revolution in many phases of electronics and, in particular, in lasers incorporating these layers. The distances are so small, comparable to the deBroglie wavelength, that the *quantum size effects* (QSE) become easily observable to us living in the macroscopic world. The ability to grow many layers of such different semiconductors on top of each other implies the ability to create artificial materials with a tailoring of characteristics. Indeed it is not clear where the revolution will stop—if ever; it appears to be limited only to the imagination and ingenuity of the researcher.

---

*This has been and will continue to be debated in the literature.

Sec. 11.6 Quantum Size Effects

One of the major considerations in a semiconductor laser theory is the number of carriers occupying the available states in the various bands. This is determined by the pumping rate, the Fermi functions, *and* the density of states. In our initial walk-through of semiconductor lasers, we assumed, implicitly, that all dimensions were huge compared to the deBroglie wavelength and thus were able to treat the allowed momentum states of the electrons (and holes) as continuous variables. This led to the density of states as a function of k being $\rho(k) = k^2 dk/\pi^2$ or, expressed in terms of energy (measured from the band edge), as

$$\rho(E)dE = \frac{1}{2\pi^2} \cdot \left\{ \frac{2m^*}{\hbar^2} \right\}^{3/2} E^{1/2} dE \qquad (11.2.8)$$

When quantum size effects are important, as they are for these ultrathin layers, the density of states changes to reflect the quantization of momentum perpendicular to the thin layer. As before, the allowed projections of the **k** vector are quantized* according to

$$k_x = \frac{\pi}{L_x}; \qquad k_y = \frac{\pi}{L_y}; \qquad \text{and} \qquad k_z = \frac{\pi}{L_z} \qquad (11.2.2)$$

If we assumed one of these dimensions, say L_z, is much smaller than the other two, then the allowed quantum states are distributed in **k** space in the manner shown in Fig. 11.14.

Only the states in Fig. 11.14 indicated by the heavy dots (●) with nonzero quantum (or "mode") numbers are allowed. It is apparent from this figure that the density of modes in the (k_x, k_y) plane is much larger than the density in the k_z direction (where only one mode is shown). As before, each mode occupies a *volume in k space* of

$$V_k = \left(\frac{\pi}{L_x}\right) \cdot \left(\frac{\pi}{L_y}\right) \cdot \left(\frac{\pi}{L_z}\right) \qquad (11.1.4)$$

but now we need to count the number of modes included as k increases by dk, keeping k_z equal to constant π/L_z (for the $q = 1$ mode). Because of the much larger density of points in the k_x, k_y plane, we treat those mode numbers as continuous variables as before and find the number of allowed states in the interval $k_\parallel$ to $k_\parallel + dk_\parallel$ (i.e., motion parallel to the thin surface).

$$dA(k_x, k_y) = \frac{\pi}{2} k_\parallel dk_\parallel \qquad (11.6.1)$$

---

*The presence of confining layers modifies the expression for k_z—see the standard quantum problem of a "particle in a box." However, we will keep our simple quantization to illustrate the effect.

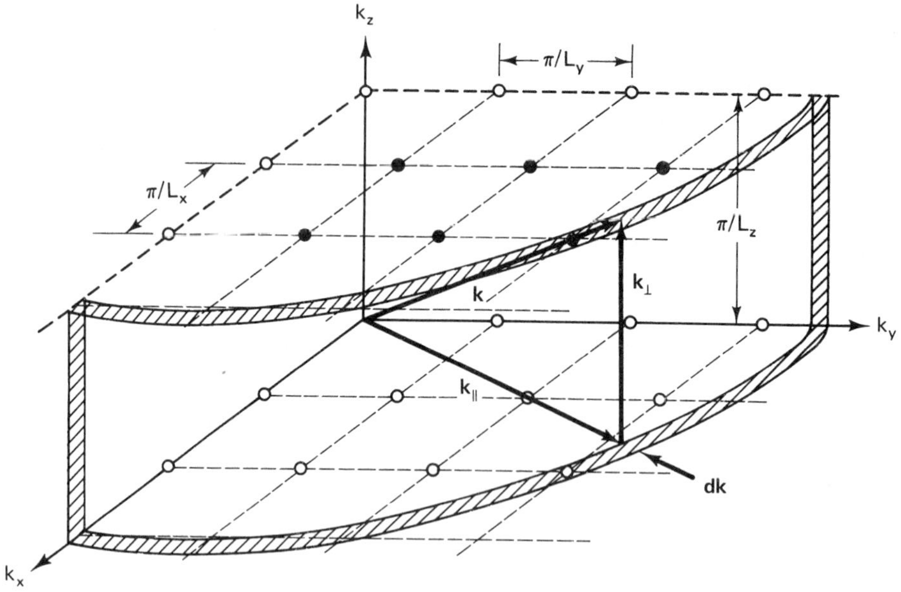

Figure 11.14 Allowed momentum vectors in a "thin" (i.e., $L_z < 200$ Å) semiconductor. The solid dots represent allowed states.

This is equal to the area density of points in the k_x, k_y plane [1 point per $(\pi/L_x)(\pi/L_y)$ in k space] times the area of 1/4 of the cylindrical base. If we also multiply by the density of points per unit of $k_z (=1/\pi L_z)$, and double the answer to allow for two spin orientations, we obtain the total number of modes between $k_\parallel$ and $k_\parallel + dk_\parallel$.

N_k = [density of points in $k_x k_y$] · (in k_z) · [2 spins] · (1/4 of cylindrical area)

or

$$N_k dk_\parallel = \left(\frac{L_x L_y L_z}{\pi^3}\right) \cdot 2 \frac{2\pi k_\parallel dk_\parallel}{4} \left(\frac{\pi}{L_z}\right) \qquad (11.6.2)$$

This is fine as it stands, but we prefer to refer to the total vector **k** not k: however, they are related by

$$k^2 = \left(\frac{\pi}{L_z}\right)^2 + k_\parallel^2 \qquad (11.6.3a)$$

and thus

$$2k\,dk = 2k_\parallel dk_\parallel \qquad (11.6.3b)$$

Sec. 11.6 Quantum Size Effects

The mode density (per unit of k) for $k_z = \pi/L_z$ is

$$\frac{N_k dk}{(L_x L_y L_z)} = \rho_k dk = \frac{1}{\pi^2} k dk \left(\frac{\pi}{L_z}\right) \tag{11.6.4}$$

Now we convert to energy as before by

$$E = \frac{(\hbar k)^2}{2m^*} \quad \text{with} \quad E > E_1 = \frac{[\hbar(\pi/L_z)]^2}{2m^*} \quad \text{i.e., } q = 1 \tag{11.6.5}$$

The density of states in the energy interval dE is

$$\rho(E)dE = \frac{1}{2\pi^2}\left(\frac{2m^*}{\hbar^2}\right)\left(\frac{\pi}{L_z}\right) dE \text{ provided } E > E_1 \tag{11.6.6}$$

Thus the density of states is a *constant* independent of energy provided E is larger than the first allowed state E_1, which in turn must be larger than the normal band edge of the semiconductor. Hence by choosing the dimension L_z, one can "design" the energy state and thus "engineer" the band gap. A simple manipulation of the formulas for the density of states indicates that this *constant* value given by (11.6.6) is contained within the usual value given by (11.2.8). This is shown in Fig. 11.15.

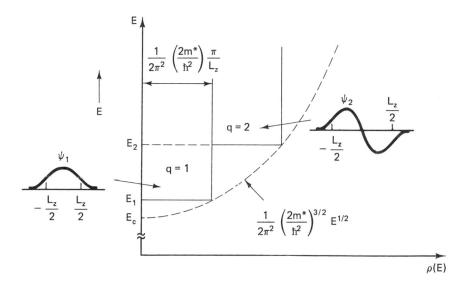

Figure 11.15 Density of states in a quantum well of thickness L_z. The lighter dashed curve is the normal density of states given by (11.2.8). The sketch indicates dependence of the wave function along z for the two subbands shown.

Now, of course, if we pick the energy high enough, one can allow k to have a projection along k_z equal to $2\pi/L_z$. This merely requires repeating the prior development with identical results except that it defines another subband that starts at $E = E_2 = \hbar^2(2\pi/L_z)^2/2m^*$ and proceeds upward at a constant value.

There are a couple of artistic issues added to Fig. 11.15 for clarity and to help understand the origin of the quantized states. For the first allowed band the ($q = 1$) wave function is just that of a simple standing ($\lambda/2$) wave, more or less going to zero at the boundaries. For $q = 2$ and the second mode, the wave function is antisymmetric and corresponds to fitting a full wavelength of the electron wave function between the boundaries. The constant density of states above the respective energies E_1 and E_2 merely reflects the idea that $k_{\parallel}$ and thus the momentum parallel to the interface can be just about anything provided L_x, L_y are sufficiently large. Note also that this stair-step–like density of states is contained within the density of states for a large semiconductor. This has an immediate consequence for a laser.

The gain coefficient of *any* laser, gas, solid state, or semiconductor, is always proportional to the inversion multiplied by how that inversion is smeared out among the various energy states (i.e., the line shape function). By restricting the number of states as a function of energy until $E > E_1$, a larger number of electrons can have the same energy within the band and thus these inverted electrons can be stimulated much more effectively by an electromagnetic wave.

One should realize that the above development is incomplete. One cannot hold a 100-Å wafer with a tweezer! It has to be grown on another substrate with still another layer grown on top of it. The beginning and terminating layers affect the wave functions of the electrons in the confining structure somewhat. Even so, there is not any law against repeating the growth procedure with some rather fantastic and complex structures being achieved. *One can engineer a band gap both in amplitude and in space!* Fig. 11.16 is one example of what has been done and points out that the combination is an artificial material with applications limited only by the imagination!

Although all of the above has focused on the conduction band, similar effects occur in the valence band but with some additional complications. Since it has both light and heavy holes (i.e., different effective masses) the positions of the subbands are also different. Transitions can occur between an electron state in the conduction band to either a light hole (lh) or a heavy hole (hh) state in the valence band with the details best left to more advanced texts.

An example of the use of the quantum size effects discussed above is illustrated in Fig. 11.17. This is intended to show the details of the active region of a diode laser with electrons being injected from the left and holes from the right. The quantum wells are so short that it is possible for the injected carrier to pass through the regions of GaAs because of the lack of collisions necessary to trap the carriers in the wells. Consequently, the aluminum concentration is slowly graded over much larger distances to aid in the trapping of the injected

Sec. 11.6 Quantum Size Effects

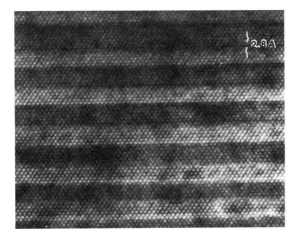

Figure 11.16 A multiple quantum well structure. This is a *t*ransmission *e*lectron *m*icroscope (TEM) display showing alternating layers of GaAs and $Al_x Ga_{1-x} As$ materials. Note that the transition between the materials occurs abruptly on an atomic scale. (Data provided by Prof. J. J. Coleman.)

carriers in the central region where the recombination takes place by transitions between allowed states in the quantum wells formed by sandwiching thin layers of GaAs between even thinner layers of AlAs. These confining layers create a potential barrier to the wave function of the electron (or hole) trapped in the GaAs well and make the calculations given above reasonably accurate. This example also illustrates the use of the fact mentioned earlier, namely, that the band gap and index of refraction vary in opposite directions as the percentage of aluminum in the alloy is changed.

There are a couple of issues to be noted about the technology and the physics of such lasers:

1. The distances are incredibly small. For instance, the 10-Å confining layers of AlAs, which provide the barrier to the electron wave function in the 50-Å GaAs layer is only ~2–unit cell dimensions thick (see Table 11.1). Nevertheless, Fig. 11.16 indicates that the transition from one material to another can be sharp on an atomic scale. There is considerable technology invested in achieving that type of growth capability.

2. The graded layers provide a convenient structure to collect the carriers injected from either side. The size of the quantum wells is so small that the injected carriers will not scatter and collect into the wells if the graded layers are not present.

3. The first allowed band above E_c is not given by (11.6.5) because the confining layers do not present an infinitely high barrier as was assumed in the calculations of this section. However, the important point to remem-

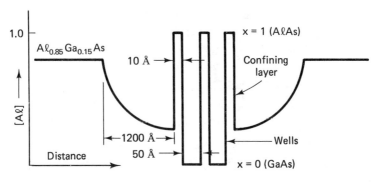

(a) Aluminum concentration at the junction

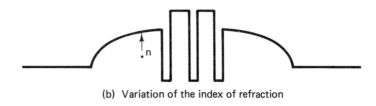

(b) Variation of the index of refraction

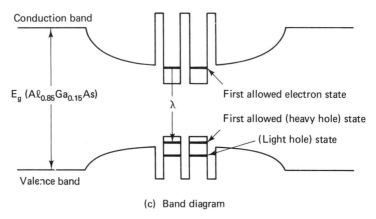

(c) Band diagram

Figure 11.17 An application of quantum size effects for junction lasers: (a) shows the typical variation of the aluminum concentration as a function of depth in the junction. The number of wells can be changed from one to many. (b) and (c) show the effect of the aluminum concentration on the index of refraction and the band gap.

ber is that the energy E_1 is a function of the layer thickness and thus *can be considered as a design variable*. Thus the wavelengths can also be *designed* (within limits).

4. The change in the aluminum concentration creates a highly desirable dielectric waveguide for the fields being amplified.

One should realize that an exact calculation of many of the physical features may not be possible, either because of incomplete knowledge of the growth parameters, the material properties, or a dependence of those properties on the size. However, the above is offered as an indication of some of the possibilities and to convey some of the excitement that such devices generate. Imagination appears to be the most serious limitation.

11.7 MODULATION OF SEMICONDUCTOR LASERS

The fact that the diode lasers are so conveniently pumped by simple current injection (mAmps and at a few volts drop) makes them the clear-cut candidate for communications. A typical light-out current-in (*L-I*) curve is shown in Fig. 11.18 with threshold currents as low as 10 to 50 mAmp and a nearly linear light output above threshold.

It takes little imagination to recognize that if the current is varied by ΔI around an operation level I_0, then the light output will also vary in a more or less synchronous fashion. If the variation is slow enough, then the two will be "in

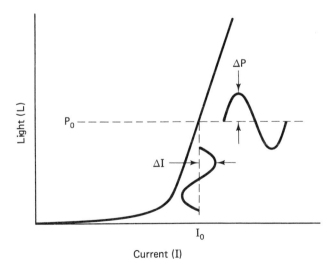

Figure 11.18 A modulated laser.

phase." However, the quality of many communication systems depends on the bandwidth, and thus our problem is to ascertain the frequency response of the diode transmitter. We do this by examining the interplay between stimulated emission and the injected carriers.

Let us now formulate the rate equations for a semiconductor laser. Because of the small physical size of these lasers, we again treat the system as a circuit element whereby we ignore the spatial variations of carrier density and photons with z. We also treat the cross-sectional overlap between the electromagnetic mode and the inverted population as a constant that only needs to be computed once.

$$\Gamma = \frac{\int_{-d/2}^{+d/2} E^2(x)dx}{\int_{-\infty}^{+\infty} E^2(x)dx} \tag{11.7.1}$$

In other words, Γ is that fraction of the photon flux of the cavity mode that overlaps the inverted population (assumed to be in the width d).

It should be clear from the previous sections that the density of the inverted population N (electrons in the band and holes in the valence band) is equal to the injected current divided by the electronic charge and the distance over which the recombination takes place. Thus the production term for the inversion becomes

$$\frac{dN}{dt} \text{(production)} = \frac{J}{ed} \tag{11.7.2a}$$

The carriers are lost spontaneously by both radiative and nonradiative processes. Let us assign a rate for the total recombination in terms of a carrier lifetime $(1/\tau_s)$. Inasmuch as it is a spontaneous decay, the rate of decrease of the inversion is proportional to the inversion multiplying that rate.

$$\frac{dN}{dt} \text{(spontaneous recombination)} = -\frac{N}{\tau_s} \tag{11.7.2b}$$

Stimulated emission depletes the inversion in direct proportion to the intensity of the stimulating wave. However, it is never very important until a laser is near or above threshold—then it is all important. Thus the stimulated recombination rate of carriers should account for the fact that the inversion density must be at least high enough to make the cavity *transparent* (i.e., the gain just about equal to the losses) before stimulated loss of carriers occurs. Furthermore, the net rate of stimulated emission will tend to drive the inversion back to its transparent or threshold condition.

$$\frac{dN}{dt} \text{(stimulated emission)} = -A(N - N_{\text{nom.}})P \tag{11.7.2c}$$

Sec. 11.7 Modulation of Semiconductor Lasers

where $N_{\text{nom.}}$ is the value of the inversion that makes the material transparent and the factor A is an abbreviation for a host of other quantities. For instance, the photons in the cavity are spread out over a cross section that is much larger than the inverted width, d, but only the field in that region does the stimulation. Consequently the optical confinement factor Γ appears as one factor in A. The others are (1) the gain coefficient per unit of inversion; (2) the group velocity of the wave composing the cavity mode; and (3) the volume of the cavity. (Surely the need for an abbreviation should be appreciated!)

Thus the total rate of change of the inversion is the sum of (11.7.2a) → (11.7.2c):

$$\frac{dN}{dt} = \frac{J}{ed} - \frac{N}{\tau_s} - A(N - N_{\text{nom.}})P \quad (11.7.3)$$

This equation is merely a system of bookkeeping of the supply and use of the carriers and how they interact with the photons.

The photons in the cavity mode have a separate equation of motion that is coupled to (11.7.3). The stimulated emission term, (11.7.2c), is the same except for a sign change.

$$\frac{dP}{dt}\text{(stimulated emission)} = A(N - N_{\text{nom.}})P \quad (11.7.4a)$$

Only a fraction of the recombination included in (11.7.2b) results in a photon, and only a fraction of that emission enters the cavity mode* whose photon number is represented by P. We hide our ignorance of the precise amount of spontaneous decay that yields a photon in this mode by assigning a fraction β of (11.7.2b) to the photon equation.

$$\frac{dP}{dt}\text{(spont.)} = \beta\frac{N}{\tau_s} \quad (11.7.4b)$$

Finally the photons are lost from the cavity at a rate dictated by the unavoidable resistive losses remaining inside the cavity as well as by coupling to the external world. As usual, the photon lifetime is used here to describe the change in P.

$$\frac{dP}{dt}\text{(coupling, internal losses)} = -\frac{P}{\tau_p} \quad (11.7.4c)$$

---

*Spontaneous emission goes into any of 4π directions with just about equal probability, but stimulated emission goes into the same direction, at the same frequency, and same polarization as the stimulating wave.

Thus the rate equation for the photons becomes

$$\frac{dP}{dt} = +A(N - N_{\text{nom.}})P + \beta\frac{N}{\tau_s} - \frac{P}{\tau_p} \qquad (11.7.5)$$

These two simple equations contain a lot of information, and it behooves us to extract as much as possible from them.

11.7.1 Static Characteristics

Let us assume a static (DC) excitation so that all time derivatives are zero. We assume $J = J_0$, $N = N_0 > N_{\text{nom.}}$ and thus $P = P_0$ (to be determined) so that (11.7.3) can be solved for the inversion density.

$$A[N_0 - N_{\text{nom.}}]P_0 + \beta\frac{N_0}{\tau_s} = \frac{P_0}{\tau_p} \qquad (11.7.6a)$$

or

$$A[N_0 - N_{\text{nom.}}] = \frac{1}{\tau_p} - \beta\frac{N_0}{P_0 \tau_s} \qquad (11.7.6b)$$

where the subscript 0 refers to the DC or steady-state values of all parameters. The two forms of (11.7.6) state something very simple but very important. Equation (11.7.6a) states that the sum of the stimulated and spontaneous rates must be equal to the photon loss rate (internal plus external), as is true for *any* laser. Equation (11.7.6b), which is a minor rewrite of (11.7.6a), points out that spontaneous emission becomes less and less important as the optical power becomes bigger. It also points out that the net gain rate, $A(N_0 - N_{\text{nom.}})$, is clamped more or less at threshold and is equal to the photon loss rate (if the last term involving spontaneous emission divided by P_0 is neglected).

Now let us substitute (11.7.6a) into (11.7.3) and solve for the photon density.

$$0 = \frac{J_0}{ed} - \frac{N_0}{\tau_s} - \frac{P_0}{\tau_p} + \beta\frac{N_0}{\tau_s} \qquad (11.7.7a)$$

or

$$P_0 = \frac{J_0 \tau_p}{ed} - (1 - \beta)\frac{N_0 \tau_p}{\tau_s} \qquad (11.7.7b)$$

It is convenient to rewrite this equation in a manner that keep tabs on the spontaneous power.

$$P_0 = \frac{\tau_p}{ed}\cdot\left[J_0 - \frac{N_0 ed}{\tau_s}\right] + \beta\frac{N_0 \tau_p}{\tau_s} \qquad (11.7.7c)$$

As mentioned earlier, N_0 is clamped more or less at threshold for lasing, and thus (11.11.7c) suggests that the diode output can be written in the following format:

$$P_0 = K(J - J_{th}) + P_{spont.} \quad (11.7.7d)$$

where the threshold current density J_{th} is $N_{nom.}\, ed/\tau_s$. Such an expression is consistent with the sharp "break" in the light (L) output vs. current (I) curves found in most lasers such as that indicated in Fig. 11.18.

There is another way of explaining this sharp threshold behavior. Below threshold, spontaneous emission goes into any of the electromagnetic modes coupling to the active region—$[(8\pi\nu^2 n^3/c^3)\,\Delta\nu] \cdot$ vol. in number. This is a huge number for even a small semiconductor: pick GaAs of volume 0.1-cm cube, $n = 3.6$, $\lambda_0 = 8400$ Å, and $\Delta\lambda = 50$ Å; the number of modes staggers the imagination—$1.17 \times 10^{+10}$! Most of those modes represent waves propagating in the wrong direction. We, living in the external world, "see" only those photons emitted in our direction and in the bandwidth of our receiving system, which is a very small fraction of the total (say 1 part in 10^3). When the laser is above threshold, the excess pumping goes into one or at most a few, perhaps 10 to 20, modes, which represent waves propagating toward us. Thus it is no wonder that we see a dramatic increase in intensity when stimulated emission takes place. (Incidentally, this same line of reasoning applies to any laser.)

11.7.2 Frequency Response of Diode Lasers

The above analysis is sufficient when the injected current is a slowly varying function of time, but under high-frequency excitation, such as at a GHz rate, we have to be concerned about the ability of stimulated emission to keep up with the rate of carrier injection.

Let us assume that the diode is biased above threshold by DC current $J_0 > J_{th}$ and an AC current $\Delta J(t)$ is added:

$$J = J_0 + \Delta J(t) \quad (11.7.8a)$$

Thus the carrier density N and photon density P should have a DC and a time-varying component.

$$N = N_0 + \Delta N(t) \quad (11.7.8b)$$

$$P = P_0 + \Delta P(t) \quad (11.7.8c)$$

where it is presumed that the time-varying components are small compared to the DC values. Now we substitute these equations into (11.7.3) and (11.7.5) to obtain

$$\frac{dN_0}{dt} + \frac{d\Delta N(t)}{dt} = \qquad (11.7.9a)$$
$$\frac{J_0}{ed} + \frac{\Delta J(t)}{ed} - \frac{N_0}{\tau_s} - \frac{\Delta N(t)}{\tau_s} - A[N_0 + \Delta N(t) - N_{\text{nom}}] \cdot [P_0 + \Delta P(t)]$$

and

$$\frac{dP_0}{dt} + \frac{d\Delta P}{dt} = \qquad (11.7.9b)$$
$$A[N_0 + \Delta N(t) - N_{\text{nom}}] \cdot [P_0 + \Delta P(t)] + \beta \frac{[N_0 + \Delta N(t)]}{\tau_s} - \frac{P_0 + \Delta P(t)}{\tau_p}$$

We neglect products of time varying quantities (i.e., $\Delta N \cdot \Delta P$) as being of second order and equate, separately, the DC variables and time-varying components. The equations involving the DC terms reproduce precisely the material covered in the previous section. The equation for the AC variables becomes

$$\frac{d\Delta N(t)}{dt} = \frac{\Delta J(t)}{ed} - \left(\frac{1}{\tau_s} + AP_0\right) \cdot \Delta N - A[N_0 - N_{\text{nom}}] \cdot \Delta P(t) \quad (11.7.10a)$$

$$\frac{d\Delta P}{dt} = A[N_0 - N_{\text{nom}}]\Delta P + AP_0 \Delta N - \frac{\Delta P}{\tau_p} + \beta \frac{\Delta N}{\tau_s} \quad (11.7.10b)$$

These can be simplified somewhat by the use of (11.7.6b) in which we neglect the spontaneous term as being small when the laser is biased well above threshold so that (11.7.10a) and (11.7.10b) become

$$\frac{d\Delta N(t)}{dt} = \frac{\Delta J(t)}{ed} - \left(\frac{1}{\tau_s} + AP_0\right) \cdot \Delta N - \frac{\Delta P}{\tau_p} \quad (11.7.11a)$$

$$\frac{d\Delta P}{dt} = \left[AP_0 + \frac{\beta}{\tau_s}\right]\Delta N \quad (11.7.11b)$$

where the first and third terms of (11.7.10) cancel on use of (11.7.6b). We can eliminate the coupling by differentiating one and substituting the other.

$$\frac{d^2\Delta N}{dt^2} + \left[\frac{1}{\tau_s} + AP_0\right]\frac{d\Delta N}{dt} + \frac{1}{\tau_p}\left[AP_0 + \frac{\beta}{\tau_s}\right]\Delta N = \frac{1}{ed} \cdot \frac{d\Delta J}{dt} \quad (11.7.12)$$

Sec. 11.7   Modulation of Semiconductor Lasers

The details of the derivation of a similar equation for ΔP are left as a problem; the answer is

$$\frac{d^2 \Delta P}{dt^2} + \left[\frac{1}{\tau_s} + AP_0\right]\frac{d\Delta P}{dt} + \frac{1}{\tau_p}\left[AP_0 + \frac{\beta}{\tau_s}\right]\Delta P = \frac{\Delta J}{ed}\left[AP_0 + \frac{\beta}{\tau_s}\right]$$
(11.7.13)

It is most informative if we form the "transfer" characteristics by letting $\Delta J(t) = \Delta J_m \exp(j\omega t)$ in (11.7.13) and solving for $\Delta P(t) = \Delta P_m \exp(j\omega t)$. The light-current transfer function becomes

$$\frac{\Delta P_m}{\Delta J_m} = \frac{\frac{1}{ed} A \left(P_0 + \frac{\beta}{\tau_s}\right)}{\left[\frac{1}{\tau_p}\left(AP_0 + \frac{\beta}{\tau_s}\right) - \omega^2\right] + j\omega \cdot \left[\frac{1}{\tau_s} + AP_0\right]}$$
(11.7.14)

This modulation response has the functional form of a second-order low-pass filter. The resonance in the transfer characteristic occurs when

$$\omega^2 = \frac{1}{\tau_s} \cdot \left[AP_0 + \frac{\beta}{\tau_s}\right] \approx \frac{AP_0}{\tau_p}$$
(11.7.15)

Equation (11.7.15) implies that the resonant frequency increases as the photon lifetime decreases, which is to be expected, and also increases as CW power increases. At higher frequencies, the response drops off as ω^{-2} or 40 dB per decade of frequency.

Fig. 11.19, taken from a publication by Lau and Yariv, illustrates a qualitative agreement with this theory. The resonance does increase with DC current (i.e., laser power) but seems to fall off faster than 40 dB/decade suggested by the analogy with a second-order circuit filter, but this is most likely caused by parasitic circuit effects. Although the theory suggests that a high average power leads to a high modulation capability, there is a practical upper limit to this approach. At too high a power, the photon flux at the output facet of the diode literally destroys the surface—it self-destructs! Other trends that are in accordance with the theory are clearly demonstrated in that paper. However, the important point is that multi-GHz modulation is possible. Let us end this chapter with the same thought as before: semiconductor lasers are clearly the optical device of choice for low power control, interrogation, and communication. Although the semiconductor laser suffers in comparison to the much larger gas and solid-state laser in terms of power/energy and beam quality, many applications do not need those extremes. Finally, the semiconductor laser naturally mates to the rest of modern electronics, and that convenience makes up for a lot of its minor shortcomings. Some of the detailed electromagnetic issues associated with semiconductor lasers are addressed in Chapter 13.

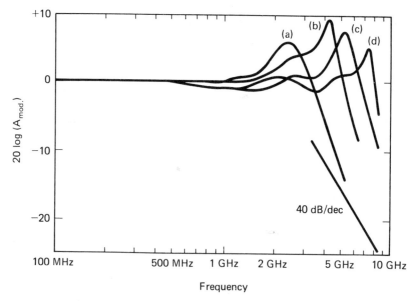

Figure 11.19 Modulation characteristics of a short-cavity (120-μm), buried-heterostructure laser as a function of bias levels: (a) 1 mW, (b) 2 mW, (c) 2.7 mW, and (d) 5 mW. (Data from Lau and Yariv.[4])

PROBLEMS

11.1. Consider an intrinsic GaAs semiconductor with $m_e^* = 0.067\, m_0$, $m_{hh}^* = 0.55\, m_0$, $m_{lh}^* = 0.067\, m_0$ and $Eg = 1.43$ eV. Optical pumping creates $5 \times 10^{18}\, \text{cm}^{-3}$ electrons in the conduction band, leaving an equal number of holes in the valence band. Assume $T = 0°\text{K}$.
  (a) What is the position of the quasi-Fermi level, F_n, relative to the conduction band? (Ans. 0.1596 eV.)
  (b) What is the position of F_p relative to E_v? (Ans. 0.0189 eV.)
  (c) What are the densities of the light and heavy holes? (Ans. $n_{hh} = 4.8 \times 10^{18}\, \text{cm}^{-3}$; $n_{lh} = 2 \times 10^{17}\, \text{cm}^{-3}$.)
  (d) What is the "speed" of the electrons in the highest filled state? (Ans. 9.14×10^7 cm/sec.)

11.2. Electron-hole pairs are created in an intrinsic GaAs semiconductor (at $0°\text{K}$) by the absorption of a focused argon-ion laser at 5145 Å. Assume a volumetric pumping rate of 10^3 W/cm^{-3} and that every absorbed photon creates one electron-hole pair. Assume that the electron-hole pairs are generated by the ion laser, recombine according to (11.4.22) with $\beta = 2 \times 10^{-10}$ cm^3/sec and come into an equilibrium such that $\partial/\partial t = 0$. (Since $T = 0$, all of the carriers are created by the optical pump.)

(a) What is the steady-state density of the electrons (or holes)? (Ans. $3.6 \times 10^{15}/\text{cm}^{-3}$.)

(b) What is the spacing $F_n - E_c$ (in eV)? (Ans. 1.28×10^{-3} eV.)

11.3. Show that (11.4.12c) is the same as (11.4.12a).

11.4. Show that the requirement of $f_n(E_b) > f_v(E_a)$ reduces to $F_n - F_p > h\nu$.

11.5. An intrinsic GaAs semiconductor is irradiated by a wave with $h\nu - E_g = 0.05$ eV. Assume k conservation and 0°K and ignore light holes ($E_g = 1.43$ eV).

(a) Identify the energy levels in the conduction and valence band that can participate in absorption or gain (i.e., find $E_b - E_c$ and $E_v - E_a$). (Ans. $\Delta E_c = 0.044$ eV; $\Delta E_v = 0.0054$ eV.)

(b) What is the minimum number of electron-hole pairs necessary to achieve optical gain at the wavelength? (Ans. $N > 7.4 \times 10^{17}$ cm^{-3}.)

11.6. Show the steps in the derivation of (11.7.11) from (11.7.9a) and (11.7.9b).

11.7. Read the article entitled "Ultra-High Speed Semiconductor Lasers" by K. Y. Lau and A. Yariv, IEEE J. Quant. Electr. QE-21, 121–137, 1985, and answer the following questions.

(a) Which figure illustrates the fact that the resonant frequency varies inversely with photon lifetime? (Ans. Fig. 5.)

(b) Use the data of this figure and some reasonable approximations to estimate the functional dependence of ω on τ_p and compare with theory.

(c) The authors estimate that intrinsic differential gain increases by cooling from $+22°$ to $-50°$ C. What is that ratio? (Ans. ~1.8; see p. 123.)

11.8. Consider the following problem sketched on the diagram below, which was inspired by the article by D. Marcuse and F. Nash, "Computer Model of an Injection Laser with Asymmetrical Gain Distribution," IEEE J. Quant. Electr. QE-18, 30–43, 1982.

(a) If α is the power loss coefficient over the length 0 to l with a gain coefficient of γ from $L + l$, what is the ratio of the power emerging from M_2 compared to that from M_1? (Express your answer in terms of mirror reflectivities $R_{1,2}$ alone.)

(b) Does this law depend on a specific saturation law or assumption of spatial uniformity?

(c) What do the authors mean by a "soft" transition?

(d) At the end of Appendix I, the authors suggest a means for including spontaneous emission in the formulation of the equation for the laser power. Carry out this suggestion for the situation shown on the diagram on p. 408.

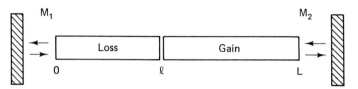

11.9. The following is the mode structure obtained from a diode laser 380 μm long. (From T. Paoli, IEEE J. Quant. Electr. *QE-9*, 267, 1973.)
(a) Find the effective index of refraction defined by

$$n(\lambda) - \lambda \, \partial n / \partial \lambda = n_{\text{eff}}.$$

(b) What is the value of the dispersion term $\partial n / \partial \lambda$ if $n = 3.6$?

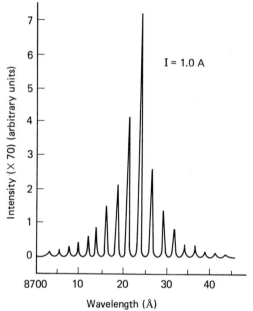

11.10. Assume that the radiation field from the laser sketched in Fig. 11.10 is given by

$$\mathbf{E} = E_0 \exp[-(y/w_y)^2] \exp[-(x/w_y)^2]\mathbf{a}_y$$

where x is perpendicular to the junction, y is parallel, $w_x = 1$ μm, and $w_y = 10$ μm.
(a) If the total power radiated by this laser were 100 mW, what is the value of E_0 (in V/cm)?
(b) Assume that the index of refraction of the semiconductor were 3.6 and that the standard transmission line formulas for the reflection or transmission coefficients were applicable (they are, almost). Find the field in the device at the output facet.
(c) Find the angles $\theta_\perp$ and $\theta_\parallel$ for this laser.

11.11. Read and report on the paper by K. Y. Liou, et al., Appl. Phys. Lett. *50*, 380, 1987.
   (a) The authors used a scanning Fabry-Perot interferometer to measure the mode hopping. If one assumes the laser mode is a δ function in frequency, what is the finesse of the instrument used?
   (b) Explain the difference between Figs. 2a and 2b. Why did the sharp peaks degrade?
   (c) What is the frequency change (in Hz) when the diode current is changed from 55 to 57 mAmp?

11.12. The following questions refer to "Fine Structure of Frequency Chirping and FM Sideband Generation of a Single-Longitudinal-Mode Semiconductor Laser under 10 GHz Direct Intensity Modulation" by C. Lin, et al., Appl. Phys. Lett. *46*(1), 12, 1985.
   (a) What are the major points covered in this paper?
   (b) Why is the FM spectra "intrinsic" to a directly modulated semiconductor laser?
   (c) Show that the carrier should be depressed to zero at an FM index of 2.4 (cf paragraph 2, p. 13). Show, by a sketch, what the spectrum would be for an index of 5.
   (d) If the carrier density varied as $N_0 + \Delta N \cos \omega t$, derive a formula for the variation of the refractive index as a function of time.

11.13 The dimensions in a semiconductor may become so small that diffusion of the carriers away from the point of birth can become a problem. The dynamics of the carrier generation and loss become

$$\frac{dN}{dt} = G - \beta N^2 - \frac{N}{\tau_D}$$

where τ_D = diffusion lifetime = 2.6 ns
β = electron-hole recombination rate = 2×10^{-10} cm^3/sec

   (a) What must be the carrier concentration for the recombination rate to exceed the loss by diffusion? (Ans. $N > 1.9 \times 10^{18}$ cm^{-3}.)
   (b) If a carrier density of 3×10^{18} cm^{-3} is to be maintained by optical pumping by an argon laser at 5145 Å, and each absorbed photon creates 1 electron-hole pair, what must be the absorbed power (per unit of volume) from the pump? (Ans. $p = 1.14 \times 10^9$ W/cm^3.)

11.14. The density of states in the conduction band of a semiconductor with all dimensions large compared to a deBroglie wavelength is given by

$$\rho(\epsilon) = \frac{1}{2\pi^2} \cdot \left[\frac{2m^*}{\hbar^2}\right]^{3/2} (\epsilon - E_c)^{1/2} \qquad \epsilon > E_c$$

If one dimension is small, say $L_z = 100$ Å, the density of states in the first allowed band becomes

$$\rho(\epsilon) = \frac{1}{2\pi^2} \cdot \left[\frac{2m^*}{\hbar^2}\right] \cdot \frac{\pi}{L_z} \qquad \epsilon > E_1 > E_c$$

where E_1 results from the quantization of the z component of the momentum, $k_z \cdot L_z = \pi$. This yields the stair-step density of states shown on the diagram below.

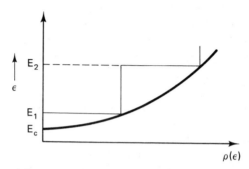

(a) If $L_z = 100$ Å, find the energy spacing $(E_1 - E_c)$ and $(E_2 - E_c)$ assuming $m_e^* = 0.067\, m_0$. (Ans. $E_1 - E_c = 56.3$ meV.)

(b) Suppose electron densities of 5×10^{17} cm^{-3} were maintained in the conduction bands of both a "large" and a "small" slab of semiconductor. Find the energy interval over which the electrons are distributed (assume $T = 0°$K for the two cases). (Ans. $\Delta E = 34.3$ meV and 17.89 meV.)

(c) The allowed energies in the valence band are given by the same relation as used in (a). Evaluate the spacing $E_v - E_1(lh)$ or $E_1(hh)$ for $m_{lh}^* = m_e^*$ and $m_{hh}^* = 0.55\, m_0$. (Ans. hh = 6.86 meV and lh = 56.3 meV.)

(d) Use the results of (a) and (c) to find the long wavelength limit of the recombination radiation between the first allowed state in the conduction band and the first heavy-hole state, assuming $E_c - E_v = 1.43$ eV. (Ans. 1.49 eV.)

(e) Assume a recombination lifetime of 1 nsec, approximate the frequency distribution of the transition by the reduced density of filled states found in (d), and find the gain coefficient of both a "large" and a "small" semiconductor with $n = p = 5 \times 10^{17}$ cm^{-3} and $T = 0°$K. (Ignore light holes.)

11.15. The spontaneous emission from a GaAs semiconductor laser can be approximated by the graph shown below. The length of the wafer is 680 μm, the index of refraction is 3.6, the facet reflectivity is 0.3; the residual absorption coefficient in the crystal is 10 cm^{-1}; and the recombination lifetime is 1 ns.

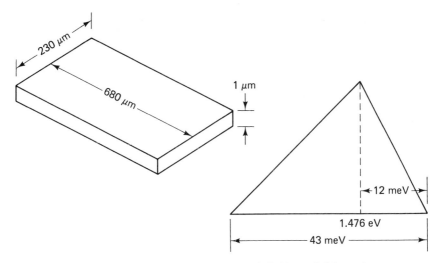

(a) What is the wavelength of peak gain? (Ans. 0.84 μm.)
(b) What is the FWHM of the gain coefficient in Hz and cm^{-1}? (Ans. 5.1×10^{12} Hz and 169 cm^{-1})
(c) What must be the inverted carrier density to bring this wafer to threshold? (Ans. 6.5×10^{15} cm^{-3})
(d) This carrier density must be sustained by some sort of pumping—carrier injection, photo pumping, or E-beam excitation. Estimate the minimum pumping power to maintain an inversion of 10^{16} cm^{-3} throughout the wafer. (Ans. 3.72 mW.)

REFERENCES AND SUGGESTED READINGS

1. It is interesting to note that three separate research groups obtained injection lasers almost simultaneously:
   (a) M. I. Nathan, W. P. Dumke, G. Burns, F. H. Dills, and G. Lasher, "Stimulated Emission of Radiation from GaAs p-n Junctions," Appl. Phys. Lett. *1*, 62, 1962.
   (b) T. M. Quish, R. J. Keyes, W. E. Krag, B. Lax, A. L. McWhorter, R. H. Redike, and H. J. Zeiger, "Semiconductor Maser of GaAs," Appl. Phys. Lett. *1*, 91, 1962.
   (c) N. Holonyak, Jr. and S. F. Bevacqua, "Coherent (Visible) Light Emission from Ga(As$_{1-x}$P$_x$)," Appl. Phys. Lett. *1*, 82, 1962.
2. See the Special Issue on Semiconductor Lasers in IEEE J. Quant. Electron. *QE-21*, No. 6, 1985. This contains 38 separate papers on various topics such as single frequency, high-speed operation, optical amplifiers high power, quantum well lasers, novel structures spectra, modeling and noise, and electron-hole recombination.
3. See also another Special Issue on Semiconductor Lasers, IEEE J. Quant. Electron. *QE-23*, No. 6, 1987. The editorial, introduction and historical papers are especially

recommended for easy reading. The first paper in the AlGaAs and Visible Lasers: "Optimization and Characterization of Index-guided Visible AlGaAs/GaAs Graded Barrier Quantum Well Laser Diodes," L. J. Maust, M. E. Givens, C. A. Zmudzinski, M. A. Emanuel, and J. J. Coleman was the basis for the discussion of Fig. 11-17.

4. K. Y. Lau and A. Yariv, "Ultra-High Speed Semiconductor Lasers," IEEE J. Quant. Electron. *QE-21*, No. 2, 1985. This was the basis of Section 11-17.
5. See IEEE J. Quant. Electron. *QE-22*, No. 9, 1986. Special Issue on Semiconductor Quantum Wells and Superlattices: Physics and Applications.
6. G. H. B. Thompson, *Physics of Semiconductor Laser Devices* (New York: John Wiley & Sons, Inc., 1980).
7. H. K. Kressel and J. K. Butler, *Semiconductor Lasers and Heterojunction LED* (New York: Academic Press, Inc., 1977).
8. S. M. Sze, *Semiconductor Devices, Physics and Technology* (New York: John Wiley & Sons, Inc., 1985). See also, S. M. Sze, *Physics of Semiconductor Devices*, 2nd ed. (New York: John Wiley & Sons, Inc., 1981), Chap. 12.
9. K. A. Jones, *Introduction to Optical Electronics* (New York: Harper & Row, 1987).
10. R. G. Hunsperger, *Integrated Optics, Theory and Techniques*, 2nd ed. (New York: Springer-Verlag, 1984), Chaps. 10–14.
11. J. I. Pankove, *Optical Processes in Semiconductors* (Englewood Cliffs, N.J.: Prentice-Hall, Inc., 1971.)
12. D. Botez and G. J. Herskowitz, "Components for Communication Systems: A Review," Proc. of IEEE *68*, 689–731, June 1980.
13. N. Holonyak, Jr., R. Kolbas, R. D. Dupris, and P. D. Dapkus, "Quantum Well Heterostructure Lasers," IEEE J. of Quant. Electron. *QE-16*, 170–180, 1980.
14. B. G. Streetman, *Solid State Electronic Devices*, 2nd ed. (Englewood Cliffs, N.J.: Prentice-Hall, Inc., 1980), Chap. 10.
15. W. Shockley, *Electrons and Holes in Semiconductors* (Princeton, N.J.: D. Van Nostrand Company, 1950).
16. M. G. Bernard and G. Duraffourg, "Laser Conditions in Semiconductors," Phys. Status Solidi *1*, 699, 1961.
17. W. P. Dumke, "Interband Transitions and Maser Action," Phys. Rev. *127*, 1559, 1962.
18. F. Stern, "Semiconductor Lasers: Theory," Vol. 1, Article B4, p. 425a; and H. Kressel, "Semiconductor lasers: Devices," in *The Laser Handbook*, Vol. 1, Article B5, p. 441, Eds. F. T. Arecchi and E. O. Schulz-DuBois, (New York: North Holland Publishing Company, 1972).
19. H. C. Casey, Jr. and F. Stern, "Concentration Dependent Absorption and Spontaneous Emission in Heavily Doped GaAs," J. Appl. Phys. *47*, 631, 1976.
20. G. Lasher and F. Stern, "Spontaneous and Stimulated Recombination Radiation in Semiconductors," Phys. Rev. *133*, A 553, 1964.
21. Zh. I. Alferov, V. M. Andreev, V. I. Korol'koy, E. L. Portnic, and D. N. Tret'yakov, Injection Properties of Ω-Al$_x$Ga$_{1-x}$AsP GaAs Heterojunctions," Friz. Tekh, Poluprov, *2*, 1016, 1968. See also Sov. Phys. Semicond *2*, 843, 1969.
22. H. C. Casey, Jr., and M. P. Panish, *Heterostructure Lasers* (New York: Academic Press, Inc., 1978).

12

Gas-Discharge Phenomena

12.1 INTRODUCTION

Much of the qualitative understanding of the gas discharge was obtained in the early 1900s (or even earlier); indeed, much of today's "modern physics" has its roots in the explanation of one phenomenon or another associated with a gas discharge.* As such, gas discharge, be it a lamp or a laser, is the "oldest" modern electronic device.

There are many exotic and complicated reasons for the performance of a gas laser, but the prosaic ones are important also—and are sometimes forgotten. A gas is an atomic system in a chaotic state. Consequently, there is very little that human beings can do to a gas that has not happened to it before and that will not automatically cure itself if the excitation is removed. (The obvious exception is the electrical initiation of a chemical reaction, in which a new compound is formed such as in the HF laser discussed in Chapter 10.) Owing to its tolerance of such mistreatment, we can inject large amounts of power into a gas (with only minimal misgivings) and expect large amounts of optical power to be generated.

Indeed, the simple fluorescent lamp is one of the most efficient light sources available; 60% of the electrical power (i.e., $V \cdot I$) is converted to optical

---

*For instance, the whole sequence of events starting with Rydberg's formula, leading to Bohr's theory of the hydrogen atom, and ultimately to Schrödinger's and Heisenberg's quantum mechanics was aimed directly at explaining the spectra emitted by a gas discharge.

power at 253.7 nm. Unfortunately, we have to convert that radiation to the visible by utilizing a phosphor; even so, the overall efficiency is 25% to 30%, better than the best laser. Sodium-vapor lamps and mercury-arc lamps are even more efficient in the production of visible radiation.

One last major advantage of a gas over a solid-state or semiconductor laser lies in its ability to be scaled to large volumes and literally kilograms of active material without straining the budget or manufacturing technology. For instance, a cylinder of CO_2 with $A = 100$ cm^2 (~4 in. square) and $l = 100$ cm long at 50 atm contains about 1 kg of an active laser medium at a cost of about 25 cents. Although dry ice is quite cheap, and thus the total cost is ridiculously low, even the cost of more exotic gases is far below that of their solid-state counterparts.

While this example points out the simplicity of scaling, it also points out the inherent problems associated with gas lasers. They tend to be big and require even larger power supplies operating at voltages and currents that are not compatible at present with modern solid-state technology.

This chapter attempts to give a primer on gas-discharge theory, so as to develop an appreciation of why some gas lasers work so well in spite of the *apparent* nonselective and deceptively simple means of pumping.

We approach a gas discharge from a phenomenological viewpoint and draw on experiment wherever possible. This enables us to go straight to an explanation with a minimal amount of theory. Our goal is to maximize the understanding of the physics for a minimum investment in arithmetic.

First, the terminal characteristics of a simple DC discharge will be described. Unfortunately, we will have to wait until the end of the chapter to give a detailed explanation of the rather strange V–I characteristics. However, it will be made clear that a discharge is *not* a simple positive resistor—if anything it is a negative resistance device that must be ballasted by external means. Failure to do so can result in disastrous consequences (to the power supply, not the gas).

Given these terminal characteristics, we then look more closely at the regions of the discharge. Because of Kirchhoff's current law and some rather obvious physical facts, we find distinct regions of a discharge: the cathode region, consisting of a cathode "dark" space, where a disproportional amount of voltage is dropped; the negative glow, where no voltage is dropped; and the positive column, which occupies the major portion of the inner electrode distance. Even though the positive column is the only *nonessential* part of a gas discharge, 95% of all gas lasers use this region to excite the gas. (The other 5% use the negative glow.)

These simple experimental facts and elementary physics will lead us to some very important conclusions: namely, that a gas discharge is very nearly space charge neutral, and that electrons *and* ions must be created at exactly the same rate as they are lost. This is called the ionization balance condition and is necessary for the *existence* of a discharge.

Sec. 12.2 Terminal Characteristics

Once the existence is established (using proof by intimidation), we then look at how the power enters the electron gas and is transferred to the neutrals, exciting the quantum levels of interest. At this point we have to delve deeper into the microscopic domain to introduce a collision cross section for various processes.

Armed with this understanding, we should be able to do a respectable job of predicting the excitation rate of the quantum levels.

12.2 TERMINAL CHARACTERISTICS

A gas discharge is a very simple device—two electrodes separated by a distance l driven by a power supply with a series ballast resistor in the manner shown in Fig. 12.1(a). For the purposes of illustrating some general characteristics, we

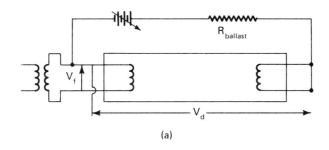

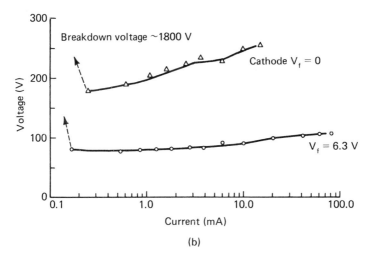

Figure 12.1 Experiment on a fluorescent lamp: (a) circuit; (b) data results.

consider a simple experiment on a very common discharge tube, a fluorescent lamp,* the results of which are plotted in Fig. 12.1(b).

There are many obvious features to be noted:

1. By no stretch of the imagination can one classify this device by a simple resistance. The voltage changes by only 20 V when the current is varied by three decades. Thus it is most logical to consider a discharge as a current-controlled device.
2. The voltage required to initiate the discharge is very high (1800 V) compared to its normal operating voltage of 80 to 100 V.
3. Whether the cathode can or cannot emit electrons by thermionic emission has a marked effect on the terminal voltage. As we will see next, the excess voltage required for the cold cathode case is dropped across a very small region adjacent to the cathode.

Although these features are peculiar to a common fluorescent lamp, the same general characteristics are present even in the most exotic gas-discharge lasers.

12.3 SPATIAL CHARACTERISTICS

If one were to measure the potential as a function of the distance between the cathode and the anode, data similar to those shown in Fig. 12.2(b) would be obtained and we would find a strong correlation between the "dark" and "bright" regions of the discharge. The potential rises very sharply near the cathode, remains more or less constant for a short distance, and rises more or less uniformly throughout the rest of the discharge length. These regions are called the cathode "dark" space,† the negative glow, and the positive column, for traditional reasons.

We can start to understand this figure and the reasons for these regions by adding some "obvious" physics in succeeding graphs. For instance, we *know* that the total current is constant independent of z, as shown in Fig. 12.2(c). We would guess that most of the current is carried by the mobile electrons, although some could be carried by the massive positively charged ions. Whatever the fraction of the current carried by each type of charge, *their sum must be a constant.*

---

*It is not our purpose here to give a complete explanation of this lamp; rather, we wish to emphasize some of the "strange" features of any discharge.

†It is not dark; rather, the visible emission is much weaker there than in the negative glow and positive column.

Sec. 12.3 Spatial Characteristics

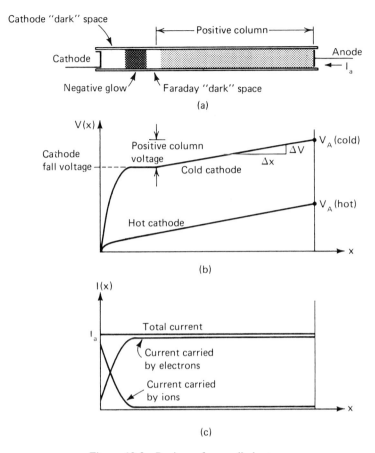

Figure 12.2 Regions of a gas discharge.

This simple fact explains the cathode regions and also dependence of the terminal voltage on the emission characteristics of the cathode. If the cathode does not emit very many electrons, the ions must carry the lion's share of the current in the CDS, as shown in Fig. 12.2(c). But, because of their large mass, this requires a large field, and thus a large cathode fall voltage (typically 100 to 400 V) is required to accelerate the ions to sufficient velocity to carry the current demanded by the circuit. If the cathode emits copious quantities of electrons through thermionic emission, the ions do not have to carry so much of the current, and the cathode fall voltage drops to roughly the ionization potential of the gas.

The negative glow is caused by those electrons that have gained a kinetic energy corresponding to a significant fraction of the cathode fall voltage. The energetic electrons slow down in the negative glow by exciting and ionizing collisions; thus this region of a discharge is an "electron beam"–produced plasma.

In the positive column, the potential varies linearly with distance, and thus the field E is a constant. The fact that the field is a constant can be utilized in Gauss's law to specify the most characteristic feature of plasma physics—*space-charge neutrality*.* Thus, while we have free charged carriers, the number density of electrons is compensated, almost perfectly, by an equal number of positive ions.

The electrons, being much lighter, respond to the presence of the electric field in a much more vigorous fashion than do the more sluggish ions. Consequently, almost all of the electrical power enters through the electrons, to be apportioned by them to the other constituents of the system. It is these other constituents, the excited neutral atoms and molecules (and sometimes excited ions), which produce the coherent radiation of the laser.

Thus it is most important that we pay close attention to the electron gas and learn how it is produced, how it gains energy from the field, and how it then transfers this energy to quantum states of a laser.

12.4 ELECTRON GAS

12.4.1 Background

In a typical gas laser, the electron density ranges from 10^{10} to 10^{13} cm^{-3}, typical of a low-pressure discharge, to 10^{15} to 10^{17} cm^{-3} for a high-pressure heavily pumped laser.[†] Obviously, we cannot afford the computational effort to follow the trajectory of each individual electron as it gains energy from the field and loses it on collisions with the more numerous neutral atoms or molecules. Thus a statistical procedure is in order where we treat the electrons as a minority gas interacting primarily with the heavy neutral atoms or molecules.

12.4.2 "Average" or "Typical" Electron

Let us follow the trajectory of a typical electron in the time interval between collisions (i.e., immediately after emerging from a collision and before it makes another, such as that shown in Fig. 12.3). In that time interval, the electron experiences only the force of the applied electric field according to Newton's laws:

---

*Only near surfaces such as the anode, cathode, or walls do space-charge effects play a role. For our purposes here, we ignore these "sheath" regions.

†In spite of these large numbers, the degree of ionization of a gas laser seldom exceeds 0.1%.

Sec. 12.4 Electron Gas

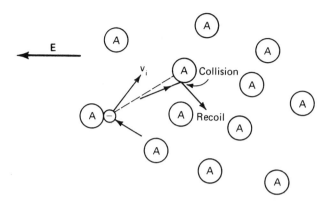

Figure 12.3 Trajectory of a typical electron.

or
$$m\frac{d\mathbf{v}}{dt} = -e\mathbf{E}$$
$$\mathbf{v} = \mathbf{v}_i - \frac{e}{m}\mathbf{E}t \triangleq \mathbf{v}_i + \mathbf{v}_{\text{ord}}$$
(12.4.1a)

where $\mathbf{v}_i$ is the electron's initial thermal velocity, which has, of course, any orientation whatsoever with respect to the "ordered" value specified by the electric field. For n_e electrons, the electrical current is

$$\mathbf{i} = \sum_{j=1}^{n_e} (-e)\mathbf{v}_j \qquad (12.4.1\text{b})$$

where each $\mathbf{v}_j$ has the format of (12.4.1a).

If we consider these equations and Fig. 12.3 for a moment, the following comments are at least palatable, if not obvious:

1. Once the electron collides with an atom, A, the recoil becomes a new "initial" velocity to be reapplied in (12.4.1).
2. The electrons collide with the atoms by virtue of their random "initial" or "thermal" velocity. As we will show shortly, (12.4.13c), the "ordered" velocity is much smaller than the rms value of the random velocity; hence, the collision rate does not depend on the field explicitly.
3. This "initial" random velocity is, for the most part, in any of 4π directions, and the collision probability is more or less independent of this direction.
4. Collisions destroy any memory of its prior trajectory and condition. Thus the ordered momentum gained from the field is converted to disordered thermal motion.

5. For n_e large in (12.4.1b) the vector sum of the initial velocities averages to zero, with a net current being conducted by the "drift" motion of the charges in the time between collisions.

Since the gain in momentum of each electron is interrupted by collisions, we modify the momentum balance equation to account for this scattering:

$$m\frac{d\mathbf{w}_d}{dt} = -e\mathbf{E} - m\mathbf{w}_d \nu_c \quad (12.4.2)$$

where ν_c is the collision frequency for momentum scattering and the $\mathbf{w}_d$ (rather than v) is used for the mean drift motion of the electrons. For steady (DC) fields, one can ignore the inertial term and obtain the drift velocity of the "average" electron.

$$\mathbf{w}_d = -\frac{e\mathbf{E}}{m\nu_c} = -\frac{e\tau}{m}\mathbf{E} \quad (12.4.3)$$

with $\tau = 1/\nu_c$ being the mean free time. The drift velocity per unit electric field is the mobility

$$\mu_e = -\frac{\mathbf{w}_d}{\mathbf{E}} = \frac{e}{m\nu_c} = \frac{e\tau}{m} \quad (12.4.4)$$

Now the collision rate of the electron in a gas depends on the density of scattering atoms, the "size" or cross section of the atom (σ_c), and the relative velocity between the atom and electron. This velocity is almost entirely caused by the random thermal speed of the electron v_{th}.

$$\nu_c = N_A \sigma_c v_{\text{th}} \quad (12.4.5)$$

The net electrical current carried by this drift is

$$\mathbf{J} = n_e e \mathbf{w}_d \quad (12.4.6)$$

By combining (12.4.3) with (12.4.6), we obtain the conductivity in terms of average electron behavior

$$\sigma = \frac{\mathbf{J}}{\mathbf{E}} = \frac{n_e e^2}{m\nu_c} \quad (12.4.7)$$

The fact that the electron gas carries current means that the *electric field* transfers energy *to* the *electron gas* with a rate

$$p_{\text{el}} \text{ (W/vol.)} = \mathbf{E} \cdot \mathbf{J} = \frac{n_e e^2}{m\nu_c}E^2 \quad (12.4.8)$$

Note that the statement of (12.4.2) has apparently broken the log-jam of mathematical equations, but now they are coming fast and furious. Let us stop here

Sec. 12.4 Electron Gas

and recapitulate our ideas before proceeding further with the question of the fate of this power.

Let us back up to (12.4.1a) and square it to obtain the kinetic energy of the electron:

$$\tfrac{1}{2} m \mathbf{v} \cdot \mathbf{v} = \tfrac{1}{2} m (|\mathbf{v}_i|^2 + 2\mathbf{v}_i \cdot \mathbf{v}_{\text{ord}} + |\mathbf{v}_{\text{ord}}|^2) \qquad (12.4.9)$$

Now $\mathbf{v}_i$ is randomly oriented with respect to the ordered velocity; hence, the second term averages out to zero for a large number of electrons. We attempt to describe the behavior of this large number by an average electron by considering the fate of the energies represented by the first and last terms of (12.4.9). This attempt is sketched in Fig. 12.4.

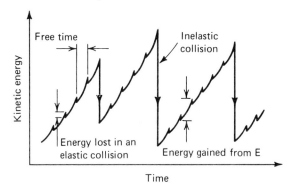

Figure 12.4 Possible energy history of an electron.

Thus the electron gains energy from the field between collisions, loses a very small fraction of its total kinetic energy in an elastic collision (nothing is perfectly elastic), and accumulates this energy until it can make an *inelastic* collision in which the *internal* quantum energy of the atom increases, thereby decreasing the kinetic energy by the corresponding difference. It is this last process that leads to the excited states for a gas laser.

With this picture in mind we can describe the rate at which the mean kinetic energy, W_e, of the electron gas is changing with time. It is increasing because of the power from the electric field and decreasing in small steps by elastic collisions and in large steps by inelastic collisions.

$$\frac{dW_e}{dt} = p_{\text{el}} - \underbrace{\nu_c \delta n_e (\tfrac{3}{2}\epsilon_k - \tfrac{3}{2}\epsilon_A)}_{\text{gas heating}} - \underbrace{\sum_j n_e \nu_{\text{inel}} \Delta W_j}_{\text{excitation}} \qquad (12.4.10)$$

where W_e = number density of electrons times the average energy of the typical electron
$W_e = n_e(\tfrac{3}{2}\epsilon_k)$
ϵ_k = a characteristic energy of the electrons (kT_e)

ϵ_A = a characteristic energy of the atoms (kT_A)
δ = fraction of the excess energy lost per elastic collision = $2m/M$
ν_{inel} = inelastic collision rate
ΔW_j = energy lost in an inelastic collision

Equation (12.4.10) contains much more in the way of definitions than science; indeed, the various terms are introduced to aid in the physical reasoning. The first term is the power entering the electron gas from the field, and the last two specify how and why the power leaves the electron gas. As Fig. 12.4 implies, each electron loses a small fraction of its excess kinetic energy, even in an elastic collision. Since there are n_e electrons with an average collision rate of ν_c, the power lost to the neutrals in the form of gas heating is the product of the density of electrons, the collision rate ν_c, and the difference between the characteristic energies of the electrons and neutrals.*

The gas excitation term follows from a similar line of reasoning. There are n_e electrons, with each electron making on the average ν_{inel} excitation collisions per second, with each event costing an energy ΔW_j. Thus the last term of (12.4.10) represents the following excitation event

$$e(\text{KE}) + A \longrightarrow A^*(\Delta W) + e(\text{KE} - \Delta W) \qquad (12.4.11)$$

and A* may be a rotational, vibrational, or electronic state.

The characteristic energies, ϵ_k, can be related to the temperature of the various gases by

$$\epsilon \Rightarrow kT \qquad (12.4.12)$$

This procedure implies that the various gases have a Maxwellian velocity distribution, an excellent approximation for the atoms but a rather poor one for the electrons, as will be seen later.

Some very important trends associated with all gas discharges can be shown by combining (12.4.8) and (12.4.10) under some very simplifying assumptions. Let us ignore the inelastic term in (12.4.10) and thus condemn the conclusions to be strictly applicable to an inefficient laser or light source, since, by definition, an insignificant amount of power is used to excite the quantum levels. By doing so, we strip away unnecessary mathematics and obtain a very important result. For a steady-state system with $dW_e/dt = 0$,

$$0 = \frac{n_e e^2}{m\nu_c} E^2 - \nu_c \delta n_e \frac{3}{2}(\epsilon_k - \epsilon_A) \qquad (12.4.13a)$$

or

$$\frac{3}{2}(\epsilon_k - \epsilon_A) = \frac{2}{\delta}\frac{1}{2}m\left(\frac{eE}{m\nu_c}\right)^2 \qquad (12.4.13b)$$

$$= \frac{2}{\delta}\left(\frac{1}{2}mw^2\right) \qquad (12.4.13c)$$

---

*If the electrons are isothermal with the neutrals, no energy is interchanged, since just as many electrons are heated by neutrals as neutrals are by electrons.

Sec. 12.4 Electron Gas

(12.4.13c) provides a promised connection between the mean energy of the electrons and the kinetic energy associated with the drift motion of the swarm. Note that δ, a very small number, appears in the denominator. Thus the random kinetic energy of the electron, $\frac{3}{2}\epsilon_k$, is much greater than the drift energy, $\frac{1}{2}mw^2$, imparted to the electrons by the field. One can state these conclusions in reverse order; namely, (12.4.13c) states that the drift motion is a very small fraction of the random kinetic energy.

The form of (12.4.13b) demonstrates another quite general trend associated with gas discharges if the explicit formula for ν_c (12.4.5) is used.

$$\frac{3}{2}(\epsilon_k - \epsilon_A) = \frac{2}{\delta}\frac{1}{2}m\left(\frac{e}{m\sigma_A v_{\text{th}}}\right)^2\left(\frac{E}{N}\right)^2 \qquad (12.4.13d)$$

We could approximate the thermal velocity by $\frac{1}{2}mv_{\text{th}}^2 = \frac{3}{2}\epsilon_k$ and find an explicit expression for ϵ_k as a function of E/N. Frankly, the arithmetic mess is not worth it but the following trend should be clear.

The characteristic energy, ϵ_k, (or temperature) of the electron gas is a monotonically increasing function of the ratio of electric field to neutral gas density:

$$\epsilon_k = f\left(\frac{E}{N}\right) \qquad (12.4.14)$$

This trend is perfectly general in spite of the approximations made.

It is interesting to apply this trend to the data shown in Fig. 12.1. Since the lamp voltage was a constant (more or less), we now know that the electron "temperature" is independent of current. (In fact, if we had measured the positive-column voltage only, we would have found that it *decreased* with current and thus the electron "temperature" does also.) This should not be construed to imply that we do not increase the power to the discharge with increasing current—we do. However, the average energy of electron gas does not change to any extent.

Even though the picture of an "average" electron is simple and appealing, it has some basic conceptual difficulties. The "average" electron does not have the energy to make the inelastic collisions essential for the excitation of the electronic quantum levels of interest in a laser. The CO_2 laser is an exception, and this fact alone accounts for the inherent efficiency (more about this later). It is the "exceptional" electron that performs the function of exciting the mercury in the fluorescent lamp, and the amazing part is that 60% to 70% of the electrical power is utilized by these "exceptional" electrons! The following example can be utilized to emphasize this conceptual difficulty as well as to give a feeling for the numbers involved in the mathematics given above.

Example: The Fluorescent Lamp of Fig. 12.1

This lamp is 40 in. long and filled with about 3 torr of argon with a small drop of mercury. The vapor pressure of mercury at normal operating temperature is about

30 mtorr. (The energy-level diagram for mercury is given in Fig. 15.1.) Since there is so little mercury and so much argon, the elastic collision rate is almost entirely controlled by the argon.

Let us consider the case when the cathode is a good thermionic emitter so that the cathode fall can be estimated to be about 15 V. Then E/N of the positive column can be found from the measured voltage at 10 mA, length, and pressure specifications (see Fig. 12.1): $E = (V - V_0)/L$; $V = 90$ V; $l = 40 \times 2.54$ cm; therefore $E = 0.74$ V/cm $= 74$ V/m. $p = 3$ torr; therefore, $N_A = 3 \times 3.56 \times 10^{16}$ cm^{-3} and $E/N = 6.91 \times 10^{-18}$ V-cm^2. To compute the drift velocity by (12.4.3), we need an estimate of the elastic collision rate ν_c, which requires, in turn, an estimate of the characteristic energy of the electrons. (This is typical of gas-discharge calculations—we need the final answer before we can start!) Let us anticipate that the characteristic energy, ϵ_k, will be close to 1.0 eV and then force the calculations to yield a self-consistent picture. With this initial guess, the thermal velocity is found from $\frac{1}{2} m v_{th}^2 = \frac{3}{2} e(\epsilon_k)$ (ϵ_k in volts); therefore $v_{th} = 7.26 \times 10^8$ cm/sec.

The elastic collision cross section in argon is a violent function of the electron energy, as exemplified by the values shown in Table 12.1. For our *rough* calculations here, we pick the value of the cross section at the characteristic energy and find $\sigma_c = 1.5 \times 10^{-16}$ cm^2 and then use the thermal velocity computed previously to estimate ν_c.

$$\nu_c = N_A \sigma v_{th} \quad [\text{Eq. (12.4.5)} = 1.16 \times 10^9 \text{ collisions/sec};$$

$$(\text{i.e., roughly 1 collision per nsec})]$$

Thus an estimate for the drift velocity is given by (12.4.3):

$$w_d = \frac{eE}{m\nu_c} = 1.12 \times 10^4 \text{ m/sec} = 1.12 \times 10^6 \text{ cm/sec}$$

As promised, the drift is much smaller than the thermal velocity and the drift energy is small indeed.

$$\frac{1}{2} m w_d^2 = 3.57 \times 10^{-4} \text{ eV}$$

If every collision were elastic, then δ in (12.4.10) would be $2m/M$, with M being the mass of the argon atom (40 amu). But we know that the fluorescent lamp does work (it

TABLE 12.1

| | Total Scattering Cross Section in Argon | | |
ϵ(V)	$\sigma_c(\times 10^{-16}$ cm$^2)$	ϵ(V)	$\sigma(\times 10^{-16}$ cm$^2)$
0.107	1.27	0.693	0.745
0.198	0.36	1.01	1.49
0.257	0.155	1.51	2.33
0.295	0.161	2.09	3.18
0.465	0.31	2.48	4.06

From Kieffer.[1]

Sec. 12.4 Electron Gas   425

does produce light), and thus there must be many inelastic collisions. To avoid evaluating the last term of (12.4.10), we pick an effective δ of five times the classical value and use the simplified form (12.4.13c). (NOTE: This is equivalent to assuming that four fifths of the electrical power is transferred from the electrons to the neutrals by processes other than elastic collisions.)

$$\frac{3}{2}(\epsilon_k - \epsilon_A) = \frac{2}{\delta}\left(\frac{1}{2}mw_d^2\right), \quad \delta \sim 5 \times \frac{2m}{M} = 1.36 \times 10^{-4}$$

$$= 5.25 \text{ eV}$$

or

$$\epsilon_k = 3.49 \text{ eV}$$

Obviously this is not consistent with our initial estimate of $\epsilon_k = 1.0$ eV. Hence, we revise our estimate to 1.5 eV and repeat the previous five calculations. Table 12.2 shows how one rapidly converges to the self-consistent *mathematical* solution for this lamp. Table 12.2 was worked out to a far greater precision than the accuracy of the theory warrants. As such, Table 12.2 should be used solely as an example for the physical thought involved and as an example of how we force mathematics to conform to the physics.

TABLE 12.2 CALCULATIONS ON LAMP ASSUMING THAT
           $\delta = 5 \times 2m/M = 1.36 \times 10^{-4}$

ϵ_k (V)	v_{th} ($\times 10^5$ m/sec)	σ_c ($\times 10^{16}$ cm^2)	ν_c ($\times 10^9$ sec^{-1})	w ($\times 10^4$ m/sec)	ϵ_d ($\times 10^{-4}$ eV)	ϵ_k (V)
1.0	7.26	1.5	1.16	1.12	3.57	3.49
1.5	8.9	2.3	2.18	6.03	1.04	1.01
1.1	7.6	1.6	1.3	1.03	2.92	2.85
1.4	8.6	2.1	1.93	0.68	1.33	1.30
1.3	8.3	1.9	1.68	0.78	1.75	1.72
1.35	8.4	1.95	1.76	0.75	1.60	1.57
1.38	8.5	2.0	1.82	0.72	1.49	1.46
1.39	8.56	2.05	1.87	0.70	1.41	1.38

As we shall see subsequently, using (12.4.3) through (12.4.14) is a gross simplification, and thus the exact numerical answers are in error. However, the trend expressed in Table 12.2 is correct; namely, the drift velocity is much smaller than the random thermal velocity; the drift energy ϵ_d is much smaller than the characteristic energy ϵ_k; and collisions, even in this low-pressure gas, are frequent indeed.

Nevertheless, the numerical values given here are typical, in spite of the gross oversimplification in theory. These numbers serve to point out the inherent weakness of the average or "typical" electron approach.

For instance, Table 12.2 indicates that a characteristic energy of ~1.39 eV will yield a self-consistent set of parameters. However, the energy of the first excited quantum state in argon, 11.54 eV, is large compared to this quantity, and thus almost no excitation is seen in the majority gas. The resonant 6^3P_1 state of mercury is 4.9 eV, and in spite of its minority status, considerable excitation is seen—indeed, it is the radiation from this state at 254 nm that is responsible for the excitation of the phosphor.

But even this energy is large compared to the characteristic energy of the electrons. Thus we are forced to conclude that the average electron does not excite the quantum levels of interest.

12.4.3 Electron Distribution Function

The example of the preceding section points out the inherent weakness of the "average" electron approach: phenomena such as electronic excitation and/or ionization require electron energies far in excess of the average of the electron gas. To compute these quantities we must specify the fraction of the electron number density that has sufficient energy to perform the various excitations of interest. That specification is the electron distribution function.

For instance, all electrons can be influenced by the electric field, and thus all electrons participate in the conduction of current. All electrons undergo elastic collisions, and thus all contribute to gas heating. But only a fraction have enough energy to excite a vibrational level, a smaller fraction can excite an electronic level in the neutral gas, and a much smaller fraction can ionize the atoms.

The defining equation for the electron distribution function is

$$\frac{dn_e}{n_e} = f(v_x, v_y, v_z)\, dv_x dv_y dv_z \qquad (12.4.15)$$

Thus f is the density of the electrons in the three-coordinate velocity space. Since the electrons have to be somewhere in this space, we also have a normalization condition:

$$\int\!\!\!\int\!\!\!\int_{-\infty}^{+\infty} f(v_x, v_y, v_z)\, dv_x dv_y dv_z = 1 \qquad (12.4.16)$$

Fig. 12.5(a) shows the geometrical interpretation of the distribution function and also serves to illustrate the conversion from velocity coordinates, (v_x, v_y, v_z) to that of speed, that is, v *without* a subscript, or kinetic energy, ϵ.

$$\frac{1}{2}mv^2 = \frac{1}{2}m(v_x^2 + v_y^2 + v_z^2) = \epsilon \qquad (12.4.17)$$

Thus, if $f(v_x, v_y, v_z)$ is specified, one goes through the mathematical gyration of converting the Cartesian velocity coordinates to spherical ones: the speed and the two angles.

As one would guess, the distribution function is almost (but not quite) spherically symmetric, and thus one need only specify its value in terms of one variable, either speed or kinetic energy. This enables one to plot the density of

Sec. 12.4 Electron Gas

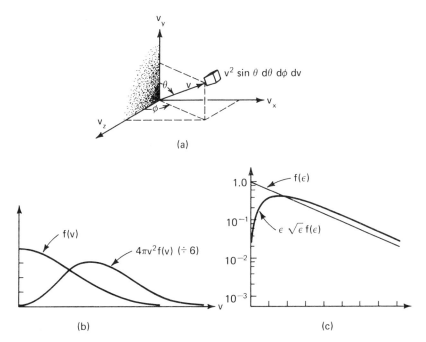

Figure 12.5 Electron distribution function: (a) the geometry and relation between the speed and the Cartesian velocities; (b) a plot of $f(v)$ and $4\pi v^2 f(v)$ *assuming* spherical symmetry; and (c) a plot of $f(\epsilon)$ and $[\epsilon f(\epsilon)]^{1/2}$ *assuming* spherical symmetry.

electrons per unit of speed or energy as shown in Fig. 12.5(b) and (c). As shown in Fig. 12.5(c), it is convenient to define a different quantity, F (of energy ϵ), to describe the fraction of electrons within an energy interval $d\epsilon$ at ϵ.

$$\frac{dn_e}{n_e} \triangleq F(\epsilon)\, d\epsilon \triangleq 4\pi v^2 f(v)\, dv \triangleq f(v_x, v_y, v_z)\, dv_x dv_y dv_z \qquad (12.4.18)$$

where the relations between v, (v_x, v_y, v_z), and ϵ are specified by (12.4.17). This assumption of spherical symmetry will return to haunt us in the next section.

To be specific and to illustrate the various methods of describing the distribution function, let us assume a spherically symmetric Maxwell-Boltzmann distribution of electron velocities:

$$f(v_x, v_y, v_z) = A \exp\left[-\frac{1}{2}\frac{m(v_x^2 + v_y^2 + v_z^2)}{kT_e}\right] \qquad (12.4.19a)$$

where kT_e = a characteristic energy of the electron gas = ϵ_k
T_e = electron "temperature"

The normalization condition, (12.4.16), evaluates the constant A, and the final form of f is

$$f(v_x, v_y, v_z) = \left(\frac{m}{2\pi kT}\right)^{3/2} \exp\left(-\frac{1}{2}\frac{m(v_x^2 + v_y^2 + v_z^2)}{kT_e}\right) \quad (12.4.19b)$$

We now use the string of equalities expressed by (12.4.17) and (12.4.18) to determine

$$F(\epsilon) = 4\pi v^2 f(v)(dv/d\epsilon)$$

with

$$v^2 = v_x^2 + v_y^2 + v_z^2; \quad v = \left(\frac{2\epsilon}{m}\right)^{1/2}; \quad \frac{dv}{d\epsilon} = \frac{1}{2}\left(\frac{2}{m}\right)^{1/2}\epsilon^{1/2}$$

or

$$F(\epsilon) = \left(\frac{1}{kT_e}\right)\frac{2}{\pi^{1/2}}\left(\frac{\epsilon}{kT_e}\right)^{1/2}\exp\left(-\frac{\epsilon}{\epsilon_k}\right); \quad \epsilon_k = kT_e \quad (12.4.20)$$

The factors preceding the exponential term come from such mundane mathematical gyrations as the normalization and the transformation from (v_x, v_y, v_z) coordinates to speed or energy coordinates. Therefore terms such as those will be present in all energy distribution functions. However, the last factor is the controlling term; it is called the *spherically symmetric* part of the distribution function, which will be denoted by the special symbol f_0.

Quite often, only this spherical symmetric part of the distribution function is given: it may be 1 at zero energy as it is here, its amplitude may be specified in terms of a particular author's favorite units, or it may be expressed in arbitrary units. Only its relative shape is important, because one can always retreat to (12.4.19a) and redo the next two equations. Some authors prefer to plot only f_0 in the manner indicated in Fig. 12.5(c) (i.e., $\log f_0$ versus ϵ). The reason should be clear. If one has a simple exponential term for this spherically symmetric part, the graph will be a straight line. If it is not a Maxwellian, it will be obvious at a glance.

Most electron energy distribution functions are not given by a simple Maxwellian, and thus the word "temperature" has no meaning whatsoever. However, the characteristic energy, ϵ_k, which takes on the value of kT_e for a Maxwellian, continues to have validity and, most important, can be measured as a function of E/N.

12.4.4 Computation of Rates

The determination of the spherically symmetric part of the distribution function is beyond the scope of this book; we will assume that someone else has done those calculations and that they are available. What can we do?

Sec. 12.4 Electron Gas

Everything that is desired to compute the excitation and transport processes in a laser (or a lamp) can be found from f_0 provided, of course, that the cross section for a particular process is known. (However, we must assume that these data are known, for without them, f_0 cannot be found.) The purpose of this section is to provide a prescription for this computation.

For any process, r, that depends primarily on the random motion of the electrons, the average of the process is computed by first identifying that rate for a single electron at a specific energy (or velocity), multiplying by the fraction of electrons that have this energy [or $F(\epsilon)\,d\epsilon$], and summing over all energies (or velocities).

$$\langle r \rangle = \int_0^\infty r(\epsilon) F(\epsilon) \, d\epsilon \tag{12.4.21a}$$

or

$$\langle r \rangle = \iiint_{-\infty}^{+\infty} r(v) f(v_x, v_y, v_z) \, dv_x dv_y dv_z \tag{12.4.21b}$$

For instance, let the rate be the elastic collision frequency, ν_c, which played such an important role in Sec. 12.4.2. Then the microscopic rate $r = \nu_c = N_A \sigma_{el}(\epsilon) v$, with $v = (2\epsilon/m)^{1/2}$. If we assume a Maxwellian distribution and assume that σ is a constant, independent of energy, then all integrations can be carried out in terms of elementary functions:

$$\langle \nu_c \rangle = \int_0^\infty \underbrace{N_A \sigma_{el} \left(\frac{2\epsilon}{m}\right)^{1/2}}_{\nu_c} \underbrace{\frac{2}{\sqrt{\pi}} \left(\frac{\epsilon}{kT_e}\right)^{1/2} \exp\left(-\frac{\epsilon}{kT_e}\right) d\left(\frac{\epsilon}{kT_e}\right)}_{F(\epsilon)\,d\epsilon} \tag{12.4.22a}$$

$$= N_A \sigma_{el} \left(\frac{8kT_e}{\pi m}\right)^{1/2} \int_0^\infty x e^{-x} \, dx \tag{12.4.22b}$$

$$= N_A \sigma_{el} \left(\frac{8kT_e}{\pi m}\right)^{1/2} = N_A \sigma_{el} \langle \bar{v} \rangle \tag{12.4.22c}$$

where $\langle \bar{v} \rangle$ is the mean speed of the assumed Maxwellian distribution, $(8kT_e/\pi m)^{1/2}$. Although the answer suffers from all the assumptions made, it is reasonably close to the correct answer.

If we request an inelastic collision rate, ν_{inel}, the format provided by (12.4.12) is correct, but the limits of integration must be changed. For instance, it would be silly to consider the rate of 1-eV electrons exciting a 5-eV electronic level; that rate is 0! Thus, given the energy level ϵ_j above the ground state, we find

$$\langle \nu_{inel} \rangle = N_A \sigma_{inel} \left(\frac{8kT_e}{\pi m}\right)^{1/2} \int_{\epsilon_j}^\infty \left(\frac{\epsilon}{kT_e}\right) \exp\left(-\frac{\epsilon}{kT}\right) d\left(\frac{\epsilon}{kT_e}\right)$$

$$= N_A \sigma_{inel} \langle \bar{v} \rangle \left(1 + \frac{\epsilon_j}{kT_e}\right) \exp\left(-\frac{\epsilon_j}{kT_e}\right) \tag{12.4.23}$$

where it has been assumed that σ_{inel} is a constant for $\epsilon > \epsilon_j$ and is 0 otherwise. That assumption and that of a Maxwellian distribution were made for the sole purpose of illustrating the procedure. In all but contrived problems, these integrals must be evaluated numerically.

Note that the probability of an inelastic collision decreases as the exponential factor, $\exp(-\epsilon_j/kT)$. Even though the inelastic cross section may be comparable to the elastic one, the rates of the two types of collisions are considerably different. For instance, we found that the elastic rate was 1.87×10^9 sec^{-1} in Table 12.2. Using the same constant cross section of 2.05×10^{-16} cm^2 above a threshold of 11.54 eV for excitation of argon atoms, a characteristic energy of 1.39 eV, we obtain $\nu_{\text{inel}} = 3.98 \times 10^6$, far smaller than the elastic value.

If we attempt to use (12.4.21b) to predict a transport quantity, say, the drift velocity, we run into a serious problem of our own making—all answers come out to be identically zero! The reason is that current is a vector quantity—a drift in a specific direction with respect to the field. In mathematical terms, (12.4.21) would tell us to multiply an odd function, the drift, by a spherically symmetric velocity distribution, an even function, and sum over velocities—a procedure that guarantees a zero result!

12.4.5 Computation of a Flux

The assumption responsible for the null result is that the distribution function is spherically symmetric, whereas, in fact, it is not. In the presence of an electric field that drives the current, the distribution function is skewed in the direction of the drift motion of the electrons. One can obtain a very good estimate of this anisotropy by using some rather transparent reasoning and some elementary vector analysis.

For a steady-state situation, the net change in the distribution function (whatever it may be) with respect to time from *all* causes is, by definition, zero. Thus the electric field accelerates the electrons in a definite direction and therefore drives the distribution function away from spherical symmetry, whereas collisions scatter the directed velocity and thus try to return the system to the isotropic state. The time rate of change of this interplay is zero:

$$\mathbf{a} \cdot \nabla_v f(v) + \nu_c (f - f_0) = 0 \qquad (12.4.24\text{a})$$

or

$$f = f_0 - \frac{\mathbf{a} \cdot \nabla_v f}{\nu_c}$$
$$= f_0 + \frac{e\mathbf{E}}{m\nu_c} \cdot \nabla_v f \qquad (12.4.24\text{b})$$

Sec. 12.4 Electron Gas

where f_0 is the spherically symmetric part of the distribution function and **a** is the acceleration from the electric field (**a** $= -e\mathbf{E}/m$ as usual). Note that the format of the second term of (12.4.24a) implies that any deviation from spherical symmetry would disappear as $\exp(-\nu_c t)$ if the electric field were suddenly removed from the plasma. This is precisely what our intuition would suggest.

Now it is a matter of patience and fortitude with vector analysis to obtain an expression for the electric current. For an electric field in the z direction, we must evaluate

$$J_z = -ne \iiint v_z f \, dv_x dv_y dv_z \qquad (12.4.25a)$$

with f provided by (12.4.24b), and since we assume that the anisotropy is small, it is permissible to replace f in (12.4.24b) by f_0 in the gradient term. It is convenient and conventional to use the electron speed v [the radial velocity coordinate of Fig. 12.5(a)] rather than the Cartesian coordinate velocities v_x, v_y, v_z in the development. Thus

$$v_z = v \cos \theta$$

$$\nabla_v f = \frac{\partial f}{\partial v} \mathbf{a}_v + \frac{1}{v} \frac{\partial f}{\partial \theta} \mathbf{a}_\theta + \frac{1}{v \sin \theta} \frac{\partial f}{\partial \varphi} \mathbf{a}_\varphi$$

$$\mathbf{a}_v \cdot \mathbf{a}_z = \cos \theta; \qquad \mathbf{a}_z \cdot \mathbf{a}_\theta = -\sin \theta; \qquad \mathbf{a}_z \cdot \mathbf{a}_\varphi = 0$$

$$dv_x dv_y dv_z = v^2 \sin \theta \, d\theta \, d\varphi \, dv$$

where $\mathbf{E} = E_z \mathbf{a}_z$. After integrating with respect to θ and ϕ, the following formula is obtained for the electrical current in terms of the isotropic part of the distribution function:

$$J_z = -\frac{4\pi}{3} \frac{ne^2}{m} \int_0^\infty \frac{E}{\nu_c(v)} v^3 \frac{\partial f_0}{\partial v} dv \qquad (12.4.25b)$$

It is interesting to note that *if* the isotropic part of the electron distribution function is given by a Maxwellian (12.4.19b) and *if* the collision frequency ν_c is independent of electron speed (a constant mean free time), then the electrical conductivity, J/E, found from (12.4.25b) is identical to the simple formula found earlier, (12.4.7). Seldom are the "if"s" satisfied; hence, the more exact theory must be used for detailed calculation. However, the simple formula (12.4.7) can be used when only rough (~ 2 times) numbers are desired.

Quite often the drift velocity has been *measured* as a function of E/N, and thus the *correct* answer is available without any effort on our part. Of course, the answer will probably be presented in a graphical format rather than by an elementary function, but it is rigorously correct. An example of this approach will be presented in Sec. 12.6.5.

12.5 IONIZATION BALANCE

For any of the foregoing theories to be useful or pertinent, one must have the electrons! Such a mundane statement is very important, because the ionization balance condition controls the E/N of a discharge, and E/N in turn controls the performance of the laser.

To appreciate the implications of such a trivial and transparent statement, consider the continuity equation for electrons:

$$\frac{dn_e}{dt} = \text{production of new electrons (and ions) by those present in a discharge}$$
$$+ \text{ production of new electrons by an externally controlled source (e.g., an electron beam, UV ionization, x-rays)}$$
$$- \text{ losses of electrons by various processes in the plasma} \quad (12.5.1)$$

To make the problem simple and transparent, we first ignore the external source and pick a simple lifetime description for the electron losses; that is, the loss is proportional to n_e divided by some lifetime τ_e (which is presumed to be known). Now the first term in (12.5.1) is nothing more than an inelastic collision of an electron with a neutral atom producing an ion and a second electron. Thus the production of new electrons is the product of n_e and the ionization frequency of the electron gas:

$$\frac{dn_e}{dt} = +\nu_i n_e - \frac{n_e}{\tau_e} \qquad (12.5.2)$$

If one takes this equation literally, it "proves" that a discharge cannot exist. For if n_e is zero, dn_e/dt is also, and thus n_e will never change from zero. However, the equation is too simple; the external source is never exactly zero, and thus there are always a few electrons around to start the process. Obviously, ν_i must be larger than $1/\tau_e$ for the electron density to grow to a respectable value from this initial value.

Having talked ourselves out of one difficulty, in the starting of a discharge, we back into another. Why and how is an equilibrium between production and loss ever established? Although one might first guess that the assumptions for (12.5.2) are again at fault, the answer usually lies with a much more practical and mundane consideration—our power supply is finite!

As more and more electrons are produced, a corresponding increase in current is passed by the discharge, the available voltage across the discharge tube drops, which lowers the E/N, which, in turn, lowers ν_i until an equality is established between the production and loss terms of (12.5.2). This self-limiting process is sketched in Fig. 12.6. At point A, the voltage across the lamp is large,

Sec. 12.5   Ionization Balance   433

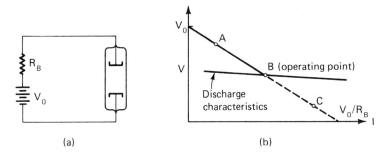

Figure 12.6   (a) Circuit. (b) Approach to equilibrium for a DC discharge.

E/N is large, and $dn_e/dt > 0$. Therefore the current increases. At C, the reverse is true and the current decreases. At B, we have a balance.

Since an ionization event is just a special type of inelastic collision event, we can use (12.4.23b) to proceed further in our analysis of the DC discharge provided that we set ϵ_j equal to the ionization energy of the gas. We set $dn/dt = 0$ in (12.5.2) and find a single equation for the characteristic electron energy.

$$N_A \sigma_i \left(\frac{8kT_e}{\pi m}\right)^{1/2} \left(1 + \frac{\epsilon_j}{kT_e}\right) \exp\left(-\frac{\epsilon_j}{kT_e}\right) = \frac{1}{\tau_e} \qquad (12.5.3)$$

Although not a simple task, (12.5.3) could be solved for this characteristic energy; the many assumptions involved make the effort somewhat futile. But the trends expressed here are most important and hold true even with the most esoteric type of discharge.

For a given gas (which specifies the ionization energy ϵ_i and σ_i) at a specified pressure (which specifies N_A) and loss rate (which may be much more complex than the simple lifetime model used here), there is a unique value of the characteristic energy which satisfies the ionization balance equation. Inasmuch as ϵ_k and E/N are related [see (12.4.14)] then

There is a unique value of E/N for a self-sustained steady-state discharge.

Thus a solution to an equation such as (12.5.1) combined with one similar to (12.4.14) yields the maximum, minimum, and *only* value of E/N permitted for a self-sustained operation. Note that current or electron density does not enter in this balance, and thus E/N is independent of current. This fact explains the flat V–I characteristic of the fluorescent lamp of Fig. 12.1. Some of these conclusions change slightly for an electron-loss process, which has a stronger electron density dependence, but it is only a minor change.

If an external source, such as an electron beam, is used to provide additional ionization, the E/N can be lowered. What happens is that the external source makes up for the decrease of internal ionization due to the lower E/N.

This role of the external source enables one to obtain the best E/N ratio for very efficient excitation of levels of the CO_2 laser.

12.6 EXAMPLE OF GAS-DISCHARGE EXCITATION OF A CO_2 LASER

12.6.1 Preliminary Information

Let us end this chapter with a practical example of laser excitation by a gas discharge, that of a high-pressure CO_2 laser in the transverse electric atmospheric (TEA) configuration shown in Fig. 12.7.

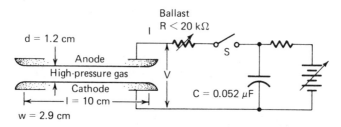

Figure 12.7 Geometry of a small high-pressure CO_2 laser discharge (as described in L. J. Denes and J. J. Lowke, Appl. Phys. Lett. *23*, 130, Aug. 1973.)

We will concentrate on the gas-discharge problem with the goal of determining what fraction of the electrical power drawn from the power supply is used to create the upper (001) state of the CO_2 molecule and how the rest of the power is apportioned to the other processes. Fortunately, this problem has been addressed with detailed experiments and theoretical calculations by the members of the Westinghouse Research Laboratory,* with the results being published in two classic papers (see Refs. 2 and 3). The purpose of this section is to understand and appreciate the significance of this work. We will *not* address the excited-state kinetics and optical coupling problem.

12.6.2 Experimental Detail and Results

The geometry shown in Fig. 12.7 is typical of those used to excite high-power lasers. The experiment consisted of charging the capacitor to a sufficiently high voltage, closing the switch (a triggered spark gap), and measuring the current and voltage.

---

*Many other workers and groups have addressed this problem. However, these two papers[2,3] are unique in the sense that they are readily available, easily read, and complete in every sense of the word, and relate theory to experiment directly.

Sec. 12.6 Example of Gas-Discharge Excitation of a CO_2 Laser

When the switch is closed, the full capacitor voltage appears across A-K. However, after a very short time interval of $\sim$1 to 2 ns, the ionization in the gap increases, which allows the current to increase, which, in turn, lowers the voltage across the gap due to the IR drop across the series resistance and the unavoidable $L(di/dt)$ drop in the wiring inductance. Eventually (in about 20 nsec), an equilibrium is established in the manner described in the preceding section (see Fig. 12.5): the current is limited such that the A-K voltage plus IR drop equals the capacitor voltage. From that time on, one can consider the experiment as a DC discharge.

If this type of procedure is repeated for various values of series resistance and total pressure, the electrical V–I characteristics of the discharge can be determined. A collection of such data is shown in Fig. 12.8, where three different mixtures are used in combination with three different total pressures.

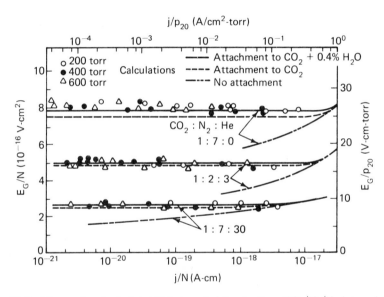

Figure 12.8 Measured and calculated V–I characteristics for three CO_2/N_2/He laser mixtures. Experimental values at three values of pressure are indicated by data points. Curves represent results of numerical calculations with varying conditions of attachment. (Reprinted from Fig. 2 of Denes and Lowke,[2] with permission of L. J. Denes and J. J. Lowke and the American Physical Society.)

Note that the V–I terminal characteristics of this exotic laser mixture are similar to those of the lowly fluorescent lamp described in Fig. 12.1, with the maintaining voltage (Ed) being independent of current (JA) over three to four orders of magnitude, in agreement with the ideas expressed by (12.6.1). Note also that the E/N depends quite critically on the mixture involved, with heavy helium concentrations requiring the lowest E/N. Thus, by (12.4.13d), we should expect the lowest average energy, ϵ_k, of the electron gas. This, as we shall see, results in the highest efficiency of excitation of the upper laser level.

These are the experimental facts of life; the purpose of the following sections is to obtain a correlation between the theory and experiment.

12.6.3 Theoretical Calculations

As has been implied many times throughout this chapter, gas-discharge *calculations* are extremely messy and time consuming, even though the theory of the calculations is straightforward. In all but contrived problems in textbooks, one must resort to the computer to obtain realistic results in a finite length of time. Fortunately, Lowke et al.[3] have published the results of their calculations, which are directly applicable to the experiment described above. The solid curves in Fig. 12.8 are one result of these calculations.

Our purpose here is to show how one arrives at these predictions and, as a bonus, to determine what fraction of the electrical power is used to excite the laser.

We will examine a specific case:

1. Total gas pressure = 400 torr
2. Mixture 1:2:3 $CO_2:N_2:He$
3. Concentration of various constituent gases
    a. $[CO_2] = 2.36 \times 10^{18}$ cm^{-3}
    b. $[N_2] \;\, = 4.72 \times 10^{18}$ cm^{-3}
    c. $[He] \;\, = 7.08 \times 10^{18}$ cm^{-3}
    d. $[N_T] \;\, = \Sigma = 1.42 \times 10^{19}$ cm^{-3}

It is important to realize that a considerable body of knowledge must be accumulated about the electron collision processes before one can even begin. For instance, if a discharge is to exist, some sort of ionization collision must take place. That collision may be with CO_2, N_2, or helium, and to compute that rate, one must multiply the rate, $\sigma_i v_e$, by the distribution function and sum over all velocities in accordance with (12.4.21). That procedure sounds simple enough except for the fact that the distribution function of the electrons is also unknown at this stage.

Thus the first step is to compute a distribution function $F(\epsilon)$, which requires the knowledge of all important inelastic and elastic collision processes. To appreciate why this knowledge is necessary, let us reconsider Fig. 12.4. The electrons move up the energy scale because of the presence of the electric field at a more or less smooth monotonic rate but take large discontinuous steps downward because of the inelastic collisions. Thus, if an inelastic collision is very probable within a given energy interval, there is a rapid depletion of electrons out of that range, and the equilibrium population of free electrons with those energies will be small. For instance, there are two very important inelastic processes that are vitally important to the excitation of the upper CO_2 laser level:

$$e + CO_2(000) \longrightarrow CO_2(001) \quad (12.6.1a)$$

$$e + N_2(v = 0) \longrightarrow N_2(v = n), \quad n = 1 \text{ to } 8 \quad (12.6.1b)$$
$$\hookrightarrow + CO_2(000) \longrightarrow CO_2(001)$$

Lowke et al.[3] account for 31 distinct inelastic processes in the computation of $F(\epsilon)$. The details of that computation are beyond the scope of this book, but the results are very pertinent and can be used with minimal effort on our part.

Typical distribution functions for various values of E/N for the $1:2:3$ mixture are shown in Fig. 12.9.* As one would expect, there are more electrons at higher energy for higher values of E/N. The curve labeled $E/N = 3 \times 10^{-16}$ (V-cm^2) illustrates a very important point made earlier—that the distribution function is *not* a Maxwellian. Indeed, the dashed curve is a Maxwellian, with the same characteristic energy ϵ_k or "electron temperature."

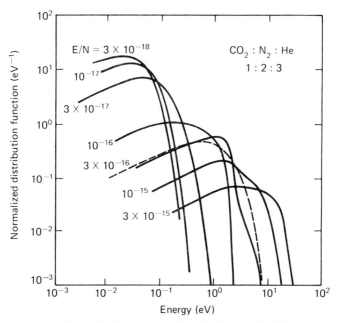

Figure 12.9 Derived distribution functions for $CO_2 : N_2 : He = 1:2:3$ for various values of E/N (V–cm^2). (Reprinted from Fig. 4 of Lowke et al.[3] with permission of J. J. Lowke, A. V. Phelps, and B. W. Irwin and the American Physical Society.)

---

*The distribution function is $F(u) = u^{1/2} f(u)$, with u being the electron energy (in eV) and normalized such that $\int_0^\infty u^{1/2} f(u) \, du = 1$.

There is a simple reason for the fact that the actual distribution has a deficiency of electrons in the 2- to 5-eV range. This is because of the very large resonant cross section for vibrational pumping of N_2 (12.6.1b). Fig. 10.15 illustrates that cross section.

Once the distribution function is determined, one can compute the electrical transport quantities in the straightforward but tedious manner indicated by (12.4.25). Obviously, this has to be done numerically, and typical results are plotted in Fig. 12.10, which gives the drift velocity, w, as a function of the parameter E/N. (Note that if we blindly trusted the simple average electron approach, the curves would be a straight line—the fact that they are not is due to the presence of the large inelastic cross sections mentioned earlier and their influence on the distribution function.)

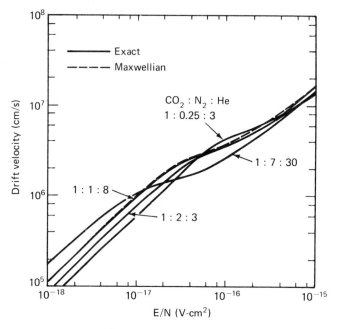

Figure 12.10 Calculated drift velocities of electrons for various gas mixtures of CO_2, N_2, and He. (Reprinted from Fig. 6 of Lowke et al.,[3] with permission of J. J. Lowke, A. V. Phelps, and B. W. Irwin and the American Physical Society.)

In fact, one should approach such a graph with a proper appreciation of prior history. The drift velocity is a measurable quantity (see e.g., Huxley and Crompton[4]), and it is from such measurements that the cross sections used in this calculation were inferred. Thus we should treat Fig. 12.10 as more than a mere calculation—almost as an experimental fact.

One also obtains the transverse diffusion coefficient in such measurements, and thus the ratio of it to the electron mobility $w/E = \mu$ can be considered as an experimental number to be accorded the same respect as the drift velocity.

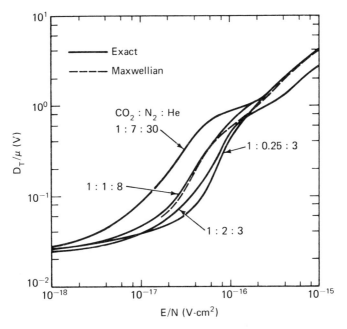

Figure 12.11 Calculated values of D_T/μ for various gas mixtures of CO_2, N_2, and He. (Reprinted from Fig. 7 of Lowke et al.,[3] with permission of J. J. Lowke, A. V. Phelps, and B. W. Irwin and the American Physical Society.)

This ratio is plotted in Fig. 12.11. The importance of this quantity can be appreciated if one recalls the Einstein relation

$$\frac{D_T}{\mu} = \frac{kT_e}{e} \quad \text{(for a Maxwellian)} \tag{12.6.2a}$$

a result valid only for a Maxwellian distribution. The fact that the true distribution is not Maxwellian implies that the words "electron temperature" are meaningless. However, the ratio D_T/μ still has the dimensions of energy (in eV) and is a general prescription for the characteristic energy of the electron gas.

$$\epsilon_k = \frac{D_T}{\mu} \quad \text{(definition of "characteristic energy")} \tag{12.6.2b}$$

12.6.4 Correlation Between Experiment and Theory

This subsection shows how the theoretical calculations of Sec. 12.6.3 can be correlated with the experimental results given in Sec. 12.6.2.

The fact that a discharge exists in quasi-CW fashion means that there must be an exact balance between the electron production and loss.

$$e + M \longrightarrow M^+ + 2e \quad \text{(production)} \tag{12.6.3a}$$

$$e + M' \longrightarrow M'' + 0 \cdot e \quad \text{(loss)} \tag{12.6.3b}$$

Whereas the ionization process indicated by (12.6.3a) is somewhat obvious for a production mechanism, the loss, (12.6.3b), encompasses a multiplicity of possibilities. One could lose the electron by attachment (creating a sluggish negative ion), the electron could recombine with a positive ion, or the electron (and ion) could disappear from the system by diffusion.

If we assume that attachment and recombination are the dominant loss process, the electron continuity equation becomes

$$0 = \frac{dn_e}{dt} = n_e \alpha w_d - n_e a w_d - \gamma n_e N_i + S_{ext} \quad (12.6.4)$$

In (12.6.4), αw_d is the ionization rate* associated with (12.6.3a); aw_d is the attachment rate per electron, implied by (12.6.3b); γ is the electron-ion recombination coefficient; and S_{ext} is an ionization process controlled by devices external to the discharge (more about this later).

Since we are dealing with a plasma, with its requirement of near-space-charge neutrality, one can set $N_i = n_e$ and abbreviate the third term of (12.6.4) as γn_e^2. Let us restrict our attention for the moment to the case of low electron densities so that γn_e^2 can be neglected and also assume that $S_{ext} = 0$. Thus (12.6.4) reduces to a trivial-appearing equation:

$$\alpha = a \quad \text{or} \quad \frac{\alpha}{N} = \frac{a}{N} \quad (12.6.5)$$

The problem hidden in this simple equation is that both α and a are functions of E/N, and thus an iterative procedure must be followed to obtain a solution.

The information necessary to obtain such a solution is contained in Fig. 12.12, which was calculated by those authors from the computed distribution function and the published cross sections for the various processes. The procedure for obtaining the E/N value for a discharge in a $1:2:3$ mixture is illustrated in Table 12.3. A first guess for E/N is 3×10^{-16} V-cm²; at this value $\alpha/N = 1.42 \times 10^{-22}$ cm² and $a/N = 4 \times 10^{-21}$ cm² from Fig. 12.12. Obviously, the loss is greater than the production, and hence we need a larger E/N. Table 12.3 illustrates how fast one can converge to a solution. Having obtained our first theoretical solution, it is time to compare it to a measurement. If one will refer back to Fig. 12.7, where E/N is plotted for this mixture, excellent agreement between theory (solid curves) and experiment (triangles, dots, and circles) is demonstrated.

Now let us examine the implications of the neglect of recombination in (12.6.4). If we identify $J = n_e e w_d$ (as usual) and neglect S_{ext} again, this equation can be cast into the following format:

$$\frac{J}{N}\left(\frac{\alpha}{N} - \frac{a}{N}\right) = \frac{\gamma}{(eW)^2}\left(\frac{J}{N}\right)^2 \quad (12.6.4) \rightarrow (12.6.6)$$

*The term α is called the first Townsend ionization coefficient.

Sec. 12.6   Example of Gas-Discharge Excitation of a CO_2 Laser   441

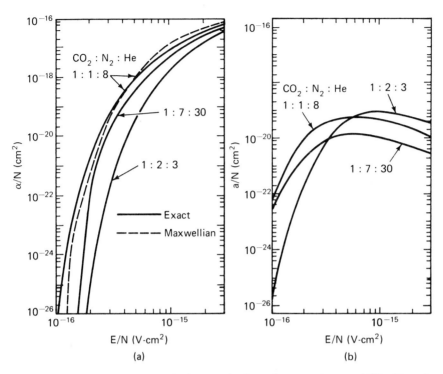

Figure 12.12 Calculated values of α/N and a/N for various gas mixtures of CO_2, N_2, and He; α is the ionization coefficient (cm^{-1}) and a is the attachment rate. The dashed curve indicates values obtained assuming a Maxwellian distribution of the 1:1:8 mixture. (Reprinted from Figs. 9 and 10 of Lowke et al.,[3] with the permission of J. J. Lowke, A. V. Phelps, and B. W. Irwin and the American Physical Society.)

Thus the solution given by Table 12.3 assumes that the right-hand side of (12.6.6) is much smaller than either term on the left. Obviously, if the current is high enough, this is no longer valid, and (12.6.6) requires that $\alpha/N > a/N$. This is the reason that the E/N required for a self-sustaining discharge increases at the high current points of Fig. 12.7. (γ was assumed to be 10^{-7} cm^3/sec for those curves.)

TABLE 12.3 SOLUTION FOR E/N IN A 1:2:3 MIXTURE

E/N (V-cm^2)	α/N (cm^2)	a/N (cm^2)	Comments
3×10^{-16}	1.42×10^{-22}	4×10^{-21}	Loss $\gg$ production
5×10^{-16}	3.6×10^{-20}	2.77×10^{-20}	Production $>$ loss
4×10^{-16}	4×10^{-21}	1.7×10^{-20}	Loss $>$ production
4.8×10^{-16}	2.66×10^{-20}	3.05×10^{-20}	Close!
4.9×10^{-16}	3.1×10^{-20}	3.1×10^{-20}	Solution!

Before we proceed to the laser performance, it is appropriate to translate some of these numbers to a more familiar and practical format. For a pressure of 400 torr, we have $N = 1.42 \times 10^{19}$ cm^{-3}; hence

$$E = \left(\frac{E}{N}\right)N = 4.9 \times 10^{-16} \times 1.42 \times 10^{19} = 6.96 \text{ kV/cm} \quad (12.6.7\text{a})$$

and the terminal voltage across the electrodes is

$$V = Ed = 6.96 \times 10^3 \text{ V/cm} \times 1.2 \text{ cm} = 8.35 \text{ kV} \quad (12.6.7\text{b})$$

For $J/N \ll 10^{-17}$ A-cm, the right-hand side of (12.6.6) can be neglected, and thus V is independent of I.

$$\left.\begin{array}{c} J < 10^{-17} \text{ A-cm} \cdot (N = 1.42 \times 10^{19} \text{ cm}^{-3}) \\ \text{or} \\ J < 142 \text{ A/cm}^2 \\ \text{Therefore} \\ I < 4.03 \text{ kA!} \end{array}\right\} \quad (12.6.7\text{c})$$

Obviously, it is not a simple task to switch such high currents and handle such high voltages.

Let us arbitrarily pick a current density far below this value so as to have confidence in our prior solution—say, $J = 1$ A/cm^2. Then

$$I = 29 \text{ A} \quad (12.6.7\text{d})$$

$$V = 8.35 \text{ kV} \quad (12.6.7\text{e})$$

and the power delivered to this $1.2 \times 10 \times 2.9$ cm^3 volume is

$$P = 242 \text{ kW} \quad (12.6.7\text{f})$$

or

$$\frac{P}{V} = 6.96 \text{ kW/cm}^3 \quad (12.6.7\text{g})$$

No wonder the laser has to be pulsed!

It is also of interest to compute the plasma parameters, such as the degree of ionization and the characteristic energy of the electrons:

$$n_e = \frac{J}{ew_d} \quad (12.6.8\text{a})$$

We use Fig. 12.9 to find $w_d = 8.9 \times 10^6$ cm/sec at $E/N = 4.9 \times 10^{-16}$ V-cm^2. Thus

$$n_e = \frac{1 \text{ A/cm}^2}{1.6 \times 10^{-19} \text{c} \times 8.96 \times 10^6 \text{ cm/sec}} \quad (12.6.8\text{b})$$

$$= 7.02 \times 10^{11} \text{ e-i pairs/cm}^3$$

Sec. 12.6 Example of Gas-Discharge Excitation of a CO₂ Laser

Degree of ionization

$$f = \frac{n_e}{N} = \frac{7.02 \times 10^{11} \text{ cm}^{-3}}{1.42 \times 10^{19} \text{ cm}^{-3}} = 4.9 \times 10^{-8} \quad (12.6.8c)$$

To say that this gas is weakly ionized is an understatement of classic proportions!

The characteristic energy of these 7×10^{11} cm^{-3} electrons in response to the voltage of 8.35×10^3 V can be found from Fig. 12.10. For $E/N = 4.9 \times 10^{-16}$ V-cm,

$$\frac{D_T}{\mu} = \epsilon_k = 1.6 \text{ eV}$$

Thus, in spite of the high voltage applied, the characteristic energy is quite low. As we shall see, we would prefer it to be lower still.

12.6.5 Laser-Level Excitation

The preceding section showed that one can predict the electrical characteristics quite accurately. Now we ask the laser question: How much of the 242 kW (12.6.7f) could one extract as optical power at 10.6 μm? In general, this is a most complex question. But it surely can never be more than that fraction used to excite the upper state: the purpose of this section is to compute that limit.

To compute this quantity, one must return to the distribution function at the E/N value for the discharge and account for the uses of the energy transferred from the electrons to the neutral atoms. Lowke et al.[3] did this accounting, and the results are reproduced in Fig. 12.13.

They kept four separate accounts, which are plotted in Fig. 12.13. Category I represents the fraction of total power used for not-too-interesting and most probably deleterious purposes, such as gas heating through elastic collisions, exciting the lower laser level, and rotational excitation. Category II is the answer to the question of this subsection. This is the fraction used to excite those quantum states that eventually feed the 001 level of CO_2. Thus the energy lost in creating a vibrationally excited N_2 molecule is included here because it can transfer its energy to CO_2 very efficiently.

Category III is the power used to excite the electronic levels of the molecules. That fraction is useless for the CO_2 laser but generally harmless otherwise. Category IV is the fraction used in ionization and thus in keeping the discharge alive! As the curves show, this fraction is negligible except at high E/N values. Even though miniscule and ignorable insofar as a power balance is concerned, it is absolutely essential for the *existence* of the discharge.

For the value of $E/N = 4.9 \times 10^{-16}$ V-cm², Fig. 12.13 predicts the following ratios:

(I) Elastic, gas heating, etc. 14.3%

(II) Upper laser level 54.3%

(III) Electronic excitation 31.4% (12.6.9)

(IV) Ionization Negligible but positive

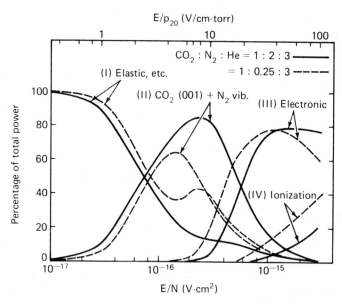

Figure 12.13 Percentage of power lost to (I) elastic collisions, rotational excitation of N_2, and excitation of bend and stretch modes of CO_2; (II) CO_2 (001) level and the first eight vibrational levels of N_2; (III) electronic excitation; and (IV) ionization. Increasing the ratio of N_2 to CO_2 increases the predicted efficiency given by (II). (Reprinted from Fig. 12 of Lowke et al.,[3] with the permission of J. J. Lowke, A. V. Phelps, and B. W. Irwin and the American Physical Society.)

Thus if we use 242-kW electric power (12.6.7f) to excite this gas mixture, 54.3% is used to create the upper state. Since the quantum efficiency of this laser is 39.9%, one could hope for an optical generation rate of

$$P_{opt} = 242 \text{ kW} \times 0.543 \times 0.399 = 52.4 \text{ kW} \qquad (12.6.10)$$

or a "wall-plug" efficiency of 21.7%.

Obviously, this is the best one could expect with optimum extraction and perfect excited-state kinetics. It is most impressive that such numbers have been approached.

There are also practical points to be gained from these numbers:

1. Although the wasted fraction 14.3% is small, the power is large: 0.143×242 kW $= 34$ kW. Thus the gas gets hot in a hurry!
2. Even if only 1% of the 52.4 kW of optical power were absorbed in the mirrors, they, too, will become hot and probably be damaged.
3. The power supply to drive this laser is not a simple device.

Again, we conclude—no wonder such lasers are pulsed.

12.7 ELECTRON BEAM SUSTAINED OPERATION

As impressive as the performance figures of the preceding section are, we see from Fig. 12.13 that we can do better. For instance, if E/N were 2.27 V-cm^2, then nearly 85% of the electrical power would go into exciting the upper laser level. Unfortunately, E/N of a *self-sustained discharge* is *not* 2.27 × 10^{-16} V-cm^2 but is 4.9 × 10^{-16} V-cm^2, as computed previously. Thus this 85% is a pipe dream unless we do things differently.

This is the role of the external source of ionization, S_{ext}, which until now has been so conveniently ignored. Its function is to create the ionization so that the applied field can transfer energy to the electrons at the optimum rate. We try to have as many electrons as possible in the proper energy interval to excite the upper laser level. This external ionization source can be an x-ray machine, a nuclear reactor, or an energetic electron beam. The latter is the most efficient, convenient, and practical and is therefore the most common method of providing this ionization. Electrons are accelerated to very high voltages—say, 100 kV—so that they can pass straight through the foil used to separate the high-pressure laser discharge cavity from the vacuum required to sustain the high voltage.

Once through the foil, the energetic electrons slow down by making many ionizing and excitation types of collisions. For instance, it takes roughly twice the ionization energy to create an electron-ion pair. But even the maximum ionization potential of any gas—that of helium is 24.5 eV—is insignificant compared to the energy of the electron beam. Thus each beam electron is capable of producing an enormous number of secondaries, and it does so. This additional controlled source is, in effect, a substitution for the internal ionization one; hence, the required E/N to effect a balance between production and loss is lowered.

This means that we have to rework the entire problem from the continuity equation onward. We start with the prescribed value of $E/N = 2.27 \times 10^{-16}$ V-cm^2 for optium laser excitation and find from Fig. 12.12 that the internal production factor is $\alpha/N = 2 \times 10^{-24}$ cm^2, whereas the loss factor is $a/N = 10^{-21}$ cm.^{2} Thus we now neglect the first term in the continuity equation, (12.6.4), and cast it into the format involving current.

$$\left(\frac{a}{N}\right)\frac{J}{eN} + \frac{\gamma}{(eW_d)^2}\left(\frac{J}{N}\right)^2 = \frac{S_{ext}}{N^2} \qquad (12.6.4) \rightarrow (12.6.11)$$

If we demand the same optical power out as before, 52.4 kW, then the amount used to create the upper state is P_{opt} divided by the quantum efficiency, or

$$P_{\text{category II}} = \frac{52.4}{0.399} = 131 \text{ kW} \qquad (12.6.12a)$$

This is 85% of the electrical power, or

$$P_{\text{elect}} = \frac{131 \text{ kW}}{0.85} = 154 \text{ kW} \quad (12.6.12b)$$

The voltage across the device is prescribed by the desired E/N.

$$V = \left(\frac{E}{N}\right) Nd$$

$$= 2.27 \times 10^{-16} \times 1.42 \times 10^{19} \times 1.2 = 3.86 \text{ kV} \quad (12.6.12c)$$

To obtain the 154 kW at this voltage means that the electrical currents must increase from the previous case (12.6.7d):

$$I = \frac{P_{\text{elect}}}{V} = \frac{154 \text{ kW}}{3.86 \text{ kV}} = 39.9 \text{ A} \quad (12.6.12d)$$

This translates to a current density of 1.38 A/cm² or $J/N = 9.72 \times 10^{-20}$ A-cm. Thus the drift velocity is found from Fig. 12.9. The plasma parameters are

$$w_d = 5.46 \times 10^6 \text{ cm/sec} \quad (12.6.13)$$

and hence the electron density is given by

$$n_e = \frac{J}{eW} = \frac{1.38}{1.6 \times 10^{-19} \times 5.46 \times 10^6}$$

$$= 1.58 \times 10^{12} \text{ e-i pairs/cm}^3 \quad (12.6.14a)$$

This has increased by a factor of 2 over that computed previously in (12.6.8). The characteristic energy is much reduced.

$$\epsilon_k = \frac{D_T}{\mu} \quad (\text{see Fig. 12.10}) = 0.91 \text{ eV} \quad (12.6.14b)$$

All parts of (12.6.12) are the desired result. Now, what must be the strength of this external source of ionization to obtain this result? To answer this, we return to (12.6.11) and use the preceding numbers to compute S_{ext}. For a typical value of $\gamma = 10^{-7}$ cm³/sec,

$$S_{\text{ext}} = N^2 \left[\left(\frac{a}{N}\right) \frac{J}{N} \frac{1}{e} + \frac{\gamma}{(ew)^2} \left(\frac{J}{N}\right)^2 \right] \quad (12.6.11)$$

$$= 3.72 \times 10^{17} \text{ e-i pairs/cm}^3/\text{sec} \quad (12.6.15)$$

If one is to use an energetic electron beam—say, 100 kV—to create this ionization, the S_{ext} can be related to the beam current density by

$$S_{\text{ext}} = n_b v_b \sum_j N_j \sigma_j = \frac{J_{\text{beam}}}{e} \sum_j N_j \sigma_j \quad (12.6.16)$$

Sec. 12.7 Electron Beam Sustained Operation

$$j = CO_2, N_2, He, \text{ the type of gas}$$

where J_{beam} is the injected current density of the high-energy beam, v_b the beam velocity $= (2eV_b/m)^{1/2}$, V_b the beam voltage, N_j the density of the jth type of gas, and σ_j the ionization cross section of that gas at the beam energy (presumed to be 100 kV). Table 12.4 gives the pertinent data for CO_2, N_2, and He.

TABLE 12.4 DATA FOR BEAM IONIZATION BY 100-kV ELECTRONS

Gas	Ionization Potential (eV)	Estimated Cross Section (cm^2)*
CO_2	13.769	3.2×10^{-18}
N_2	15.576	1.9×10^{-18}
He	24.586	4.2×10^{-19}

*Data extrapolated from Kieffer.[1]

Equating (12.6.16) to (12.6.15) and using Table 12.4 with the appropriate densities of the gases, we find that

$$J_b = 3 \times 10^{-3} \text{ A/cm}^2 \qquad (12.6.17\text{a})$$

or

$$I_b = 88 \text{ mA} \qquad (12.6.17\text{b})$$

The beam power is

$$P_{\text{beam}} = V_b I_b = 8.8 \text{ kW} \qquad (12.6.18)$$

Thus our optimized laser wall plug efficiency is

$$\eta = \frac{P_{\text{opt}}}{P_{\text{discharge}} + P_{\text{beam}}} = \frac{52.4 \text{ kW}}{154 \text{ kW} + 8.8 \text{ kW}} = 32.2\% \qquad (12.6.19)$$

The examples of these two sections illustrate the potential power of high-pressure gas lasers. The CO_2 system is not the only gas laser capable of producing these extreme power levels; however, it does prove the main points of this chapter:

1. A gas discharge laser is a very simple device, two electrodes with a rather-poor-quality vacuum between them. A precise description requires an enormous investment in prior experimental work and definite familiarity with computational techniques.
2. However, the ideas involved are very simple. Hopefully, this chapter will inspire you to examine texts devoted exclusively to gas-discharge theory.
3. If nothing else, you should realize that a gas laser is capable of producing enormous powers—in some cases at respectable efficiencies.

PROBLEMS

12.1. Consider a 1-cm-radius discharge tube filled uniformly with an e-i pair density of 10^{13} cm^{-3}. If all the electrons disappeared and thus uncovered the positive ionic charge, what would be the potential of the wall with respect to the center? (Ans. 4.52 MV; obviously this means that the electrons have a vanishingly small probability of disappearing at a different rate than that of the ions.)

12.2. Show that the maximum electrical current density that can be passed by a system without space-charge neutralization is given by

$$J = \frac{4}{9}\epsilon_0 \left(\frac{2e}{m}\right)^{1/2} \frac{V^{3/2}}{d^2}$$

where the anode-to-cathode spacing is d and the potential difference is V (the Child-Langmuir law).

12.3. Suppose that an electron collides "elastically" with a stationary heavy neutral of mass M. Show that if the scattering is isotropic, the electron loses, on the average, a fraction $2m/M$ of its energy per collision.

12.4. Suppose that one suddenly removed the voltage across the lamp of Table 12.2. Estimate how long it will take the electron energy to decay from 1.38 eV to that of the neutrals (0.026 eV). (Assume elastic losses only.)

12.5. The current represented by the *random* motion of the electrons is huge compared to the circuit current. For instance, consider the parameters of a fluorescent lamp in common use: $I = 0.25$ A, $D = 1$ in.; $N_e = N_i = 10^{12}$ cm^{-3}. Assume that the electron distribution is Maxwellian at $kT/e = 1.4$ eV and show that the random current is given by

$$J_r = \frac{N_e e}{4}\left(\frac{8kTe}{\pi m}\right)^{1/2}$$

Evaluate for the conditions given.

12.6. Suppose that the spherical symmetric part of the distribution function is given by

$$f_0(v) = A/[1 + (v/v_0)^N]$$

where v_0 is some characteristic *speed* and A the value to be determined from the normalization condition. (a) Evaluate A; (b) plot $F(\epsilon)$ as a function of energy.

12.7. Show that if ν_c is independent of velocity and f_0 is Maxwellian, (12.4.7) is the correct value for the conductivity.

12.8. Evaluate the percentage error involved in using (12.4.3) for the drift velocity by evaluating (12.4.25b) for the following dependence of ν_c on velocity.

$$\nu_c(v) = \nu_0 \left(\frac{v}{v_0}\right)^m, \quad m = -3, \ldots, +3$$

12.9. Can you give a simple explanation of why elastic collisions are always much more frequent than inelastic ones?

12.10. Repeat the reasoning discussed in Sec. 12.4.5 for the case where the plasma density is nonuniform. Find an expression analogous to (12.4.25b) for the flux of electrons in the direction of the spatial gradient. That coefficient is called the diffusion coefficient.

12.11. Suppose that the power supply voltage in Fig. 12.1 has an open circuit voltage of 300 V in series with a ballast resistor of 20 kΩ. What is the operating current and lamp voltage for (a) a hot cathode; (b) a cold cathode?

12.12. Use the data of Figs. 12.10 and 12.11 to plot the ratio of the drift energy to the characteristic energy.

12.13. Plot the predicted operating E/N vs. electron-beam current for the case discussed in Fig. 12.7. Vary the beam current from 0 to 0.1 A/cm^2.

Problems on the excitation of the CO_2 laser. (Some of these questions *require* that you read the original articles; hopefully, you will do this in *any* case.)

12.14. Suppose that the ballast resistor in Fig. 12.6 was 100 Ω and the capacitor was charged to 20 kV. Ignore the initial phases of the discharge described in the text taking place for times less than 50 nsec. Use the case discussed in the text of a 1:2:3 mixture at 400 torr.

(a) Sketch the time variation of the current through the discharge. (Assume that the triggered spark gap opens when the current drops below 20 A, and neglect the voltage drop across S.) What is the duration of the pulse? (Ans: $\tau = 9.12 \ \mu s$.)

(b) (1) How much energy is stored in the capacitor at $t = 0$? (Ans. 10.4 joules.)

(2) Estimate how much of this energy is transferred to the gas. (Ans. 4.19 joules.)

(3) How much energy remains stored in the capacitor? (Ans. 2.79 joules.)

(4) How much energy is lost in the 100-Ω ballast resistor? (Ans. 3.42 joules.)

(5) What is the peak power dissipated in the ballast? (Ans. 1.36 MW.)

(c) How much of the energy delivered to the discharge is used
  (1) To excite the N_2 vibrational levels and the 001 state of CO_2? (Ans. 2.28 joules.)
  (2) To heat the gas via elastic collision and excitation of the lower laser level? (Ans. 0.599 joules.)
(d) If one assumes that all the energy of (c2) goes into random translational motion of the gas molecules, what is the temperature rise at the end of the current pulse? (Ans. $\Delta T = 58.6°K$.)
(e) Assume that an equilibrium is established between the 001 state of CO_2 and the $v = 1$ state of N_2 and the energy of (c1) resides in these states. Assume a perfect match in energy between the two states.
  (1) How much is in the CO_2 system? (Ans. 0.76 joules.)
  (2) How much is in the N_2 system? (Ans. 1.52 joules.)
  (3) What percentage of the CO_2 and N_2 molecules are in the 001 state or $v = 1$ state? (Ans. 19.7%.)
(f) The energy stored in the CO_2 system can be extracted at a rate limited only by the photon-buildup time (see Chapter 9 on gain switching), whereas that stored in the N_2 system is limited by the collisional transfer rate between N_2 and CO_2. Assume a gain switched pulse of 3-nsec duration and a transfer rate of 10^7 sec^{-1}. Assume further that the lower state decays with a lifetime of 10 nsec.
  (1) Estimate the energy in the gain switch pulse. (Ans. $0.76/2 = 0.38$ joules.)
  (2) Estimate the energy in the long tail representing the transfer from N_2 to CO_2. (Ans. $1.52 + 0.38$ joules $= 1.90$ joules.)
  (3) Sketch the output of this laser. (Ans. A peak of 0.127 MW followed by a long tail [more or less triangular] with a slope of 10^7 sec^{-1} containing the 1.9 joules of [f2].)

12.15. Suppose the electron gas of a fluorescent lamp was described by a Maxwellian distribution function with a characteristic energy $kT_e/e = 1.5$ eV whereas the Hg$^+$ has a temperature of 300°K. Any insulated object in contact with this plasma (i.e., a glass wall or a floating metallic probe) must attain a negative potential with respect to the bulk in order that equal numbers of positive and negative charges arrive at the surface. Compute this "floating potential" difference.

12.16. Assume that the spherically symmetric part of the distribution function of the electron gas can be approximated by

$$f_0(v) = A\left[1 - \frac{v}{v_0}\right] \text{ for } v < v_0; \quad 0 \text{ otherwise}$$

(a) Evaluate the parameter A from the normalization requirement.
(b) Relate the parameter v_0 to the average energy of the electron gas.

Now assume there is an electric field E that gives the electrons energy, and they in turn transfer their excess energy to the neutrals (density of 10^{17} cm^{-3}) by three types of collisions:

Elastic: $\sigma_c = 10^{-16}$ cm^2; $G = 10^{-4}$
Excitation: $\sigma_x = 10^{-17}$ cm^2; threshold energy of 5 eV = energy lost per collision
Ionization: $\sigma_i = 10^{-18}$ cm^2; threshold energy of 10 eV = energy lost per collision

(c) Use (12.4.25) and the definition of current, $\mathbf{J} = ne\mathbf{w}$ to compute the drift velocity, $\mathbf{w}$, as a function of E/N. Plot these results in the manner of Fig. 12.10.
(d) Use the energy balance equation to relate the characteristic energy of the electrons to E/N. Plot the results in the same manner as was used for Fig. 12.11.
(e) Compute the fractional power used for the various processes as a function of E/N and plot the results in the manner used for Fig. 12.13.

REFERENCES AND SUGGESTED READINGS

1. L. J. Kieffer, "A Compilation of Electron Collision Cross Section Data for Modeling Gas Discharge Lasers," JILA Information Center Report No. 13, Sept. 1973.
2. L. J. Denes and J. J. Lowke, "V-I Characteristics of Pulsed CO_2 Laser Discharges," Appl. Phys. Lett. *23*(3), 130–132, Aug. 1973.
3. J. J. Lowke, A. V. Phelps, and B. W. Irwin, "Predicted Electron Transport Coefficients and Operating Characteristics of CO_2-N_2-He Laser Mixtures," J. Appl. Phys. *44*(10), 4664–4471, 1973.
4. L. G. H. Huxley and R. W. Crompton, *The Diffusion and Drift of Electrons in Gases* (New York: John Wiley & Sons, Inc., 1974).
5. W. L. Nighan, "Electron Energy Distributions and Collision Rates in Electrically Excited N_2, CO, and CO_2," Phys. Rev. *A2*, 1989–2000, 1970.

GENERAL GAS DISCHARGE REFERENCES

John F. Waymouth, *Electric Discharge Lamps* (Cambridge, Mass.: The M.I.T. Press, 1971).
A. von Engel, *Ionized Gas*, 2nd ed. (London: Oxford-at-the-Clarendon Press, 1965).
James D. Cobine, *Gaseous Conductors* (New York: Dover Publications, Inc., 1958).

S. C. Brown, *Introduction to Electrical Discharges in Gases* (New York: John Wiley & Sons, Inc., 1966).

B. E. Cherrington, *Gaseous Electronics and Gas Lasers* (Oxford: Pergamon Press, 1979).

G. F. Weston, *Cold Cathode Glow Discharge Tubes* (London: ILIFFE Book Ltd., 1968).

B. M. Smernov, *Physics of Weakly Ionized Gases,* translated from Russian. (Moscow: Mir Publishers, 1981).

B. Chapman, *Glow Discharge Processes* (New York: John Wiley & Sons, Inc., 1980).

13

Advanced Topics in Electromagnetics of Lasers

13.1 INTRODUCTION

By now one should realize that a significant part of the laser electronics resides in its electromagnetic description. We have tried to avoid unnecessary complications so as to appreciate the simplicity of laser operation by stimulated emission of the electromagnetic wave. However, there are some obvious holes in our simplistic approach, especially when one considers a semiconductor system. A few of these deficiencies and questions that should be addressed are

1. Not all gas, dye, and solid-state lasers use "stable" resonators. Indeed, the question of stability does not even arise for semiconductor lasers.
2. Is there a "general" approach that can handle any cavity configuration?
3. What are the fields inside a semiconductor laser, and how do they couple to the outside world? This is a significant question, for if the field is at one location and the inverted population at another, little stimulated emission will take place.
4. Some lasers use a feedback scheme that is "distributed" through the medium. Our simple *round-trip gain* for threshold and *round-trip phase shift* for oscillation frequency specifications take on a more complicated appearance in these circumstances but are still applicable.

5. Arrays can be constructed in which the fields from individual lasers can be added coherently so as to maximize the power into a given direction. In principle, one should be able to electronically control this direction just as phased-array antennas are controlled in the microwave spectrum.

All of the above are electromagnetic problems associated with lasers and are addressed below. While electromagnetic problems are the focus of this chapter, do not forget the other issues of lasers, stimulated emission, saturation, pumping schemes, and so on, that are still operative.

13.2 UNSTABLE RESONATORS

13.2.1 General Considerations

Although the use of stable resonators was instrumental in the development of lasers, they do have a disadvantage of having a very *small* mode volume. Consequently, it is difficult to pack enough excited atoms into this volume to generate high power in the laser oscillator. Furthermore, it is difficult to restrict oscillation to the $TEM_{0,0}$ mode unless we insert apertures to increase the diffraction losses on the higher-order modes. Even this is a nontrivial task; we must adjust the *size* and *position* of the aperture with respect to an *unknown* axis of the cavity.

Thus we naturally look fondly at the edges of the unstable regime, where spot sizes at the various mirrors tend to become very big (see Fig. 5.5 for the hemispherical cavity). The spot size at the spherical mirror goes to infinity as $d \to R$. Unfortunately, we can expect the accuracy of the theory of Gaussian beams to degenerate rapidly as one approaches the edges of the stability diagram and be totally inapplicable in the unstable regime.

A valuable alternative to this was provided by Siegman.[1] He reasoned that we could consider the field between mirrors of an unstable resonator to be that of a limited extent *spherical* wave whose size is determined by the aperture of the mirror.

For instance, consider the geometry shown in Fig. 13.1(a), obviously unstable by our previous analysis. If we assume a spherical wave to originate from the point P_1 as in Fig. 13.1(a) (undefined at this state) with a size determined by the aperture a_1, a considerable fraction of this wave misses M_2. However, this incident wave does illuminate the mirror M_2 more or less uniformly. Consequently, it will generate a limited-extent spherical wave whose extent is dictated by a_2 apparently originating from a point P_2, as indicated in Fig. 13.1(b). Obviously, P_2 is the image of the source at P_1. We have a self-consistent picture if the point P_1 is also the image of the source at P_2. The part of the wave that leaks around the mirrors can be considered useful output, and the fraction

Sec. 13.2 Unstable Resonators

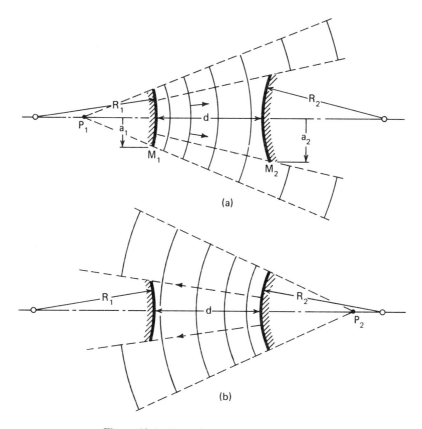

Figure 13.1 Wave fronts in an unstable cavity.

that makes a complete round-trip determines the required gain of the laser medium.

Let us now try to translate these ideas into an analytic description. Consider the geometry shown in Fig. 13.2 and postulate the existence of a set of points, P_1 and P_2, that act as virtual sources (or the object) for the waves that illuminate the other mirror. It will be convenient to measure all distances in units of the mirror spacing, d; that is, P_1 is located $r_1 d$ from M_1 and $(r_1 + 1)d$ from M_2. Thus P_1 is the virtual source of radiation impinging on M_2. The reflected wave appears to originate from the object point P_2. Using standard mirror formulas relating the image and object distances yields

$$\left. \begin{array}{c} \dfrac{1}{(r_1 + 1)d} - \dfrac{1}{r_2 d} = \dfrac{2}{R_2} \\[2mm] \dfrac{1}{r_2} - \dfrac{1}{r_1 + 1} = 2(g_2 - 1) \end{array} \right\} \quad (13.2.1)$$

or

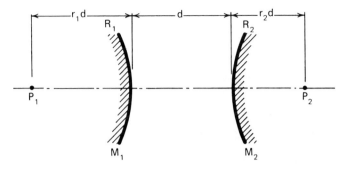

Figure 13.2 Unstable cavity.

where

$$g_2 = 1 - \frac{d}{R_2} \tag{13.2.2}$$

A self-consistent picture is obtained if the new source at P_2 has its image point at the original P_1.

$$\left.\begin{array}{c}\dfrac{1}{(r_2 + 1)d} - \dfrac{1}{r_1 d} = \dfrac{2}{R_1} \\[2ex] \dfrac{1}{r_1} - \dfrac{1}{r_2 + 1} = 2(g_1 - 1)\end{array}\right\} \tag{13.2.3}$$

where

$$g_1 = 1 - \frac{d}{R_1} \tag{13.2.4}$$

We must now solve these equations simultaneously to find these characteristic points $r_1 d$ and $r_2 d$. Note that if the r's are positive numbers, these points lie *outside* the cavity; a negative number would indicate that a source point lies on the reflecting side of the mirror. Perseverance with the arithmetic leads to

$$r_1 = \frac{[1 - (g_1 g_2)^{-1}]^{1/2} - 1 + g_1^{-1}}{2 - g_1^{-1} - g_2^{-1}} \tag{13.2.5a}$$

$$r_2 = \frac{[1 - (g_1 g_2)^{-1}]^{1/2} - 1 + g_2^{-1}}{2 - g_1^{-1} - g_2^{-1}} \tag{13.2.5b}$$

Having found that a self-consistent picture is possible, let us turn to the question of losses. Recall that we postulated a limited extent spherical wave originating at the points P_1 and P_2 with the angular extent being determined by the mirrors M_2 and M_1, respectively. The fraction of power reflected from M_2 that originated

Sec. 13.2 Unstable Resonators

at M_1 is [see Fig. 13.3(a)]

$$\Gamma_2 = \frac{\text{solid angle of } M_2 \text{ with origin at } P_1}{\text{angular extent of wave originating at } M_1}$$

$$= \frac{\pi a_2^2/4\pi(r_1 + 1)^2 d^2}{\pi a_1^2/4\pi r_1^2 d^2} \tag{13.2.6}$$

and for that portion reflected from M_1 [see Fig. 13.3(b)]:

$$\Gamma_1 = \frac{\text{solid angle of } M_1 \text{ with origin at } P_2}{\text{angular extent of wave originating at } M_2}$$

$$= \frac{\pi a_1^2/4\pi(r_2 + 1)^2 d^2}{\pi a_2^2/4\pi r_2^2 d^2} \tag{13.2.7}$$

Thus the fraction of the power that survives a round-trip is

$$\Gamma^2 = \Gamma_1 \Gamma_2 = \left[\frac{r_1 r_2}{(r_1 + 1)(r_2 + 1)}\right]^2 \tag{13.2.8}$$

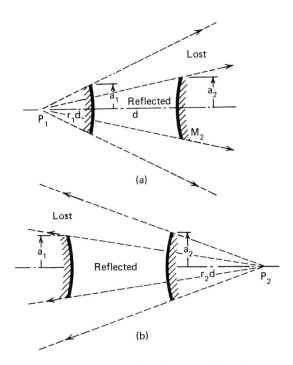

Figure 13.3 Power lost in an unstable cavity.

Note that this is the important parameter insofar as laser oscillators are concerned. [The single-pass gain must be such that net power gain exceeds the loss (i.e., $G\Gamma > 1$); see (0.1).] Note also that this expression is independent of the mirror sizes within the context of this first-order theory. At first glance, this may seem strange, but recall the physical situation being studied.

If the size of the mirror M_2 in Fig. 13.3 were made larger, less of the power would leak out that end. But this reduction would be compensated by the increased angle of the radiation impinging on M_1 from point P_2. Fig. 13.4 illustrates a case where we should think before blindly plugging into formulas. The physical size of the mirror M_2 has nothing whatsoever to do with the radiation impinging on M_1. We should use the aperture defined by the incident radiation onto M_2 as it is shown, rather than the physical size.

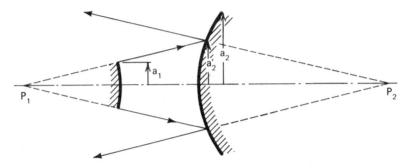

Figure 13.4 Extreme case of an unstable resonator.

The example shown in Fig. 13.5, of a cavity with a finite aperture medium between the mirrors, is even more practical than the foregoing case. If we assume that the edges of the medium (a discharge tube or a laser rod) are perfectly absorbing, then the cone angles are dictated by the combination of the mirrors, gain medium, and geometry. Here we see the use of a "beam slicer" to extract useful power output and to define the effective aperture of M_1. It is obvious from this figure that one should minimize the amount of power dumped uselessly into the walls of the active medium.

Let us return to the formula for the round-trip transmission (13.2.8) and express the result in terms of the g parameters by utilizing the relations between r_1 and r_2 and g_1 and g_2 (13.2.5a) and (13.2.5b). The mean one-way transmission* coefficient through the cavity can be expressed as

$$\Gamma = \pm \frac{r_1 r_2}{(r_1 + 1)(r_2 + 1)}$$

*The mean power lost per pass is related to Γ by
$$L = 1 - \Gamma$$

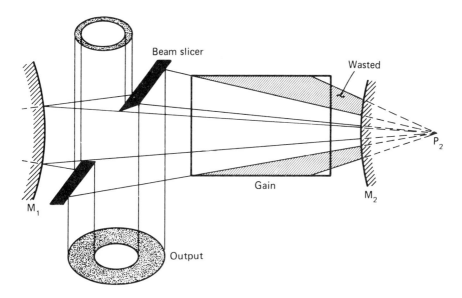

Figure 13.5 Laser using an unstable resonator.

or:
$$\Gamma = (\pm)\frac{1 - [1 - (g_1g_2)^{-1}]^{1/2}}{1 + [1 - (g_1g_2)^{-1}]^{1/2}} \qquad (13.2.9)$$

Note that if $g_1g_2 > 0$, the quantity involving the square root is less than 1. Hence, the upper choice in sign is applicable in order for the transmission coefficient to be positive and less than 1. If $g_1g_2 < 0$, the negative sign must be chosen to obtain sensible answers for the transmission coefficient. Thus one has a *positive* and a *negative* branch of unstable resonators. We can solve for the contours of equiloss and plot these on a stability diagram:

positive branch $\qquad\qquad$ negative branch

$g_1g_2 > 1 \qquad\qquad\qquad g_1g_2 < 0$

$$\Gamma = \frac{1 - [1 - (g_1g_2)^{-1}]^{1/2}}{1 + [1 - (g_1g_2)^{-1}]^{1/2}} \qquad \Gamma = \frac{[1 - (g_1g_2)^{-1}]^{1/2} - 1}{[1 - (g_1g_2)^{-1}]^{1/2} + 1}$$

Hence,

$$[1 - (g_1g_2)^{-1}]^{1/2} = \frac{1 - \Gamma}{1 + \Gamma} \qquad [1 - (g_1g_2)^{-1}]^{1/2} = \frac{1 + \Gamma}{1 - \Gamma}$$

Solving for g_1g_2 in terms of the loss, we obtain

$$\text{(a)} \; g_1g_2 = \frac{(1 + \Gamma)^2}{4\Gamma} \qquad \text{(b)} \; g_1g_2 = -\frac{(1 - \Gamma)^2}{4\Gamma} \qquad (13.2.10)$$

Thus the equiloss contours are hyperbolas on the stability diagram. The positive branch corresponds to the first and third quadrants (Fig. 13.6), whereas the negative branch is in the second and fourth quadrants. If, for example, we wanted a cavity with a power loss per pass of 25%, then $\Gamma = 0.75$ and

$$g_1 g_2 = \frac{(1.75)^2}{4 \times 0.75} = 1.0208 \qquad (+ \text{ branch})$$

$$g_1 g_2 = -\frac{(0.25)^2}{4 \times 0.75} = -0.0208 \qquad (- \text{ branch})$$

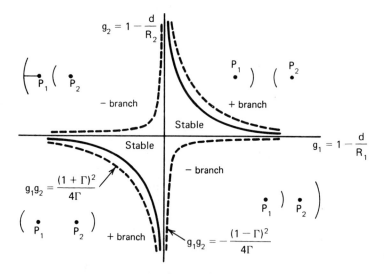

Figure 13.6 Equiloss contours for the unstable cavities.

13.2.2 The Unstable Confocal Resonator

Consider the problem of an unstable confocal resonator shown in Fig. 13.7, where the mirrors have a common focal point at a distance $d/2$ to the left of M_1. Thus $|R_1| = d_0$ and $R_2 = 3d_0$, to have a common focal point.

We can return to the previous formulas for r_1 and r_2 to find an indeterminate relationship of the form $r_1 = N/D = 0/0$. This difficulty can be resolved by recognizing that perfect alignment in spacing is impossible. Thus $d = d_0(1 + \delta)$, where $d_0 = |R_1|$, and then take the limit as our accuracy improves by letting δ approach zero.

We can circumvent all the dull arithmetic by looking at the physics of the problem and using some reasonable guesswork. If we guess that point P_1 is at the focal point of M_2, the image P_2 is at $-\infty$. Now the image of P_2 in mirror M_1 is

Sec. 13.2 Unstable Resonators

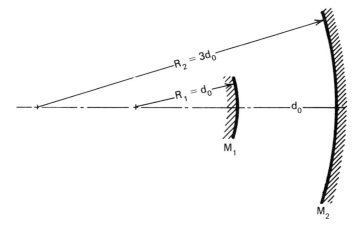

Figure 13.7 Confocal resonator.

the focal point—hence, we have achieved a self-consistent picture of the virtual sources of the radiation within the cavity. Thus the picture is as shown in Fig. 13.8. Within the context of this first-order theory, the radiation should come out in the form shown. Even though the first-order theory predicts a uniformly illuminated annular region, we know that this is impossible. Diffraction effects will round off the corners and fill up the center region. Indeed, an article by Frieberg, Chenausky, and Buczek[2a]* shows that the far-field pattern closely resembles a Gaussian beam.

Incidentally, Fig. 13.9 illustrates the "burn" pattern when a 14-kW CO_2 laser using a confocal unstable resonator (at the University of Illinois) is dumped into a bucket of sand (thereby making a low grade of glass). It is obvious that Fig. 13.8 closely resembles this burn pattern.

The level of sophistication of understanding of the fields in unstable resonators has advanced considerably since the publication of this first-order theory. There are obvious oversimplifications but, amazingly, it does give a good "feel" for the cavity and a reasonably good approximation for the loss per pass. For the case analyzed in this section:

$$\Gamma_2 = 1$$

$$\Gamma_1 = \frac{\pi a_1^2}{\pi(3a_1)^2} = \frac{1}{9}$$

Therefore

$$\Gamma = (\Gamma_1\Gamma_2)^{1/2} = \frac{1}{3}$$

---

*See also Frieberg et al.,[2b] Fig. 7.

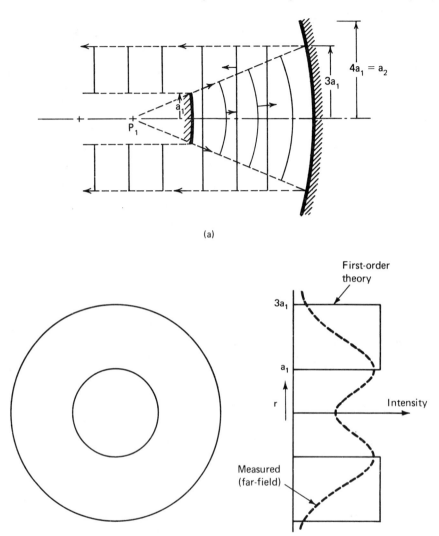

Figure 13.8 (a) Radiation pattern for the cavity of Fig. 13.7. (b) and (c) Field distribution.

Thus there is a loss of 66% per pass in this cavity. The power gain per pass should be

$$G\Gamma > 1 \quad \text{or} \quad G > 3 \text{ or } 4.77 \text{ dB}$$

Many lasers have small-signal gains sufficient to meet this criterion.

Sec. 13.3   Integral Equation Approach to Cavities   463

Figure 13.9 Burn pattern from a 14-kW CO_2 laser using an unstable resonator. The "sand" is incandescent because of the intense beam. (Photograph by S. Hutchison.)

13.3 INTEGRAL EQUATION APPROACH TO CAVITIES

13.3.1 Mathematical Formulation

Until now, we have used three approaches to optical cavities, which are listed below in order of sophistication along with some obvious difficulties:

1. Treat the fields inside the cavity as *uniform plane waves*. However, there are two problems with this approach: (a) We cannot permit an infinite cross section, so how "big" should it be? (b) Once we have answered (a), then how do we account for the part that misses a finite size mirror?

2. Use Hermite-Gaussian beam modes, but these are only applicable for stable cavities and even for those, diffraction effects as a result of the fields at the edges of mirrors are ignored.

3. Use the self-consistent "image" theory of unstable resonators discussed in Sec. 13.2. That theory only applies for unstable cavities and even there predicts sharp edges to the field distribution. Furthermore, it completely ignores the phase of the fields.

There is another approach that avoids the above problems at the expense of mathematical complexity and almost certain absence of analytic solutions except for special circumstances. However, it is perfectly general. In fact, it was the first serious theoretical approach to optical cavities, and the results, which predicted high Q cavities, even for an open resonator such as two plane mirrors, had a tremendous impact on laser research.

The task is to find a self-consistent solution to the Helmholtz equation for each Cartesian component of the field.

$$\nabla^2 E + k^2 E = 0 \qquad (13.3.1)$$

Now we might try to solve (13.3.1) by analytic means or at the last resort by numerical analysis, but for open cavities the boundary conditions are not well defined. For partial differential equations, this represents a critical lack of information. Thus we search for an integral equation formulation that is equivalent to (13.3.1) under which we can hide our ignorance.

In principle, we can always obtain a *formal* solution to (13.3.1) by use of Green's function G, which is the response of the system at the point $\mathbf{r}$ due to a unit source at the point $\mathbf{r}_0$. Green's function obeys an equation similar to (but worse in complexity than) (13.3.1).

$$\nabla^2 G + k^2 G = -\delta(x - x_0)\delta(y - y_0)\delta(z - z_0) \qquad (13.3.2)$$

If (13.3.1) is multiplied by G, (13.3.2) by E, and the difference integrated over the volume of the cavity, then the following is obtained:

$$\iiint [G\nabla^2 E - E\nabla^2 G]\,dV = \iiint E\delta(\mathbf{r} - \mathbf{r}_0)\,dV \qquad (13.3.3)$$

Now we apply some dry mathematics from vector calculus.

$$\nabla \cdot (a\mathbf{A}) \equiv a\nabla \cdot \mathbf{A} + \mathbf{A} \cdot \nabla a \qquad (13.3.4a)$$

and

$$\iiint \nabla \cdot \mathbf{B}\,dV \equiv \oiint \mathbf{B} \cdot \mathbf{n}\,dA \qquad (13.3.4b)$$

We apply (13.3.4a) twice: assigning $\mathbf{A}$ to be ∇E and then ∇G with a being G and E, respectively, and then recognizing that the bracket in (13.3.3) can be expressed as the divergence of another vector $\mathbf{B}$; thus the result can be converted from a volume integral to one over the surface bounding the cavity. The property of the δ function evaluates the field at the interior point $\mathbf{r}_0$.

$$E(\mathbf{r}_0) = \oiint [G\nabla E - E\nabla G] \cdot \mathbf{n}\,dA \qquad (13.3.5)$$

where $\mathbf{n}$ is a unit vector normal to the surface enclosing the cavity. This is an important formula in itself. It states that the field at any point inside the cavity is related to G and E on the surfaces bounding the system. However, it does not

Sec. 13.3 Integral Equation Approach to Cavities

appear that much practical progress has been made: we do not know E anywhere, much less on the surfaces, and the task of finding G appears to be worse!

The utility of (13.3.5) lies in our ability to inject some physical insight into this dry bit of mathematics and obtain an approximate form for G. Further, the form of (13.3.5) suggests that an iterative procedure (on a computer) might be fruitful: guess at E, evaluate the integrals, compare the result to the guess, correct the guess, and start again. Follow the loop until the guess is self-consistent (or computer funds are exhausted). This is precisely what was and is done. However, first let us obtain a logical but approximate form for G, which follows from its physical definition—the field from a unit source located at (x_0, y_0, z_0). For free space that is just a spherical wave of unit amplitude

$$G = \frac{1}{4\pi |\mathbf{R}|} e^{-j\phi(r,r_0)} \qquad (13.3.6)$$

The term $\phi(r, r_0)$ is the phase shift experienced by the wave in propagating from (x_0, y_0, z_0) to (x, y, z) (or conversely) and $|\mathbf{R}| = |\mathbf{r} - \mathbf{r}_0|$ is the distance from $\mathbf{r}_0$ to $\mathbf{r}$.

Now comes the uninspiring task of evaluating the gradients in (13.3.5) and applying it to the geometry shown in Fig. 13.10. This figure is a generic optical cavity (shown with two flat mirrors, but it could be any slightly curved surface) with the vectors $\mathbf{r}$, $\mathbf{r}_0$, and $\mathbf{R}$ indicated. A number of issues are to be noted:

1. We are interested in computing the field at M_2 in terms of the field at M_1.
2. Thus $\mathbf{r}_0$ is on M_2 and $\mathbf{r}$ is on M_1, in accordance with (13.3.5).
3. The mirror sizes $a_{1,2}$ are small compared to the spacing d, and hence the angle θ is small.

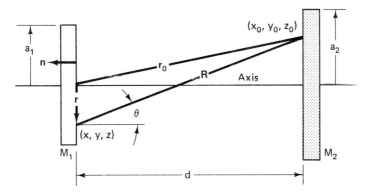

Figure 13.10 An optical cavity illustrating the geometry and coordinates used for the integral equation.

For transverse fields in the immediate vicinity of M_1, propagating (mostly) in the z direction, we have

$$E = f(x,y)e^{-jkz} \tag{13.3.7a}$$

Hence

$$\mathbf{n} \cdot \nabla E = (-\mathbf{a}_z) \cdot (-jk\mathbf{a}_z)E = +jkE \tag{13.3.7b}$$

The gradient of G involves a differentiation of the product of the $1/R$ (the amplitude factor) and the exponential term representing the phase shift $\phi(r,r_0)$ going from $\mathbf{r}_0$ to $\mathbf{r}$. The latter is overwhelmingly dominant because that phase goes through a complete cycle (2π) every time the distance changes by λ_0, whereas the distance $|R|$ experiences just a small fractional change of the order of $\sim \lambda/d \ll 1$.

For small angles, the phase term ϕ can be expressed as $(k \cos \theta)(z_0 - z)$, and thus ∇G is given by

$$\nabla G = jk \cos \theta \, \mathbf{a}_z \left(\frac{1}{4\pi R}\right) e^{-j\phi(r,r_0)} \tag{13.3.8a}$$

Thus

$$\mathbf{n} \cdot \nabla G = -jk \cos \theta \left(\frac{1}{4\pi R}\right) e^{-j\phi(r,r_0)} \tag{13.3.8b}$$

Combining (13.3.7b) and (13.3.8b) with (13.3.5) yields

$$E(r_0) = jk \iint_{M_1} \frac{E(r)(1 + \cos \theta) e^{-j\phi(r,r_0)} \, dA}{4\pi R} \tag{13.3.9}$$

One more approximation is in order: the distance R from any point on M_1 to the arbitrary point r_0 does not deviate significantly from d for a reasonably small mirror. Further, $\cos \theta \sim 1$, and we obtain

$$E(r_0) = \frac{j}{\lambda(z - z_0)} \iint_{M_1} E(r) e^{-j\phi(r,r_0)} \, dA \tag{13.3.10a}$$

r_0 can be anywhere inside the cavity such that $|z - z_0|$ is large compared to the "size" of M_1 so as to ensure the validity of our approximations. In particular, it is valid for $|z - z_0| = d$; that is, the point z_0 is on M_2; and thus

$$E(M_2) = \frac{j}{\lambda d} \iint E(r) e^{-j\phi(r,r_0)} \, dA \tag{13.3.10b}$$

This is our fundamental result, and it is the basic equation for numerical analysis. Of course, the identification of which mirror is 1 (or 2) is completely arbitrary as is the identification of the variables r and r_0—the labeling can be interchanged

Sec. 13.3 Integral Equation Approach to Cavities

at will. For a simple cavity shown in Fig. 13.10, (13.3.10) must be applied twice with an interchange of labels: once going from 1 to 2, and once more in going from 2 back to 1.

13.3.2 The Fox and Li Results[5]

It is just a matter of programming, computer time, and patience to obtain the characteristic field distributions of various cavity configurations using (13.3.10). A.G. Fox and T. Li[5] started with the easiest geometry of all—two identical strips of width $2a$ and infinitely long (in and out of paper in Fig. 13.11). The initial guess was simply a uniform field with a constant phase over the strip, zero otherwise. Fig. 13.12 is a reproduction of their Fig. 5 showing the resulting pattern with the field normalized to be 1 at the center of the strip mirror, after one transit. Note that there are some rather violent gyrations for this case. After 300 transits, the field settles down and starts looking more or less like a Gaussian beam.

The lower part of this figure illustrates the phase of the fields on one of the mirrors referenced to that at the center. The initial wave was started with constant phase independent of x. For every "bump" seen in the top for the amplitude, there is a corresponding gyration in phase for one transit. After 300 transists, and having achieved a self-consistent reproduction, the phase is just about as smooth as the amplitude with the field at edge lagging that at the center. In other words, even though the mirrors are flat, the phase front is *not;* that is, the mirrors are not surfaces of constant phase.

Given the fields, it is just a matter of computing the fraction of the intensity intercepted by each mirror and identifying the remainder (the part that misses) as the diffraction loss for the cavity. This fraction, taken from Fig. 8 of Fox and Li[5] is shown in Fig. 13.13 and confirms our intuition: large mirrors separated by reasonable distances have small losses. Now, however, it is given a scientific measure.

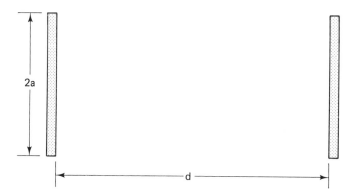

Figure 13.11 The "strip" mirror system analyzed by Fox and Li.[5]

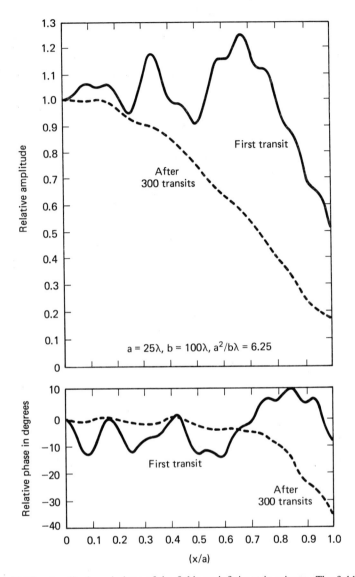

Figure 13.12 Amplitude and phase of the fields on infinite strip mirrors. The fields were started with an uniform amplitude $E(x) = E_0$; $|x| < a$ (zero otherwise) and with a constant phase. (From Fox and Li.[5])

13.3.3 The Confocal Resonator[6]

There is only one geometry that yields an analytic answer with a minimum number of approximations: this is the confocal geometry consisting of two identical "square" mirrors with a common radius of curvature $R = b$, which are separated by a distance $d = b$, and thus share a common focal point as shown in Fig. 13.14.

Sec. 13.3 Integral Equation Approach to Cavities

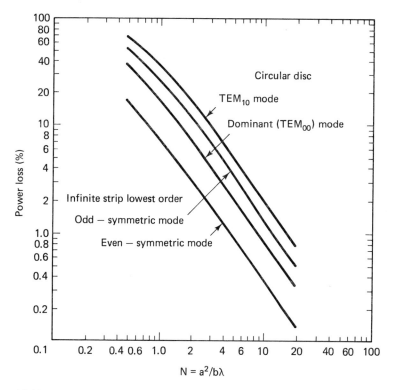

Figure 13.13 Losses in an open cavity per transit for a strip mirror of width $2a$ and a circular disk of radius a. (From Fox and Li,[5] Fig. 8.)

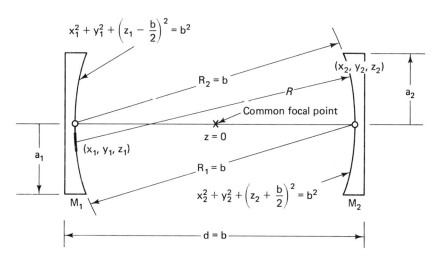

Figure 13.14 The confocal resonator.

Also shown in Fig. 13.14 are the equations describing the surfaces of the two mirrors, which are needed in order to compute the phase shift of a wave originating from dx_1, dy_1, at M_1 and propagating to x_2, y_2 on M_2.

$$\phi(2,1) = k[(x_2 - x_1)^2 + (y_2 - y_1)^2 + (z_2 - z_1)^2]^{1/2} = kR \qquad (13.3.11)$$

After considerable uninspiring arithmetic, one finds that the distance R from (x_1, y_1, z_1) to (x_2, y_2, z_2) can be approximated by

$$R = b - \frac{x_1 x_2}{b} - \frac{y_1 y_2}{b} \qquad (13.3.12)$$

which is equivalent to approximating the spherical mirror surface by a parabolic one. Thus (13.3.10) can be written as

$$E(x_2, y_2) = \frac{k}{2\pi b} e^{-j(kb - \pi/2)} \iint_{M_1} E(x_1, y_1) e^{jkx_1 x_2/b} e^{jky_1 y_2/b} \, dx_1 dy_1 \qquad (13.3.13)$$

where the multiplicative factor of j in (13.3.10) has been expressed as $\exp(j\pi/2)$ and combined with the axial phase shift kb.

It is very important to keep in mind the "game plan." We are computing the field on M_2 in terms of that on M_1; which in turn can be computed in terms of that on M_2. If we had two dissimilar mirrors, one would have to complete the loop and demand self-consistency for the round-trip. For the case considered here, we can cheat a bit and merely require that the field at M_2 be a scaled replication of the field at M_1. The format of (13.3.13) suggests factoring the field into a product. That is,

$$E(x_1, y_1) = f(x_1) g(y_1) \qquad (13.3.14a)$$

and

$$E(x_2, y_2) = [\sigma_x f(x_2)][\sigma_y g(y_2)] \qquad (13.3.14b)$$

where σ_x, σ_y are complex constants to be determined. Equation (13.3.13) can be factored and rewritten

$$\sigma_x f(x_2) = \left[\frac{k}{2\pi b} e^{-j(kb - \pi/2)}\right]^{1/2} \int_{-a_1}^{+a_1} f(x_1) e^{jkx_1 x_2/b} \, dx_1 \qquad (13.3.15)$$

with a corresponding equation for $g(y)$:

$$\sigma_y g(y_2) = \left[\frac{k}{2\pi b} e^{-j(kb - \pi/2)}\right]^{1/2} \int_{-a_1}^{-a_1} g(y_1) e^{jky_1 y_2/b} \, dy_1 \qquad (13.3.16)$$

Let us define some normalized quantities. Let

$$N = \frac{a^2}{b\lambda} = \text{Fresnel number} \qquad (13.3.17)$$

Sec. 13.3 Integral Equation Approach to Cavities

$$C = 2\pi N \qquad (13.3.18a)$$

$$X_{1,2} = \frac{x_{1,2}}{a}\sqrt{C} \qquad (13.3.18b)$$

Then (13.3.15) can be written in terms of normalized variables:

$$\sigma_x f(X_2) = \left[\frac{ka^2}{2\pi bC}\right]^{1/2} \{e^{-j(kb-\pi/2)}\}^{1/2} \int_{-\sqrt{C}}^{\sqrt{C}} f(X_1) e^{jX_1 X_2}\, dX_1 \qquad (13.3.19)$$

In (13.3.10) through (13.3.19) there is a "hint" of a Fourier transform creeping into view; it becomes most evident if (13.3.17) and (13.3.18) are used for the first prefactor of (13.3.19).

$$\frac{ka^2}{2\pi bC} = \frac{1}{2\pi}$$

Hence, (13.3.19) appears as a finite Fourier transform:

$$\sigma_x f(X_2) = \{e^{-j(kb-\pi/2)}\}^{1/2} \left\{\frac{1}{\sqrt{2\pi}} \int_{-\sqrt{C}}^{\sqrt{C}} f(X_1) e^{jX_1 X_2}\, dX_1\right\} \qquad (13.3.20)$$

If $C \to \infty$, then it *is* a simple Fourier transform and the solution for $f(X_1)$ is a Gaussian. In other words, suppose $f(X_1) = \exp - (X_1^2/2)$. Then (13.3.20) can be manipulated as follows:

$$I = \frac{1}{\sqrt{2\pi}} \int_{-\infty}^{\infty} f(X_1) e^{jX_1 X_2}\, dX_1 = \frac{1}{\sqrt{2\pi}} \int_{-\infty}^{\infty} \exp\left[-\left(\frac{X_1^2}{2} - jX_1 X_2\right)\right] dX_1$$

The exponent can be expressed as a perfect square plus an extra term.

$$\frac{X_1^2}{2} - jX_1 X_2 = \left(\frac{X_1}{\sqrt{2}} - j\frac{X_2}{\sqrt{2}}\right) + \frac{X_2^2}{2} = \frac{U^2}{2} + \frac{X_2^2}{2} \qquad (13.3.21a)$$

where

$$U = X_1 - jX_2 \qquad (13.3.21b)$$

Thus the integral becomes

$$I = e^{-X_2^2/2}\left\{\frac{1}{\sqrt{2\pi}} \int_{-\infty}^{\infty} e^{-u^2/2}\, du\right\} \qquad (13.3.22)$$

Now there is a key issue that is in danger of being obscured by the smoke of mathematics. We assumed a field varying as $\exp - X_1^2/2$ on M_1 and found a field identical to it on M_2 (since X_1 and X_2 are equivalent variables merely expressing a distance along x at the two mirrors). Because we choose a symmetrical cavity, the return trip will obviously reproduce the original field and hence we have identified a characteristic *mode* of the cavity. The assumption of

$C \to \infty$, that is, infinite mirror size, leads to $|\sigma_{x,y}| = 1$. (For finite mirrors, the amplitude would be less than 1.) The total field is given by

$$E(X_2, Y_2) = e^{-j(kb - \pi/2)} e^{-(x_2^2 + y_2^2)/2} \tag{13.3.23}$$

Now it is appropriate to remove the normalizations on the spatial variables:

$$\frac{X_1^2}{2} = \frac{x_1^2 C}{2a^2} = \frac{x_1^2}{2a^2} \cdot \frac{2\pi a^2}{b\lambda} = \frac{x_1^2}{b\lambda/\pi} \equiv \frac{x_1^2}{w_s^2}$$

where $(13.3.24)$

$$w_s^2 = \frac{b\lambda}{\pi} = \left(\frac{2b}{k}\right)$$

The parameter w_s is the spot size (e^{-1}) of the Gaussian beam at the spherical mirror and is *precisely* the result that was obtained in Chapter 5 for this geometry.

Resonance can be found directly from the phase of (13.3.23).

$$kb - \frac{\pi}{2} = q\pi$$

$$\nu = \frac{c}{2d}\left(q + \frac{1}{2}\right) \tag{13.3.25}$$

where the fact that $b = d$ has been used. This can be compared directly with (6.5.4.) for $m = p = 0$:

$$\nu = \frac{c}{2d}\left[q + \frac{1 + m + p}{\pi} \cos^{-1}(g_1 g_2)^{1/2}\right] \tag{6.5.4}$$

Since

$$g_1 = 1 - \frac{d}{R_1} = 0 = g_2 = 1 - \frac{d}{R_2} \qquad \therefore \cos^{-1}(g_1 g_2)^{1/2} = \frac{\pi}{2}$$

and

$$\nu = \frac{c}{2d}\left(q + \frac{1}{2}\right)$$

and thus we arrive at precisely the same result for the frequency.

For finite-sized mirrors, some of the field from M_1 "misses" M_2 and represents a power loss. A numerical evaluation of (13.3.20) leads to the fact that $|\sigma_{x,y}| < 1$. Fox and Li also evaluated this loss, and their results are plotted in Fig. 13.15. Note that the loss is very small even for a low Fresnel number. Fig. 13.16 shows the loss (per pass) as a function of the $g_{1,2}$ parameters going from a stable cavity to the unstable version covered earlier, both treated with the integral equation approach. The dashed curves for the unstable cavity represent the loss predicted by the simple geometric optics approach of Sec. 13.3.

Sec. 13.3   Integral Equation Approach to Cavities

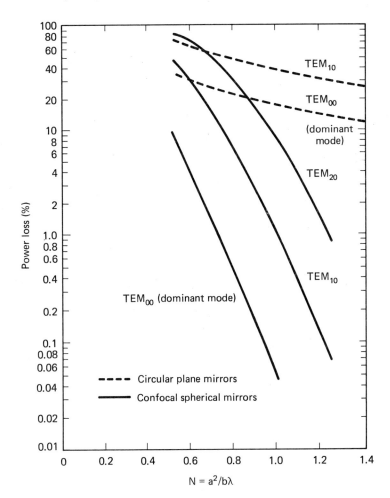

Figure 13.15   Power loss per transit vs. $N = a^2/b\lambda$ for confocal spherical mirrors. (Dashed curves for circular plane mirrors are shown for comparison.) (From Fox and Li.[5])

The integral equation formulation for optical cavities is a very powerful tool. It obviously requires considerably more mathematics than does the simple Gaussian beam optics of earlier chapters, but as the above calculation indicates, the answers are identical, as they should be, for cases that can be analyzed by either approach. This should inspire confidence in both approaches. Unfortunately, the integral equation approach most often requires a computer solution similar to that used by Fox and Li. Only in contrived cases, such as the confocal geometry, can an analytic solution be obtained. But let us face it, computer time is cheap, and thus we have a quite general approach to solving for the modes in *any* cavity.

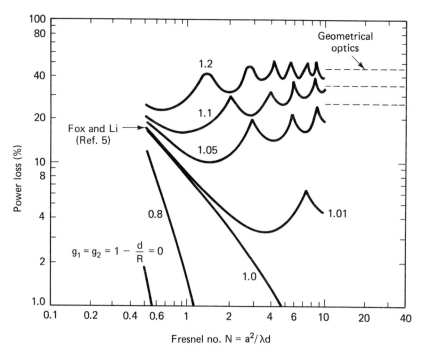

Figure 13.16 Loss per bounce vs. Fresnel number for stable and unstable resonators. (From Siegman.[1])

13.4 SEMICONDUCTOR CAVITIES

The cavities used for semiconductor lasers are characterized by transverse dimensions comparable to the wavelength being amplified, and hence a waveguide approach is most often used in the analysis. The vertical geometry of a typical stripe laser is shown in Fig. 13.17. The injected current is usually confined to a "stripe" whose width is much larger than the depth over which the recombination of the carriers takes place.

As mentioned in Chapter 11, one can substitute Al for Ga in the GaAs lattice, which raises the band gap and simultaneously lowers the index of refraction but keeps the crystal structure more or less identical. For various choices of x in $Al_xGa_{1-x}As$ and various layer thicknesses, one can have from 0 to 4 heterojunctions for the simultaneous confinement of the injected carriers and the electromagnetic wave. Most of the essential features of the wave guidance can be obtained by analyzing the simpler case of a three region waveguide, named 1, 3, and 5 for historical reasons, shown in Fig. 13.17. We search for fields that propagate in the $\pm z$ direction as $\exp\pm(\gamma z)$ and obey the wave equa-

Sec. 13.4 Semiconductor Cavities   475

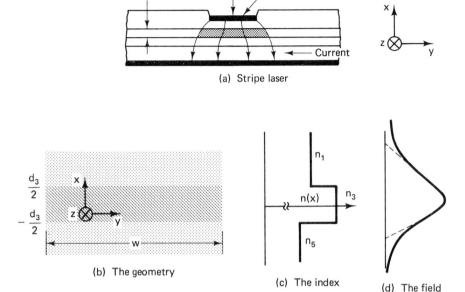

Figure 13.17 (a) The geometry of a stripe laser. (b) The asymmetric slab waveguide representative of many semiconductor lasers. (c) A sketch of the mode in the slab.

tion. There are two general classes of such fields, TE or TM, which are of primary interest:

$$\nabla_t^2 E_z + \left[\gamma^2 + \frac{\omega^2}{c^2} n^2(x,y)\right] E_z = 0 \quad (H_z = 0 \text{ or TM}) \quad (13.4.1a)$$

$$\nabla_t^2 H_z + \left[\gamma^2 + \frac{\omega^2}{c^2} n^2(x,y)\right] H_z = 0 \quad (E_z = 0 \text{ or TE}) \quad (13.4.1b)$$

where

$$\nabla_t^2 = \frac{\partial^2}{\partial x^2} + \frac{\partial^2}{\partial y^2}$$

The index of refraction can be continuous functions of x and y as implied by (13.4.1) or discontinuous as implied by Fig. 13.17.

Both types of modes obey the same type of scalar wave equation, and thus some general conclusions can be obtained by examining the characteristics of the solution for the various regions. We, of course, are primarily interested in the case $\gamma = j\beta$, that is, wave motion in the z direction with size of the mode being bounded in the transverse directions. If we neglect the variation of the fields in the y direction for a moment, then (13.4.1a) and (13.4.1b) can be expressed as

$$\frac{\partial^2 \Psi}{\partial x^2} + \left[\frac{\omega^2}{c^2}n^2(x,y) - \beta^2\right]\Psi = 0 \qquad (13.4.2)$$

This equation states a very important point about either mode if the desired goal of obtaining a *guided* field distribution of such lasers is remembered. We prefer a high peak transverse electric field in the immediate vicinity of the recombination region d_3 and then dropping to zero for $|x| \to \infty$. This characteristic is achieved if $\beta^2 < (\omega n_3/c)^2$ in the central region, which results in a standing wave or trigonometric type of solution there, and $\beta^2 > (\omega n_{1,5}/c)^2$ for the exterior regions, which leads to the exponential tails on the modes. Thus the phase constant β divided by $\omega/c = k_0$ lies between two extremes for these desired trapped (i.e., nonradiating) modes.

$$n_{1,5} < \frac{\beta}{k_0} < n_3 \qquad (13.4.3)$$

Values of β/k_0 less than n_1 or n_5 (or both) lead to waves propagating or radiating in the x direction and thus are not modes guided along z and suitable for the laser.

Although the presence or absence of a z component of an electric or magnetic field characterizes the mode as TM or TE, respectively, most prefer to work with the transverse components E_x or E_y (or H_y, H_x), which are more easily visualized in terms of intensity. The relationships between the transverse fields and the z component were given earlier (4.4.1) and are repeated below using γ instead of $j\beta$ for the propagation constant and $k = \omega n/c$.

$$\mathbf{E}_t = -\frac{1}{\gamma^2 + k^2}[\gamma \nabla_t E_z - j\omega\mu\, \mathbf{a}_z \times \nabla_t H_z] \qquad (13.4.4a)$$

$$\mathbf{H}_t = \frac{1}{\gamma^2 + k^2}[-j\omega\varepsilon_0 n^2 \mathbf{a}_z \times \nabla_t E_z - \gamma \nabla_t H_z] \qquad (13.4.4b)$$

The following analysis closely parallels that of a symmetric slab of Chapter 4, but now we face up to the fact that seldom is $n_1 = n_5$, which makes the arithmetic somewhat tedious. However, the importance of such lasers dictates that the effort be expended to extract all the information possible for the asymmetric slab typical of the lasers.

13.4.1 TE Modes

For fields uniform in the y direction in Fig. 13.17 ($\partial/\partial y = 0$), the component E_y satisfies

$$\frac{\partial^2 E_y}{\partial x^2} + \left(\gamma^2 + \frac{\omega^2}{c^2}n^2_{1,3,5}\right)E_y = 0 \qquad (13.4.5)$$

Sec. 13.4 Semiconductor Cavities

in each of the three regions. We anticipate a wave-guided z along with $\gamma = j\beta$ and its amplitude vanishing at $|x| \to \infty$. Thus

$$E_y^{(1)} = A_1 \exp\left[-h_1\left(x - \frac{d_3}{2}\right)\right] \quad (13.4.6a)$$

$$E_y^{(3)} = A_3 \cos h_3 x + B_3 \sin h_3 x \quad (13.4.6b)$$

$$E_y^{(5)} = A_5 \exp\left[+h_5\left(x + \frac{d_3}{2}\right)\right] \quad (13.4.6c)$$

where

$$h_1^2 = \beta^2 - (\omega n_1/c)^2 \quad (13.4.7a)$$

$$h_3^2 = (\omega n_3/c)^2 - \beta^2 \quad (13.4.7b)$$

$$h_5^2 = \beta^2 - (\omega n_5/c)^2 \quad (13.4.7c)$$

If β/k_0 satisfies the inequality of (13.4.3), then the right-hand sides of (13.4.7a) to (13.4.7c) are positive real.

Equations (13.4.6a) to (13.4.6c) can be rewritten to emphasize the continuity of the fields at the boundaries $x = \pm d_3/2$.

$$E_y^{(1)} = A\left\{\cos\left(\frac{h_3 d_3}{2} - \phi\right) \exp\left[-h_1\left(x - \frac{d_3}{2}\right)\right]\right\} \quad (13.4.8a)$$

$$E_y^{(3)} = A[\cos(h_3 x - \phi)] \quad (13.4.8b)$$

$$E_y^{(5)} = A\left\{\cos\left(\frac{h_3 d_3}{2} + \phi\right) \exp\left[+h_1\left(x + \frac{d_3}{2}\right)\right]\right\} \quad (13.4.8c)$$

This form also emphasizes the fact that the field need not be "centered" at $x = 0$ either as a symmetric or an asymmetric function. A sketch of the E_y field as a function of x is shown in Fig. 13.18.

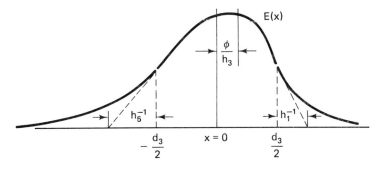

Figure 13.18 A sketch of the dominant TE_0 mode.

Although the form of (13.4.8) guarantees the continuity of E_y at $x = \pm d_3/2$, one must also match the tangential components of $H_z^{(1,3,5)}$ along these same planes. Combining (13.4.4b) and (13.4.1b) yields

$$H_z = -\frac{1}{j\omega\mu_0} \frac{\partial E_y}{\partial x} \qquad (13.4.9)$$

Thus

$$H_z^{(1)} = \frac{h_1}{j\omega\mu_0}\left\{A\cos\left(\frac{h_3 d_3}{2} - \phi\right)\exp\left[-h_1\left(x - \frac{d_3}{2}\right)\right]\right\} \qquad (13.4.10a)$$

$$H_z^{(3)} = \frac{h_3}{j\omega\mu_0}[A\sin(h_3 x - \phi)] \qquad (13.4.10b)$$

$$H_z^{(5)} = -\frac{h_5}{j\omega\mu_0}\left\{A\cos\left(\frac{h_3 d_3}{2} - \phi\right)\exp\left[+h_5\left(x + \frac{d_3}{2}\right)\right]\right\} \qquad (13.4.10c)$$

The continuity of H_z at the boundaries leads to

$$\left(\frac{h_3 d_3}{2} - \phi\right) = \tan^{-1}\left(\frac{h_1}{h_3}\right) = A \qquad (13.4.11a)$$

$$\left(\frac{h_3 d_3}{2} + \phi\right) = \tan^{-1}\left(\frac{h_5}{h_3}\right) = B \qquad (13.4.11b)$$

Thus the propagation constant γ, which is hidden in the h parameters (13.4.7), determined implicitly by the sum of (13.4.11a) and (13.4.11b) or

$$\tan(h_3 d) = \frac{\tan A + \tan B}{1 - \tan A \tan B} = \frac{h_3(h_1 + h_5)}{h_3^2 - h_1 h_5} \qquad (13.4.12)$$

The difference between (13.4.11b) and (13.4.11a) yields an expression for ϕ:

$$\tan 2\phi = \frac{h_3(h_5 - h_1)}{h_3^2 + h_1 h_5} \qquad (13.4.13)$$

13.4.2 TM Modes

The equation for H_x of the TM modes is identical to (13.4.5), and thus its solution has the same format as (13.4.6a) through (13.4.6c) with the same definitions of $h_{1,2,3}$ as given by (13.4.7a) through (13.4.7c). However, different multiplication factors appear in the secular equation (13.4.12) and for ϕ in (13.4.13) because of the different indices of refraction. Using (13.4.4):

$$E_z = \frac{1}{j\omega\varepsilon_0 n^2} \frac{\partial H_y}{\partial x} \qquad (13.4.14)$$

and matching H_y and E_z at $x = \pm d_3/2$, one obtains

$$\tan(h_3 d_3) = \frac{\dfrac{h_3}{n_3^2}\left[\dfrac{h_1}{n_1^2} + \dfrac{h_5}{n_5^2}\right]}{\left(\dfrac{h_3}{n_3^2}\right)^2 - \left(\dfrac{h_1}{n_1^2}\right)\left(\dfrac{h_5}{n_5^2}\right)} \tag{13.4.15}$$

and

$$\tan 2\phi = \frac{\dfrac{h_3}{n_3^2}\left[\dfrac{h_5}{n_5^2} - \dfrac{h_1}{n_1^2}\right]}{\left(\dfrac{h_3}{n_3^2}\right)^2 + \left(\dfrac{h_1}{n_1^2}\right)\left(\dfrac{h_5}{n_5^2}\right)} \tag{13.4.16}$$

13.4.3 Physical Comparison of TE and TM Modes

Although there is a difference in the propagation constant and ϕ between the two configurations, that is a numerical problem rather than a physical phenomenon that is easily detected with a minimum of sophisticated instruments. Both modes are superpositions of plane waves internally reflected at the boundaries at $x = \pm d_3/2$. The distinguishing, easily detected, physical feature is the difference in polarization of the fields as shown in Fig. 13.19. The optical electric field is in the y direction for the TE modes and lies in the xz plane for the TM orientation. Most semiconductor lasers oscillate in the TE modes because the reflectivity of a cleaved facet for that orientation is higher than for the TM case. To compute this reflectivity is a rather formidable task because the radiation field is composed of a distribution of plane waves to match the internal mode with the radiating field at the semiconductor facet. That is a task reserved for more advanced texts. Many use the standard Fresnel formula in conjunction with Fig. 13.19 to estimate the reflectivity. If the angle θ there were $0°$, then the standard transmission line formula applies:

$$R = \left[\frac{n_3 - 1}{n_3 + 1}\right]^2 \approx 0.32 \text{ for } n_3 = 3.6 \text{ (TE and TM modes)}$$

However, Fig. 13.20 suggests that there is the possibility of Brewster's angle occurring for the TM orientation but not for the TE case. At Brewster's angle ($\theta \cong 16°$ in Fig. 13.19), the reflectivity would be zero, if the waves were uniform plane waves. Because the fields are *not* uniform plane waves, zero reflectivity does not occur. But this line of reasoning does indicate that the TM reflectivity and thus the feedback are always less than those of the TE modes. This is borne out by detailed calculations and is shown in Fig. 13.20.

480 Advanced Topics in Electromagnetics of Lasers Chap. 13

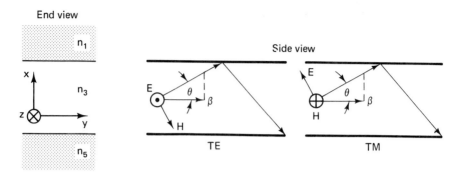

Figure 13.19 A comparison between TE and TM slab waveguide modes showing the difference in polarization.

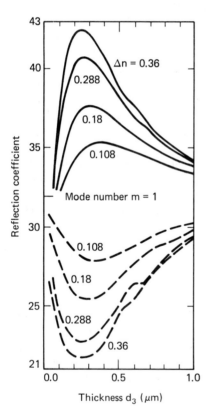

Figure 13.20 The facet reflection coefficient for the TE (solid) and TM (dashed) modes. (From Ikegami.[20])

13.5 GAIN GUIDING: AN EXAMPLE

The previous section is an example of the use of a spatial variation of the index of refraction to create a waveguide to confine the wave that is propagating in the z direction. The index can have discontinuous jumps as was considered previously (cf Fig. 13.17) or be a continuous variable, as was considered in Chapter 4 for fibers. For most heterostructure lasers, there is always an index variation in the direction perpendicular to the plane of the junction. Spatial variation of the gain along the plane of the junction can also provide guidance of a mode, and this is the problem to be addressed here. The practical situation is sketched in Fig. 13.21 for a heterostructure laser.

To make the following arithmetic tolerable, we assume the fractional concentrations of aluminum of the confining p and n layers are chosen to generate a smooth variation of dielectric constant in the x direction as sketched

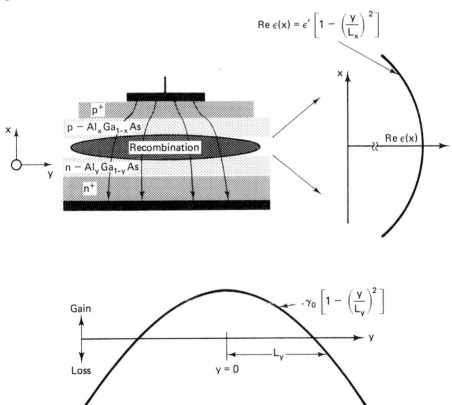

Figure 13.21 The assumed variation of the dielectric constant and gain in a heterostructure laser.

in Fig. 13.21(b). Alternatively, we can interpret the parabolic variation as a simple continuous approximation to the discontinuous jumps considered in the previous section, and thus L_x is an adjustable fitting parameter. Obviously, the extent of the field in x must be much less than L_x to avoid the absurdity of $\epsilon(x)$ becoming negative.

The gain coefficient is due to the recombination of the carriers, as discussed in Chapter 11, and it will reflect the local current density at the plane of the junction. Because of the small width of the recombination region, we neglect variations of the gain coefficient with x but approximate its spatial variation with y by

$$\gamma(y) = \gamma_0\left(1 - \frac{y^2}{L_y^2}\right) \qquad (13.5.1)$$

Here we acknowledge the idea that the gain coefficient can be negative (i.e., absorption) for sufficiently large values of y greater than L_y. We can anticipate that both γ_0 and L_y are functions of the current through the diode.

Thus the electromagnetic problem consists of finding solutions to the wave equation with a complex inhomogeneous dielectric constant of the form:

$$\epsilon(x, y) = \epsilon_0\left[\epsilon'\left(1 - \frac{x^2}{L_x^2}\right) + j\epsilon''\left(1 - \frac{y^2}{L_y^2}\right)\right] \qquad (13.5.2)$$

Hence the wave equation becomes

$$\nabla^2 \begin{Bmatrix} E \\ H \end{Bmatrix} + \frac{\omega^2}{c^2}\left[\epsilon'\left(1 - \frac{x^2}{L_x^2}\right) + j\epsilon''\left(1 - \frac{y^2}{L_y^2}\right)\right]\begin{Bmatrix} E \\ H \end{Bmatrix} = 0 \qquad (13.5.3a)$$

If the medium were uniform and of infinite extent, and *If* we were dealing with uniform plane waves propagating as $\exp[(\gamma_0/2) - j\beta)]z$, then one would relate the field gain coefficient $(\gamma_0/2)$ and phase constant β with the dielectric parameters (ϵ', ϵ'') according to

$$\beta^2 = \frac{\omega^2}{c^2}\epsilon' = \frac{\omega^2}{c^2}n_r^2 \qquad (13.5.4a)$$

$$\gamma_0\beta = \frac{\omega^2}{c^2}\epsilon'' \qquad (13.5.4b)$$

Hence it is appropriate to rewrite (13.5.3a) with those abbreviations:

$$\nabla^2\begin{Bmatrix} E \\ H \end{Bmatrix} + \left[\beta^2\left(1 - \frac{x^2}{L_x^2}\right) + j\gamma_0\beta\left(1 - \frac{y^2}{L_y^2}\right)\right]\begin{Bmatrix} E \\ H \end{Bmatrix} = 0 \qquad (13.5.3b)$$

Now we follow a procedure similar to that used in Chapter 3: assume that the fields vary as

$$(E,H) = \psi(x,y,z)e^{(\gamma_0/2 - j\beta)z} \qquad (13.5.5)$$

Sec. 13.5 Gain Guiding: An Example

where $\dfrac{\partial^2 \psi}{\partial z^2} \ll 2j\beta \dfrac{\partial \psi}{\partial z}$ as was argued in Chapter 3. The wave function ψ obeys the following equation with the additional assumption $|\gamma_0| \ll |\beta|$, which is always true except for pathological cases.

$$\nabla_t^2 \psi - 2j\beta \frac{\partial \psi}{\partial z} - \frac{\beta^2 x^2}{L_x^2}\psi - j\gamma_0 \beta \frac{y^2}{L_y^2}\psi = 0 \qquad (13.5.6)$$

One should not fall into the trap of thinking that (amplified) uniform plane waves have been assumed with a growth of $\exp(\gamma_0 z/2)$ and a phase change of $\exp(-j\beta z)$. Not so. Those terms are factors in the expression for the field but the wave function ψ depends on z, and thus the total modal growth and phase change remain to be determined.

Now the procedure becomes identical (rather than merely similar) to Chapter 3. Assume that ψ can be expressed as

$$\psi(x,y,z) = \left\{\exp\left[-j\left(P_y(z) + \frac{\beta x^2}{2q_x(z)}\right)\right]\right\}\left\{\exp\left[-j\left(P_y(z) + \frac{\beta y^2}{2q_y(z)}\right)\right]\right\} \qquad (13.5.7)$$

where this form emphasizes that the wave function is expressed as a separated product of two functions $\psi_x(x,z) \cdot \psi_y(y,z)$. We need not go through the standard steps of separating the variables, because the unknown functions, $P_x(z)$, $P_y(z)$, $q_x(z)$, and $q_y(z)$, are already identified and appear in the exponent. Equation (13.5.7) is substituted into (13.5.6), and terms involving x^2, x^0, y^2, and y^0 are grouped together.

$$\left\{\beta^2\left[\frac{q_x' - 1}{q_x^2} - \frac{1}{L_x^2}\right]x^2 - \beta\left[\frac{j}{q_x} + 2P_x'\right]x^0 \right.$$
$$\left. + \beta^2\left[\frac{q_y' - 1}{q_y^2} - j\frac{\gamma_0/\beta}{L_y^2}\right]y^2 - \beta\left[\frac{j}{q_y} + 2P_y'\right]y^0\right\} = 0 \qquad (13.5.8)$$

All factors of the powers of x or y must be separately equal to zero, which thus yields ordinary differential equations for the complex beam parameters $q_{x,y}$ and $P_{x,y}(z)$. However, we are interested in a guided mode, not freely expanding waves as were considered in Chapter 3. Hence, we also require that $q_x'(z) = q_y'(z) = 0$; that is, the complex beam parameter, q, should be independent of z. From the coefficient of x^2 in (13.4.13) (with $q_x' = 0$), we obtain the following:

$$q_x^2 = -L_x^2 \quad \text{or} \quad \frac{1}{q_x} = -j\frac{1}{L_x} \qquad (13.5.9a)$$

From the coefficient of x^0:

$$P_x'(z) = -j\frac{1}{2q_x}; \qquad -jP_x(z) = +j\frac{z}{2L_x} \qquad (13.5.9b)$$

From the coefficient of y^2 (with $q'_y = 0$):

$$\frac{1}{q_y^2} = -j\frac{\gamma_0}{\beta L_y^2} \quad \text{or} \quad \frac{1}{q_y} = \left(\frac{\gamma_0}{2\beta}\right)^{1/2}\frac{1}{L_y}(1 - j1) \quad (13.5.9c)$$

and from the coefficient of y^0:

$$P'_y(z) = \frac{-j}{2q_y}; \quad -jP_y(z) = -\left(\frac{\gamma_0}{2\beta}\right)^{1/2}\frac{z}{2L_y} + j\left(\frac{\gamma_0}{2\beta}\right)^{1/2}\frac{z}{2L_y} \quad (13.5.9d)$$

Now the physical interpretation of the complex beam parameter is the same as that assigned in Chapter 3; that is,

$$\frac{1}{q} \equiv \frac{1}{R} - j\frac{\lambda}{\pi w^2}$$

The significant changes are that we have demanded that the radius of curvature of the phase front, R, and the spot size, w, *be independent of z*, and both parameters be allowed to be different in the x and y direction. With this in mind, we can reassemble the field:

$$\frac{E(x\,y\,z)}{E_0} = \left\{\exp-\left(\frac{\beta x^2}{2L_x}\right)\right\}$$

$$\times \left\{\exp\left[-\frac{\beta y^2}{2L_y}\left(\frac{\gamma_0}{2\beta}\right)^{1/2}\right]\right\} \times \left\{\exp\left[-j\frac{\beta y^2}{2L_y}\left(\frac{\gamma_0}{2\beta}\right)^{1/2}\right]\right\}$$

$$\exp\left\{-j\left[\beta - \frac{1}{2L_x} - \frac{1}{2L_y}\left(\frac{\gamma_0}{2\beta}\right)^{1/2}\right]z\right\}$$

$$\times \left\{\exp\left[\frac{\gamma_0}{2} - \left(\frac{\gamma_0}{2\beta}\right)^{1/2}\frac{1}{2L_y}\right]z\right\} \quad (13.5.10)$$

There are various parts of this equation that are easily discernible:

1. The first line of (13.5.9) obviously describes a Gaussian beam variation in the x direction with a spot size w_x^2 given by

$$w_x^2 = \frac{2L_x}{\beta} \quad (13.5.11a)$$

with the wave front being planar (i.e., $R_x = \infty$) along x.

2. The first factor of the second line of (13.5.10) describes a Gaussian beam with a spot size given by

$$\frac{y^2}{w_y^2} \equiv \frac{\beta y^2}{2L_y}\left(\frac{\gamma_0}{2\beta}\right)^{1/2} \quad \text{or} \quad w_y^2 = \frac{2\sqrt{2}\,L_y}{(\gamma_0\beta)^{1/2}} \quad (13.5.11b)$$

Sec. 13.5 Gain Guiding: An Example

The quadratic variation of the phase with y, the second factor of line 2 of (13.5.10), indicates that the wave front is curved with the radius of curvature given by

$$\exp\left(-j\frac{\beta y^2}{2R}\right) = \exp\left[-j\frac{\beta y^2}{2L_y}\left(\frac{\gamma_0}{2\beta}\right)^{1/2}\right]$$

or (13.5.11c)

$$R = L_y\left(\frac{2\beta}{\gamma_0}\right)^{1/2}$$

3. The third line merely indicates that the wave propagates with a modal phase constant given by

$$\beta_{\text{modal}} = \beta - \frac{1}{2L_x} - \left(\frac{\gamma_0}{2\beta}\right)^{1/2}\frac{1}{2L_y} \quad (13.5.11d)$$

with β given by (13.5.4a). The factors "correcting" β are the result of the penetration of the field into regions of lower dielectric constant (for x variation) and into regions of lower gain—even loss—for the y variation.

4. The net *power* gain coefficient for this *mode* is also reduced from the peak value of γ_0 because of the penetration of the field into the loss region.

$$\gamma_{\text{modal}} = \gamma_0 - \left(\frac{\gamma_0}{2\beta}\right)^{1/2}\frac{1}{L_y} \quad (13.5.11e)$$

Example

Let us choose some typical numerical values so as to appreciate the physical significance of the previous development.

Let $\lambda_0 = 8400$ Å, $n = 3.6$; $\gamma_0 = 100$ cm^{-1}; $L_y = 5 \times 10^{-4}$ cm; $L_x = 10^{-3}$ cm

$$\therefore \beta = \frac{2\pi n}{\lambda_0} = 2.69 \times 10^5 \text{ rad/cm}$$

Hence $w_x = 0.862$ μm by (13.5.11a). The real index of refraction decreases from 3.6 to $3.6(1 - 0.0074) = 3.573$ in this "spot-size" distance. This calculation points out that it does not take much of a change in the index to guide the wave. The field is described by a Gaussian with a "spot size" in the y direction, $w_y = 5.22$ μm, from (13.5.11b).

The gain coefficient defined by (13.4.16) changes quite significantly over this spot size, decreasing from a peak value of 100 cm^{-1} at $y = 0$ to become *negative* (i.e., absorptive) at $y = w_y$. Because of the field penetrating into the absorptive regions, the effective (or modal) gain coefficient is given by (13.5.11e) to be $\gamma_{\text{modal}} = 72.75$ cm^{-1}. The situation is sketched in Fig. 13.22(a) where the field is plotted directly below the assumed spatial variation of the gain coefficient. The positive value of γ_{modal} reflects the fact that the peak field lies near $y = 0$ where the gain is largest. The tails of the field

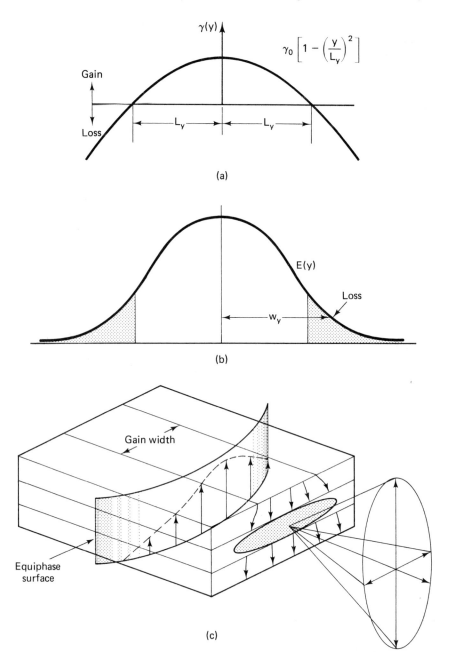

Figure 13.22 (a) The variation of the gain along the plane of the junction and the resulting electric field of the mode. (b) The equiphase surface for a gain-guided laser.

Sec. 13.6   Distributed Feedback

penetrate into the lossy region and thus account for the reduction of the modal gain coefficient.

The radius of curvature of the phase front is given by (13.5.11c) to be $R = 366.9$ μm. The fact that the radius is not infinity indicates that the wavefront is *curved* in the y direction (but not in the x direction). In other words, the beam is *astigmatic*. This is sketched in Fig. 13.22. In order to evaluate the far-field radiation pattern from such lasers, one must account for both the amplitude and phase variation over the output facet.

The astigmatism can be observed to us in the outside world by the radiation pattern. The Gaussian beam in the x direction has the $z = 0$ plane located at the output facet, and therefore the standard laws from Chapter 3 for the expansion into free space are applicable. However, the wave front is *curved* in the y direction as it arrives at the output facet and thus some additional work is involved in computing the free space radiation. One needs to compute the y variation of the complex beam parameter on the *air* side of the output facet mirror by using the *ABCD* matrix for a dielectric-air interface for that purpose. The spot size w_y will be the same in the air as in the laser; however, the radius of curvature will be different. This makes the beam appear to originate from a plane behind the output mirror, which in turn increases the beam spread in the y direction. The details are left for a problem (naturally).

13.6 DISTRIBUTED FEEDBACK

In most of our work, we have made the tacit assumption of simple mirrors for feedback elements. While this is prevalent, there are good reasons for distributing the feedback, although we will have to wait to appreciate them. As a start, let us convince ourselves that it is an interesting problem.

Consider the system shown in Fig. 13.23, in which a strong field is propagating to the right in a medium composed of periodic small reflections. To the zeroth-order approximation, this strong signal propagates more or less as if the medium were uniform and ignores the small reflections. At each of the planes a, b, c, d, ..., the wave propagating in the positive z direction generates a "small" wave propagating back to the left with the total "reflected" field being composed of all the individual reflections. The problem with this zeroth-order picture is that the total field propagating in the negative direction does not remain small if the phase delay between the individual small wave packets is chosen correctly. For instance, the wave generated at the plane b lags the corresponding field generated at place a by $k\Lambda$ radians. When the b wave propagates back to position a, it experiences another $k\Lambda$ phase delay, yielding a total phase delay of b with respect to a (at the plane a) of $k(2\Lambda)$. Thus the total field propagating to the left in Fig. 13.23 is

$$E^-(z=a) = \underbrace{\Delta\rho E_0}_{a} + \underbrace{\Delta\rho E_0 e^{-jk2\Lambda}}_{b} + \underbrace{\Delta\rho E_0 e^{-jk4\Lambda}}_{c} + \underbrace{\Delta\rho E_0 e^{-jk6\Lambda}}_{d} + \cdots$$

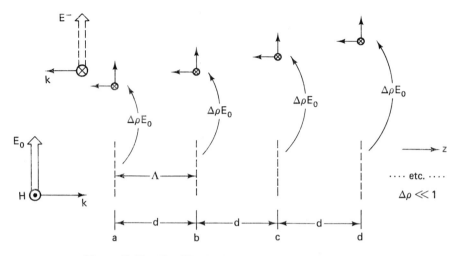

Figure 13.23 "Small" reflections from a periodic discontinuity.

If $2k\Lambda$ is an integral number of 2π radians, the individual small wavelets add and, with N planes of reflections, E^- is $N\Delta\rho E_0$. For N large, this reverse wave could be comparable to the incident one. Obviously, we are "manufacturing" energy in a completely passive system, a neat (but impossible) trick—a consequence of our simplifying zeroth-order approximation. Thus we must resort to a more precise analysis.

However, this simple picture does provide us with a motivation for proceeding. Note that this feedback is frequency selective, since only those frequencies satisfying the condition

$$k(2\Lambda) = \frac{\omega}{c} n(2\Lambda) = m2\pi$$

or (13.6.1)

$$\Lambda = m\frac{\lambda}{2} \quad \text{with} \quad \lambda = \frac{\lambda_0}{n} \quad (n = \text{index of refraction})$$

generate components which add in-phase, and thus result in a wave of significant amplitude. The condition expressed by (13.6.1) is the same as found for Bragg scattering of electrons from the periodic arrangement of atoms in a crystal. The integer, m, corresponds to the various orders; in what follows, we will pick the simplest case of $m = 1$.

Such a feedback system can be realized in a semiconductor by etching a periodic structure along the axis of the diode and then regrowing a different material on top of the array. Fig. 13.24(a) shows an idealized version of that published by Nakomura et al.[22] It is a fact of life that the index of refraction goes

Sec. 13.6   Distributed Feedback

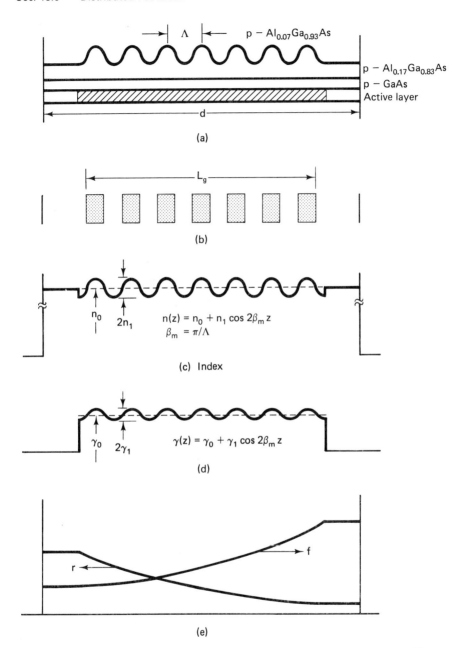

Figure 13.24 (a) Geometry for a distributed feedback semiconductor laser (after Yariv[22]); (b), (c), and (d) are sketches of the material density, index, and inversion with z; and (e) is a first-order guess as to the variation of the *f*orward and *r*everse waves inside the cavity.

down as the aluminum fraction goes up, and thus the guiding properties and the effective width of the guide are both periodic functions of z. Now the phase constant of the guided modes is found from the matching of the fields at these transverse boundaries (cf Sec. 13.5), and thus the effective dielectric constant is also a periodic function of z as shown in Fig. 13.24(c). One should also expect the gain coefficient to have a similar variation as shown in Fig. 13.24(d).

Let us focus our attention on the wave equation describing the variation of the field with respect to z and hide our ignorance of the details of the field matching in the transverse plane by using the periodic representation of n and γ shown in Fig. 13.24. We also travel the same logic path relating the gain coefficient and the index to the complex dielectric constant as was used in the previous section (13.5.4) to obtain

$$\frac{d^2E}{dz^2} + \left[\frac{\omega^2}{c^2}(n_0 + n_1 \cos 2\beta_m z)^2 + j\frac{\omega}{c}n_0(\gamma_0 + \gamma_1 \cos 2\beta_m z)\right]E = 0$$

(13.6.2)

Let us abbreviate in the same manner as before:

$$\beta_0 = \frac{\omega}{c}n_0 \qquad (13.6.3a)$$

$$g_0 = \frac{\gamma_0}{2} \qquad (13.6.3b)$$

$$g_1 = \frac{\gamma_1}{2} \qquad (13.6.3c)$$

$$\kappa = \frac{\omega}{c}\frac{n_1}{2} + j\frac{g_1}{2} \qquad (13.6.3d)$$

where the use of a *field* amplification factor g_0 is used rather than $\gamma_0/2$ to avoid endless factors of 2. After assuming "small" amplitudes of perturbation from a uniform medium so that $n_1 \ll n_0$, $g_0 \ll \beta$, $g_1 \ll g_0$, one obtains

$$\frac{d^2E}{dz^2} + [\beta_0^2 + j2g_0\beta_0 + 4\beta_0\kappa \cos 2\beta_m z]E = 0 \qquad (13.6.4)$$

A few comments about the abbreviations are in order. The quantity β_0 would be the phase constant of the wave if everything were uniform ($\kappa = 0$). The quantity $\beta_m = \pi/\Lambda$ is a measure of the *mechanical* change in the material properties—it is not an electrical phase shift of anything—it just has the same dimensions as β_o.

The solution to all wave equations can be represented by a superposition of waves propagating in the two directions—most of the time we consider them one at a time. Here, however, a wave going in the $+z$ direction generates one

Sec. 13.6 Distributed Feedback

going in the reverse $(-z)$ and conversely, as was suggested by our zeroth-order analysis in Fig. 13.23. Thus we look for a combination of *forward* and *reverse* waves as a solution of (13.6.4).

$$E(z) = f(z)e^{-j\beta_m z} + r(z)e^{j\beta_m z} \tag{13.6.5}$$

Here we are attempting to play somewhat the same game as was used for Gaussian beams in Chapter 3 and in the previous section. The exponential factors of (13.6.5) represent a first-order guess of a plane wave with a phase constant synchronous with the mechanical periodicity, but with the functions $f(z)$ and $r(z)$ "correcting" that guess. We substitute this assumed format into (13.6.4), use the exponential form for the cosine function, and keep track of terms with common exponential factors.

$$\frac{d^2 E}{dz^2} = [-\beta_m^2 f - 2j\beta_m f' + f'']e^{-j\beta_m z} + [-\beta_m^2 r + 2j\beta_m r' + r'']e^{j\beta_m z}$$

$$[\ ]E = [\beta_0^2 + 2jg_0\beta_0]f e^{-j\beta_m z} + [\beta_0^2 + 2jg_0\beta_0]r e^{j\beta_m z}$$

$$+ 4\beta_0 \kappa \left[\frac{(e^{j2\beta_m z} + e^{-j2\beta_m z})e^{-j\beta_m z}}{2}\right]f \tag{13.6.6}$$

$$+ 4\beta_0 \kappa \left[\frac{(e^{j2\beta_m z} + e^{-j2\beta_m z})e^{j\beta_m z}}{2}\right]r$$

So far we are exact; now comes the first set of approximations. We assume that the second derivative terms are small and neglect terms that vary rapidly in space such as $\exp \pm j3\beta_m z$. Setting each factor multiplying $\exp \pm j\beta_m z$ equal to zero yields

$$\{[(\beta_0^2 - \beta_m^2) + 2jg_0\beta_0]f - 2j\beta_m f' + 2\beta_0 \kappa r\}e^{-j\beta_m z} = 0 \tag{13.6.7a}$$

$$\{[(\beta_0^2 - \beta_m^2) + 2jg_0\beta_0]r + 2j\beta_m r' + 2\beta_0 \kappa f\}e^{j\beta_m z} = 0 \tag{13.6.7b}$$

A further approximation is in order: we need not distinguish between β_0 and β_m unless the difference occurs. Hence, let $\beta_0 = \beta_m + \delta$ with δ being the phase shift per unit of length in *excess* of the mechanical value $\beta_m = \pi/\Lambda$.

$$\beta_0^2 - \beta_m^2 = (\beta_0 + \beta_m)(\beta_0 - \beta_m) \cong 2\beta_m \delta$$

Equation (13.6.7) reduces to a more manageable form:

$$f'(z) - (g_0 - j\delta)f(z) = -j\kappa r(z) \tag{13.6.8a}$$

$$r'(z) + (g_0 - j\delta)r(z) = j\kappa f(z) \tag{13.6.8b}$$

Let us stop here for a moment and ensure that we have not lost the physics in the cloud of mathematics. If the "ripples" of Fig. 13.24 were zero, then the forward and reverse waves are decoupled and $\kappa = 0$. Then (13.6.8a) is easily solved: $f^0 = A \exp[(g_0 - j\delta)z]$. This is exactly what one would expect with the wave growing as $\exp(g_0 z)$ and experiencing a phase shift given by $\exp[-j(\beta_m + \delta)z]$. Thus we can proceed with some confidence.

Now we combine (13.6.8a) with (13.6.8b) to obtain uncoupled equations for f and r:

$$\begin{Bmatrix} f'' \\ r'' \end{Bmatrix} - p^2 \begin{Bmatrix} f \\ r \end{Bmatrix} = 0 \tag{13.6.9a}$$

where

$$p^2 = \kappa^2 + (g_0 - j\delta)^2 \tag{13.6.9b}$$

Now before we rush to solve these, we should stare intently at Fig. 13.24 so as to anticipate the solution. If there were *no* coupling between the waves (i.e., $\kappa = 0$), then one would expect the forward wave f to propagate in the active medium as $\exp + p(z + L/2)$ and start with an amplitude determined by the reflection of the reverse wave at mirror M_1. Similarly, the reverse wave, r, should propagate as $\exp - p(z - L/2)$ and start with an amplitude determined by the reflection at M_2. Thus it is appropriate to pick a solution that expresses those ideas even in the case where $\kappa \neq 0$. Inasmuch as the forward wave is being "generated" by the reverse wave, its mathematical description must include some of the characteristics of a wave propagating in the negative z direction.

$$f(z) = A e^{p(z+L/2)} + B e^{-p(z-L/2)} \tag{13.6.10a}$$

Equation (13.6.8a) is used to generate a corresponding equation for $r(z)$:

$$r(z) = \frac{f' - (g_0 - j\delta)f}{-j\kappa} \tag{13.6.8a}$$

or

$$r(z) = \frac{[p - (g_0 - j\delta)]}{-j\kappa} A e^{p(z+L/2)} + \frac{[p + (g_0 - j\delta)]}{j\kappa} B e^{-p(z-L/2)} \tag{13.6.10b}$$

Now the fields are forced to be self-consistent by applying the boundary conditions at the mirrors M_1 and M_2:

$$E^+(z = -d/2) = \sqrt{R_1}\, E^-(z = -d/2) \tag{13.6.11a}$$

$$E^-(z = +d/2) = \sqrt{R_2}\, E^+(z = +d/2) \tag{13.6.11b}$$

Now the fields merely propagate in the regions outside the distributed feedback, and this just represents a phase delay of $2\beta_0(d/2 - L/2)$ for the incident wave going from $L/2$ to $d/2$, another equal phase delay for the wave coming back to $L/2$. This is easily taken into account by replacing the field reflection coefficients at the mirrors by a transformed value at $z = \pm L/2$.

$$f(z = -L/2) = \Gamma_1 r(z = -L/2) \tag{13.6.12a}$$

$$r(z = +L/2) = \Gamma_2 f(z = +L/2) \tag{13.6.12b}$$

Sec. 13.6 Distributed Feedback

with*
$$\Gamma_{1,2} = \sqrt{R_{1,2}} e^{-j[\beta_0(d-L)+\beta_m L]} \quad (13.6.12c)$$

The extra phase factor, $\exp - j\beta_m L$, arises because we have removed a phase factor of $\exp -j\beta_m z$ in going from the *fields* to the functions $f(z)$ and $r(z)$; compare with (13.5.5).

Applying (13.6.12) yields two equations for the constants A and B:

$$A\left\{1 + \Gamma_1 \frac{[p - (g_0 - j\delta)]}{j\kappa}\right\} \\ + Be^{pL}\left\{1 - \Gamma_1 \frac{[p + (g_0 - j\delta)]}{j\kappa}\right\} = 0 \quad (13.6.13a)$$

$$Ae^{pL}\left\{\Gamma_2 + \frac{[p - (g_0 - j\delta)]}{j\kappa}\right\} \\ + B\left\{\Gamma_2 - \frac{[p + (g_0 - j\delta)]}{j\kappa}\right\} = 0 \quad (13.6.13b)$$

In order to avoid a trivial solution, $A = B = 0$, the determinant of the coefficients must be zero. This leads to a rather formidable equation:

$$e^{2pL} = \left\{\frac{\Gamma_1 + \Gamma_2 + \Gamma_1\Gamma_2\left[\dfrac{p - (g_0 - j\delta)}{j\kappa}\right] - \left[\dfrac{p + (g_0 - j\delta)}{j\kappa}\right]}{\Gamma_1 + \Gamma_2 + \dfrac{[p - (g_0 - j\delta)]}{j\kappa} - \Gamma_1\Gamma_2\dfrac{[p + (g_0 - j\delta)]}{j\kappa}}\right\} \quad (13.6.14)$$

This is even worse than it appears: the parameter p is complex; so also is κ and, of course, $\Gamma_{1,2}$. Let us make sure it represents something sensible before presenting its solution.

Example 1: Weak coupling ($\kappa \to 0$)

Assume κ is real, which means that the ripples in Fig. 13.24 are entirely the result of variations in the effective index of refraction. Assume also that κ is small compared to g_0. Then we obtain a simple expression for the complex parameter p.

$$p = p' - jp'' \approx (g_0 - j\delta) + \frac{\kappa^2}{2(g_0 - j\delta)} \quad (13.6.15a)$$

$$\therefore p - (g_0 - j\delta) \approx \frac{\kappa^2}{2(g_0 - j\delta)} \quad (13.6.15b)$$

$$p + (g_0 - j\delta) \approx 2(g_0 - j\delta) \quad (13.6.15c)$$

---

*Any resemblance to transmission line theory or to Smith chart manipulations should not be a surprise; in any case it is intentional. To convince yourself that the extra phase does belong, assume a uniform gain and dielectric medium between $\pm L/2$ and 0 otherwise. Choose a field A at some point and propagate it around the loop and demand self-consistency. The reflection coefficients for f and r, which are only part of the fields, are given by (13.6.12c).

Equation (13.6.14) becomes

$$e^{2pL} = \frac{-2(g_0 - j\delta) + j\kappa(\Gamma_1 + \Gamma_2) + \dfrac{\kappa^2 \Gamma_1 \Gamma_2}{(g_0 - j\delta)}}{-2\Gamma_1\Gamma_2(g_0 - j\delta) + j\kappa(\Gamma_1 + \Gamma_2) + \dfrac{\kappa^2}{(g_0 - j\delta)}} \qquad (13.6.16)$$

For $\kappa \to 0$, (13.6.16) assumes a very simple form.

$$e^{2pL} = e^{2(g_0 - j\delta)L} = \frac{+1}{\Gamma_1 \Gamma_2} \qquad (13.6.17a)$$

This should be very familiar to all, but let us make sure that we appreciate the familiarity. For this example, the quantity $p = (g_0 - j\delta)$; $2g_0 L = \gamma_0 L$ the integrated gain (for intensity); $\delta = (\beta_0 - \beta_m)$; the factor +1 (one) can be expressed as $\exp(-j2q\pi)$; and $\Gamma_{1,2}$ is given by (13.6.12c). Thus (13.6.17a) becomes

$$e^{\gamma_0 L} e^{-j2\delta L} = \frac{e^{-j2q\pi}}{\sqrt{R_1 R_2} e^{-2j(\beta_0 d - L + \beta_m L)}} \qquad (13.6.17b)$$

Equating magnitude and phase of (13.6.17b) yields

$$\gamma_0 L = \ln \frac{1}{\sqrt{R_1 R_2}} \quad \text{(for the amplitude part)} \qquad (13.6.18a)$$

and

$$2\delta L + 2[\beta_0(d - L) + \beta_m L] = q2\pi \quad \text{(for the phase part)}$$

inasmuch as $\delta = \beta_0 - \beta_m$; $\beta_0 = \omega n_0/c$; and $\omega = 2\pi\nu$, the phase part leads to

$$\nu = \frac{c}{2n_0 d} q \qquad (13.6.18b)$$

Equation (13.6.18a) is, of course, the threshold condition for a simple laser and (13.6.18b) is the frequency of oscillation, both in standard from.

• • •

Thus (13.6.14) contains both the threshold condition and the frequency of oscillation, both being buried in the fact that all of the parameters are complex.

Example 2: κ real and small compared to g_0 and $R_1 = R_2 = 0$

In this case, all of the feedback is provided by the "ripples" in the index of refraction and $\Gamma_1 = \Gamma_2 = 0$. The field factors f and r must build up from a zero amplitude at $\pm L/2$. Equation (13.6.14) becomes

$$e^{2pL} = -\frac{p + (g_0 - j\delta)}{p - (g_0 - j\delta)} \qquad (13.6.19)$$

Sec. 13.6   Distributed Feedback

Multiply by $p + (g_0 - j\delta)$ in the numerator and denominator, use the fact that $p^2 - (g_0 - j\delta)^2 = \kappa^2$, and take the square root with the following result:

$$e^{pL} = \frac{p + (g_0 - j\delta)}{\pm j\kappa}$$

Form the reciprocal of this equation, multiply numerator and denominator by $p - (g_0 - j\delta)$ identify κ^2 as before, and add to the above, so as to obtain an equation equivalent to (13.6.19).

$$e^{pL} - e^{-pL} = \frac{2p}{\pm j\kappa} \tag{13.6.20a}$$

If κ is small, the gain must be high to establish oscillation. Hence, the first exponential term is much larger than the second, and the self-consistent condition becomes

$$e^{pL} = \frac{2p}{\pm j\kappa} \tag{13.6.20b}$$

Thus the required gain to establish oscillation for a given detuning parameter δ is found from a (numerical) solution to

$$e^{p'L} = \frac{2[(p'L)^2 + (p''L)^2]^{1/2}}{\kappa L} \tag{13.6.21a}$$

with

$$(p' - jp'')^2 = \kappa^2 + (g_0 - j\delta)^2 \tag{13.6.21b}$$

The detuning δ, or frequency of oscillation, is found from the phase of (13.6.20b):

$$\exp[-j(p''L) + j(q + 1/2)\pi] = \exp[+j\tan^{-1}(p''L/p'L)]$$

or

$$(q + 1/2) = \frac{p''L}{\pi} + \frac{1}{\pi}\tan^{-1}\left(\frac{p''L}{p'L}\right) \tag{13.6.22}$$

where the factor j has been converted to $\exp(j\pi/2)$ and extra factors of ± 1 expressed as $\exp(jq\pi)$. Numerical methods must be used to solve these equations for the required gain and frequency, and the procedure used is as follows.

One expects a solution close to the Bragg frequency given by

$$2\frac{\omega_B}{c}n_0\Lambda = m \cdot 2\pi$$

or

$$\nu_B = m\left[\frac{c}{2n_0\Lambda}\right] \tag{13.6.1}$$

Thus

$$\frac{\delta L}{\pi} = \frac{(\beta_0 - \beta_m)L}{\pi} = \frac{\nu - \nu_B}{c/(2n_0L)}$$

which is a convenient horizontal axis to use.

One first computes the required integrated gain (g_0L) from (13.6.21a) as a function of δL to generate a curve similar to Fig. 13.25(a). Then one evaluates the right-hand side of (13.6.22) and plots that as a function of $\delta L/\pi$ as shown in Fig. 13.25(b). When that normalized phase crosses a value $(q \pm 1/2)\pi$ we have achieved a simultaneous solution to the real and imaginary parts of (13.6.20).

For the example illustrated in Fig. 13.25, $\kappa L = 0.5$ and this intersection occurs at $\delta L/\pi = \pm 0.67$. At this value of the detuning, the required integrated gain is $g_0L = 2.625$ or a single pass gain of 22.8 dB.

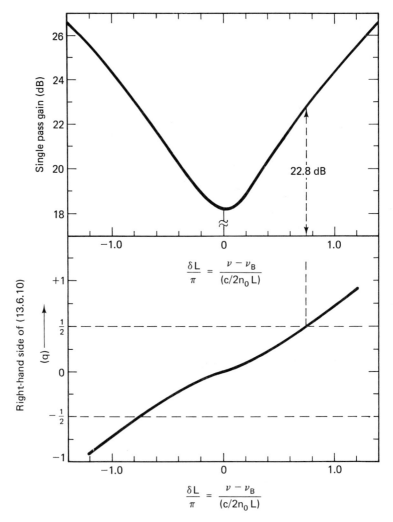

Figure 13.25 Gain in dB and phase shift of a distributed feedback laser; the parameter $\kappa l = 0.5$ for this example.

Note that oscillation is not at the Bragg frequency, which has the lowest required gain, but at two frequencies symmetrically placed with respect to that value. If we had allowed ripples in the gain, then κL would be complex and the perfect asymmetry (with respect to $\delta = 0$) in the phase curve of Fig. 13.25(b) would be destroyed and thus one of the modes would have a lower threshold than the other. Quite often, a 90° phase shifting section (i.e., $\lambda/4$) is incorporated into the middle of the structure and this modification gets rid of the 1/2 term in (13.6.22). As a result, oscillation will occur close to the Bragg frequency, and there will be considerable discrimination against the other modes oscillating. In any event, there is definitely a frequency selectivity in this feedback scheme. One may be dismayed by the mathematical complexity of distribution, but the laser has no problem in solving these equations—it does so in a nanosecond or so.

13.7 LASER ARRAYS

13.7.1 System Considerations

One cannot increase the excitation to any one given laser cavity to arbitrarily high values so as to generate high optical power and expect it to survive! There are many laser physics issues that negate this approach and practical considerations that forbid it. For instance, if one attempts to load a CO_2 with too much power, the "waste heat" raises the rotational temperature, which in turn lowers the gain. In semiconductor lasers, the internal optical electric field may become so large that it literally self-destructs because of optical breakdown at any minor imperfection at the output facet. (This presumes, of course, that the device did not melt because of its waste heat.)

One approach to obtaining high power is to utilize the coherence properties of lasers to "add" the fields generated by many parallel sources, each of which could be working at its optimum operating point. A similar situation is often used in the microwave domain with a master oscillator driving many (N) amplifiers with each of those exciting an element in an array antenna. Ideally, one can obtain N times the power radiated by a single element but into the pattern of the much larger array, which implies a much narrower radiated beam. As a bonus, the beam pattern can be electronically scanned by controlling the relative phase between the elements. There has been a major effort to adapt this older microwave technology to the optical domain with the majority of the effort being directed toward semiconductor arrays, but some work has also been devoted to coupling the larger CO_2 or chemical lasers also.

13.7.2 The Semiconductor Laser Array—Physical Picture

The problem to be addressed (slowly) is shown in Fig. 13.26(a) and differs from the microwave scheme in an important issue: there are N oscillators, not amplifiers, which are coupled and radiate at the output aperture. Each laser

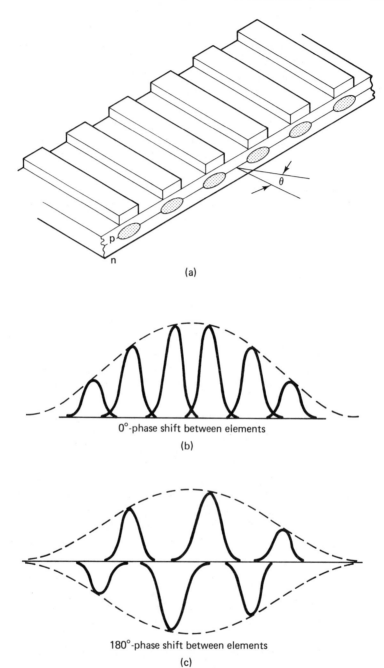

Figure 13.26 (a) A typical 6-stripe semiconductor array. (b) An in-phase field distribution. (c) The same field elements as (b) but with a 180° phase shift between adjacent stripes.

oscillates at the same frequency (or wavelength) and is locked in a specific phase relationship with respect to each other, which need not be zero or even the same between adjacent stripes. The theoretical questions to be addressed are "What are the allowed *phases* of the coupled oscillators, and, if more than one solution is possible, which is the most desirable?"

It is as important to realize that the question is real and makes a significant difference to us in the outside world as it is to answer it with pages of mathematics. Suppose that each stripe produces a nice smooth mode profile with a central amplitude of each laser under another smooth envelope as shown in Fig. 13.26(b). In order to compute the far field or radiation pattern of this array, the observable, one needs the specification of the relative phase of each element in addition to the amplitude specification. If, for instance, all oscillators had a common phase (which can be assigned to be zero), then the radiation from this array would be a maximum along the axis because each differential source at $\pm x$ would contribute an equal-in-phase radiated field there. If, however, the phase of each adjacent oscillator were differed by 180°, the field along the central axis would be 0 (zero!) because of the phase cancellation of contributions from each symmetrically placed but out-of-phase element. If we avoid the mathematics for a moment, we would guess that the far-field radiation pattern of the in-phase distribution would be a single lobe with an angle determined (roughly) by the bigger envelope, whereas the 180° distribution would be a two-lobe pattern with a zero on the axis. Detailed calculation using antenna theory will verify that guess. Obviously, one would prefer the in-phase distribution, but alas, a "vote" is not taken on the matter: the laser physics *decides* the issues. Unless special techniques are used, the laser prefers the out-of-phase distribution because the total field is minimized in the regions between the stripes, which are absorptive.*

Inasmuch as there are $N = 6$ elements in the example shown in Fig. 13.26, we also suspect that there are $N = 6$ different combinations of phase and amplitude assignments among the elements. These are the so-called "supermodes"—the coherent combination of the individual modes of each strip. The next section describes those combinations.

13.7.3 The Supermodes of the Array

The laser array must oscillate at a common frequency (wavelength) so that a coherent addition of the fields is possible; hence each laser must couple to and lock with a particular phase relationship to its adjacent neighbor. Because of this coupling, the propagation constant of the modes in the array may differ somewhat from that of an isolated element, but not by much. Further, we expect the transverse field configuration of the individual element to be hardly changed

---

*There is no current in the gaps; therefore the semiconductor is not inverted; and thus there is absorption.

from that of normal mode for an isolated stripe as found in Sec. 13.5, which is expressed in symbolic form by

$$\mathbf{E}_1 = E_1(x,y)e^{-j\beta_0 z} \tag{13.7.1}$$

where the fact that there is gain in the stripe is ignored in the interest of simplicity and β_0 is the propagation constant for an isolated element. We presume that (13.7.1) satisfies the *uncoupled wave equation* in all of its glory and gory detail with the appropriate boundary conditions. With nearest neighborhood coupling, the amplitude of the field for the lth element obeys

$$\frac{dE_l}{dz} = -j\beta_0 E_l - j\kappa E_{l+1} - j\kappa E_{l-1} \tag{13.7.2}$$

with l running from 1 to N and κ indicating how much of the neighbor is injected into the lth cavity. The functional dependence of the fields on the transverse coordinates has also been suppressed so as to concentrate on the amplitudes of each stripe. Our task is to find a nontrivial solution to the *entire* array, which can be expressed in matrix form by

$$\frac{d}{dz}\begin{bmatrix} E_1 \\ E_2 \\ \cdot \\ \cdot \\ \cdot \\ \cdot \\ E_N \end{bmatrix} = -j \begin{bmatrix} \beta_0 & \kappa & 0 & \cdots & & 0 \\ \kappa & \beta_0 & \kappa & 0 & \cdots & 0 \\ 0 & \kappa & \beta_0 & \kappa & \cdots & 0 \\ \cdot & & & \kappa & \cdots & \cdots \\ \cdot & & & & & \cdot \\ 0 & \cdots & & \kappa & \beta_0 & \kappa \\ 0 & \cdots & & & \kappa & \beta_0 \end{bmatrix} \begin{bmatrix} E_1 \\ E_2 \\ \cdot \\ \cdot \\ \cdot \\ \cdot \\ E_N \end{bmatrix} \tag{13.7.3a}$$

or

$$\frac{d}{dz}\overline{\overline{\mathbf{E}}} = -j\overline{\overline{\mathbf{M}}} \cdot \overline{\overline{\mathbf{E}}} \tag{13.7.3b}$$

where $\overline{\overline{\mathbf{E}}}$ is a column matrix and $\mathbf{M}$ is a tridiagonal square array. If we assume that the fields can be expessed as

$$E_l^{(k)} = A_l^{(k)} e^{-j\beta_k z} \tag{13.7.4}$$

then (13.7.2) becomes a simple eigenvalue problem.

$$-j \begin{bmatrix} \Delta\beta & \kappa & 0 & \cdots & & 0 \\ \kappa & \Delta\beta & \kappa & \cdots & & 0 \\ 0 & \kappa & \Delta\beta & \kappa & \cdots & \cdot \\ \cdot & \cdot & \cdot & \cdot & \cdot & \cdot \\ \cdot & \cdot & \cdot & \cdot & \cdot & \cdot \\ 0 & \cdots & & \kappa & \Delta\beta & \kappa \\ 0 & \cdots & & & \kappa & \Delta\beta \end{bmatrix} \begin{bmatrix} A_1^{(k)} \\ A_2^{(k)} \\ \cdot \\ \cdot \\ \cdot \\ \cdot \\ A_N^{(k)} \end{bmatrix} = 0 \tag{13.7.5}$$

Sec. 13.7 Laser Arrays

with $\Delta\beta = \beta_0 - \beta_k$. For a nontrivial solution (i.e., $A_l^{(k)} \neq 0$), the determinant must vanish, and this procedure guarantees N solutions for the propagation constants in the array with each solution corresponding to a different amplitude distribution among the elements.

One can solve the above equations without recourse to the beauties and subtleties of matrix theory by recognizing the close analogy between difference and differential equations. Consider one of the equations of (13.7.5):

l^{th} row:
$$(\beta_0 - \beta_k)A_l^{(k)} + \kappa(A_{l-1}^{(k)} + A_{l+1}^{(k)}) = 0$$

Guess that
$$A_l^{(k)} = A_0^{(k)} \sin l\theta \tag{13.7.6}$$

where θ is a parameter to be determined. Expand the coefficients A_{l-1} and A_{l+1} by the standard trigonometric formulas, add, and multiply by κ as dictated by (13.7.6):

$$A_{l-1}^{(k)} = A_0^{(k)} [\sin l\theta \cos\theta - \cos l\theta \sin\theta]$$
$$A_{l+1}^{(k)} = A_0^{(k)} [\sin l\theta \cos\theta + \cos l\theta \sin\theta]$$
$$\kappa(A_{l+1}^{(k)} + A_{l-1}^{(k)}) = 2\kappa A_0^{(k)} \sin l\theta \cos\theta$$

Insert the guess of (13.7.6) into the lth row of (13.7.5) and cancel the common factors of $A_0^{(k)} \sin l\theta$:

$$(\beta_0 - \beta_k)A_0^{(k)} \sin l\theta = -2\kappa A_0^{(k)} \sin l\theta \cos\theta$$
$$(\beta_0 - \beta_k) = -2\kappa \cos\theta \tag{13.7.7}$$

This equation should be interpreted as determining the propagation constants β_k in terms of the parameter θ, which is not known but which can be found from a bit of physics: the array is the same whether viewed from the front or back and thus the Nth element must look like the first if you go to the other side of the page.

$$A_N^{(k)} = A_0^{(k)} \sin N\theta = A_1^{(k)} = A_0^{(k)} \sin\theta$$
$$N\theta = k\pi - \theta$$

or
$$\theta = \frac{k\pi}{N+1} \tag{13.7.8}$$

The pertinent data for the various combinations of fields of the supermodes are now available:

$$A_l^{(k)} = A_0^{(k)} \sin l\left\{\frac{k\pi}{N+1}\right\} \tag{13.7.9}$$

$$\beta_k = \beta_0 + 2\kappa \cos\left[\frac{k\pi}{N+1}\right] \tag{13.7.10}$$

It is easy to show that there are only N values of $k = 1, 2, \ldots N$ that yield different field configurations, each with a slightly different propagation constant. The laser oscillates at a frequency such that the round-trip phase shift, $\beta_k 2d = q2\pi$, for each supermode configuration. If β_0 is approximated by $\omega n/c$, then the equation for frequency of the kth supermode is

$$\nu_k = \frac{c}{\lambda_0}\left[1 - \frac{\kappa\lambda_0}{\pi n}\cos\left(\frac{k\pi}{N+1}\right)\right] \qquad (13.7.11)$$

where the axial mode number q is approximated by $2nd/\lambda_0$, and λ_0 is the free-space wavelength from each isolated element. For $\kappa\lambda_0/2\pi n$ small, the supermodes generate a range of wavelengths centered around λ_0.

$$\frac{\Delta\lambda}{\lambda_0} = \frac{\lambda_{k=N} - \lambda_{k=1}}{\lambda_0} = \frac{\kappa\lambda_0}{\pi n}\left[\cos\left(\frac{\pi}{N+1}\right) - \cos\left(\frac{N\pi}{N+1}\right)\right] \qquad (13.7.12)$$

Fig. 13.27 illustrates the "bottom line" of these calculations—the field distribution for an $N = 5$ array.

As mentioned earlier, the physics of the laser controls which of these supermodes will be chosen, and unfortunately, the 180° phase shift between stripes, $k = 5$, is usually the answer because it has the highest gain-to-loss ratio since the total field is a minimum in the regions between the stripes. One can coax the laser into the other supermodes by external feedback or careful control of the current injection. It should be clear, however, that the far-field pattern of each of these supermodes is considerably different.

13.7.4 The Radiation Pattern

The field radiated by this array can be found from the application of (13.3.10b) of this chapter:

$$\mathbf{E}(r_0) = \frac{j}{\lambda_0 d}\int_{x_1}\int_{y_1}\mathbf{E}(x_1, y_1)e^{-j\phi(2,1)}dx_1 dy_1 \qquad (13.3.10b)$$

where the limits of integration extend over the entire aperture of the array, and $\phi(2,1)$ is the phase shift from the points on the facet to the observation point. It is not much "fun" to make these calculations, and the only saving grace comes from making the far-field approximation. (Fortunately, the semiconductor laser, even an array, is so small that almost everything is in the far field.) Then the far-field pattern is just the Fourier transform of the aperture distribution, and all of the prior history of it can be brought to bear on this problem. From an operational standpoint, one merely collects the radiation with a lens with its focal plane at that of the array. The output of the lens is the desired Fourier transform.

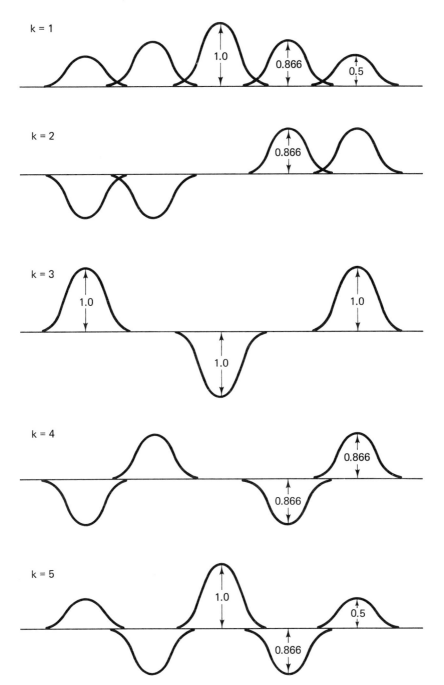

Figure 13.27 The supermode field distribution for $N = 5$.

PROBLEMS

13.1. Read the article by A. E. Siegman and R. Arrathoon, "Modes in Unstable Optical Resonators and Lens Waveguide," IEEE J. Quantum Elect. *QE-3*, No. 4, 156–163, April 1967. Answer the following questions by simple equations, a sketch, or a paragraph.
   (a) What is the parameter M and how is it related to the geometry of Fig. 13.8?
   (b) Show how our (13.3.10a) and (13.3.10b) lead to Equation 4 of the Siegman and Arrathoon paper. Why are the subscripts on g no longer necessary?
   (c) Why is "the region centered around a point $y - x/M$" so important in Equation 5? What is a "Fresnel integral"?
   (d) Show the "obvious" development of Equation 8 from Equation 7.
   (e) What is the meaning of the Fresnel number N?
   (f) Verify the phase-shift curve plotted for $N = 2.7$ in Fig. 11 using the geometrical analysis.
   (g) If a cavity were specified by $L = 50$, cm, $g = 1.10$, $N_{eq} = 2.5$, what is the size of the mirrors and the radius of curvature? Compare the loss predicted by the geometric analysis to that plotted in Fig. 11.

13.2. Consider the unstable resonator shown in the accompanying diagram. The reflection coefficient of R_1 is 0.98, that of R_2 is 0.95.

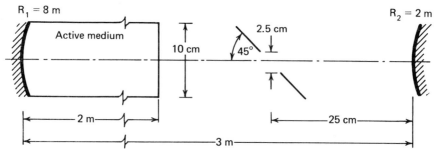

   (a) Find the location of points P_1 and P_2 and show them on a sketch of the optical cavity.
   (b) Use Siegman's simple approach to sketch the field variation within the optical cavity. Find and label all pertinent dimensions. Show the wave front coming from the beam slicer.
   (c) What is the mean loss per pass through the empty cavity?
   (d) What is the gain required of this laser?

13.3. Consider a laser with a confocal unstable resonator as shown on the accompanying diagram. Sketch the amplitude of the field at $z = 0^-$ and $z = d$ and describe why this field pattern fits the definition of a mode. What should be the gain coefficient of the active cell to obtain oscillation?

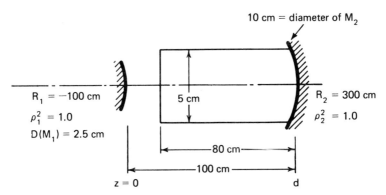

13.4. In a simple laser, oscillation builds up from noise provided by the spontaneous emission into the cavity mode with the electric field growing with time because of the gain in the cavity, and decaying because of the passive coupling to the external world. Thus the amplitude of the qth mode obeys the following:

$$\frac{dE_q}{dt} + \frac{1}{2}\left[\frac{1}{\tau_p} - 1\frac{f\gamma_0(\nu_q)(c/n)}{1 + |E_q(t)|^2/E_s^2}\right]E_q(t) = 0$$

where $E_s^2/2\eta_0 = I_s$ = the saturation intensity, γ_0 = small-signal gain coefficient, and f is the fraction of the cavity filled with the active medium.

Assume that the inversion is created instantaneously at $t = 0$ with an initial mode intensity that is 10^{-5} of its final value. As a consequence, it takes a significant amount of time (in units of the photon lifetime τ_p) to reach the steady state value of $I/I_s \approx \gamma/\alpha - 1$.

(a) Plot the time evolution of the mode intensity from its initial value to 95% of the final value for the gain-to-loss ratios, $f\gamma_0(c/n)\tau_p$, of 1.75, 2.25, and 4. Plot the intensity normalized to final value on a logarithmic scale, with the time normalized to the photon lifetime on a linear scale.

(b) Why is the factor of 1/2 present in this differential equation?

13.5 The starting point for the analysis of a finite aperture mirror system is (13.3.10b). If one applies this expression to nonconfocal geometry with $d \neq R$ and expresses the field as a product in the form $E = u(x) \cdot v(y)$, then the integral equation becomes

$$\sigma u(x_2) = \int_{-a}^{a} K(x_1, x_2) u(x_1) dx_1$$

where

$$K(x_1, x_2) = \sqrt{\frac{j}{\lambda d}} \cdot \exp\left\{-j\frac{k}{2d}[g(x_1^2 + x_2^2) - 2x_1 x_2]\right\}$$

and $g = 1 - d/R$ (same as in Chapter 2), $k = 2\pi/\lambda$, d = mirror spacing, R = radius of curvature of the two mirrors, and σ contains the attenuation and phase shift experienced by the wave in propagating from M_1 to M_2. This is Eq. 74 in the paper by H. Kogelnik and T. Li, "Laser Beams and Resonators," Appl. Opt. 5, 1550–1567, 1967. Show the essential steps in arriving at this answer from (13.3.10).

13.6 Even though we "know" that a Gaussian beam is an excellent approximation for the fields inside a finite aperture confocal resonator, let us follow the wave from M_1 to M_2 with a field pattern of a similar but different choice and apply the integral (13.3.19) where $N = a^2/\lambda d$ = Fresnel number (Note: $d = R_1 = R_2$), $C = 2\pi N$, and $X_{1,2} = (x_{1,2}/a)C^{1/2}$. Use the following expression for the field on M_1:

$$F(x_1) = \cos^2\left[\frac{\pi x_1}{2w_s}\right] \text{ and assume } N = \pi/2;\ a/w_s = 2;\ x_1 < w_s$$

(a) Make a careful sketch of the field on the surface of M_1.
(b) Find and sketch the field at M_2 with the same scale as that used for (a). Show also the part that missses the edge of M_2.
(c) Estimate the power lost per transit for this "beam."

13.7 The field on M_2 in Problem 13.6 has the same general functional shape as that on M_1, but it extends over a much larger area. Thus we have *not* found a self-reproducing field configuration for the cavity. Let us remedy that problem by letting the ratio a/w_s to be equal to some factor—call it 2m.

(a) Find the value of m that equalizes the width (FWHM) of the fields at the two mirrors.
(b) Plot the ratio $E^2(x)/E^2(0)$ on the mirror surfaces.
(c) *Estimate* the fractional loss per round-trip.

13.8 The mirrors of the confocal cavity shown on the diagram below have a tapered field reflection coefficient given by

$$\rho(r) = \rho_0 \exp[-(tr)^2]$$

where r = distance from the axis

ρ_0 = field reflection coefficient at $r = 0$ (use 0.98 for a numerical value)

t = a taper rate that can be expressed as a multiple of the normal spot size on the mirrors; $t = m/w_s$

w_s = spot size on $M_{1,2}$ (for $t = 0$) and is equal to $(2b/k)^{1/2}$

(a) Use the integral equation formulation of the problem to find a self-consistent set of fields reflected from each mirror. (Assume $a \to \infty$.)

The functional form of the solution is a Gaussian with a spot size w: you are to find the ratio $(w/w_s)^2$.

(b) Find the single pass gain, G, for the laser to oscillate.

(c) Plot the relative output *intensity* (i.e., $\mathbf{E}^2$) at M_2 as a function of the normalized distance (r/w_s) assuming the transmission $= 1 - \rho^2(r)$. Label the position of peak intensity.

(d) Use the above results and the expansion laws for a Gaussian mode to sketch the variation of the fields *inside* the cavity. Identify the planes where the wave front of the positive and negative going waves are planar and the spot sizes are a minimum. Evaluate w_0^2 at those planes.

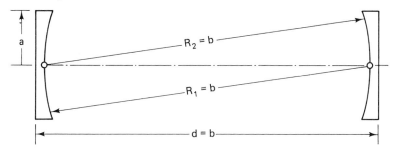

13.9. Derive the equation that determines the propagation constant $\gamma_0 = j\beta$ and the asymmetry angle ϕ for the TM modes in the geometry shown below.

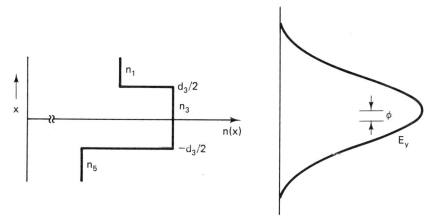

13.10. In the example considered for gain guiding, the following parameters were chosen: $\lambda_0 = 8400$ Å; $n = 3.6$; $\gamma_0 = 100$ cm^{-1}; $L_x = 10$ μm; $L_y = 5$ μm; where the length parameters refer to the parabolic gain and index profiles. The spot sizes and the radius of curvature of the mode had the following values: $w_x = 0.862$ μm; $w_y = 5.22$ μm; $R_y = 366.9$ μm;

$\gamma_{net} = 72.75$ cm^{-1}, for the field parameters *inside* the semiconductor cavity shown below.

(a) If the cavity length was 500 μm, the facet reflectivity $R_{1,2} = 0.36$, and the current density was 1000 amp/cm^2, what is the gain coefficient?

(b) Assume that the gain coefficient scales directly with current and compute the threshold current.

(c) The parameters given above describe the field inside the laser; the observable is the external field. Use the *ABCD* law for Gaussian beams to compute the following:
  (1) The beam spot sizes and the radius of curvature on the air side of the facet.
  (2) Find the far-field divergence angle of the beam as it spreads in the x and y directions. Assume $\theta/2 = dw/dz$ (in radians).
  (3) If the beam were *not* astigmatic, $R_y = \infty$, but with the same spot sizes, what would be the angle θ_y?

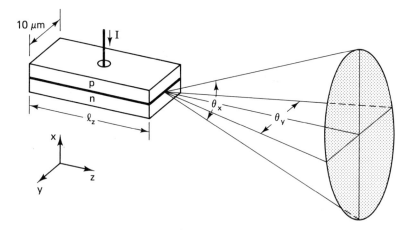

13.11. Read the article by D. D. Cook and F. R. Nash, "Gain Induced Guiding and Astigmatic Output Beam of GaAs Lasers," *J. Appl. Phys.* **46**, 1660–1672, 1975. The following questions refer to this paper.

(a) Show the development of Equations 2.2 and 2.3 from 2.1.
(b) Derive Equations 2.27 and A7 using the *ABCD* law rather than by using the procedure followed.
(c) Explain (with words and a few equations) the logic of Equation 2.11.
(d) Which equation is the *integral* definition of the modal gain, and why does it make sense in terms of stimulated emission and absorption?
(e) What is the origin of Equation C1?
(f) Explain the notation and development of Equation B5.

13.12. Consider the coupled triangular array of lasers shown in the diagram below. Assume that the propagation constant of an isolated channel is β_0 and there is a coupling coefficient of κ between each element.
  (a) Write the system of equations describing the propagation of a supermode of the array.
  (b) Find the characteristic values for all of the allowed values of the propagation constant for the supermodes.
  (c) Indicate the relative amplitudes of the fields for these supermodes.

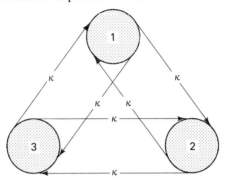

13.13. Find the propagation constants and the relative field amplitudes for the $N = 4$ linear array.

REFERENCES AND SUGGESTED READINGS

1. A. E. Siegman, "Unstable Optical Resonators for Laser Applications," Proc. IEEE *53*, 277–287, Mar. 1965.
2. (a) R. J. Freiberg, P. P. Chenausky, and C. J. Buczek, "An Experimental Study of Unstable Confocal CO_2 Resonators," IEEE J. Quant. Electron. *QE-8*, 882–892, Dec. 1972.
   (b) R. J. Freiberg, P. P. Chenausky, and C. J. Buczek, "Asymmetric Unstable Traveling-Wave Resonators," IEEE J. Quant. Electron. *QE-10*, 279–289, Feb. 1974.
   (c) A. Siegman and R. Arrathoon, "Modes in Unstable Optical Resonators and Lens Waveguides," IEEE Quant. Electron. *QE-3*, 156, 1967.
3. Yu. A. Ana'Ev, "Unstable Resonators and Their Applications," Sov. J. Quant. Electron. *1*, 565–586, 1972.
4. H. A. Haus, *Waves and Fields in Optoelectronics* (Englewood Cliffs, N.J.: Prentice-Hall, Inc., 1984).
5. A. G. Fox and T. Li, "Resonant Modes in a Maser Interferometer," Bell Syst. Tech. J. *40*, 453–488, Mar. 1961.
6. G. D. Boyd and J. P. Gordon, "Confocal Multimode Resonator for Millimeter through Optical Wavelength Masers," Bell Syst. Tech. J. *40*, 489–508, Mar. 1961.
7. G. D. Boyd and H. Kogelnik, "Generalized Confocal Resonator Theory," Bell Syst. Tech. J. *41*, 1347–1369, July 1962.

8. H. Kogelnik, "Imaging of Optical Modes-Resonators with Internal Lenses," Bell Syst. Tech. J. *44*, 455–494, Mar. 1963.
9. G. H. B. Thompson, *Physics of Semiconductor Laser Devices* (New York: John Wiley & Sons, Inc., 1980).
10. H. K. Kressel and J. K. Butler, *Semiconductor Lasers and Heterojunction LED* (New York: Academic Press, Inc., 1977).
11. H. C. Casey Jr. and M. P. Panish, *Heterostructure Lasers* (New York: Academic Press, Inc., 1978).
12. R. G. Hunsperger, *Integrated Optics: Theory and Technology*, Springer Series in Optical Science (New York: Springer-Verlag, 1984).
13. D. D. Cook and F. R. Nash, "Gain-induced Guiding and Astigmatic Output of GaAs Lasers," J. Appl. Phys. *46*, 1660–1672, 1975.
14. F. R. Nash, "Mode Guidance Parallel to the Junction Plane of Double-Heterostructure GaAs Lasers," J. Appl. Phys. *44*, 4696, 4707, 1973.
15. E. Kapon, J. Katz, and A. Yariv, "Supermode Analysis of Phase-Locked Arrays of Semiconductor Lasers," Optics Letters, Vol. 10, No. 4, 1984.
16. H. Kogelnik and C. V. Shank, "Stimulated Emission in a Periodic Structure," Appl. Phys. Lett. *18*, 152–154, 1971.
17. H. Kogelnik and C. V. Shank, "Coupled Wave Theory of Distributed Feedback Lasers," J. Appl. Phys. *43*, 2327–2335, 1975.
18. W. Streiffer, D. R. Scifres, and R. D. Burnham, "Coupled Wave Analysis of DFB and DBR Lasers," IEEE J. Quant. Electron. *QE-13*, 134–140, 1977.
19. T. Ikegami, "Reflectivity at Facet and Oscillation Mode in Double-Heterostructure Injection Lasers," IEEE J. Quant. Electron. *QE-8*, 470–476, 1972.
20. S. Chinn, "Effects of Mirror Reflectivity in a Distributed Feedback Laser," IEEE J. Quant. Electron. *QE-9*, 574–580, 1973.
21. M. Nakamura, K. Aiki, J. Umceda, and A. Yariv, "CW Operation of Distributed-Feedback GaAs-GaAlAs Diode Lasers at Temperatures Up To 300 K," Appl. Phys. Lett. *27*, 403, 1975.
22. H. Haus and C. V. Shank, "Antisymmetric Taper of DFB Laser," IEEE J. Quant. Electron. *QE-12*, 534, 1976.
23. S. L. McCall and P. M. Platzman, "An Optimized $\pi/2$ DFB Laser," IEEE J. Quant. Electron. *QE-21*, 1899, 1985.
24. H. Kogelnik and C. V. Shank, "Coupled-Wave Theory of DFB Laser," J. Appl. Phys. *43*, 2327, 1972.
25. W. H. Steier, *The Laser Handbook*, Vol. 3, Article A5, Ed. M. L. Stitch (New York: North Holland Publishing Company, 1979).

14

Quantum Theory of the Laser: An Introduction

14.1 INTRODUCTION

We have managed to wade through a lot of the physics of a laser without bothering about the mathematical description of the quantum interaction between the field and the atoms. The key simplification, which enabled us to proceed as far as we have, was the introduction of the Einstein A and B coefficients and the heavy reliance on the rate equations. However, there are limitations to this approach that are not obvious unless one examines the quantum problem in greater detail, a task to which we turn. As a bonus, we will obtain a better picture and understanding of the scope and limitations of the Einstein A and B coefficients.

In particular, we want to be more sophisticated in our approach to the A coefficient. Hence, we will discuss the mathematical formulation of the problem of predicting the transition rate from quantum mechanics. However, a lot of the notation one encounters in the literature is related to the classical model of an atom. Consequently, we will quickly run through this type of material to see the similarities when we get to the more exact model.

14.2 THE CLASSICAL MODEL OF AN ATOM

14.2.1 The Antenna Problem

Consider an elementary dipole antenna shown in Fig. 14.1. Every elementary text in electromagnetic theory computes the power radiated by this antenna.

$$i(t) = I_0 \cos \omega t$$

$$P_{\text{rad}} = \left(\frac{\mu_0}{\epsilon_0}\right)^{1/2} \cdot \frac{\pi}{3} \cdot \frac{(I_0 d)^2}{\lambda_0^2} \tag{14.2.1}$$

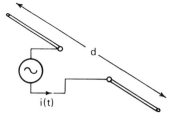

Figure 14.1 Dipole antenna.

Making the substitution for $\lambda_0 = 2\pi c/\omega$, we obtain*

$$P_{\text{rad}} = \frac{1}{3}\left(\frac{1}{4\pi\epsilon_0}\right)\frac{\omega^2}{c^3}(I_0 d)^2 \tag{14.2.2}$$

Now the current $I = I_0 \cos \omega t$ can be related to the charge crossing an imaginary surface intersecting the antenna by

$$\frac{dQ}{dt} = I_0 \cos \omega t \quad \therefore Q = \frac{I_0}{\omega} \sin \omega t = Q_0 \sin \omega t$$

Therefore

$$P_{\text{rad}} = \frac{1}{3}\left(\frac{1}{4\pi\epsilon_0}\right)\frac{\omega^4}{c^3}(Q_0 d)^2 \tag{14.2.3}$$

If we identify a group of randomly phased oscillating dipole antennas with N atoms, the total power is

$$P_{\text{rad}} = \frac{16}{3}\left(\frac{1}{4\pi\epsilon_0}\right)\frac{\pi^4 \nu^4}{c^3}|\mu|^2 N \tag{14.2.4a}$$

$$\mu = Q_0 d = \text{electric dipole moment}$$

*In (14.2.2) and many equations to follow, the quantity $1/4\pi\epsilon_0$ is kept as a separate term to facilitate the comparison to other references using the CGS system of units. In that case, this factor is equal to unity (1).

Sec. 14.2 The Classical Model of an Atom

But recall Einstein's approach to this same collection of atoms (or antennas) radiating spontaneously with random phase so as to add powers rather than fields.

$$P_{\text{rad}} = h\nu \frac{dN}{dt} = h\nu A_{21} N \tag{14.2.4b}$$

Equating the two expressions we obtain a "classical" expression for the A coefficient:

$$A_{21} = \frac{16}{3}\left(\frac{1}{4\pi\epsilon_0}\right)\frac{\pi^4 \nu^3}{c^3}\frac{|\mu|^2}{h} \tag{14.2.5}$$

One could argue that this has no bearing on the problem, but the answer is quite close to the "exact" quantum-mechanical calculation. It does point out the necessity of computing the quantum value of the electric dipole moment for allowed transitions. Similar relations can be found for magnetic dipole and electric quadrupole transitions.

14.2.2 The "Bound" Electron

Let us take another "classical" viewpoint of the atom as that having a bound active electron with a characteristic frequency ω_0 related to the atomic restoring forces and also having a damping constant $1/\tau$ caused by the radiation that it must emit as it relaxes to a lower energy state. Thus the electron obeys a differential equation such as

$$\ddot{x} + \frac{1}{\tau}\dot{x} + \omega_0^2 x = 0 \tag{14.2.6}$$

The solution is

$$x(t) = X_0 e^{-t/2\tau} \cos \omega_0 t$$

assuming that $1/\tau \ll \omega_0$ with the initial conditions that $x(t) = X_0$ at $t = 0$. The total energy of this active electron is the sum of its kinetic plus potential energies.

$$W_{\text{KE}} = \frac{1}{2}m\dot{x}^2 = \left(\frac{1}{2}mX_0^2\omega_0^2\right)e^{-t/\tau}\cos^2 \omega_0 t$$

$$W_{\text{PE}} = \frac{1}{2}m\omega_0^2 X^2 = \left(\frac{1}{2}mX_0^2\omega_0^2\right)e^{-t/\tau}\sin^2 \omega_0 t \tag{14.2.7}$$

$$W_{\text{TE}} = W_{\text{PE}} + W_{\text{KE}} = \frac{1}{2}m\omega_0^2 X_0^2 e^{-t/\tau}$$

Thus the term $1/\tau$ has the interpretation of an energy decay rate.

$$-\frac{dW}{dt} = \frac{W}{\tau}$$

Evaluating this expression at $t = 0$ when the initial energy is $m\omega_0^2 X_0^2/2$ and equating this power to the "classical" rate of radiation (14.2.3) (per oscillator) yields

$$\frac{1}{\tau}\left(\frac{1}{2}m\omega_0^2 X_0^2\right) = \frac{1}{3}\left(\frac{1}{4\pi\epsilon_0}\right)\frac{\omega_0^4}{c^3}|eX_0|^2 \tag{14.2.8}$$

where eX_0 = dipole moment. Solving for $1/\tau$ yields another classical expression for the A coefficient.

$$\frac{1}{\tau} = A_{21} = \frac{2}{3}\left(\frac{1}{4\pi\epsilon_0}\right)\frac{\omega^2}{c^3}\frac{e^2}{m} = \frac{8\pi^2}{3}\left(\frac{1}{4\pi\epsilon_0}\right)\frac{\nu^2}{c^3}\frac{e^2}{m} \tag{14.2.9}$$

14.2.3 The "Driven" Oscillator

We can also inquire into the response of this damped harmonic oscillator to an externally applied electromagnetic field $e(t) = E_0 \exp(j\omega t)$:

$$\ddot{x} + \frac{1}{\tau}\dot{x} + \omega_0^2 x = -\frac{e}{m}E_0 e^{j\omega t}$$

which yields the forced response

$$x(t) = -\frac{e}{m}\frac{E_0 e^{j\omega t}}{(\omega_0^2 - \omega^2) + j\frac{\omega}{\tau}} \tag{14.2.10}$$

If we had a collection of N oscillators (atoms), the response of the collection is just N times the response of the average one. Thus the polarization $\mathbf{P} = Nq\mathbf{x}$ can be inserted into Maxwell's equations so as to compute the reaction of these oscillators on the field:

$$\nabla \times \mathbf{H} = \epsilon_0 n^2 \frac{\partial \mathbf{E}}{\partial t} + \frac{\partial \mathbf{P}}{\partial t} = j\omega\epsilon_0 n^2 \mathbf{E}_0 e^{j\omega t} + N(-e)\dot{\mathbf{x}}(t)$$

$$= \left\{j\omega\epsilon_0 n^2 + \frac{Ne^2}{m}\frac{j\omega}{(\omega_0^2 - \omega^2) + j\frac{\omega}{\tau}}\right\}\mathbf{E}_0 e^{j\omega t} \tag{14.2.11}$$

where it is assumed that the atoms are embedded in a medium with an index of refraction n. Factoring the term $j\omega E_0$ from this expression yields the atoms'

Sec. 14.2 The Classical Model of an Atom

contribution to the complex susceptibility, or the total complex relative dielectric constant.

$$\nabla \times \mathbf{H} = j\omega\epsilon_0[n^2 + \chi'_a - j\chi''_a]\mathbf{E}_0 e^{j\omega t} \quad (14.2.12a)$$
$$= j\omega\epsilon_0[\epsilon' - j\epsilon'']\mathbf{E}_0 e^{j\omega t} \quad (14.2.12b)$$

where

$$\chi'_a = \frac{Ne^2}{m\epsilon_0} \frac{\omega_0^2 - \omega^2}{(\omega_0^2 - \omega^2)^2 + \left(\frac{\omega}{\tau}\right)^2} = \epsilon' - n^2 \quad (14.2.12c)$$

$$\chi''_a = \frac{Ne^2}{m\epsilon_0} \frac{(\omega/\tau)}{(\omega_0^2 - \omega^2)^2 + \left(\frac{\omega}{\tau}\right)^2} = \epsilon'' \quad (14.2.12d)$$

If we assume that the *fields* vary as $\exp -jkz$, we can relate the complex propagation constant $k = k' - jk''$ to the dielectric properties defined by (14.2.12)

$$\mathbf{k} \cdot \mathbf{k} = (k')^2 - (k'')^2 + j(k')(-2k'') = \left(\frac{\omega}{c}\right)^2 (\epsilon' - j\epsilon'') \quad (14.2.13)$$

The real part of the propagation constant k' is always much bigger than the imaginary part, the field attenuation or gain constant, k'', and thus we equate the real terms of (14.2.13) to obtain

$$(k')^2 = \left(\frac{\omega}{c}\right)^2 [n^2 + \chi'_a]$$

or

$$k' = \frac{\omega}{c} n \left[1 + \frac{\chi'_a}{2n^2}\right] \quad (14.2.14)$$

where we assume that the active atoms' contribution to real dielectric constant is small compared to that contributed by the host lattice.

Equating the imaginary parts of (14.2.13) and recognizing that the factor $2k''$ is what has been called the power gain coefficient $\gamma(\omega)$ throughout the book yields an expression for it in terms of the imaginary part of the susceptibility.

$$k'(-2k'') \triangleq k'\gamma(\omega) = -\frac{\omega^2}{c^2}\chi''_a \quad (14.2.15a)$$

Using (14.2.14) for k' and ignoring $\chi'_a/2n^2$ as small compared to 1 converts (14.2.15a) into a very important but simple expression.

$$\gamma(\omega) \approx -k' \cdot \frac{\chi''_a}{n^2} \quad (14.2.15b)$$

Equation (14.2.15b) is very important. The fact that we used a classical analysis of the atom to obtain the susceptibility is beside the point. The main issue is that the gain coefficient is related to χ_a'' by this simple relationship.

Equation (14.2.15b) becomes a very familiar one when (14.2.12d) is used for χ_a'', the result converted to frequency units by $\omega = 2\pi\nu$ and the usual approximation $\omega + \omega_0 \approx 2\omega_0$ is made.

$$\gamma(\omega) = -\frac{\omega}{c}n\frac{\chi_a''}{n^2} = -\frac{\pi}{n}\frac{Ne^2}{mc}\left(\frac{1}{4\pi\epsilon_0}\right)\left[\frac{1/\tau}{(\omega_0 - \omega)^2 + (1/\tau)^2}\right] \quad (14.2.16)$$

Define

$$\Delta\nu = \frac{1}{2\pi\tau} \quad (14.2.17a)$$

and

$$g(\nu) = \frac{\Delta\nu}{2\pi[(\nu_0 - \nu)^2 + (\Delta\nu/2)^2]} \quad (14.2.17b)$$

Then

$$-\gamma(\nu) = \frac{\pi}{n}\frac{Ne^2}{mc}\left(\frac{1}{4\pi\epsilon_0}\right)g(\nu) \quad (14.2.17c)$$

In the early days of spectroscopy of gases, this equation seemed to work, almost. One could *measure* an integrated absorption:

$$A_b = \int[-\gamma(\nu)]d\nu = \frac{\pi}{n}\left(\frac{Ne^2}{mc}\right)\left(\frac{1}{4\pi\epsilon_0}\right)\binom{\text{integrated}}{\text{absorption}} \quad (14.2.18)$$

Unfortunately, it appears that only a fraction f_{12} of the active electron* participated in the absorption. Thus (14.2.17c) was corrected by replacing N with $N_1 f_{12}$ with the upper state presumed to be zero.

We *know* the correct answer to (14.2.18) from Chapter 7, and this relates f_{12} to the A coefficient (and conversely).

$$A_b = \int[-\gamma(\nu)]d\nu = \frac{\pi}{n}N_1 f_{12}\left(\frac{e^2}{mc}\right)\left(\frac{1}{4\pi\epsilon_0}\right) \quad \text{by (14.2.18)}$$

$$\stackrel{\Delta}{=} \frac{\lambda_0^2}{8\pi n^2}A_{21}\left(\frac{g_2}{g_1}\right)N_1 \quad \text{by (7.7.2b)}$$

Thus

$$A_{21} = 8\pi^2 n\frac{\nu^2}{c^3}\left(\frac{e^2}{m}\right)\frac{g_1}{g_2}f_{12}\left(\frac{1}{4\pi\epsilon_0}\right) \quad (14.2.19a)$$

*The fraction f_{12} is aptly named—the absorption oscillator strength with the term $g_1 f_{12}/g_2$ in (14.2.19a) being called the emission oscillator strength.

Sec. 14.2   The Classical Model of an Atom   517

Note that this is just $(3ng_1/g_2)f_{12}$ times the "classical" A coefficient defined by (14.2.9). For numerical values, we have*:

$$A_{21} = 6.67 \times 10^{-5} \left[\frac{1}{\lambda_0(\text{meters})}\right]^2 \frac{g_1}{g_2} f_{12} n \qquad (14.2.19\text{b})$$

Equation (14.2.19) is correct—it was forced to be! Now one can massage some of the prior ones to obtain their correct value for populations in both states. The key is to substitute letter symbols according to the following sequence:

$$N \longrightarrow N_1 f_{12} \longrightarrow -\frac{g_1}{g_2} f_{12}\left[N_2 - \frac{g_2}{g_1}N_1\right] \qquad (14.2.20)$$

with f_{12} defined by (14.2.19). The real part of the susceptibility χ'_a can be related to the imaginary part irrespective of the above shenanigans by dividing (14.2.12c) by (14.2.12d) and making the usual assumption $\omega + \omega_0 \approx 2\omega_0$.

$$\frac{\chi'_a}{\chi''_a} = 2(\omega_0 - \omega)\tau = 2(\nu_0 - \nu) \cdot 2\pi\tau$$
$$= \frac{(\nu_0 - \nu)}{\Delta\nu/2} \qquad (14.2.21\text{a})$$

Now use (14.2.15b):

$$\frac{\chi'_a}{n^2} = \left(\frac{\nu_0 - \nu}{\Delta\nu/2}\right)\frac{\chi''_a}{n^2} = -\left(\frac{\nu_0 - \nu}{\Delta\nu/2}\right)\left(\frac{\gamma(\nu)}{k'}\right) \qquad (14.2.21\text{b})$$

This relation is only valid for a Lorentzian line shape, but it is very close for any reasonably well-behaved "bell-shaped" distribution if $\Delta\nu$ is interpreted as the FWHM of the gain or absorption coefficient.

14.2.4 Dispersion: Mode Pulling

It should be clear that the classical analysis can be forced to yield the correct answer by proper identification of the coefficients multiplying the response of the active electrons.[†] One of the benefits of the analysis is the recognition that the phase shift through an active medium, k', also depends on the proximity of the wave's frequency to the center of the transition. This was expressed by (14.2.14), which can be combined with (14.2.21b):

$$k' = \frac{\omega}{c} n\left(1 + \frac{\chi'_a}{2n^2}\right) = \frac{\omega}{c} n\left[1 - \left(\frac{\nu_0 - \nu}{\Delta\nu}\right)\frac{\gamma(\nu)}{k'}\right] \qquad (14.2.22)$$

*Most texts omit n from this relation since they are dealing with gases where $n \approx 1$. Its presence hints at the fact that spontaneous emission in an anistropic medium, such as ruby, is also anistropic.

[†] The only weak link is the identification of $1/\tau$ as the damping rate from radiation in (14.2.6). The quantum approach will teach us that $1/\tau$ should be assigned to phase-changing collisions. For our purposes here, we need only that $\Delta\nu = (1/2\pi\tau)$, that is, (14.2.17a).

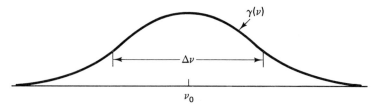

(a) Typical frequency dependence of the gain coefficient

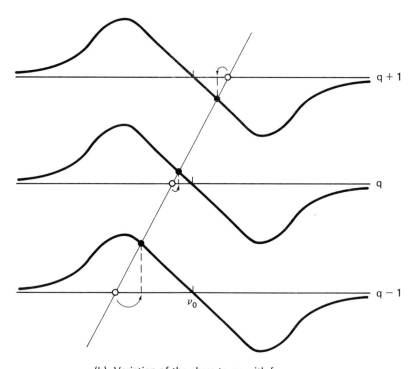

(b) Variation of the phase terms with frequency

Figure 14.2 Graphic solution for cavity resonance.

Thus the phase constant is less than or greater than $\omega n/c$ depending on whether $\nu < \nu_0$ or $\nu > \nu_0$, (assuming $\gamma(\nu)$ is positive, that is, gain). This is called "anomalous dispersion" and is manifested in a "pulling" of the oscillation frequency off a cold cavity resonance and toward line center. This is easily demonstrated by the following analysis.

The *active* cavity resonance ν_q is determined by the usual condition: RTPS $= q \cdot 2\pi$ with the phase shift per unit of length given by (14.2.22):

Sec. 14.2 The Classical Model of an Atom

$$\frac{\omega}{c} 2nd - \left(\frac{\nu_0 - \nu}{\Delta \nu}\right) \gamma(\nu) \, 2d = q \cdot 2\pi$$

$$\left(\frac{2nd}{c}\right) \nu = \left[q + \left(\frac{\nu_o - \nu}{\Delta \nu}\right)\left(\frac{\gamma(\nu) \cdot 2d}{2\pi}\right)\right] \qquad (14.2.23)$$

This is not an easy expression to solve for ν_q since $\gamma(\nu)$ is a rather complex expression. Let us use a graphic approach rather than an analytic one so as to obtain the essential features. This approach is sketched in Fig. 14.2 where both sides of (14.2.23) are plotted against frequency. The left-hand side is just a straight line of slope $(2nd/c)$. The right-hand side is a series of frequency-independent lines, differing by the integer values of q, plus the term related to the integrated gain coefficient times the length. Fig. 14.3(a) is a sketch of $\gamma(\nu)$ and is placed above the graphic solution to aid in the mental construction of frequency variation of the added term in (14.2.23). Note that if $\gamma(\nu) > 0$ and $\nu < \nu_0$, then that term is positive; for $\nu > \nu_0$, the reverse is true. Where the straight line intersects the solid curves are the possible oscillation frequencies; where it intersects the dashed horizontal lines represents the passive cavity resonance (i.e. $\gamma(\nu) = 0$). Note that modes are always pulled *toward* line center ν_0.

The pulling is not a large absolute frequency shift compared to ν_0, but it is annoying if one contemplates a coherent heterodyne detection scheme in an

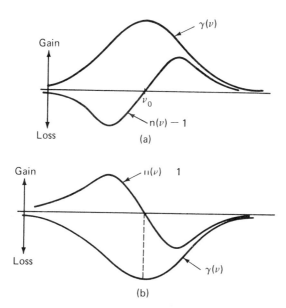

Figure 14.3 Anomalous disperson in the vicinity of a transition with (a) gain $[\gamma(\nu) > 0]$, or (b) absorption $[\gamma(\nu) < 0]$.

optical communication link. If the laser's gain fluctuates, there is an associated frequency shift in addition to any amplitude changes. A numerical evaluation of this problem is left as a problem for the student.

14.3 QUANTUM VIEWPOINT OF THE INTERACTION OF THE ATOM WITH A CLASSICAL FIELD

Up until now we have considered the interaction from a phenomenological-rate-equation point of view, with the coefficients related to the Einstein A and B terms. Although this approach is intuitively satisfying, it does have some deficiencies that are not apparent at this time. Further, we have not indicated how these terms can be connected to the fundamental description of the atom. The rate-equation approach deals with an average atom, whereas there are effects in which an entire group of atoms acts in concert. Thus we must first appreciate the reaction of a single atom to an impressed electromagnetic field.

14.3.1 General Formulation

We will approach this interaction from a semiclassical viewpoint whereby the atom is quantized and the field has its classical interpretation and obeys Maxwell's equations. The first item to be discussed is the effect of the field on the quantum picture of the atom. Consider an isolated atom; the energy levels E_n and the wave functions Ψ_n are found from a solution to

$$\mathbf{H}_{op} \Psi = j\hbar \frac{\partial \Psi}{\partial t} \tag{14.3.1a}$$

or

$$\left(-\frac{\hbar^2}{2m}\nabla^2 + V\right)\Psi = j\hbar \frac{\partial \Psi}{\partial t} \tag{14.3.1b}$$

We assume that somebody else has solved this problem to provide us with a complete set $\{\Psi_n\}$ of suitable normalized orthogonal wave functions.

$$\Psi_n(r,t) = u_n(r)e^{-j(E_n/\hbar)t} \tag{14.3.2a}$$

where

$$\int \Psi_m^* \Psi_n \, d^3x = \delta_{m,n} \tag{14.3.2b}$$

and E_n = energy of the nth state. Transitions occur between these states such that $\hbar\omega_{m,n} = E_m - E_n$ with a spontaneous rate, which, for a large collection of atoms, is called the Einstein A coefficient. Our first goal is to relate this rate back to the roots found in the quantum description of the atom.

Sec. 14.3 Quantum Viewpoint of the Interaction of the Atom with a Classical Field

Unfortunately, there are serious roadblocks to be overcome. Equation (14.3.1a) is for an isolated atom, whereas the real case involves many more-or-less identical atoms that interact with each other, other foreign atoms or systems, or external fields. There is always some electromagnetic energy present—the so-called "zero-point" fluctuations—which stimulates the spontaneous decay of one state into another. In order to obtain a value of A from first principles, it is necessary to quantize the fields, a procedure that involves much more mathematics than the level of this book warrants. We will retain the electromagnetic field as a classical quantity, and focus our attention on the rate at which absorption or stimulated emission causes transitions between the states, which is the issue of major concern anyway!

Now let us lay out our game plan for the rest of the chapter. We first pick one atom, examine its response to an external applied electric field, and guess that it is representative of the collection. This procedure allows one to derive the Einstein coefficient B_{12} (or B_{21}) and thus obtain $A_{21} = B_{21}(8\pi h\nu^3/c^3)$. This analysis will point out an inherent weakness in the rate equation viewpoint: it is still a guess that one atom represents the collection. In practice, we do not know enough about any system to specify all of the details anyway, and thus a statistical approach is necessary, which goes by the name, "the density matrix." For the simple laser interactions considered in this book, the density matrix approach leads to precisely the same answer for laser gain equation and saturation behavior as was found in Chapter 8, but with a considerably more rigorous foundation. However, it introduces a formalism that will handle much more complex problems than those considered in this book.

14.3.2 Perturbation Solution of Schrödinger's Equation

The presence of an external field adds an extra potential energy term to the total energy operator of (14.3.1). For $\mathbf{E} = E_{0x} \cos \omega t \, \mathbf{a}_x$, the added potential energy from a displacement of an electron is given by

$$v = -e\phi \qquad \phi = -\int \mathbf{E} \cdot d\mathbf{l}$$

or

$$v = ex \, E_{0x} \cos \omega t \qquad (14.3.3)$$

Now we have the job of solving the "new" equation

$$\underbrace{\left[-\frac{\hbar^2}{2m}\nabla^2 + V(r)\right.}_{\mathbf{H}_0} + \underbrace{v\vphantom{\frac{\hbar^2}{2m}}\right]}_{\mathbf{H}'}\Psi = j\hbar\frac{\partial\psi}{\partial\tau} \qquad (14.3.4)$$

In the "old days" (before lasers) one was quite justified in assuming that the internal Hamiltonian operator $\mathbf{H}_0 \gg \mathbf{H}'$ since the internal atomic fields were huge compared to anything imaginable (at that time).* Now, however, it is conceivable to interrogate with fields available from lasers that makes **H'** significant compared to **H₀**, and thus the original solution to (14.3.1) is not applicable at all. This causes all sorts of problems; transitions can occur between states that cannot be identified with the isolated atom. We avoid these problems by making the usual assumption that $H_0 \gg H'$ and thus restrict our attention to a weak field interaction (but that does not make the above problem go away).

We recognize that the solution to (14.3.1) has many of the characteristics agreeing with experiments, and thus we suspect that the solution to (14.3.4) can be expressed in terms of the isolated unperturbed atoms' wave functions.

$$\Psi = \sum_n a_n(t) u_n e^{-j(E_n/\hbar)t} \tag{14.3.5}$$

This can be done for any function since the set $\{\psi_n\}$ is *complete*. We are guided by the idea that the new system will bear a strong resemblance to the isolated one. Now the square of the coefficient of the state n, $|a_n(t)|^2$, is, of course, the probability of the system being in this particular state. Thus we substitute the infinite series into Schrödinger's equation:

$$\sum_n \left[\left(-\frac{\hbar^2}{2m} \nabla^2 + V \right) + v \right] a_n(t) u_n e^{-j(E_n/\hbar)t} = j\hbar \sum \frac{\partial}{\partial t} [a_n(t) e^{-j(E_n/\hbar)t}] u_n$$

$$= \sum E_n a_n(t) u_n e^{-j(E_n/\hbar)t} + j\hbar \sum \dot{a}_n(t) u_n e^{-j(E_n/\hbar)t} \tag{14.3.6}$$

Combining the first term on the right with the first parenthesis on the left leads to

$$\sum \left[-\frac{\hbar^2}{2m} \nabla^2 + V - E_n \right] u_n a_n e^{-j(E_n/\hbar)t} + \sum a_n(t) v u_n e^{-j(E_n/\hbar)t}$$

$$= j\hbar \sum \dot{a}_n(t) u_n e^{-j(E_n/\hbar)t} \tag{14.3.7}$$

The first series is zero since the functions u_n satisfy the original Schrödinger's equation. Now we eliminate the summation on the right by multiplying by $[u_m \exp -j(E_m/\hbar)t]^*$, that is, the complex conjugate of the mth wave function, integrating over all space, and using the orthogonality of the wave functions (14.3.2b). This procedure picks out the mth coefficient on the right because $\int u_m^* u_n \, dx = \delta_{m,n}$.

$$j\hbar \, \dot{a}_m(t) = \sum_n a_n(t) v_{m,n} e^{j\omega_{m,n} t} \tag{14.3.8}$$

---

*For instance, the field experienced by an electron in the first Bohr orbit is $e/4\pi\epsilon_0 a_0^2$ and with $a_0 = 0.526$ Å the electric field is 5.2×10^{11} V/m.

Sec. 14.3 Quantum Viewpoint of the Interaction of the Atom with a Classical Field

where the transition frequency $\omega_{m,n}$ is given by the difference in energy levels.

$$\hbar \omega_{m,n} = E_m - E_n \qquad (14.3.9)$$

and the matrix element $v_{m,n}$ is an abbreviation for

$$v_{m,n} = eE_{0x}\left[\int u_m^* x u_n \, d^3x\right] \cos \omega t$$

$$= \mu_{m,n} E_{0x} \cos \omega t \qquad (14.3.10)$$

and

$$\mu_{m,n} = e\langle x_{m,n}\rangle$$

As promised in Sec. 14.2, we obtain a result proportional to this electric dipole moment $\mu_{m,n}$. Note that the matrix element $v_{m,n}$ is zero unless v depends on space as it does in (14.3.10). Even if v does depend on space, the integral can still be zero. As we will see, such situations correspond to "forbidden" transitions.*

So far we are "exact," but the utility of all this arithmetic depends on making some useful approximations to solve the infinite set of coupled equations. Let us look at only two states, 1 and 2, as indicated in Fig. 14.4, and assume that the others do not affect the occupation probabilities $|a_{1,2}(t)|^2$ to any appreciable extent. Thus our infinite set of coupled equations reduces to just two:

$$\frac{da_2}{dt} = a_1 \frac{\mu_{21x} E_{0x}}{2j\hbar}\left[e^{j(\omega+\omega_{21})t} + e^{-j(\omega-\omega_{21})t}\right]$$
$$+ a_2 \frac{\mu_{22x} E_{0x}}{2j\hbar}\left[e^{j\omega t} + e^{-j\omega t}\right] \qquad (14.3.11)$$

$$\frac{da_1}{dt} = a_1 \frac{\mu_{11x} E_{0x}}{2j\hbar}\left[e^{j\omega t} + e^{-j\omega t}\right]$$
$$+ a_2 \frac{\mu_{12x} E_{0x}}{2j\hbar}\left[e^{j(\omega-\omega_{21})t} + e^{-j(\omega-\omega_{21})t}\right] \qquad (14.3.12)$$

——————— 2

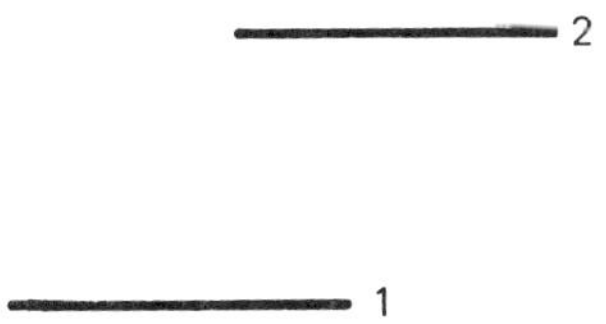

Figure 14.4 Energy-level scheme. ——————— 1

---

*Our analysis includes electric-dipole-field interactions only; other interactions, such as magnetic dipole and electric quadrapole, are possible but are much weaker than the electric dipole. Although those transitions are significant for lasers (the iodine laser at 1.3 μm is a magnetic dipole transition), they will not be covered here.

Terms of the form μ_{11} and μ_{22} are proportional to the expectation value of the atomic electron, being located at a distance $\langle x_{11} \rangle$ from the center. In other words, this is a *permanent dipole moment*. We will restrict our attention to systems where $\mu_{11} = \mu_{22} = 0$. We also neglect antiresonant terms, $\exp[\pm j(\omega + \omega_{21})t]$ because the average value of such a term is zero on a time scale of $(\omega - \omega_{21})^{-1}$. Thus the starting point for any serious analysis of the effect of the external field on a given state is

$$\frac{da_2}{dt} = a_1 \frac{\mu_{21x} E_{0x}}{2j\hbar} \exp[-j(\omega - \omega_{21})t] \qquad (14.3.13a)$$

$$\frac{da_1}{dt} = a_2 \frac{\mu_{12x} E_{0x}}{2j\hbar} \exp[+j(\omega - \omega_{21})t] \qquad (14.3.13b)$$

14.4 DERIVATION OF EINSTEIN COEFFICIENTS

One could spend a good fraction of one's life obtaining various approximate solutions to (14.3.13a) and (14.3.13b) under various circumstances. Probably the easiest to start with is the case where the system is prepared in state 1 at $t = 0$ when a weak electric field with a frequency close to the transition frequency is turned on. Our initial conditions are

$$a_1(t = 0) \sim 1, \qquad a_2(t = 0) \sim 0 \qquad (14.4.1)$$

We anticipate that this field causes the atom to undergo a transition from 1 to 2 in the manner shown in Fig. 14.5. Thus the atom is in state 1 and starting to make a transition to 2 by the *absorption* of energy from the electromagnetic wave. We examine the growth of a_2 for *short enough times* so that the approximation $a_1(t = 0) = a_{10} \sim 1$ is still valid (so as to simplify the mathematics).

$$|a_2(t)|^2 = a_{10}^2 \left[\frac{\langle\mu_{21}\rangle_x E_{0x}}{2\hbar}\right]^2 \left[\frac{\sin\left(\frac{\omega_{21} - \omega}{2}\right)t}{\frac{\omega_{21} - \omega}{2}t}\right]^2 t^2 \qquad (14.4.2)$$

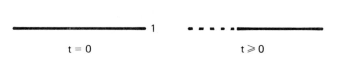

Figure 14.5 Effect of applying a field to an atom.

Sec. 14.4 Derivation of Einstein Coefficients

Let us do a bit of abbreviation. Let

$$\Omega = \frac{\langle \mu_{21} \rangle_x E_{0x}}{2\hbar} \quad (14.4.3)$$

Then, the population in state 2 becomes

$$\therefore |a_2(t)|^2 = a_{10}^2 \, \Omega^2 \left[\frac{\sin\left(\frac{\omega_{21} - \omega}{2}\right)t}{\left(\frac{\omega_{21} - \omega}{2}\right)t} \right]^2 t^2 \quad (14.4.4)$$

Now the bracketed factor oscillates violently if $\omega_{21} - \omega = \Delta\omega \neq 0$ and thus $|a_2(t)|^2$ remains small. If $\Delta\omega \approx 0$, then sinc$(x) = (\sin x/x) \approx 1$ and the occupation probability grows as t^2, a prediction that should bother us considerably. Why?

Recall Einstein's approach to this same problem, identify $|a_{10}|^2$ as N_1 and $|a_2(t)|^2$ as $N_2(t)$, and solve for the population $N_2(t)$ with the same approximations used to obtain (14.4.4) namely, $N_1 \sim N_{10}$, a constant.

$$\frac{dN_2}{dt} = B_{12} N_1 \rho_\nu \, g(\nu) \quad \text{or} \quad N_2 \simeq N_{10} B_{12} \rho_\nu \, g(\nu) t \quad (14.4.5)$$

Note that the rate equation approach would predict a linear growth with time whereas the time-dependent perturbation approach predicts a t^2 variation. *This should bother you!*

The origin of this apparent discrepancy is that we have insisted on a perfectly sharp transition given by $\hbar\omega_{21} = E_2 - E_1$, whereas for a collection of atoms, there is always a distribution of transition frequencies, which we have called the line-shape function $g(\nu_{21})$ and which is normalized.

$$\int_0^\infty g(\nu_{21}) d\nu_{21} = 1 \quad (14.4.6)$$

Thus to obtain a transition rate representative of a collection of atoms, we must prorate the response of one atom, (14.4.4), by the probability that the group of atoms has that center frequency, $g(\nu_{21})d\nu_{21}$. The average occupation probability becomes

$$\langle |a_2(t)|^2 \rangle = |a_{10}|^2 \Omega^2 \int_0^\infty \left[\frac{\sin \pi(\nu_{21} - \nu)t}{\pi(\nu_{21} - \nu)t} \right]^2 t^2 \, g(\nu_{21}) d\nu_{21} \quad (14.4.7)$$

Let us assume that the sinc(x) term is very sharply peaked about $\nu_{21} = \nu$ compared to a much slower variation of $g(\nu_{21})$ in this same frequency neighborhood. Thus we evaluate $g(\nu_{21})$ at $\nu_{21} = \nu$, pull it outside the integral, shift the interval of integration to $\pm\infty$, and massage the remainder of the integrand to look like a tabulated function:

$$\langle |a_2(t)|^2 \rangle = |a_{10}|^2 \, \Omega^2 \, g(\nu) t \left\{ \frac{1}{\pi} \int_{-\infty}^{+\infty} \left[\frac{\sin x}{x} \right]^2 dx \right\} \qquad x = \pi(\nu_{21} - \nu)t$$
(14.4.8)
$$= |a_{10}|^2 \, \Omega^2 \, g(\nu) t$$

We have now avoided the pitfall noted earlier and have obtained a result appealing to our intuition, the rate equations, and Einstein's approach. The only remaining task is to identify B_{12} by comparing (14.4.5) with (14.4.8).

$$\underbrace{N_2(t) = N_{10} B_{12} \rho_\nu \, g(\nu) t}_{(14.4.5)} \Longleftrightarrow \underbrace{\langle |a_2(t)|^2 \rangle = |a_{10}|^2 \left[\frac{\langle \mu_{21} \rangle_x E_{0x}}{2\hbar} \right]^2 g(\nu) t}_{(14.4.8)}$$

Now the energy density (at ν) is related to the electric field of the wave by

$$\rho_\nu = (\epsilon_0 E_{0x}^2)/2$$

and $\langle \mu_{21} \rangle_x^2$ can be related to the total (squared) dipole moment found in unpolarized light

$$\langle \mu_{21} \rangle_x^2 = \langle \mu_{21} \rangle^2 / 3$$

Thus

$$B_{12} = \frac{2}{3} \frac{\pi^2}{\epsilon_0} \frac{\langle \mu_{21} \rangle^2}{\hbar^2} \qquad (14.4.9)$$

We could have started this section and assumed the atom in state 2 beginning a transition to 1 because of the stimulus of the field. This merely means an interchange of subscripts $2 \to 1$, and thus (14.4.9) is also valid for the stimulated emission coefficient B_{21} (for equal degeneracies).

As mentioned earlier, we do not obtain the A coefficient without quantizing the field. However, we do know that A_{21} and B_{21} are related:

$$\frac{A_{21}}{B_{21}} = \frac{8\pi h \nu^3}{c^3} \qquad (7.3.10b)$$

Thus

$$A_{21} = \frac{8\pi h \nu^3}{c^3} B_{21} = \frac{16\pi^3}{3\epsilon_0} \left(\frac{\nu}{c} \right)^3 \frac{\langle \mu_{21} \rangle^2}{\hbar}$$

$$= \frac{64\pi^4}{3} \left(\frac{1}{4\pi\epsilon_0} \right) \left(\frac{\nu}{c} \right)^3 \frac{\langle \mu_{21} \rangle^2}{\hbar} \qquad (14.4.10)$$

Thus we have now achieved one goal of this chapter: to relate the Einstein coefficients to the most fundamental parameter of the atom, the wave function. Only in the simplest of cases, hydrogen atoms (not molecules), can this be carried through to the bitter end. If we were only interested in this particular goal, it is doubtful whether all this arithmetic is worth the result. However, this

Sec. 14.5 Time-Dependent Populations: the Rabi Approach

section serves to introduce the next one, which points out the limitations of the rate-equation approach.

14.5 TIME-DEPENDENT POPULATIONS: THE RABI APPROACH

Let us return to (14.3.13a) and (14.3.13b) and allow for the asynchronous behavior of the occupation probabilities $a_1(t)$ and $a_2(t)$; that is, if $a_2(t)$ increases, $a_1(t)$ decreases to keep $|a_1(t)|^2 + |a_2(t)|^2 = 1$. We keep our previous abbreviation and write

$$\dot{a}_2(t) = -j\Omega a_1(t) e^{j(\Delta\omega)t} \qquad (14.3.13a)$$

$$\dot{a}_1(t) = -j\Omega a_2(t) e^{-j(\Delta\omega)t} \qquad (14.3.13b)$$

where

$$\Omega = \frac{\langle\mu_{21}\rangle_x E_{0x}}{2\hbar} \qquad \text{(the Rabi frequency)} \qquad (14.5.1a)$$

$$\Delta\omega = \omega_{21} - \omega \qquad \text{(frequency detuning)} \qquad (14.5.1b)$$

Differentiate (14.3.13b) and use (14.3.13a) for $a_2(t)$ so as to obtain a single equation for $a_1(t)$.

$$\ddot{a}_1(t) + j\Delta\omega \dot{a}_1(t) + \Omega^2 a_1(t) = 0 \qquad (14.5.2)$$

If one assumes a solution of the form $a_1 = C \exp j\alpha t$, then the parameter α has two roots:

$$\alpha_{1,2} = -\left(\frac{\Delta\omega}{2}\right) \pm \sqrt{\left(\frac{\Delta\omega}{2}\right)^2 + \Omega^2} \qquad (14.5.3)$$

and $a_1(t)$ is given by

$$a_1(t) = C_1 e^{j\alpha_1 t} + C_2 e^{j\alpha_2 t} \qquad (14.5.4)$$

where the constants C_1 and C_2 are to be evaluated from the initial conditions picked to be the same as previously chosen: $a_1(0) \simeq 1$, $a_2(0) \approx 0$; and thus $\dot{a}_1(0) \approx 0$ by (14.3.13b). Applying these conditions, we obtain

$$a_1(t) = \frac{\alpha_1 e^{j\alpha_2 t} - \alpha_2 e^{j\alpha_1 t}}{\alpha_1 - \alpha_2}$$

Suffering through some uninspiring arithmetic leads to

$$a_2(t) = j\Omega e^{-j(\alpha_1 + \alpha_2/2)t} \left[\frac{\sin\dfrac{(\alpha_2 - \alpha_1)t}{2}}{\dfrac{(\alpha_2 - \alpha_1)t}{2}} \right] t \qquad (14.5.5)$$

Now if we reinsert the values for α_1 and α_2 in terms of $\Delta\omega$ and the Rabi frequency, Ω, and formulate $|a_2(t)|^2$ we obtain a familiar result:

$$|a_2(t)|^2 = \Omega^2 \left\{ \frac{\sin\left[\left(\frac{\Delta\omega}{2}\right)^2 + \Omega^2\right]^{1/2} t}{\left[\left(\frac{\Delta\omega}{2}\right)^2 + \Omega^2\right]^{1/2} t} \right\}^2 t^2 \quad (14.5.6a)$$

This equation contains a lot of physics so it is appropriate to dissect it in every way possible. First of all, note that it looks quite similar to (14.4.4) but does not have the restriction of being limited to small times such that $|a_2(t)| \ll 1$. If we cancel the t^2 term and rewrite (14.5.6a), the equation becomes easier to interpret:

$$|a_2(t)|^2 = \frac{\Omega^2}{\left(\frac{\Delta\omega}{2}\right)^2 + \Omega^2} \sin^2\left[\left(\frac{\Delta\omega}{2}\right)^2 + \Omega^2\right]^{1/2} t \quad (14.5.6b)$$

Thus the occupation probability $|a_2(t)|^2$ starts at zero, as assumed; increases to a level that depends on the frequency detuning $\Delta\omega$ *and* the field strength parameter, $\Omega = \mu E_{0x}/2\hbar$; and "flops" back and forth between the extremes at an angular frequency given by $[(\Delta\omega/2)^2 + \Omega^2]^{1/2}$. If the frequency detuning $\Delta\omega$ is zero, the flopping frequency is the Rabi value (14.5.1a).

This variation is sketched in Fig. 14.6. Note that this time dependence is completely at odds with the rate equation approach, which would indicate that the time derivative of N_2 would vanish when $|a_2(t)|^2 = |a_1(t)|^2$, that is $N_2 = N_1$. Not so! This points out a serious limitation to the rate equation approach—it will not handle the "flopping" of the populations. However, we have yet to put in the statistics of the interaction between the active and foreign atoms, and, when all that smoke clears, one recovers the Einstein approach, but it is not obvious at this time.

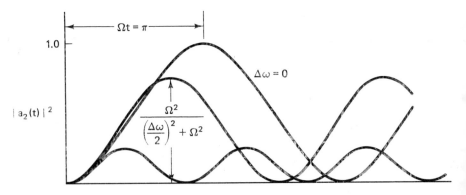

Figure 14.6 Time evolution of population in state 2.

Sec. 14.5 Time-Dependent Populations: the Rabi Approach

Let us again formulate an average atom response by summing over the line shape as was done previously (14.4.7):

$$\langle |a_2(t)|^2 \rangle = \Omega^2 \int_0^\infty \left\{ \frac{\sin[\pi^2(\nu_{21} - \nu)^2 + \Omega^2]^{1/2}t}{[\pi^2(\nu_{21} - \nu)^2 + \Omega^2]^{1/2}} \right\} g(\nu_{21})\, d\nu_{21} \quad (14.5.7)$$

Under some circumstances, one can evaluate this as was done previously. If, for instance, the field were weak such that $\Omega \ll \Delta\omega/2 = \pi(\nu_{21} - \nu)$, then the $(\sin^2 x)/x^2$ term "samples" $g(\nu_{21})$ in a narrow frequency interval around $\nu_{21} \approx \nu$. As the field becomes bigger, $\Omega > \Delta\omega/2$, and its effects are felt by a much larger spread in frequencies given by

$$\nu_{21} = \nu \pm \frac{\Omega}{\pi} \quad (14.5.8)$$

Let us take a typical case to obtain a feeling for the extent of this power broadening:

$$\text{Let } \mu = e\langle x \rangle = 1 \text{ debye} = 3.33 \times 10^{-30} \text{ C-m}$$

$$\therefore \langle x \rangle = 2.08 \times 10^{-11} \text{m} = 0.208 \text{ Å}$$

which is just about as big as one can expect for a displacement of an electron from its equilibrium value.

Let us pick a modest intensity of 100 W/cm²

$$E_{0x}^2 / 2\eta_0 = 100 \text{ W/cm}^2$$

$$\therefore E_{0x} = 275 \text{ V/cm} = 27.5 \text{ kV/m}$$

Hence

$$\frac{\Omega}{2\pi} = \frac{\mu_{21} E_{0x}}{2h} = 138 \text{ MHz}$$

In other words, the field can cause transitions over a broadened portion of the spectrum. This is the origin of what is called "power broadening." This power broadening is large compared to the "natural" broadening process (i.e., from spontaneous emission) but still small compared to a Doppler width for visible wavelengths for the numerical case considered above.

However, let us go to an obviously strong field from a 1 MW laser focused to a 1-μm spot size (area = 10^{-8} cm²). Then $E_{0x} = 2.75 \times 10^{10}$ V/m and $\Omega/2\pi = 69$ THz. If we compute an energy spread associated with this power broadening, $\hbar\Omega = \Delta E$, then

$$\frac{\Delta E}{e} = \frac{\hbar \Omega}{e} = \frac{\mu E_{0x}}{2} = \frac{\langle x \rangle E_{0x}}{2} = 0.286 \text{ eV}$$

Thus what started out to be a nice sharp transition to be interrogated by only those frequencies close to ν_{21} is now affected by radiation detuned from $\nu_{21} = \nu$

by as much as at 0.286 eV. One should not be surprised therefore to find a host of "nonlinear" optical phenomena occurring with these intense fields. We need a related but different formalism to handle this case, the density matrix description that is discussed in the next section.

14.6 THE DENSITY MATRIX

14.6.1 Definition

Maxwell's equations describe a laser quite well, provided we compute the polarization **P** according to the rules of quantum mechanics. This involves computing the time evolution of the wave function leading to the electric polarization as the case of concern to us here. Let us recall the procedure from Sec. 14.3. We assumed that the wave function for *one* of the atoms could be represented by

$$\Psi(r,t) = \sum_n C_n(t) u_n(r) \qquad (14.6.1a)$$

where $u_n(r)$ is any arbitrary set of complete orthonormal functions, but we prefer them to be solutions to the time-independent Schrödinger's equation.

$$\left[-\frac{\hbar^2}{2m} \nabla^2 + V \right] u_n = E_n u_n \qquad (14.6.1b)$$

If we desired the expectation value of some operator **A**, we would compute

$$\langle A \rangle = (\Psi, \mathbf{A}\Psi) = \int \Psi^* \mathbf{A} \Psi \, d^3x \qquad (14.6.2a)$$

a compact expression that balloons to a large one when written out in detail.

$$\langle A \rangle = \int \left\{ \sum_m C_m^* u_m^*(r) \cdot \hat{\mathbf{A}} \left[\sum_n C_n(t) u_n(r) \right] \right\} d^3x$$

$$= \sum_m \sum_n C_m^*(t) \left[\int u_m^*(r) \, \hat{\mathbf{A}} \, u_n(r) \, d^3x \right] C_n(t) \qquad (14.6.2b)$$

The quantity in the brackets [ ] is the matrix element $A_{m,n}$ of the operator **A**. Thus

$$\langle A \rangle = \sum C_m^*(t) A_{m,n} C_n(t) \qquad (14.6.2c)$$

Now let us break for a moment and touch base with the real world. We will be interested in the dipole moment $\langle \mu \rangle = e \langle x \rangle$ because the polarization is given by, $P = N \langle \mu \rangle$, and that in turn goes into Maxwell's equations. Thus, if it makes you more comfortable, use the letter symbol $\langle \mu \rangle$ instead of $\langle A \rangle$. However, since most texts keep the discussion general so as to be applicable to all cases such as

Sec. 14.6 The Density Matrix

those dealing with electric quadrupole, magnetic dipole transitions we will also continue to use **A**.

From (14.6.2a), we note that the bilinear products, $C_m^*(t)\, C_n(t)$, which control the time dependence of $\langle A \rangle$ for a single atom. When we have a collection of N atoms, each with a slightly different parameter (say velocity), which we identify by the parameter s, we multiply (14.6.2c) by the probability that the class s can exist, $\mathcal{P}_s$, and sum over s.

$$\overline{\langle A \rangle_s} = \sum_s \mathcal{P}_s \langle A \rangle_s \qquad (14.6.3a)$$

where the subscript on $\langle A \rangle_s$ denotes the expectation value of the operator for the sth class of atoms, and the bar signifies the average over that distribution. This class s can be a classical variable, velocity in the case of a Doppler distribution, or a quantum statistic such as the phase distribution between the different atoms. More to the point is that sufficient information is not available to specify a wave function for each atom anyway, and thus we must use a statistical approach.

Now we "tag" all of our previous definitions of Ψ and $C(t)$ with a (prior) superscript s to indicate that these quantities depend on which part of the distribution is being considered. It is not necessary to tag $A_{m,n}$ since we have implicitly assumed atoms with identical stationary wave functions.

Thus

$$\langle A \rangle_s = \sum_{m,n} {}^{(s)}C_m^*(t)\, A_{m,n}\, {}^{(s)}C_n(t) \qquad (14.6.2d)$$

and the statistical average becomes

$$\overline{\langle A \rangle_s} = \sum_s \sum_{m,n} \mathcal{P}_s\, {}^{(s)}C_m^*(t)\, {}^{(s)}C_n(t)\, A_{m,n} \qquad (14.6.3b)$$

Define

$$\rho_{n,m} = \sum_s \mathcal{P}_s\, {}^{(s)}C_m^*(t)\, {}^{(s)}C_n(t) \equiv \overline{C_m^* C_n} \qquad (14.6.4)$$

The array $\rho_{n,m}$ is called the *density matrix*. With it, one can compute the expectation value of *any* operator in a rather simple fashion:

$$\overline{\langle A \rangle} = \sum_n \sum_m \rho_{n,m}\, A_{m,n} \qquad (14.6.3c)$$

Now by the rules of matrix multiplication

$$(AB)_{k,l} = \sum_m A_{k,m}\, B_{m,l}$$

Equation (14.6.3c) becomes

$$\overline{\langle A \rangle} = \sum_n (\rho A)_{n,n} \stackrel{\Delta}{=} tr(\rho A) \tag{14.6.5}$$

where tr = trace = sum of the diagonal elements of the matrix (ρA).

Now $N \cdot \rho_{n,n}$ is the probability of N_n atoms being in state n, and because all of the atoms have to be in some state, we have

$$\sum_n \rho_{n,n} = tr(\rho) = 1 \tag{14.6.6}$$

Another useful fact about the density matrix $\rho_{n,m}$ is that it is Hermitian; that is, because

$$\rho_{n,m} \stackrel{\Delta}{=} \overline{C_m^* C_n} \tag{14.6.4}$$

therefore

$$\rho_{m,n} \stackrel{\Delta}{=} \overline{C_n^* C_m} = \rho_{n,m}^* \tag{14.6.7}$$

Let us focus on the electric dipole moment operator $\mu = -ex$ as an example. Thus the matrix elements $\mu_{m,n}$ are given by

$$\mu_{m,n} = -e \int u_m^*(\mathbf{r}) x u_n(\mathbf{r}) d^3x$$

If we further restrict our attention to a 2-level atom such that $\mu_{11} = \mu_{22} = 0$ (no permanent dipole moment), then, according to the density matrix prescription, the dipole moment of the distribution is given by

$$\overline{\langle \mu \rangle} = \sum_{n,m} \rho_{n,m} \mu_{m,n}$$

$$= \underbrace{\rho_{11}\mu_{11}}_{0} + \rho_{12}\mu_{21} + \rho_{21}\mu_{12} + \underbrace{\rho_{22}\mu_{22}}_{0}$$

$$= \mu(\rho_{12} + \rho_{21}) \tag{14.6.8}$$

since the phases of u can always be chosen such that $\mu_{21} = \mu_{12} = \mu$. Thus the polarization P goes into Maxwell's equations is given by

$$P = N\mu_{21}(\rho_{12} + \rho_{21}) = N\mu_{21}(\rho_{21}^* + \rho_{21}) \tag{14.6.9}$$

Let us take an example that illustrates the significance of the parameter s. It will also illustrate the insight provided by the density matrix approach that is not available otherwise.

Recall the definition of $\rho_{n,m}$ (14.6.4) and compute the polarization by (14.6.9). Suppose $\mathcal{P}_s$ represents a distribution of *phases* between 0 and 2π of state m with respect to time as kept by state n.

Sec. 14.6 The Density Matrix

Let $\mathcal{P}_s = P(\phi_s)e^{j\phi_s}$ where the probability is normalized: $\int_0^{2\pi} P(\phi_s)d\phi_s = 1$

If $P(\phi_s) = \delta(\phi - \phi_0)e^{j\phi}$, that is, a well-defined phase relationship for all atoms, then

$$\rho_{n,m} = \int \delta(\phi - \phi_0)e^{j\phi}C_m^*C_n \, d\phi = e^{j\phi_0}C_m^*C_n$$

Contrast this with the case where the phases are *uniformly* distributed between 0 and 2π.

Let

$$\mathcal{P}_s = \frac{1}{2\pi}e^{j\phi_s}$$

Hence

$$\rho_{n,m} = C_m^*C_n\left[\frac{1}{2\pi}\int_0^{2\pi} e^{j\phi_s}d\phi_s\right] \equiv 0$$

Thus there would be *no* polarization if the phases of the bilinear products were uniformly distributed over 2π radians. Thus we have now discovered something that is very important to lasers.

Anything that tends to randomize the phase of state m *with respect to* n *reduces the polarization.*

This is not an obvious fact from anything discussed to date. It was alluded to in Chapter 7 in the discussion of pressure broadening, but now we have a demonstration of its validity.

14.6.2 Equation of Motion for the Density Matrix

It is just a matter of patience using time derivatives to obtain an equation describing the time variation of ρ. We start with the definition of $\rho_{n,m}$:

$$\rho_{n,m} \stackrel{\Delta}{=} \overline{C_m^*(t)C_n(t)} \qquad (14.6.4)$$

Differentiate with respect to time (and drop the bar to save a bit in notation).

$$\frac{\partial \rho_{n,m}}{\partial t} = C_m^*\frac{\partial C_n}{\partial t} + C_n\frac{\partial}{\partial t}C_n\frac{\partial C_m^*}{\partial t} \qquad (14.6.10)$$

Now we retreat to Schrödinger's equation, use the definitions of $C(t)$, and the orthogonality of u_n to obtain an expression for the time derivatives of $C(t)$.

That is,

$$\Psi = \sum C_n(t) u_n(r) \quad \text{[defines } C_n(t)\text{]}$$

with

$$j\hbar \partial \Psi / \partial t = \mathbf{H}\Psi \quad \text{(Schrödinger's equation)}$$

Thus

$$j\hbar \sum_n \frac{\partial C_n(t)}{\partial t} u_n(r) = \sum C_n(t) \, \mathbf{H} \, u_n(r) \quad (14.6.11)$$

Multiply by $u_m^*(r)$ and use orthogonality of the spatial part of the wave functions $u(r)$ to obtain an expression for the time dependence of $C_m(t)$.

$$j\hbar \frac{\partial C_m(t)}{\partial t} = \sum_n C_n H_{m,n} \quad (14.6.12)$$

where $H_{m,n}$ represents the matrix elements of the total energy operator, internal plus any external fields. Thus

$$\frac{\partial C_n}{\partial t} = \frac{1}{j\hbar} \sum_k C_k H_{n,k} \quad (14.6.13a)$$

and

$$\frac{\partial C_m^*}{\partial t} = -\frac{1}{j\hbar} \sum_k C_k^* H_{m,k}^* \quad (14.6.13b)$$

where the dummy index k is introduced to represent the summing operation demanded by (14.6.12). Now the total energy operator is Hermitian, that is, $H_{m,n} = H_{n,m}^*$, which enables us to simplify (14.6.13) when written out in terms of (14.6.13a) and (14.6.13b):

$$\frac{\partial \rho_{n,m}}{\partial t} = C_m^* \frac{\partial C_n}{\partial t} + C_n \frac{\partial C_m^*}{\partial t} \quad (14.6.10)$$

$$= \frac{1}{j\hbar} \sum_k C_m^* C_k H_{n,k} - C_k^* C_n H_{k,m}$$

$$= \frac{1}{j\hbar} \sum_k (\rho_{k,m} H_{n,k} - \rho_{n,k} H_{k,m}) \quad (14.6.14a)$$

Some prefer to write (14.6.14a) in terms of the Poisson bracket symbol:

$$\frac{\partial \rho_{n,m}}{\partial t} = \frac{1}{j\hbar} [\rho, H]_{n,m} \quad (14.6.14b)$$

and thus (14.6.14a) defines the expansion of Poisson brackets.

14.6.3 The Density Matrix Equations for a Two-Level System

Let us again consider the case of an interaction of an external field $E(t)$ with an atom and assume $\mu_{11} = \mu_{22} = 0$. Now the total energy operator consists of a part representing the internal forces of the atom (or system) H_0 and a small perturbation $H' = -\mu E(t)$ where $H_0 \gg H'$.

Now we carefully write out the terms required by (14.6.14a), assuming as before that only states 2 and 1 are significantly affected by the external field. Let $n = 2$ and $m = 1$.

$$\frac{\partial \rho_{21}}{\partial t} = \frac{-j}{\hbar}\left\{\underbrace{\rho_{11}H_{21} - \rho_{21}H_{11}}_{k=1} + \underbrace{\rho_{21}H_{22} - \rho_{22}H_{21}}_{k=2}\right\} \quad (14.6.15a)$$

Now

$$H_{11} = \int u_1^* \mathbf{H}_0 u_1 \, d^3x$$

but

$$\mathbf{H}_0 u_1 = E_1 u_1$$

$$\therefore H_{11} = E_1, \text{ the energy of state 1}$$

Likewise

$$H_{22} = E_2, \text{ the energy of state 2}$$

Finally,

$$H_{21} = \int u_2^*[-exE_x(t)]u_1 \, d^3x$$

$$= -\mu_{21x}E_x(t)$$

Thus (14.6.15a) becomes

$$\frac{\partial \rho_{21}}{\partial t} = -j\frac{\mu_{21x}E_x(t)}{\hbar}(\rho_{22} - \rho_{11}) - j\omega_{21}\rho_{21} \quad (14.6.16)$$

where

$$E_2 - E_1 = \hbar\omega_{21}$$

Now let $n = 2$ and $m = 2$

$$\frac{\partial \rho_{22}}{\partial t} = \frac{1}{j\hbar}\underbrace{[\rho_{12}H_{21} - \rho_{21}H_{12}}_{k=1} + \underbrace{\rho_{22}H_{22} - \rho_{22}H_{22}]}_{k=2 \text{ (equals 0)}} \quad (14.6.17a)$$

Now $\rho_{1,2} = \rho_{2,1}^*$ since ρ is Hermitian. One can always pick the basis function u_n to be real and thus $H_{12} = H_{21} = -\mu_{21x} E_x(t)$.

$$\frac{\partial \rho_{22}}{\partial t} = \frac{1}{j\hbar}[\rho_{21}^* - \rho_{21}][-\mu_{21x} E_x(t)] \qquad (14.6.17b)$$

or

$$\frac{\partial \rho_{22}}{\partial t} = -j\frac{\mu_{21x} E_x(t)}{\hbar}[\rho_{21} - \rho_{21}^*] \qquad (14.6.18)$$

for $n = 1$ and $m = 1$

$$\frac{\partial \rho_{11}}{\partial t} = j\frac{\mu_{21x} E_x(t)}{\hbar}[\rho_{21} - \rho_{21}^*] \qquad (14.6.19)$$

If we subtract (14.6.19) from (14.6.18), we obtain an equation for the *difference* $\rho_{22} - \rho_{11}$, which when multiplied by the number of atoms, N, represents the population inversion $N_2 - N_1$.

$$\frac{\partial}{\partial t}(\rho_{22} - \rho_{11}) = -j\frac{2\mu_{21x} E_x(t)}{\hbar}[\rho_{21} - \rho_{21}^*] \qquad (14.6.20)$$

It is probably good to back off and survey the damage caused by all this mathematics. Let us keep in mind the reason for "draining this swamp." We want to compute the polarization P, which is related to the density matrix by

$$P_x = N\mu_{21x}(\rho_{12} + \rho_{21}) = N\mu_{21x}(\rho_{21}^* + \rho_{21}) \qquad (14.6.9)$$

The off-diagonal elements ρ_{12} and ρ_{21} are related to the applied field *and* the diagonal elements by (14.6.16), and those in turn are controlled by the product of the field and the off-diagonal elements (14.6.20). Thus we are faced with a nonlinear problem. For instance, suppose $E(t) \approx e^{j\omega t}$ and we start by assuming $\rho_{22} - \rho_{11} =$ a constant in (14.6.16). Then $\rho_{2,1}$ will also have this same time dependence. But now (14.6.20) indicates that $(\rho_{22} - \rho_{11})$ is not constant but varies as a *product* of $E(t)$ and $\rho_{21}(t)$, which implies that our assumption of a constant difference $\rho_{22} - \rho_{11}$ is *not* quite correct. We will have to talk our way out of this problem later.

Before we attempt a simultaneous solution to these equations, we need to fold in some connection to the real world. The atoms in this ensemble do collide with each other and other foreign atoms or systems, and the states do decay with time. Thus there is a need for such decay terms—a task to which we now turn.

14.6.4 Relaxation Terms in the Density Matrix

We have already seen in Sec. 14.6.1 that any type of phase randomizing event decreases the polarization by decreasing ρ_{21}. Thus we fold into (14.6.16) a rate of destruction of ρ_{21} by such collisions in terms of a time constant T_2 and the dynamic equation becomes

Sec. 14.6 The Density Matrix

$$\frac{\partial \rho_{21}}{\partial t} + \left[\frac{1}{T_2} + j\omega_{21}\right]\rho_{21} = -j\frac{\mu_{21x}E_x(t)}{\hbar}(\rho_{22} - \rho_{11}) \quad (14.6.12) \rightarrow (14.6.21)$$

We also fold in the fact that the population difference, $\rho_{22} - \rho_{11}$, is being maintained at some equilibrium value $(\rho_{22} - \rho_{11})^0$ by some pumping process, which works against a natural decay rate $1/\tau$.

$$\frac{\partial}{\partial t}(\rho_{22} - \rho_{11}) + \frac{(\rho_{22} - \rho_{11}) - (\rho_{22} - \rho_{11})^0}{\tau}$$

$$= -j\frac{2\mu_{21x}E_x(t)}{\hbar}(\rho_{21} - \rho_{21}^*) \quad (14.6.22)$$

We now see that we have the possibility of a serious notation crisis. The quantity τ is *not* the decay rate of a state, but it is the decay rate of the population *difference*. Some references use the rate $1/T_1$ for the decay of the inversion and call such phenomena "T_1 processes." Unfortunately, it is easy to fall into the trap of assigning T_1 as the lifetime of state 1. Neither T_1 nor τ can be assigned directly to either state 1 or 2. However, a most serious notation problem arises with respect to the rate $1/T_2$; *T_2 is not the lifetime of state 2; it is the lifetime of the phase of state 2 with respect to state 1*. The rate $1/T_2$ represents the *decay* of the polarization as a result of phase-interrupting processes (elastic collisions with other gases or with phonons in a solid).

14.6.5 The Polarization Current

The necessary information to compute the polarization, $P = N\mu(\rho_{21}^* + \rho_{21})$, is contained in the "equation of motion" for ρ_{21} (14.6.21) and for the population difference $\rho_{22} - \rho_{11}$ given by (14.6.22). We have already noted that there is a problem with this set of equations. If we assume ρ_{21} to vary as $E(t)$ per (14.6.21), this forces the difference, $\rho_{22} - \rho_{11}$, to vary as $E(t)$ multiplied by $\rho_{21}(t)$, which must be used on the right side of (14.6.22) and so on. We avoid this seemingly endless merry-go-round by taking a harmonic balance of the two equations and solving for the leading terms. We start with the field.

$$\text{Let } E_x(t) = E_{0x} \cos \omega t = \frac{E_{0x}}{2}[e^{j\omega t} + e^{-j\omega t}] \quad (14.6.23)$$

and assume a slow variation of the population difference $(\rho_{22} - \rho_{11})$ on the time scale of the angular frequency ω^{-1}. Now we can anticipate that only part of the field, $E_0 e^{-j\omega t}$, causes an appreciable effect on ρ_{21}. This is easily verified by considering the homogeneous solution of (14.6.21) for the case where E is suddenly clamped to zero; the solution for ρ_{21} is

$$\rho_{21}(t) = \rho_{21}(0) e^{-j\omega_{21}t} e^{-t/T_2}$$

Thus only the term $E_0 e^{-j\omega t}$ of (14.6.23) "rotates" in near synchronism with the natural response. The other part of the cosine, $e^{+j\omega t}$, rotates in the opposite (phasor) sense and has little impact on ρ_{21}. This logic path is sometimes referred to as the rotating wave approximation.

Let us assume a slowly varying *amplitude* for ρ_{21} along with a fast time scale synchronous with $E(t)$; that is, let

$$\rho_{21} = \sigma_{21}(t) e^{-j\omega t} \quad (14.6.24a)$$

and

$$\rho_{12} = \sigma_{12}(t) e^{+j\omega t} \quad (14.6.24b)$$

We also know that $\rho_{2,1} = \rho_{12}^*$; hence we know too, that

$$\sigma_{21}(t) = \sigma_{12}^*(t) \quad (14.6.24c)$$

Now we make a careful harmonic balance of (14.6.21) and (14.6.22). If ρ_{21} and ρ_{21} vary as the driving frequency ω according to (14.6.21), then the product term $E(t)(\rho_{21} - \rho_{21}^*)$ has a slowly varying or "zero frequency" component and an additional variation at 2ω that appears in the expression for $\rho_{22} - \rho_{11}$. If one chases this harmonic reasoning around the loop a few times, then it becomes clear that odd harmonics, ω, 3ω, 5ω, and so on appear in the expression for ρ_{21}, and even harmonics 0, 2ω, 4ω, and so on appear in the difference $(\rho_{22} - \rho_{11})$. Taking only the first terms of the harmonic sequence for each quantity and using (14.6.24), we obtain

$$\left\{ \frac{d}{dt} \sigma_{21} + \left[\frac{1}{T_2} + j(\omega_{21} - \omega) \right] \sigma_{21} \right\} e^{-j\omega t}$$
$$= -j \left[\frac{\mu_{21x} E_{0x}}{2\hbar} e^{-j\omega t} + \frac{\mu_{21x} E_{0x}}{2\hbar} e^{+j\omega t} \right] (\rho_{22} - \rho_{11}) \quad (14.6.25)$$

The last term is neglected in the rotating wave approximation, and that step enables one to cancel common factors of $e^{-j\omega t}$.

$$\frac{d\sigma_{21}}{dt} + \left[\frac{1}{T_2} + j(\omega_{21} - \omega) \right] \sigma_{21} = -j\Omega(\rho_{22} - \rho_{11}) \quad (14.6.26)$$

where $\Omega = \mu_{21x} E_{0x}/2\hbar$, the Rabi flopping frequency. Substituting (14.6.24a) and (14.6.24b) into (14.6.22) and keeping only zero frequency terms yield

$$\frac{d}{dt}(\rho_{22} - \rho_{11}) + \frac{(\rho_{22} - \rho_{11}) - (\rho_{22} - \rho_{11})^0}{\tau} = -2j\Omega(\sigma_{21} - \sigma_{21}^*) \quad (14.6.27)$$

Let us pick the simplest of all cases—a steady state so that $d/dt = 0$. Thus (14.6.26) can be solved for σ_{21} in terms of the difference in populations.

Sec. 14.6 The Density Matrix

$$\sigma_{21} = \frac{\Omega(\rho_{22} - \rho_{11})}{(\omega_{21} - \omega) + j\frac{1}{T_2}} \quad (14.6.28a)$$

Hence

$$\sigma_{21}^* = \frac{\Omega(\rho_{22} - \rho_{11})}{(\omega_{21} - \omega) - j\frac{1}{T_2}}$$

and

$$\sigma_{21} - \sigma_{21}^* = -j\frac{2\Omega}{T_2}\left[\frac{1}{(\omega_{21} - \omega)^2 + \left(\frac{1}{T_2}\right)^2}\right](\rho_{22} - \rho_{11}) \quad (14.6.28b)$$

Now we substitute (14.6.28b) into (14.6.27) and solve for the population difference in terms of the equilibrium value.

$$\frac{(\rho_{22} - \rho_{11}) - (\rho_{22} - \rho_{11})^0}{\tau} = -\frac{4\Omega^2}{T_2}\left[\frac{1}{(\omega_{21} - \omega)^2 + \left(\frac{1}{T_2}\right)^2}\right](\rho_{22} - \rho_{11})$$

or

$$(\rho_{22} - \rho_{11}) = \frac{(\rho_{22} - \rho_{11})^0}{1 + 4\Omega^2 \tau T_2\left[\frac{(1/T_2)^2}{(\omega_{21} - \omega)^2 + (1/T_2)^2}\right]} \quad (14.6.29)$$

We have seen this equation before—back in Chapter 8 where saturation was first discussed. Remember, $\rho_{22} - \rho_{11}$ represents the population difference, and saturation is a reduction of that quantity as the stimulating field E (contained in Ω) becomes bigger. The resemblance becomes more striking if we define a normalized line shape function to be unity at line center as we did in Chapter 8.

$$\overline{g}(\omega) = \frac{(\Delta\omega/2)^2}{(\omega_{21} - \omega)^2 + (\Delta\omega/2)^2}$$

$$= \frac{(\Delta\nu/2)^2}{(\nu_{21} - \nu)^2 + (\Delta\nu/2)^2} = \overline{g}(\nu) \quad (14.6.30a)$$

where

$$\frac{\Delta\omega}{2} = \frac{1}{T_2} \quad \text{or} \quad \Delta\nu = \frac{1}{\pi T_2} \quad (14.6.30b)$$

With these definitions, (14.6.29) can be rewritten

$$(\rho_{22} - \rho_{11}) = \frac{(\rho_{22} - \rho_{11})^0}{1 + 4\Omega^2 \tau T_2 g(\nu)} \quad (14.6.31)$$

and we obviously have a term that looks like $[1 + I/I_s\, g(\nu)]^{-1}$ as found in Chapter 8.

Now we use this solution for $(\rho_{22} - \rho_{11})$ in (14.6.28a) to find an expression for σ_{21} (and σ_{21}^*). What we really need is the polarization P given by (14.6.9), which is related to σ_{21} by

$$P = N\mu_{21x}(\rho_{21} + \rho_{21}^*) \quad \text{(real quantity)} \quad (14.6.9)$$

$$= N\mu_{21x}(\sigma_{21} e^{-j\omega t} + \sigma_{21}^* e^{j\omega t}) \quad \text{by (14.6.24)}$$

$$\therefore P_x = N\mu_{21x}[2(\text{Re }\sigma_{21})\cos \omega t + 2(\text{Im }\sigma_{21})\sin \omega t] \quad (14.6.32)$$

Previously we had defined a complex susceptibility:

$$P_x = \epsilon_0(\chi' - j\chi'')E_{0x} e^{j\omega t} \quad (14.6.33a)$$

where $E_0 e^{j\omega t}$ was a phasor representation of a real field $E_{0x}\cos \omega t$. In terms of real functions,

$$P_x(t) = \epsilon_0 E_{0x}\{\chi' \cos \omega t + \chi'' \sin \omega t\} \quad (14.6.33b)$$

and thus

$$\epsilon_0 \chi' E_{0x} = N\mu_{21x} \cdot 2(\text{Re }\sigma_{21}) \quad (14.6.34a)$$

$$\epsilon_0 \chi'' E_{0x} = N\mu_{21x} 2(\text{Im }\sigma_{21}) \quad (14.6.34b)$$

Now it is just a simple but painful task to substitute (14.6.31) into (14.6.28a), rationalize, and identify real and imaginary parts of σ_{21}. After considerable work, one obtains

$$\chi' = -\frac{\mu_{21x}^2 T_2}{\epsilon_0 \hbar} \frac{(\omega_{21} - \omega)(T_2)^2}{1 + 4\Omega^2 \tau T_2 + (\omega_{21} - \omega)^2 (T_2)^2} \Delta N^0 \quad (14.6.35)$$

$$\chi'' = \frac{-\mu_{21x}^2 T_2}{\epsilon_0 \hbar} \frac{1}{1 + 4\Omega^2 \tau T_2 + (\omega_{21} - \omega)^2 T_2^2} \Delta N^0 \quad (14.6.36)$$

where

$$\Delta N^0 = N(\rho_{22} - \rho_{11})^0$$

Now recall that we had established quite firmly that the gain coefficient $\gamma(\nu)$ is related to χ'' by

Sec. 14.6   The Density Matrix

$$\gamma(\nu) = -\frac{k'}{n^2}\chi'' \qquad (14.6.37)$$

and thus we have finally arrived at a derivation of the saturated gain coefficient that has its roots firmly planted in quantum theory. If one works with the total dipole moment (for unpolarized light) $[\mu_{21x}]^2 = \mu_{21}^2/3$, one obtains

$$\gamma(\omega) = \frac{\omega}{c}\frac{1}{n}\frac{\mu_{21}^2}{3\epsilon_0}\frac{T_2}{\hbar}\left[\frac{\Delta N_0}{1 + 4\Omega^2\tau T_2 + (\omega_{21} - \omega)^2 T_2^2}\right] \qquad (14.6.38a)$$

or

$$\gamma(\omega) = \frac{\frac{\omega}{c}\frac{\mu_{21}^2 T_2}{3n\epsilon_0\hbar}\overline{g}(\omega)\Delta N^0}{1 + 4\Omega^2\tau T_2\overline{g}(\omega)} \qquad (14.6.38b)$$

where $\overline{g}(\omega)$ is the line shape normalized to 1 at line center and given by (14.6.30). The numerator of (14.6.38b) is the *small*-signal gain coefficient $\gamma_0(\omega)$ (i.e., when $\Omega \to E_{0x} \to 0$), and the denominator expressed the saturation behavior. If we recall the relationship between the A coefficient and dipole moment μ_{21} as given by (14.4.10), then one can show that (14.6.38) is precisely the same as that found using the rate equations with the A and B coefficients as was done in Chapter 8, provided we interpret τ and T_2 correctly; that is, T_2 is related to the line broadening and τ is *lifetime of the inversion*.

Inasmuch as the above density matrix approach used considerably more mathematics with precisely the same answer, one can ask, "Was this trip necessary"? The answer is yes.

At the very minimum, one should now appreciate the fact that the simple rate equations will handle many laser phenomena quite adequately. The above analysis has identified a very important interpretation to the homogeneous line width, $\Delta \nu_h = 1/\pi T_2$. Previously, we had more or less treated $\Delta \nu_h$ in a rather disdainful fashion, whereas now we should have developed a reasonable respect for it.

The density matrix approach also generates the complex Lorentzian in a natural fashion. This concept was very useful for describing mode pulling in this chapter and for computing the field transfer function of a laser medium in Chapter 9.

The most important reason for this "trip" is to identify a formal procedure for addressing other problems. Although the mathematics of the density matrix is long and tedious, it will handle the more complicated problems such as multiphoton effects, Raman scattering, and other nonlinear phenomena with the same formalism (but with more tedious algebra).

PROBLEMS

14.1. If a bound electron undergoes simple harmonic motion at a frequency ω_0, it is being accelerated and thus radiates power at the frequency ω_0. Therefore the amplitude of the simple harmonic motion must be damped because of that radiation, since the total energy of system (bound electron + radiation) must be conserved.

(a) Use such arguments to show that the equation of motion for the bound electron is

$$\ddot{z} + \omega_0^2 z - \frac{1}{6\pi\epsilon_0}\frac{e^2}{mc^3}\dddot{z} = 0$$

(b) Assume a displacement of the form $z = d \exp[(1/2\tau) + j\omega_0]t$; neglect all terms involving powers of $1/\tau$ greater than 1, and show that the radiation damping is given by

$$\frac{1}{\tau} = \frac{1}{6\pi\epsilon_0}\frac{e^2\omega_0^2}{mc^3}$$

14.2. (a) Show that (14.2.12a) can be derived from (14.2.12b) by applying the Kramers-Kronig relations (see Appendix II).

(b) The simplified form of the Lorentzian—say (14.2.15)—does *not* satisfy the requirements for the strict applicability of Kramers-Kronig relations. Why?

14.3. Use the perturbation theory of Sec. 14.4 to compute the A coefficient for the $3p \to 2s$ transition in hydrogen. (Ans. $A = 2.245 \times 10^7$ sec^{-1}.) What is the wavelength of the radiation? (Ans. $H_\alpha = \lambda_0 = 6562.86$ Å.)

14.4. Consider a particle bound in a one-dimensional potential well [i.e., $V(x) = 0$; $0 < x < d$; $V(x) = \infty$ otherwise]. Compute the transition probabilities between the quantum levels.

14.5. The one-dimensional well of Problem 14.4 with a finite potential barrier is a good approximation for the case of an electron in the conduction band between two layers with a higher band gap. In Chapter 11 we assumed that barrier to be infinite and thus the wave function went to zero at the barrier and k_z was equal to π/L_z (11.2.2). Assume a barrier of 0.5 eV, $L_z = 200$ Å, and an effective mass of 0.067 m_0 for the electron; compute the energy of the first allowed state and the value of k_z.

14.6. The power radiated by a classic magnetic dipole, $\mathbf{m} = I\,\mathbf{A}$, is

$$P_{\text{rad}} = \frac{1}{2\pi} \cdot \frac{\omega^4}{c^4} \cdot \sqrt{\frac{\mu_0}{\epsilon_0}} \cdot \mathbf{m}^2$$

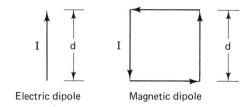

Electric dipole   Magnetic dipole

(a) Find an expression for the classical A_m coefficient for magnetic dipole radiation.

(b) Evaluate the ratio of the A coefficient derived in (a) to that of an electric dipole assuming the above geometry for each. Assume $d = 1$ Å, $\lambda_0 = 10{,}000$ Å, and the same current for each.

14.7. Use the data provided in Chapter 10 concerning the ruby system to estimate the amount of pulling on a mode that would have been located 1 GHz away from line center. Assume an inversion sufficient to sustain a gain coefficient of 0.05 cm^{-1} (at line center), oscillation on R_1 line at 6943 Å with $E \perp$ to c axis, and a temperature of 300°K. Approximate the line shape by a Lorentzian.

14.8. Compute the effect of atomic dispersion on the group velocity of an optical pulse propagating in an atomic medium with the frequency center of the pulse coinciding with the atomic resonance for both a normal and an inverted population. Express the group velocity as a function of the peak loss (or gain) coefficient for the case of a Lorentzian line. Ignore hole burning and assume that the spectrum of the pulse is narrow compared to the width of the line.

14.9. The line shape of the helium-neon laser at 6328 Å is Doppler broadened with a width of ~ 1.9 GHz. The stimulated emission cross section is 5×10^{-13} cm^2; the upper state ($3s_2$, $g_2 = 1$) has a lifetime 55 ns; the A_{21} coefficient is 6.56×10^6 sec^{-1}; and the lower state ($2p_4$, $g_1 = 2$) has a lifetime of 19.3 ns. The homogeneous line width can be estimated to be 10 MHz. Because $\Delta \nu_h \ll \Delta \nu_D$, the laser can oscillate on any of those cavity modes in which the gain coefficient exceeds the loss. Assume an inversion of two times threshold and a cavity length of 100 cm.

(a) Find the saturation intensity (in W/cm^2).

(b) What is the spacing in MHz between the modes?

(c) If the intensity of the mode at line center were eight times the saturation intensity, what is the width of the hole burnt in the inversion?

(d) Find the dipole moment of this transition (in coulomb-meters).

(e) If the intensity of the mode at line center were 0.5 W/cm^2, what is the Rabi flopping frequency (in MHz)?

14.10. If one includes decay out of the states coupled to the field, the occupation probabilities obey the following differential equations:

$$\frac{da_2}{dt} = -\frac{a_2}{2T_2} - j\Omega a_1 \exp(j\Delta\omega t)$$

$$\frac{da_1}{dt} = -\frac{a_1}{2T_1} - j\Omega a_2 \exp(-j\Delta\omega t)$$

where Ω = Rabi flopping frequency and $1/\tau_{2,1}$ are the spontaneous decay rates to levels other than 2 or 1. Assume equal decay rates and the normal population condition: $a_2(0) = 0$ and $a_1(0) = 1$. Find the time dependence of $|a_2(t)|^2$.

14.11 The density matrix formulation of the field-atom interaction yielded the following expression for the imaginary part of the susceptibility χ'', assuming homogeneous broadening.

$$\chi'' = \frac{(\mu_{21})^2 T_2}{3\epsilon_0 \hbar} \cdot \frac{\Delta N_0}{1 + (\omega_{21} - \omega)^2 T_2^2 + 4\Omega^2 T_2 \tau}$$

where $\Delta N_0 = N(\rho_{11} - \rho_{22})_0$, which corresponds to $N_1 - N_2$

Ω = Rabi flopping frequency

$(\mu_{21x})^2 = (\mu_{21})^2/3$

Show that this expression and (14.6.37) lead to the same expression for the (saturated) gain coefficients as was derived from the rate equations (8.3.12) provided one uses the correct interpretation of the factors T_2 and τ. In particular, manipulate the expression to show that it contains
(a) The homogeneous line width $\Delta\nu_h$
(b) The Lorentzian line shape

$$g(\nu) = \frac{\Delta\nu_h}{2\pi\left[(\nu_0 - \nu)^2 + \left(\frac{\Delta\nu_h}{2}\right)^2\right]}$$

(c) The stimulated emission cross section $\sigma(\nu_0) = A_{21} \cdot \frac{\lambda^2}{8\pi^2} \cdot \frac{2}{\Delta\nu_h}$
(d) The saturation intensity I_s and the homogeneous saturation law
(e) The "hole" width $\Delta\nu_H$
(f) The coefficient for spontaneous emission A_{21}

14.12. A convenient interpretation of the parameter $1/\tau$ of the density matrix approach is that it represents the rate of recovery of the population difference ($\Delta\rho$) caused by stimulated emission after the stimulating field is removed.

$$\frac{\partial(\rho_{22} - \rho_{11})}{\partial t} + \frac{(\rho_{22} - \rho_{11}) - (\rho_{22} - \rho_{11})^0}{\tau} = 2j\Omega(\sigma_{21} - \sigma_{21}^*)$$

where

$$\Omega = \frac{eE_{0x}}{2\hbar}$$

Thus if E_{0x} is suddenly clamped to zero, then

$$(\rho_{22} - \rho_{11}) - (\rho_{22} - \rho_{11})^0 = \Delta\rho \exp(-t/\tau)$$

or

$$\left.\frac{\partial(\rho_{22} - \rho_{11})}{\partial t}\right|_{t=0} = \frac{\Delta\rho}{\tau}$$

Use this last relationship and the rate equations for the two-level system shown below to relate τ to the lifetimes of the states. Consider two extreme cases for the branching ratio of state 2: $\phi = 0$, that is, hardly any of state 2 decays spontaneously to state 1; and for $\phi = 1$ all of 2 decays to 1 but do *not* assume $\tau_1 \ll \tau_2$. You may assume that $P_2 \tau_2 > P_1 \tau_1$.

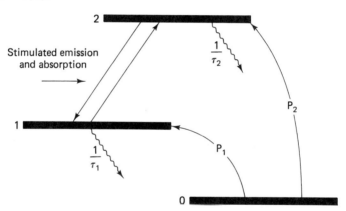

14.13. The frequency dependence of the real part of the complex susceptibility for a homogeneously broaded line is given by

$$\frac{\chi'}{n^2} = A_{21} \cdot \frac{\lambda_0^3}{16\pi^2 n^3} \cdot \left[\frac{g_2}{g_1} N_1 - N_2\right] \cdot \left\{\frac{\nu_0 - \nu}{\pi\left[(\nu_0 - \nu)^2 + \left(\frac{\Delta\nu_h}{2}\right)^2\right]}\right\}$$

If the distribution of center frequencies for a Doppler line were approximated by a "square" distribution:

$$p(f) = \frac{1}{\Delta \nu_s} \text{ for } \nu_0 - \Delta \nu_s/2 < f < \nu_0 + \Delta \nu_s/2$$

find the susceptibility for the Doppler-broadened transition using these approximations.

14.14. The experimental data presented in the paper by P. W. Smith "Mode Selection in Lasers," Proc. IEEE, page 428, Fig 13, 1972, on the homogeneous line width of a $He^3:Ne^{20}$ laser line at 6328 Å can be represented by an equation of the form:

$$\Delta \nu_h \text{ (MHz)} = 33.3 + 111.1p$$

where p is the pressure (in torr) of the 7 : 1 gas mixture. Other data for this transition in a typical laser application are as follows: $A_{21} = 6.56 \times 10^6 \text{ sec}^{-1}$; $A_2 = \Sigma A_{2j} = 12.8 \times 10^6 \text{ sec}^{-1}$; $A_1 = 51.2 \times 10^6 \text{ sec}^{-1}$; $T = 500°K$.

(a) Should the graph of $\Delta \nu_h$ go through zero at zero pressure? Was that an experimental error? If not, what should be the zero pressure intercept?

(b) What is T_2 for a pressure of 3 torr? What is the lifetime of the upper state under these circumstances?

(c) At what pressure would the homogeneous line width be equal to the Doppler width?

(d) What is the electric dipole moment for the $3s_2 - 2p_4$ transition? Express the result in debyes (1 D = 3.33×10^{-30} m).

(e) What is the saturation intensity for a pressure of 3 torr?

(f) What is the Rabi flopping frequency (in hertz) for a stimulating wave with an intensity of 100 W/cm²?

REFERENCES AND SUGGESTED READINGS

1. G. Herzberg, *Atomic Spectra and Atomic Structure,* 2nd ed. (New York: Dover Publications, Inc., 1944).
2. A. C. G. Mitchell and M. W. Zemansky, *Resonance Radiation and Excited Atoms* (New York: Cambridge University Press, 1971), especially Chap. 3.
3. R. M. Eisberg, *Fundamentals of Modern Physics* (New York: John Wiley & Sons, Inc., 1961), Chap. 9.
4. E. Merzbacker, *Quantum Mechanics* (New York: John Wiley & Sons, Inc., 1971), Chaps. 19 and 20.
5. A. Yariv, *Quantum Electronics,* 2nd ed. (New York: John Wiley & Sons, Inc., 1975), Chaps. 2 and 3.
6. W. S. Chang, *Principles of Quantum Electronics* (Reading, Mass.: Addison-Wesley Publishing Company, Inc., 1969), Chaps. 5 and 6.
7. A. Maitland and M. H. Dunn, *Laser Physics* (Amsterdam: North-Holland Publishing Company, 1969), Chaps. 2 and 3.

8. M. O. Scully and M. Sargent III, "The Concept of the Photon," Phys. Today, 38–47, Mar. 1972.
9. A. Matveyev, *Principles of Electrodynamics* (New York: Reinhold Publishing Corp., 1966), Chap. 7.
10. G. Herzberg, *Spectra of Diatomic Molecules* (Princeton, N.J.: D. Van Nostrand Company, 1950).
11. A. van der Ziel, *Solid State Physical Electronics,* 2nd ed. (Englewood Cliffs, N.J.: Prentice-Hall, Inc., 1968), Chap. 2.
12. R. L. White, *Basic Quantum Mechanics* (New York: McGraw-Hill Book Company, 1966), Chap. 11.
13. Willis E. Lamb, "Theory of Optical Maser," Phys. Rev. *A134,* A1429–1450, June 15, 1964.
14. See also some of the collected papers in *Laser Theory,* Ed. Frank S. Barnes (New York: IEEE Press, 1972). Part I, Historical Papers, is especially recommended.
15. M. Sargent, III, M. Scully, and W. Lamb, Jr., *Laser Physics* (Reading, Mass.: Addison-Wesley Publishing Company, Inc., 1974).
16. R. H. Pantell and H. E. Puthoff, *Fundamentals of Quantum Electronics* (New York: John Wiley & Sons, Inc., 1969).
17. R. P. Feynman, R. B. Leighton, and M. Sands, *The Feynman Lectures on Physics,* Vol. III (Reading, Mass.: Addison-Wesley Publishing Company, Inc., 1965).
18. U-Fano, "Description of States in Quantum Mechanics by Density Matrix and Operator Techniques," Rev. of Mod. Phys. *20,* 74–93, 1957.

15

Spectroscopy of Common Lasers

15.1 INTRODUCTION

This chapter briefly reviews the notation commonly used to describe the quantum state of atoms and molecules. Most students will have had some introduction to atomic spectra so only a brief résumé will be given here. However, many will not have had an exposure to molecular vibrational-rotational structure; hence, a more detailed explanation is given. In any case, our goal is to learn the *language* of spectroscopic notation in a minimum of time with a minimum of mathematics. For a more detailed explanation of the underlying principles, one should consult Refs. 1 through 4.

This chapter will provide you with more than a set of dry rules, regulations, selections, rules and formulas. We will also try to examine those issues about various lasers that can only be appreciated after the spectroscopy is understood. Sections devoted to those special issues are marked with an asterisk (*).

15.2 ATOMIC NOTATION

15.2.1 Energy Levels

With the exception of atomic hydrogen (and possibly helium), an atom composed of z protons and z electrons is much too complicated to expect an analytic solution for the quantum levels. However, various perturbation and coupling schemes have evolved that have proved to be remarkably accurate in predicting the trends and rules of the spectra. Actually, the process worked in reverse order: the energy levels and facts about the spectra were known from experiment, which, in turn, led to schemes reproducing the known answer.

One of the most successful is the LS (or Russel-Saunders) coupling scheme, in which a quantum state of an atom is labeled in the following manner:

$$Nl^k\ {}^{2S+1}L_J \qquad (15.2.1)$$

where L is the total orbital angular momentum quantum number, S the total spin angular momentum, and J the magnitude of the total angular momentum. $\mathbf{J} = \mathbf{L} + \mathbf{S}$ according to the vector model covered in Herzberg.[1] By convention, letter symbols are used for L according to the following scheme:

L	S	P	D	F	G	H	I	$\cdots$
Value of L	0	1	2	3	4	5	6	

The number N denotes the orbit number (in the Bohr sense), and l the angular momentum states of the last k active electrons. Quite often l^k is omitted.

For instance, the ground state of mercury is $6\ {}^1S_0$ and the first five excited states are $6\ {}^3P_0$, $6\ {}^3P_1$, $6\ {}^3P_2$, $6\ {}^1P_1$, and $7\ {}^3S_1$. According to the table above, we have the following information provided:

$6\ {}^1S_0$: $N = 6, L = 0, J = 0, S = 0$ (read "six singlet-S-zero")

$6\ {}^3P_0$: $N = 6, L = 1, J = 0, S = 1$ (read "six triplet-P-zero")

$6\ {}^3P_1$: $N = 6, L = 1, J = 1, S = 1$ etc.

$6\ {}^3P_2$: $N = 6, L = 1, J = 2, S = 1$

$6\ {}^1P_1$: $N = 6, L = 1, J = 1, S = 0$

$7\ {}^3S_1$: $N = 7, L = 0, J = 1, S = 1$

In an energy-level diagram, states with different multiplicities, $2S + 1$, are separated (along the $L = 0$ or the S column) and various states with common L are arranged in columns. Fig. 15.1, for mercury, illustrates these conventions. Also shown are some transitions that illustrate the selection rules discussed below.

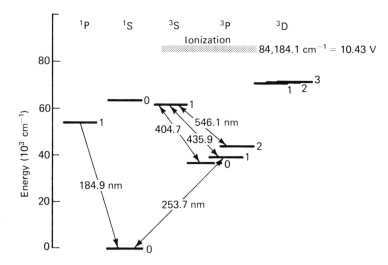

Figure 15.1 Partial energy–level diagram for mercury. (Data from Moore,[5] Vol. 3.)

15.2.2 Transitions—Selection Rules

The theory behind the selection rules is buried in reams of mathematics—see Chapter 14 for a small part of it—and for every "rule" there appears to be an "exception."* However, a good starting point is to commit the following three to memory:

$$\Delta J = \pm 1, 0 \quad \text{but} \quad J = 0 \leftarrow\!\!\!|\!\!\!\rightarrow J = 0 \quad (15.2.2)$$

Thus J can change by ± 1 or 0 (but not 2, 3, etc.), and transitions between states with $J = 0$ are forbidden. This is a "rule" that has few exceptions.

$$\Delta L = \pm 1 \quad (15.2.3)$$

A well-known laser transition in iodine, $5^2 P_{1/2} \rightarrow 5^2 P_{3/2}$, at $\lambda = 1.315$ μm violates this rule, since $\Delta L = 0$. Such a transition is a magnetic dipole transition.

$$\Delta S = 0 \quad (15.2.4)$$

In other words, transitions involving a spin change are forbidden. This rule seems to be the first to fall! In the light elements such as helium, it is rigorously obeyed, but more and more deviations are found in the heavier elements. For instance, the 253.7-nm transition in mercury violates this rule and yet is one of the strongest lines in the spectra. Indeed, it is that transition that is used to excite the phosphor in a fluorescent lamp.

*The word "exception" is not really appropriate, but it fits the situation for now. The rules are given for electric dipole transitions, whereas the exceptions do not fit that category; see the discussion following (15.2.3).

Sec. 15.3 Molecular Structure—Diatomic Molecules   551

As an example, the $6\ ^3P_1 \rightarrow 6\ ^1S_0$ transition is allowed according to (15.2.2) ($\Delta J = +1$), is allowed according to (15.2.3) ($\Delta L = +1$), but is forbidden according to (15.2.4). Obviously the transition does take place—proving that atoms cannot read or obey rules. Note that two of the first three excited states, $6\ ^3P_2$ and $6\ ^3P_0$, do not have a radiative transition back to the ground state. For the first, $\Delta J = 2$, and $J = 0 \rightarrow J = 0$ for the second. Such states are called metastable.

15.3 MOLECULAR STRUCTURE—DIATOMIC MOLECULES

15.3.1 Preliminary Comments

When one combines two atoms to form a diatomic molecule, such as H_2, N_2, or CO, the system has quantum states associated with *rotation* and *vibration* in addition to the *electronic* structure. In other words, the atoms may vibrate with respect to each other and rotate about an axis—all while the electrons in the combined system can undergo changes also.

To a first approximation, one assumes that these three types of "motion" are independent of each other. To be more precise, one assumes that the wave function can be factored into a product of rotational, vibrational, and electronic wave functions. The advantage of this approximation is that each motion can be treated separately and then combined in the end. This also assumes that the energies associated with the different types of motion are additive.

If two atoms, A and B, are separated by a long distance, they retain their separate identities with their associated electronic structure and literally ignore one another. If we move them closer together, various forces come into play to either attract or to repel the other atom. At last, the horizontal axis on an energy-level diagram means something—we plot this interaction potential as a function of separation of the nucleii, as shown in Fig. 15.2. In Fig. 15.2, two different interaction potentials are shown: the solid curve represents the potential energy for a bound molecular state, and the dashed one represents a repulsive state.

As shown in this figure, the minimum energy is when the two atoms are separated by a distance r_e. As indicated in Fig. 15.2, the value of this minimum depends on whether one of the atoms is in an electronic excited state; for now, let us focus on the case where both are in the ground state. While the two atoms are in the potential well, we have a bound diatomic molecule, with a binding— called the dissociation energy—of the difference between $V(r = \infty)$ and the lowest allowed energy state near the "equilibrium" distance r_e.

As soon as we start talking about distances of a few angstroms, typical of r_e, it is not possible to identify a fixed position as an equilibrium position of two atoms, for this would require zero velocity. The uncertainty principle forbids such absolute determinacy. Rather, we must solve for energy levels in this more

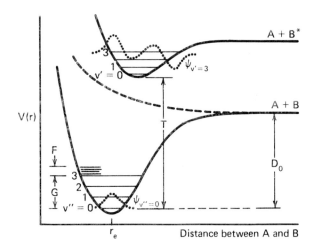

Figure 15.2 Potential-energy diagram for a hypothetical molecule AB. By convention, the vibrational energy is denoted by the letter symbol G, rotational energy by F, and electronic energy by T; for the CO molecule, $r_e = 1.128322$ Å.

or less parabolic potential well. Every elementary text in quantum mechanics solves that problem, so we will just use one of the major results of that analysis:

> The energy levels are more or less uniformly spaced in the well, with the minimum energy level lying above the minimum of the well.

These levels are the vibrational levels that correspond to a classical vibration of the two nuclei bound together by a spring. If the well were exactly parabolic, the energy levels would be spaced by a multiple of $\hbar$ times the vibration frequency, $\omega_{\text{vib.}}$, and the minimum would be $(\hbar\omega_{\text{vib.}}/2)$ above the minimum of the parabola. However, it is not a parabola, so some modification must be made to the above.

Thus we have a bound vibrating molecule. It takes little imagination to recognize that some more energy could be tied up in rotation. Thus in a given electronic manifold, specified by $V(r)$, there is vibrational and rotational energy in addition to the electronic value. We can have transitions of the following types:

1. Rotational: the vibrational quantum number and the electronic state do not change.
2. Vibrational-rotational (VR): the electronic state does not change.
3. Electronic: everything changes.

Pure rotational transitions occur in the far infrared to microwave portion of the spectrum (λ = cm to 15 μm); VR transitions occur in the 20 to 2 μm region; and electronic ones are in the range 1 to 10 eV. In the material given below, we

Sec. 15.3 Molecular Structure—Diatomic Molecules

show how to compute the various levels given the molecular data. In view of the complexity of (3), only the procedure for naming the states will be given.

15.3.2 Rotational Structure and Transitions

The minimum data necessary to specify the rotational energy levels are the rotational constant B_e and how this constant depends on the particular vibrational level to yield B_v.

$$B_v = B_e - \alpha_e\left(v + \frac{1}{2}\right) \quad (15.3.1)$$

where v is the vibrational quantum number and B_e and α_e are part of the specification of the molecule. Then the energy level of the rotational state, F, depends on B_v and the angular momentum quantum number J according to*

$$F(J) = B_v J(J + 1) \quad (15.3.2)$$

Rotational transitions occur between adjacent states according to $\Delta J = \pm 1$ ($\Delta J = 0$ means that there was not a transition for this case). Labeling this transition by the J value *of the lower state* yields the following formula for the transition frequencies (measured in cm^{-1}):

$$F(J + 1) - F(J) = \bar{\nu}(J) = 2B_v(J + 1) \quad (15.3.3)$$

Typical values for B_e and α_e (for the ground electronic state of CO)[6] are

$$B_e = 1.931271 \text{ cm}^{-1}, \qquad \alpha_e = 0.017513 \text{ cm}^{-1}$$

15.3.3 Thermal Distribution of the Population in Rotational States*

It is obvious from these typical numbers that the spacing between rotational levels is very small compared to energies of random motion of the gas molecules ($kT = 208.5$ cm^{-1} at 300°K). Consequently, collisions between molecules are very effective in establishing a thermal population of the rotational states. The population in a rotational state J of vibrational manifold v is dictated by Boltzmann statistics according to

$$\frac{N_{v,J}}{N_v} = \frac{(2J + 1) \exp\{-[hcB_v J(J + 1)/kT]\}}{\sum_{J=0}^{\infty}(2J + 1) \exp\{-[hcB_v J(J + 1)kT]\}} \quad (15.3.4)$$

where $(2J + 1)$ is the degeneracy of the quantum state J. The numerator of (15.3.4) is the statistical weight of the rotational state J times the Boltzmann factor, whereas the denominator is the sum of similar factors for all states in this

---
*There is a small correction to (15.3.2) of the form, $-D_v J^2(J + 1)^2$, but it is usually neglected. See Herzberg[2] for more detail.

vibrational manifold. Thus a vibrational population N_v is apportioned among many—say, 50 to 100—different rotation states. With very little error, one can approximate the infinite series by an integral:

$$\sum (2J + 1) \exp\{-[hcB_v J(J + 1)/kT]\}$$

$$\rightarrow \int (2J + 1) \exp\{-[hcB_v J(J + 1)/kT]\}dJ = \frac{kT}{hcB_v} \quad (15.3.5)$$

$$\therefore \frac{N_{v,J}}{N_v} = \frac{hcB_v}{kT}(2J + 1) \exp\left[-\frac{hcB_v J(J + 1)}{kT}\right] \quad (15.3.6)$$

This function is plotted in Fig. 15.3 for the $v = 0$ state in the CO molecule using the data given previously. Note that there is very little population in the low and very high values of J; most of the populations appear at energies at $\sim kT$.

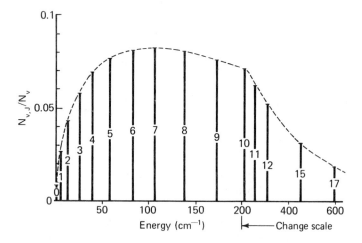

Figure 15.3 Rotational distribution of the population in the $v = 0$ state of CO at 300°K.

At first glance, one might think that there was a built-in population inversion between some of the rotational states—say, between $J = 4$ and $J = 3$. There is, in the sense of strict numbers, but it does not help build a laser. The quantity that counts insofar as the laser gain is the number divided by the degeneracy:

$$\gamma = \sigma\left(N_2 - \frac{g_2}{g_1}N_1\right) = \sigma g_2\left(\frac{N_2}{g_2} - \frac{N_1}{g_1}\right)$$

This difference is always negative for all values of J for a thermal distribution.

15.3.4 Vibrational Structure

The vibrational energy level, $G(v)$, can be computed from

$$G(v) = \omega_e\left(v + \frac{1}{2}\right) - \omega_e x_e\left(v + \frac{1}{2}\right)^2 + \omega_e y_e\left(v + \frac{1}{2}\right)^3 + \cdots \quad (15.3.7)$$

where ω_e, $\omega_e x_e$, and so on are all measured constants of a molecule in a particular electronic state. In the ground electronic state of CO, these constants[5] are

$$\omega_e = 2169.8233 \text{ cm}^{-1}$$
$$\omega_e x_e = 13.2939 \text{ cm}^{-1}$$
$$\omega_e y_e = 1.57 \times 10^{-5} \text{ cm}^{-1}$$

There is a temptation to factor ω_e from the expression, but since the products, $\omega_e x_e$ and $\omega_e y_e$, ..., are always specified, it is a futile exercise.

One should recognize (15.3.7) as a Taylor series expansion that converges very rapidly, inasmuch as the higher-order coefficients decrease dramatically. The first term is the energy level of a particle in a parabolic potential well; the remainder is the correction for the fact that $V(r)$ in Fig. 15.2 is not a parabola. The quantity $c\omega_e$ is roughly the vibrational frequency of the classical mass-spring system. Thus the heavier the atoms, the smaller the vibrational frequency. For example, ω_e for molecular hydrogen is 4395.2 cm^{-1}, that of D_2 is 3118.4 cm^{-1}, and the ratio is $2^{1/2}$ to within 0.3%.

The total energy of a rotation state J in a vibrational manifold v is given by the sum (15.3.7) and (15.3.2):

$$E(v, J) = G(v) + F(J) \quad (15.3.8)$$

15.3.5 Vibration-Rotational Transitions

These transitions are between rotational states of adjacent vibrational manifolds, according to the following selection rules:

$$\Delta v = \pm 1$$
$$\Delta J = \pm 1, 0, \quad J = 0 \;\;\leftrightarrow\!\!\!|\;\; J = 0 \quad (15.3.9)$$

Transitions $\Delta v = 2$ do occur but are much weaker.

If the J value of the lower state is one greater than the upper, the family of transitions is called the P branch; if it is the same, the family is called the Q branch; and if it is less, it is called the R branch.

P branch: $\Delta J = -1$, $J_{upper} = J_{lower} - 1$
Q branch: $\Delta J = 0$, $J_{upper} = J_{lower}$ (15.3.10)
R branch: $\Delta J = +1$, $J_{upper} = J_{lower} + 1$

In most diatomic molecules, the Q branch is *not* present in the VR spectra (NO is the only exception). In homonuclear diatomic molecules, such as H_2, N_2, O_2, and so on, none of the VR transitions are observable. The classical reason for this last fact is that no amount of stretching of two identical molecules can form an electric dipole. Consequently, *all* vibrational levels of homonuclear molecules are metastable. That fact plays a major role in the spectacular performance of the CO_2/N_2 laser.

It is more a matter of perseverance than intelligence to apply (15.3.9) to compute the wave number of the transitions. If we label the transition by the lower-state J values, we have

$$\bar{\nu} = G(v') - G(v'') + F(J') - F(J'')$$

For the P branch, $J' = J'' - 1$, or

$$\bar{\nu}_{P(J)} = \bar{\nu}_0 - (B_{v'} + B_{v''})J + (B_{v'} - B_{v''})J^2 \quad (15.3.11)$$

For the R branch, $J' = J'' + 1$, or

$$\bar{\nu}_{R(J)} = \bar{\nu}_0 + 2B_{v'} + (3B_{v'} - B_{v''})J + (B_{v'} - B_{v''})J^2 \quad (15.3.12)$$

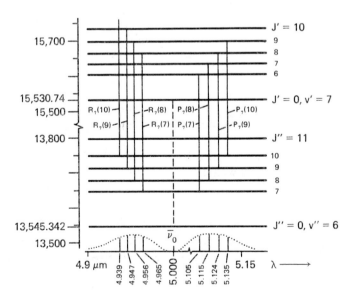

Figure 15.4 Example of the VR spectra of CO on the $7 \to 6$ band.

Sec. 15.3   Molecular Structure—Diatomic Molecules   557

where the single prime represents the upper state, the double prime represents the lower, and $\bar{\nu}_0$ is the "center" of the band ignoring rotation.

$$\bar{\nu}_0 = G(v') - G(v'') \qquad (15.3.13)$$

An example of VR transitions in CO is shown in Fig. 15.4. We also introduce the shorthand notation used in lasers to name a transition: for example, $P_7(10)$ means that it is in the P branch, originates at $v = 7$, $J = 9$ and terminates on $v = 6$, $J = 10$.

15.3.6 Relative Gain on P and R Branches— Partial and Total Inversions*

Whereas gas collisions are very effective in maintaining an equilibrium distribution among rotational states, they are far less so for vibrational states. Indeed, in an electric discharge laser, the population in vibration states tends to equilibrate at the electron temperature, which can be many times (10 to 50) the gas temperature. Moreover, it is quite common to speak of three temperatures, translational, rotational, and vibrational, to describe the thermal distribution of atoms in those states. This leads to some important consequences for lasers:

1. The gain on P branch from a given J level is always higher than the R branch originating at the same upper level.
2. One can always obtain gain on the P branch for some value of J even though the total number of molecules in the upper vibrational state is *less* than that in the lower state provided that $T_v > T_R$. This is called a *partial inversion*.

Both statements can be proved by a patient examination of the laser-gain equation:

$$\gamma(\nu) = \sigma(\nu)\left(N_2 - \frac{g_2}{g_1}N_1\right) = \sigma(\nu)g_2\left(\frac{N_2}{g_2} - \frac{N_1}{g_1}\right) \qquad (7.5.16)$$

We identify the upper state 2 with the rotational state J' of the v' vibrational state, J'', v'' with that of the lower state 1, and recall that $g_2 = 2J' + 1$, $g_1 = 2J'' + 1$. Substituting (15.3.6) into the laser-gain equation leads to a rather long and painful expression:

$$\gamma(\nu) = \sigma(\nu)(2J' + 1)\left\{N_{v'}\frac{hcB_{v'}}{kT_r}\exp\left[-\frac{hcB_{v'}J'(J'+1)}{kT_r}\right]\right.$$
$$\left. - N_{v''}\frac{hcB_{v''}}{kT_r}\exp\left[-\frac{hcB_{v''}J''(J''+1)}{kT_r}\right]\right\} \qquad (15.3.14)$$

This equation looks much worse than it is. Let us simplify by ignoring any differences in the rotational constants (i.e., let $B_{v''} = B_{v'} = B$) and expressing the gain for the P and R branches in terms of the J value for the *upper state* (even though a transition is usually labeled by the J value of the lower state).

For the P branch, $J' = J$, $J'' = J + 1$:

$$\frac{\gamma_P}{N_{v'}\sigma(\nu)} = \frac{hcB}{kT_r}(2J + 1)\left\{\exp\left[-\frac{hcBJ(J + 1)}{kT_r}\right]\right.$$
$$\left. - \frac{N_{v''}}{N_{v'}}\exp\left[-\frac{hcB(J + 1)(J + 2)}{kT_r}\right]\right\} \quad (15.3.15a)$$

For the R branch, $J' = J$, $J'' = J - 1$:

$$\frac{\gamma_R}{N_{v'}\sigma(\nu)} = \frac{hcB}{kT}(2J + 1)\left\{\exp\left[-\frac{hcBJ(J + 1)}{kT_r}\right]\right.$$
$$\left. - \frac{N_{v''}}{N_{v'}}\exp\left[-\frac{hcB(J - 1)(J)}{kT_r}\right]\right\} \quad (15.3.15b)$$

These functions are plotted in Fig. 15.5 for different values of the vibrational inversion ratio $N_{v'}/N_{v''}$. Note that at a common J value of the upper state the gain of the P branch is always higher than that of the R branch, and furthermore that gain can still be achieved on the P branch even when $N_{v'}/N_{v''} < 1$.

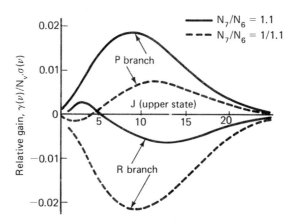

Figure 15.5 Relative gain on the $7 \rightarrow 6$ transition of CO for different values of the vibrational population ratio.

If $N_{v'}/N_{v''} < 1$, gain is obtained over only a fraction of the rotational states—hence its name, *partial inversion;* if the ratio is greater than 1, it is called a *total inversion*. While one can always obtain mathematical gain on high rotational states in the P branch, the value of the gain becomes vanishingly small because of the small population at high J values. Thus, unless special precautions are taken, a laser will always tend to oscillate on the P branch at the expense of the R-branch inversion.

15.4 ELECTRONIC STATES IN MOLECULES

15.4.1 Notation

The notation for the electronic manifold of a molecule is similar to that for an atom, with capital Greek letters replacing the letters. In addition, the *lowest* electronic manifold is given the name X followed by spectroscopic formula for the configuration, with the multiplicity $2S + 1$ preceding the L value as a superscript. For CO, the ground state is $X\,^1\Sigma$. As indicated, capital Greek letters, $\Sigma, \Pi, \Delta, \ldots$, are used rather than $S, P, D, \ldots$.

The excited states with the *same* multiplicity are then given the names A, $B, C, \ldots$, followed by spectroscopic formulas, as we ascend in energy. The excited state with *different* multiplicities are given the names, $a, b, c, \ldots$. The sole exception to this naming procedure is for N_2, where the roles of the capital and lowercase letters are interchanged (see Fig. 15.6).

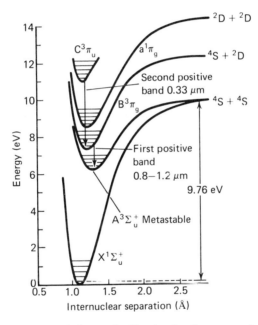

Figure 15.6 Energy-level diagram for N_2, showing the common laser transitions.

There is such a rich variety of rules for electronic transitions that even a summary will not be attempted here. Suffice it to say that there are transitions between different vibrational and rotational quantum numbers within each electronic state and that the vibrational and rotational constants depend on the electronic state. Because of the smallness of the typical rotational constants, there is a transition within every angstrom or so within a band.

15.4.2 The Franck-Condon Principle

There is one issue associated with changes in electronic states of a molecule that is very important but is simple in concept and easy to understand. The Franck-Condon principle states that all electronic processes, such as electronic transitions and electron-molecule inelastic collisions, must take place along a vertical path (i.e., r is a constant).

The classical explanation for this is that the bound electrons can readjust their orbits instantaneously compared to the more sluggish vibrational motion of the nucleii.

Thus the most probable absorption path in Fig. 15.2 is from $v'' = 0$ in the electronic ground state to $v' = 3$ in the excited electronic state. The same is true for inelastic electron collisions. Note that to dissociate the molecule into two free atoms, A and B, we must provide sufficient energy to go vertically from $v = 0$ to the repulsive electronic state (shown as dashed curve in Fig. 15.2). From that point, the two atoms will fly apart, sharing the excess energy.

If the molecule were in $v' = 0$ of the excited state, it would most probably radiate back to $v'' = 3$, maybe to $v'' = 2$, but not to 0 or 1, because this would require a considerable change in "position" of the molecule during the transition.

This principle is most important in all types of lasers: gas, liquid (dye), and solid state. In the semiconductor lasers, the rule is called the "$\Delta k = 0$" or "k selection rule" and explains why indirect band-gap semiconductors such as Ge and Si do not lase, whereas a direct band-gap material, such as GaAs, does.

15.4.3 Molecular Nitrogen Lasers*

A partial energy-level diagram of molecular nitrogen laser is shown in Fig. 15.6. As noted above, nitrogen is an exception to the rules for naming the levels, and thus the A, B, and C states have different multiplicities (a triplet) than does the ground state ($X\ ^1\Sigma_g^+$). In a light gas such as N_2, the rule forbidding intersystem crossing is obeyed (i.e., singlet $\leftarrow\!\!|\!\!\rightarrow$ triplet), and thus the lowest $A\ ^3\Sigma_u^+$ state is metastable against radiation back to the ground level.

Not only is the lowest laser level metastable, the lifetime of the B state is *longer* than that of the C state. Thus, at first glance, almost everything is wrong, yet it lases rather spectacularly on the $C \rightarrow B$ band.

Part of the reason for its performance in spite of the lifetime odds against it is because of the favorable excitation route from X to C, allowed by the Franck-Condon principle of the preceding section. For instance, most of the excitation in a nitrogen discharge would be from the $v = 0$ vibrational level in the $X\ ^1\Sigma_g^+$ state. Since the "equilibrium" position, r_e, of the C state is close to that of the ground state, electron impact excitation will proceed along that path rather than to the B or A states (see Table 15.1). For instance, if an electron has 15 eV of energy (sufficient to excite all the states of Table 15.1), then by experiment, excitation to the C state is twice as likely as to the B or A state. (See Ref. 8 for more details.)

TABLE 15.1 DATA ON N_2

State	T (cm^{-1})	ω_e (cm^{-1})	$\omega_e x_e$ (cm^{-1})	r_e (Å)	τ
$X\ ^1\Sigma_g^+$	0	2359.61	14.456	1.094	∞
$A\ ^3\Sigma_u^+$	50,206.0	1460.37	13.891	1.293	seconds
$B\ ^3\Pi_g$	59,626.3	1734.11	14.47	1.2123	10 μsec
$C\ ^3\Pi_u$	89,147.3	2035.1	17.08	1.148	40 nsec

However, the real critical issue associated with the nitrogen laser has nothing to do with spectroscopy, atomic physics, stimulated emission, or any other esoteric subject but has most everything to do with the speed of the electrical discharge *circuit*. It has to be capable of switching very high voltages (15 to 40 kV is typical) with a very fast rise time ($\leq$10 ns). (That is a nontrivial job!) The speed is required to sustain the electron "temperature" at a high enough value to take advantage of the favorable excitation route allowed by the Franck-Condon principle.

If all the provisos are met, the N_2 lases quite spectacularly in the near UV on the *C-B* band, in the pulsed mode obviously. Gains are so high (50 to 75 dB/m are typical) that cavities are superfluous. One mirror is used to merely not "waste" the photons from one end.

PROBLEMS

15.1. Refer to Table 10.3. What electronic selection rule forbids the emission on the $3s_2 \rightarrow 2p_9$; $2p_4 \rightarrow 1s_3$? (Ans. $\Delta J = \pm 1, 0$.)

15.2. Refer to Fig. 10.13. Which selection rule is violated for the 5145-Å and 5287-Å laser lines in the argon ion? (Ans. $\Delta S = 0$)

15.3. Why do all rare gases have two metastable states?

15.4. Use Moore[5] to name and evaluate the energies of the metastable states of the rare gases. Present your answer in the following format:

Gas	Spectroscopic Name	Energy Level (cm^{-1})	Energy Level (eV)
He			
Ne			
Ar			
Kr			
Xe			

15.5. Compute and verify the VR spectra of the $7 \to 6$ spectra of CO shown in Fig. 15.4.

15.6. Make a plot similar to Fig. 15.5 for the $7 \to 6$ transition in CO for two different rotational temperatures, $T_r = 150°K$ and $T_r = 300°K$. Does the CO laser benefit from cooling? Is this a general rule, or is it peculiar to CO?

15.7. Use Herzberg[2] to find the data necessary to analyze the $3 \to 2$ band of the HF laser.

15.8. Refer to Table 10.5 for data for the CO_2 laser.
   (a) Construct a table of wave numbers for the P-branch transition in the 10.4-μm band.
   (b) Suppose that there was a total inversion for this band and that the gain on the $P(22)$ transition was 0.5%/cm. What is the ratio $N_{v'}/N_{v''}$? (Assume Doppler broadening at 300°K.)

15.9. Compute the wavelength of the iodine laser. (Use Moore[5] for energy levels.)

15.10. Fig. 15.5 implies that there is a minimum value of J before positive gain is observed on a VR transition. If one expresses the vibrational population by a Boltzmann factor of the form

$$\frac{N_{v'}}{N_{v''}} = \exp\left[-\frac{G(v') - G(v'')}{kT_v}\right]$$

then the vibrational temperature T_v can be positive or negative corresponding to the types of inversions.
   (a) What is that correspondence?
   (b) Show that one can always obtain gain provided that J is larger than j_{min} given by

$$J > J_{min} = \frac{G(v') - G(v'')}{2B}\frac{T_r}{T_v} - 1 \quad \text{[Eq. (2-11) of Polanyi}^7\text{]}$$

15.11. The nominal wavelength of the resonance transition in mercury that excites the phosphor in a common fluorescent lamp is 2537 Å. Actually there are 10 separate lines within 0.05 Å of this nominal value because of the dependence of the center frequency on the isotope and because of the splitting induced by the interaction of the nuclear spins (of the odd isotopes 199 and 201) with the electronic states. The table below shows the mass of the isotope, its relative abundance, the nuclear spin I, the shift from the center of the Hg(198) transition, and the relative contribution (i.e., intensity) of each transition.

Isotope	Abundance	Spin I	Shift (cm⁻¹)	Intensity
196	0.1%	0	+0.137	0.1%
198	9.89	0	0.0	9.89
200	23.77	0	−0.160	23.77
202	29.27	0	−0.337	29.27
204	6.85	0	−0.511	6.85
199	16.45	1/2 (A)	−0.514	5.48
		(B)	−0.160	10.96
201	13.67	3/2 (a)	−0.489	6.84
		(b)	−0.023	4.56
		(c)	+0.229	2.28

These lines are Doppler broadened around their center frequencies corresponding to a gas temperature of 90°C. Construct the spontaneous emission profile and present the results in graph format. The following questions are meant as an aid for the construction of the graph. (See Halstead and Reeves[9] and Anderson et al.[10])

(a) What is the Doppler width of the Hg(198) transition? Is this width large or small compared to the shifts in the table above? Is it worthwhile worrying about the different widths for each transition?

(b) Plot the spectral distribution of intensity if the lines were a delta function at the respective center frequencies. Identify the spikes by the isotopes and/or the hyperfine component, that is, 201(a) and so on. Are there any natural groupings of transitions that are separated by less than a Doppler width and thus could be considered as one transition? What are the relative strengths of the groups and what are the center frequencies of each?

(c) Use the approximations suggested by (b) to construct a careful graph of the emission profile of the 2537-Å line.

 (1) At what wave number (relative to the center of the 198 transition) does the spontaneous emission reach a peak?
 (2) What is the FWHM of the composite line shape?
 (3) Express the amplitudes of the peaks by a proportionality sequence L:M:N and so on.

REFERENCES AND SUGGESTED READINGS

1. G. Herzberg, *Atomic Spectra and Atomic Structure* (Englewood Cliffs, N.J.: Prentice-Hall, Inc., 1937; New York: Dover Publications, Inc., 1944).
2. G. Herzberg, *Spectra of Diatomic Molecules* (Princeton, N.J.: D. Van Nostrand Company, 1950).
3. G. Herzberg, *Infrared and Raman Spectra* (Princeton, N.J.: D. Van Nostrand Company, 1966).

4. E. U. Condon and G. H. Shortly, *The Theory of Atomic Spectra* (Cambridge, England: Cambridge University Press, 1957).
5. C. Moore, *Atomic Energy Levels*, NSRDS-NBS-35, Vols. 1–3 (Washington, D.C.: U.S. Dept. of Commerce, 1971).
6. P. Krupenie, *The Band Spectra of Carbon Monoxide*, NSRDS-NBS-5 (Washington, D.C.: U.S. Dept. of Commerce, 1966).
7. J. C. Polanyi, "Vibrational-Rotational Population Inversion," Appl. Opt., Suppl. 2 Chem. Lasers, 109–127, 1965.
8. L. A. Newman and T. A. DeTemple, "Electron Transport Parameters and Excitation Rates in N_2," J. Appl. Phys, *47*, 1912–1915, 1976.
9. J. A. Halstead and R. A. Reeves, "Time Resolved Spectroscopy of the Mercury $6\,^3P_1$ State," J. Phys. Chem. *85*, 2777, 1981.
10. J. B. Anderson, J. Maya, M. W. Grossman, R. Laqueskenko, and J. F. Waymouth, "Monte Carlo Treatment of Resonant-radiation Imprisonment in Fluorescent Lamps," Phys. Rev. A, *31*, 2968, 1985.

16

Detection of Optical Radiation

16.1 INTRODUCTION

Detection of optical radiation is a topic that has a long history, preceding the invention of the laser by many years.* However, the communication capabilities of the laser have spurred renewed interest in making the detectors faster, more sensitive, and more convenient and versatile. This chapter discusses the characteristics of some of the common quantum detectors and the origin of noise in various systems.

We discuss first the manner in which these detectors convert a photon flux into an electrical current and ignore the question of noise. However, the latter question is most important and we will invest considerable effort in describing the origin of the noise. We then return to specific detector classes and discuss their use in a typical signal-processing environment.

*In fact, Einstein's explanation of the photoelectric effect can be considered as the start of the wide acceptance of the concept of a photon, but detectors go even further back. Calibrated optical detectors had to be in existence so that the blackbody spectra could be *measured* and *then* explained by Planck.

16.2 QUANTUM DETECTORS

The term "quantum detector" implies that there is a direct correspondence between the number of photons absorbed and the number of electrical carriers (electrons or holes) generated and subsequently used in the circuit. Note that this restriction eliminates thermal detectors* from our consideration.

The ratio of electrical current generated (carriers per second) to photons per second absorbed is the quantum efficiency:

$$\eta_{qe} = \frac{\text{number of carriers generated}}{\text{number of photons absorbed}} \quad (16.2.1a)$$

$$= \frac{i/e}{P_{\text{opt.}}/h\nu} \quad (16.2.1b)$$

assuming, of course, that every carrier generated is collected by the circuit. In a *p-n* diode, carriers generated by the absorption of photons beyond a diffusion length of the junction merely recombine and do not contribute to the circuit current.

Let us take some examples to illustrate the characteristic of some common detectors.

16.2.1 Vacuum Photodiode

Probably the easiest detector to analyze and to visualize this conversion of a photon flux into an electrical current is the vacuum photodiode shown in Fig. 16.1. Indeed, it was this device that gave convincing "proof" of the existence of a photon.

If the photon energy, $h\nu$, is larger than the photoelectric work function of the cathode, current will flow; if not, *nothing* is obtained irrespective of the bias voltage or the intensity of the incoming beam. (If the intensity is large enough, it can *heat* the photocathode, which will then emit thermionic electrons. Under such extreme circumstances, a quantum detector becomes a thermal one.) A typical photo-response curve is shown in Fig. 16.1(b). If the wavelength is too long, the photoelectric emission process ceases. This is a function of the cathode material and its preparation; the theory is most involved and is not appropriate to be discussed here. If the wavelength is too short, the transmission of the vacuum window degrades, and this fact accounts for the high-frequency cutoff.

The photons pass through the vacuum window and impinge on the photocathode and generate η_{qe} electrons, which in turn are attracted to and collected by the anode to complete the electric circuit. The bias voltage, V_k, is usually quite high—say, 300 to 5000 V—to eliminate the possibility that the negative space

---

*Such as heating water with the photons and measuring the temperature rise with a thermometer. Although such a detector is not of a "high tech" variety, it is easily calibrated.

Sec. 16.2 Quantum Detectors

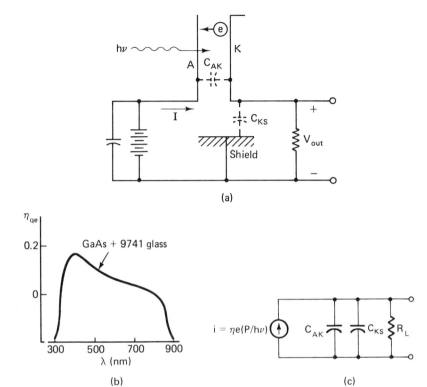

Figure 16.1 Vacuum photodiode: (a) bias circuitry; (b) typical variation of η_{qe}; and (c) equivalent circuit.

charge of the emitted electrons would limit the external current. For this same reason, the *A-K* gap distance is also made as small as practical given the constraint that the ouput capacitance should also be small. The anode will be of the form of a highly transparent metallic grid (or maybe just a few wires) so that few photons are intercepted by it.

There are a few practical issues brought out in Fig. 16.1(a) and (c). First of all, it is the *current* that is generated by the photons; hence it is logical to consider a Norton equivalent circuit for this device. The output voltage therefore is a function of the load resistance. The speed of the device is usually determined by the time constant RC_T and not by the transit time of the electrons from cathode to anode, which is exceedingly fast. Sometimes, the cathode is biased negatively and the load is placed in the anode circuit, which eliminates the cathode-shield capacitance from the circuit. Even though the DC current can only flow in one direction, $A \rightarrow K$, we will be concerned with the high-frequency components of the individual electron impulses. Therefore one can also apply this equivalent circuit for the high-frequency component of the impulses.

Vacuum photodiodes have been largely supplanted by the much more convenient solid-state detectors (smaller, faster in some cases, more sensitive, and requiring less formidable power supplies [if at all] for bias). They still are used when the optical signal is very large, large enough that focusing onto a typical solid-state detector would result in damage. If the voltage is high enough, amperes of photoelectron current can be obtained. Obviously, the time duration must be small enough so that the tube dissipation does not destroy the anode. However, this section serves to introduce the premium detector of all types—the photomultiplier.

16.2.2 The Photomultiplier

The major use of the vacuum photoelectric effect is in conjunction with the most perfect current amplifier in existence, the secondary emission amplifier. This amplifier is based on the experimental fact that many materials will emit, on the average, δ new electrons for every electron impinging on them. If the kinetic energy of the incident electron is high enough—typically 100 to 200 V— then δ is greater than 1 and one obtains essentially "noise-free" current amplification.

The combination of the vacuum-photocathode technology with anodes (referred to as dynodes) constructed from materials selected to enhance δ yields a photomultiplier in the configuration shown in Fig. 16.2.

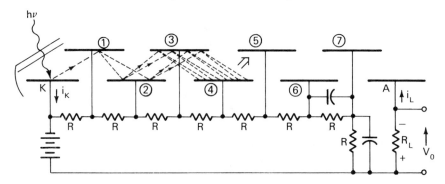

Figure 16.2 Seven-stage photomuliplier.

If one electron is emitted by the photocathode, it impinges on the first anode, called dynode 1. This element emits δ new electrons, and if the voltages are chosen correctly, those δ electrons head for the second dynode, which emits δ^2 electrons, and so on. It is obvious that if one has N dynodes with a secondary emission ratio of δ each, the current gain is

$$G = \frac{i_L}{i_K} = \delta^N \tag{16.2.2}$$

If $N = 12$ and $\delta = 4.0$, then G is huge, 1.68×10^7 (or 144 dB!).

Sec. 16.2 Quantum Detectors

It must be emphasized that this gain is essentially "noise free." Detailed considerations of noise will be covered later in the chapter, but note here that *nothing* flows through the output resistor unless that first electron starts down the multiplier chain. That is not true for any other amplifier. For instance, the common AM or FM radio is an amplifier with gains comparable to that of a photomultiplier. The "hiss" heard when it is not tuned to a station is an output with no input—noise. As we will see, that does not happen with a photomultiplier. The photomultiplier is extremely sensitive and can reach the limit of detecting single photons. To illustrate its capabilities consider the following example.

Suppose that a distance source at $\lambda = 5000$ Å reached our receiver (the photomultipler) with 20 photons per microsecond. The incoming power is thus 7.95×10^{-12} W. If the gain were as computed above and the quantum efficiency of the photocathode were $\eta_{qe} = 0.15$, the load current would be

$$i_L = G\eta e\left(\frac{P}{h\nu}\right) \quad (16.2.3)$$

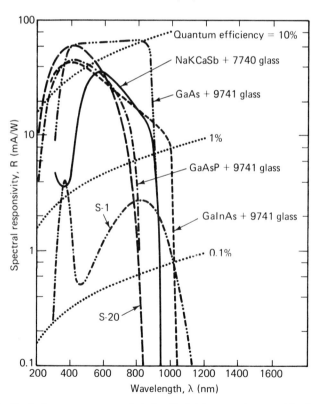

Figure 16.3 Typical response curves of photocathode/glass envelope combinations. Note that the response is plotted as current per watt and not as current per photon. (Data from Fig. 10.5 of *RCA Electro-Optics Handbook.*[5])

or 8.06×10^{-6} A, a value that is readily measured. If $R_L = 1$ MΩ, then $V_L = 8$ V, a large signal in response to 7.95 picowatts!

The dynode resistor chain in Fig. 16.2 is chosen to allot each dynode its proper share of a single power supply voltage. The capacitors are placed on the last few stages, where the dynode currents are the largest, to ensure that the voltage difference remains constant in the presence of a large pulse of photoelectrons.

Many of the problems associated with vacuum photodiodes are also present in photomultiplers; they are large and fragile and require high-voltage power supplies. However, photomultipliers hold a commanding lead in sensitivity. The most serious limitation of both is a fundamental one—the vacuum photoelectric effect ceases for $\lambda \geqslant 1.2$ μm. This is shown in Fig. 16.3 for various photocathode materials (and glass envelopes). Thus one must turn to solid-state detectors (or thermal ones) for radiation at longer wavelengths.

16.3 SOLID-STATE QUANTUM DETECTORS

16.3.1 The Photoconductor

An elementary schematic for a photoconductive element used in a detector circuit is shown in Fig. 16.4. The element is chosen such that its "dark" resistance is much larger than the load R, and thus most of the bias voltage appears across the detector chip. Depending on the type of semiconductor material used and its doping, the incoming radiation might ionize a donor, creating an excess electron in the conduction band for an n-type semiconductor, an excess hole in the valence band by promoting an electron to an acceptor level in a p-type, or an electron-hole pair by band-to-band transitions.

In any case the optical signal changes the number of free carriers in the chip and thus its resistance drops, allowing more of the bias voltage to appear across the load.

Once the excess electron (or hole) is created by the absorption of a photon,

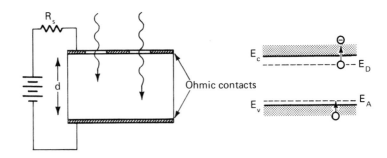

Figure 16.4 Photoconductive detector.

Sec. 16.3 Solid-State Quantum Detectors

it will drift under the influence of the field toward the appropriate contact and current flows in the external circuit. Since this is a majority-carrier conduction process, the other contact emits another carrier as soon as the first is collected. Thus current continues to flow until the carrier recombines with the donor (or acceptor).

If one defines τ_0 as being the carrier lifetime, the number of excess carriers, ΔN, generated in this detector in response to a suddenly applied optical signal is given by a solution to

$$\frac{d \Delta N}{dt} = \eta \left(\frac{P_{\text{abs.}}}{h\nu} \right) u(t) - \frac{\Delta N}{\tau_0} \tag{16.3.1}$$

where η is the quantum efficiency and $u(t)$ the step function. The solution is

$$\Delta N = \eta \left(\frac{P_{\text{abs.}}}{h\nu} \right) \tau_0 \left[1 - \exp\left(-\frac{t}{\tau_0}\right) \right] \tag{16.3.2}$$

As each carrier moves across the gap, d, it conducts a current $qv_d/d = q/\tau_d$, where τ_d is the time for the carrier to cross the gap. Thus the current in the external circuit is

$$i = q\eta \left(\frac{P_{\text{abs.}}}{h\nu} \right) \left(\frac{\tau_0}{\tau_d} \right) \left[1 - \exp\left(-\frac{t}{\tau_0}\right) \right] \tag{16.3.3}$$

Note that the factor τ_0/τ_d can be interpreted as a photoconductive gain. If $\tau_0 \gg \tau_d$, the peak current is much larger than q times the carrier generation rate.

However, this advantage is a double-edged sword! If the carrier lifetime is long, the time-response factor in (16.3.3) is slow. If one attempts to preserve a fast time response and gain, the drift time τ_d must be short. But there is a limit here also. The distance cannot be less than the absorption length of the photons—otherwise, little optical power is absorbed, and thus few excess carriers are generated.

There are other limitations and difficulties with this type of detector—a penalty we must pay for the ability to detect longer wavelengths. If the donor (or acceptor) level-to-conduction (or valence) band gap is small in order to detect a small photon energy, these donors are easily ionized by virtue of the thermal environment. For instance, Hg goes into Ge as an acceptor at 0.09 eV; hence, the longest wavelength that it can detect is 14 μm. At room temperature, $T = 300°K$, most of the acceptors would be ionized, creating a large quantity of holes and lowering the resistance of the chip. Consequently, most photoconductive detectors are cooled to liquid-nitrogen (77°K) or liquid-helium (4.2°K) temperatures. Thus what started out to be an inexpensive small chip, say, 2 mm² in area, now requires a bulky Dewar and an expensive and delicate vacuum window* to allow the IR radiation to irradiate the detector element.

---

*Most common materials, such as glass, are highly attenuating for wavelengths longer than 3 μm.

16.3.2 The Junction Photodiode

Many of the limitations and difficulties previously noted are alleviated by utilizing the highly developed semiconductor technology of p-n junctions with all of the attendant varieties, the simple p-n junction, the p-i-n diode, avalanche photodiode (APD), and the phototransistor. These devices are rugged, sensitive, fast, easily produced, easily biased, and obviously compatible with the concept of integrated optics, where an entire system, receiver, multiplexer, and transmitter are to be placed on a single chip of semiconductor material. Our attention will be limited to the first three types of detectors since these are the workhorses of the industry.

Let us imagine the processes that take place in the formation of the reverse-biased p-n junction photodiode shown in Fig. 16.5. We start with two neutral materials, doped to be p- and n-type, which are to be brought together to form the junction. At the instant of contact, electrons flow from the n region to the p region, leaving behind the immobile donor ions. Similarly, holes flow from p to n, leaving behind the negatively charged acceptors. Thus an internal space-charge field builds up to prevent further gross migration of the carriers.

Note the direction of this space-charge field. If a minority carrier in the n region—a hole—would wander into this space-charge region, it would immediately be accelerated by this field and the hole would *drift* to the p region. Similarly, if a minority carrier on the p side—an electron—wanders into the space-charge region, it will be accelerated by the field and drift to the n side. Both processes contribute to the "drift" current crossing the junction, which is, of course, balanced at zero bias by an opposite current because of the diffusion of majority carriers against the retarding field.

The addition of an external reverse bias greatly reduces the diffusion of majority carriers across the junction. However, the wanderings of the *minority* carriers in the two regions still proceed as usual—some still have the audacity to diffuse from the neutral material into the space-charge region and are thus swept across the junction. As Streetman[1, p. 151] notes, the drift current is not so much "how fast the carriers are swept across the junction, but how often" they arrive there to be accelerated. In other words, the drift current is equal to how often these minority carriers are generated within a diffusion length of the transition region or within that region itself.

If we are provided access to the diode by an optical window, we can change this generation rate by the incoming photons and thus obtain a current proportional to the *absorbed* photons. However, it should be noted that only those electron-hole pairs generated in the transition region or within a diffusion length contribute to the current. If we can neglect recombination in this depletion region, the current is given by

$$I = e\eta \left(\frac{P_{\text{abs.}}}{h\nu} \right) \quad (16.3.4)$$

Sec. 16.3 Solid-State Quantum Detectors 573

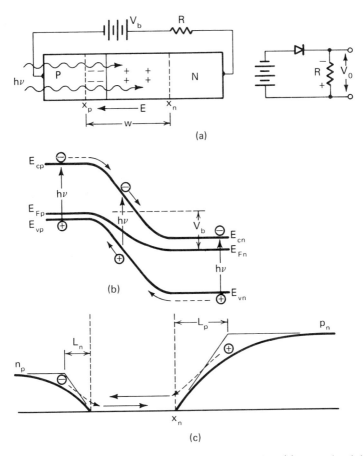

Figure 16.5 Solid-state photodiode: (a) reverse-biased *p-n* junction; (b) energy-band diagram; and (c) distribution of minority carriers.

Now it is important to remember that only the minority carriers that are created within a diffusion length of the transition region contribute to the photocurrent. Those electron-hole pairs created deeply in the neutral *n* and *p* region recombine before they ever diffuse to the junction to be influenced by the field and hence transport charge. This points out a major failing of the simple *p-n* junction diode—that one must wait for a relatively slow process, diffusion, to transport a minority carrier to the junction. Thus one could anticipate a rather slow time response if most of the carriers are generated in the neutral regions.

This is especially serious for indirect semiconductors such as silicon, in which the optical absorption coefficient is small. Since the transition width is small compared to the diffusion length, most of the pair generation takes place in the neutral regions, and thus we must depend on this slow process. This difficulty can be avoided by utilizing the *p-i-n* structure, which is discussed in Sec. 16.3.3.

Even though (16.3.4) appears to be very similar to the multiplicitive factor in (16.3.3), there are major differences in the physics. There is no possibility of photoconductive gain in a diode. Most of the applied bias appears across the depletion layers. Thus only those pairs formed within that region or that can diffuse to that region can be influenced by the field and carry electrical current.

One can utilize the solid-state physics of *p-n* junctions in a way other than the photoconductive mode described above. For instance, the optical photons generate an excess number of electron-hole pairs in the regions near the junction, and this contribution to the drift current is in addition to that caused by the thermal generation of electron-hole pairs. If the device is open circuited so that the net current is zero, the built-in field must be reduced to allow more diffusion current to flow. If I_0 is the normal reverse current of the diode as a result of thermal generation of electron-hole pairs and I_{pc} is the photocurrent produced according to (16.3.4), then the open circuit voltage is given by

$$V_{oc} = \frac{kT}{e} \ln\left(\frac{I_{pc}}{I_0} + 1\right) \qquad (16.3.5a)$$

This type of operation is called the *photovoltaic* mode and follows directly from the usual diode equation in the presence of optical generation of carriers:

$$I = I_0\left[\exp\left(\frac{eV}{kT}\right) - 1\right] - I_{pc} \qquad (16.3.5b)$$

Usually, the photoconductive mode is preferred for a variety of practical reasons. First, note that the current-to-optical power relation is linear for the photoconductive mode (16.3.4) but logarithmic for the photovoltaic one. If one wants speed of response, the choice is overwhelmingly in favor of the photoconductive mode.

This is because the depletion-layer capacitance caused by the space-charged layers of Fig. 16.5(a) appears in parallel with the load R_L. If the diode is reverse biased, then $C \propto (V)^{-1/2}$ (for an abrupt junction) and the inherent time constant of the detector $\tau = RC$ is minimized. On the other hand, this capacitance is quite large at the low voltages implied by photovoltaic operation. Incidentally, the voltage given by (16.3.5a) *cannot* be greater than the contact potential ($V_0 \sim E_g$) between *p* and *n*, irrespective of the implication that it can by (16.3.5b) (see Streetman[1] for more details). For the photovoltaic mode, we should also use a large R to ensure an open-circuit condition for (16.3.5); hence, the detector time constant is seriously degraded.

16.3.3 The *p-i-n* Diode

In all the equations so far, we have carefully insisted on using the *absorbed* optical power and knowing *where* it is absorbed. Obviously, if the photon passes through the detector, it cannot generate an electron-hole pair. The problem is shown in Fig. 16.6(a), where incoming radiation passes around the contacts through the bulk *p* region to be absorbed (partially) in and around the depletion region. The part of the radiation that is shadowed by the contacts does nothing; that absorbed deep in the *p* region and *n* regions does next to nothing (since the minority carrier so generated has a very small probability of reaching the junction before recombining); only that absorbed in the depletion layer or within a diffusion length of it is useful. Since the diffusion lengths are usually much larger than the width of the depletion region, most of the useful absorption and pair production are in the neutral region, and we must depend on the slow diffusion process for transport of the current.

We would prefer for the carriers to be generated where the field is large, so that the charge transport (i.e., velocity) is due to the fast drift rather than the slow diffusion. This is accomplished by adding an intrinsic (or at least a high-resistivity) region between the *p* and *n* layers, as shown in Fig. 16.6(b). This is

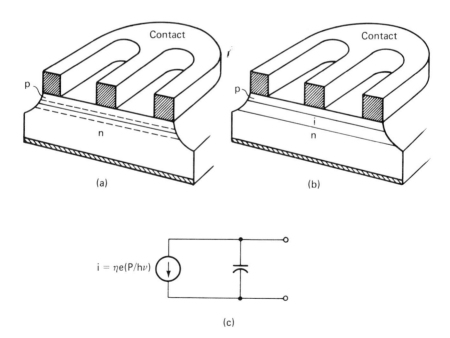

Figure 16.6 *p-n* and *p-i-n* diode: (a) *p-n* junction detector; (b) *p-i-n* detector; and (c) equivalent circuit.

called a *p-i-n* diode. Most solid-state detectors for lasers in the visible to near-IR region are of this variety. If the intrinsic layer is thick compared to the optical absorption length, most of the photocurrent will be generated in the intrinsic layer, where the field is largest. Thus any carriers generated there are immediately swept to the appropriate contact. There are other advantages, also. The intrinsic layer separates the depletion-layer charges of Fig. 16.5(a) and thus lowers the capacitance.

16.3.4 The Avalanche Photodiode

One can utilize the fact that the optically produced electrons (and holes) can create secondary pairs as they drift across the junction and thus provide an avalanche of free carriers. Diodes using this effect are called avalanche photodiodes. These second-generation carriers move under the influence of the field and can create a third generation; all generations contribute to the external current. If M new pairs are generated for each primary pair created by the photon, (16.3.4) must be multiplied by this factor, thus greatly enhancing the sensitivity of the device. Typical values of M range from 20 to 100.

One cannot use arbitrarily large values of M, because, as will be discussed later, this multiplication does contribute excess noise. Furthermore, if the electron-hole pair created by the optical photon avalanches, so does the pair created by thermal processes. Thus thermal runaway is a distinct possibility at high values of M.

16.4 NOISE CONSIDERATIONS

If one insists on large optical signals, there is really no problem with detection. Indeed, a crude standard detector for a high-powered CO_2 laser was the number of firebricks burned per second. Obviously, other "standards" can also be used. In this section, let us make sure that we have identified the problem with the detection of weak optical signals before we pull out our mathematical guns to attack the problem.

For weak signals, there are a whole host of statistical issues that must be addressed. For instance, we have indicated that the photomultiplier is able to detect ~20 photons in a microsecond or an optical energy of 8×10^{-12} W. However, if the source is a distant laser, the statistical fluctuation of the number of photons into the solid angle of our detector must be taken into account. Furthermore, the quantum efficiency of the photon cathode, η, and the secondary emission ratio, δ, are statistical *averages,* not hard and fast numbers. To illustrate this last issue, consider a "Gedanken" experiment, using a computer, to predict the generation of electrons emitted from the cathode.

Sec. 16.4 Noise Considerations

Assume a source emitting an on-off square wave with a precise number of 20 photons within the envelope of the on time. If the quantum efficiency were 100%, then we would have 20 "spikes" of current, which would then be amplified by the secondary emission amplifier. Since the gain was assumed to be 1.68×10^7 we now have $20 \times 1.68 \times 10^7$ "spikes" of current through our load resistor, each carrying a charge of 1.6×10^{-19} coulombs. If the time interval were 1 μsec, this represents an average current of 53.7 μamp. If the load resistor were 1 kΩ, then the *average* output voltage would be 53.7 mV, with the "spikes" being much larger. If there were no electrons emitted during the off time, one would have an unambiguous indication of the presence of these photons. This is shown in Fig. 16.7(a).

The situation becomes less clear for $\eta < 1$ as shown in Fig. 16.7(b) and (c). The quantum efficiency is a statistical quantity with the average value as shown on the side. For $\eta = 0.5$, we cannot get half of an electron; we get one half of the time—on the average. As a consequence our nice square-wave envelope became rather ragged due to *the statistical nature of the detection process*. The situation becomes much worse for $\eta = 0.2$— indeed, the current "spikes" remind one of a noise generator!

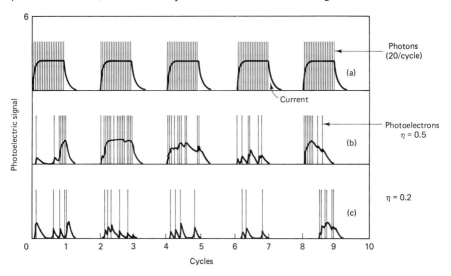

Figure 16.7 A computer experiment demonstrating the statistical nature of the conversion of photons into an electrical current (shown solid). See text for added detail.

Fig. 16-7 was generated by a computer using its random number generator so as to emit an electron according to the statisctics demanded by the quantum efficiency specification. If the source had a statistical fluctuation in number of photons during the on time, or there was thermionic emission from the cathode, these would also appear as "spikes" in the current to be further amplified by the dynode chain. These spikes could appear anywhere and destroy the infinite contrast of the "data" implied by Fig. 16-7. If the detector were a solid-state device, such as an APD, then the thermal generation of *e-h* pairs would also contribute a statistical current that acts as a fluctuation of the baseline of the output. In any case, note that the pulses do not "look" the same; sometimes they appear to replicate the envelope of the photons; others are a rather poor representation. Our task is to quantify the concept of a signal-to-noise ratio. But it should be clear that the *statistical nature of converting a photon into an electron generates its own brand of noise* in addition to any statistical fluctuation of the source. It is impossible to assign a partial blame to the detector, and hence both causes are usually lumped into a category called *quantum noise*.

Although we have used the photomultiplier as a convenient example, the same considerations apply to any quantum detector. The only problem with the others is that external "noisy" amplifiers must be used to amplify the output current to the level used here, and these amplifiers contribute their own bit of noise, to muddy the waters even further. But, in *all* cases, *each* photoelectron (or hole) produces a *current impulse* of the form

$$i = e\frac{v}{d} = \frac{e}{\tau}, \qquad 0 \le t \le \tau = \frac{d}{v} \qquad (16.4.1)$$

where v is the free electron velocity of a vacuum photodiode or the drift velocity in a semiconductor and d the distance over which it must travel before being collected, giving a lifetime τ. Please note that τ is quite small, typically less than 10^{-9} sec; thus these impulses have a nearly "flat" frequency spectral content. This fact plays a major role in the source of noise in optical detectors.

16.5 THE MATHEMATICS OF NOISE

The most essential bit of mathematics pertinent to noise consideration is the Fourier transform pair, which relates the spectral content of a video signal to its time response. Thus, if $v(t)$ is given, then $V(\omega)$ is related to it by

$$V(\omega) = \int_{-\infty}^{+\infty} v(t) e^{-j\omega t} \, dt \qquad (16.5.1)$$

and, of course, the reverse sequence can be followed to find $v(t)$ if $V(\omega)$ is given:

$$v(t) = \frac{1}{2\pi} \int_{-\infty}^{+\infty} V(\omega) e^{+j\omega t} \, d\omega \qquad (16.5.2)$$

In practice, we are forced to measure the signals over a finite sampling interval T, and thus we can assume $v(t)$ to be zero (or, more properly, not observed) outside the interval $-T/2 \le t \le T/2$. We indicate this finite time slot by a subscript:

$$V_T(\omega) = \int_{-T/2}^{+T/2} v(t) e^{-j\omega t} \, dt \qquad (16.5.3\text{a})$$

Sec. 16.5 The Mathematics of Noise

and

$$v(t) = \frac{1}{2\pi} \int_{-\infty}^{+\infty} V_T(\omega) e^{+j\omega t} \, d\omega \tag{16.5.3b}$$

Since $v(t)$ is real, we also know that $V_T(\omega) = V_T^*(-\omega)$.

Let us suppose that this video signal $v(t)$ is driving an amplifier with an input resistance R; we wish to compute the average power transferred to it. We shall now proceed to show that this power can be computed in the time domain or in the frequency domain by suitable manipulations of (16.5.3a) and (16.5.3b). We first multiply the voltage times the current and average over the sampling time T:

$$\langle P \rangle = \frac{1}{T} \int_{-T/2}^{+T/2} v(t) i(t) \, dt = \frac{1}{R} \left[\frac{1}{T} \int_{-T/2}^{+T/2} v(t) v(t) \, dt \right] \tag{16.5.4a}$$

Substitute (16.5.3b) into (16.5.3a) for the second $v(t)$:

$$\langle P \rangle = \frac{1}{R} \left\{ \int_{-T/2}^{+T/2} v(t) \left[\frac{1}{2\pi} \int_{-\infty}^{+\infty} V_T(\omega) e^{+j\omega t} \, d\omega \right] dt \right\} \tag{16.5.4b}$$

Now interchange the order of integration and identify the complex conjugate of $V_T(\omega)$.

$$\langle P \rangle = \frac{1}{R} \left\{ \frac{1}{2\pi T} \int_{-\infty}^{+\infty} V_T(\omega) \underbrace{\left[\int_{-T/2}^{+T/2} v(t) e^{+j\omega t} \, dt \right]}_{V_T^*(\omega)} d\omega \right\} \tag{16.5.4c}$$

$$= \frac{1}{R} \left[\frac{1}{2\pi T} \int_{-\infty}^{+\infty} |V_T(\omega)|^2 \, d\omega \right] \tag{16.5.4d}$$

Now since $V_T(\omega) = V_T^*(-\omega)$, the quantity $|V_T(\omega)|^2$ must be even with respect to frequency—hence we need only consider positive ω. Thus our principal result is

$$\langle P \rangle = \frac{1}{R} \left[\int_0^{\infty} \frac{|V_T(\omega)|^2}{\pi T} \, d\omega \right] \tag{16.5.4e}$$

The integrand, $|V_T(\omega)|^2/\pi T$, can be interpreted as the energy per radian-frequency interval $d\omega$ (for a 1-Ω resistor) and is referred to as the spectral density function $S_T(\omega)$.

$$S_T(\omega) = \frac{1}{\pi T} |V_T(\omega)|^2 \tag{16.5.5}$$

Even though ω is the most convenient variable for theoretical calculations, the frequency, ν (Hz), is the most common system specification. Thus we define $S_T(\nu)$ to yield the same answer as (16.5.4e):

$$\int S_T(\nu)\, d\nu \triangleq \int S_T(\omega)\, d\omega \qquad (16.5.6a)$$

or

$$S_T(\nu) = 2\pi S_T(\omega) = \frac{2}{T}|V_T(\omega)|^2 \qquad (16.5.6b)$$

One can attribute a circuit function to the mathematical quantity $S_T(\nu)\,\Delta\nu$ by defining an equivalent voltage generator capable of generating the correct average power in the load resistor when each frequency component is summed over the bandwidth of the postdetector system. If one had started our analysis with a current signal $i(t)$, another definition of $S_T(\omega)$ would result, with $|I_T(\omega)|^2 R$ replacing $(|V_T(\omega)|^2/R)$ in (16.5.4e), and we are naturally led to an equivalent current generator. Since $v = iR$, this is no more complicated than the standard technique of replacing a Thévenin equivalent voltage generator by a Norton circuit involving current sources. These equivalent circuits are shown in Fig. 16.8.

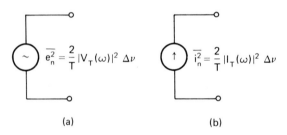

(a)   (b)

Figure 16.8 Equivalent-circuit representations of $S_T(\nu)\,\Delta\nu$.

Please note: It is the *power-generating capacity* of the generators that is specified, and this capacity, in turn, depends on the impedance level and bandwidth of the circuit elements placed on the output. The generator is not an entity in itself—*it is defined to give the right answer*—(16.5.4e) (or its counterpart in terms of current).

An example will point out the subtle implications contained in (16.5.4e). Consider the voltage signal produced on the output of the photomultiplier in the preceding section in response to the emission of a photoelectron. Obviously, the display as shown in Fig. 16.6 is the combined effect of the response of the detector *and* the circuitry (i.e., the distributed capacitance). If we looked more carefully at this pulse with a perfect display circuit—one without capacitance—we would observe an extremely short pulse, typically 1 nsec or so. As implied by (16.4.1), current only flows while the carrier is in transit (*for any detector*) and τ is very small. If we neglect the change in velocity of the carrier in transit, the transform of the current is

Sec. 16.5 The Mathematics of Noise

$$I_T(\omega) = \int_{-\tau/2}^{+\tau/2} i(t)e^{-j\omega t}\,dt = \int_{-\tau/2}^{+\tau/2} e\left(\frac{v}{d}\right) e^{-j\omega t}\,dt$$

$$= e\left(\frac{v}{d}\right)\left[\frac{e^{j(\omega\tau/2)} - e^{-j(\omega\tau/2)}}{2j\omega\tau/2}\right]\tau \qquad (16.5.7)$$

$$= e\left[\frac{\sin(\omega\tau/2)}{\omega\tau/2}\right]$$

since $\tau = d/v$. Thus the process of detecting the optical photons produces a current pulse whose frequency components are spread out over a whole band according to the spectral density function $S_T(\nu)$:

$$S_T(\nu) = \frac{2e}{T}\left[\frac{\sin(\omega\tau/2)}{\omega\tau/2}\right]^2 \bigg|_{\tau\to 0} \longrightarrow \frac{2e}{T} \qquad (16.5.8)$$

Equation (16.5.8) must be multiplied by G for the photomultiplier and τ interpreted as the pulse duration of the output current. Quite often we neglect the pulse duration and take the limit of (16.5.8) as $\tau \to 0$ and thus replace $(\sin x/x)^2$ by 1. This is the spectral density function of the detector without modification by the external circuitry.

The important point to remember was stated above, but it is repeated here for emphasis:

The very process of detecting a photon generates power over the whole band of (video) frequencies.

This is the critical point in the discussion of shot noise in Sec. 16.6.1.

Now, of course, we seldom deal with only one such pulse or video signal $i(t)$; rather, we deal with a sequence of signals such as that depicted in Fig. 16.7. Thus, if $\overline{N}$ is the average pulse *rate*, we would have $\overline{N}T$ pulses in the time interval T, and each pulse would be expressed by $i(t - t_j)$, where t_j is the time origin for each pulse.

$$i = \sum_{j=1}^{\overline{N}T} ef(t - t_j) \qquad (16.5.9)$$

where the function $f(t)$ expresses the time behavior of one such event and e is the charge transported. It is important to realize what we can or cannot specify about this output.

We can surely specify the average or DC current produced by this detector—we measure it (with an ammeter). Thus, if the output consists of $\overline{N}$ pulses per second in our sampling interval T, the DC current is merely

$$I_{DC} = \frac{1}{T}\int_{-T/2}^{+T/2} i\,dt = \frac{1}{T}(\overline{N}Te) = \overline{N}e \qquad (16.5.10)$$

But we *cannot* predict when the pulses will arrive*—only the average rate of arrival given by $\bar{N}$. Thus the Fourier transform of the output current must reflect this ignorance.

$$I_T(\omega) = \sum_{j=1}^{\bar{N}T} e \int_{-T/2}^{+T/2} f(t - t_j) e^{-j\omega t} \, dt$$
$$= \sum_{j=1}^{\bar{N}T} eF(\omega) e^{-j\omega t_j} \qquad (16.5.11)$$

where $F(\omega)$ is the transform of a single event. To compute the spectral distribution of the power contained in (16.5.11), we need to form the product $I_T(\omega) I_T^*(\omega)$:

$$|I_T(\omega)|^2 = \left[e \sum_{j=1}^{\bar{N}T} F(\omega) e^{-j\omega t_j} \right] \left[e \sum_{k=1}^{\bar{N}T} F^*(\omega) e^{+j\omega t_k} \right] \qquad (16.5.12a)$$

When $j = k$, the exponential factors cancel and the summation yields $\bar{N}T$.

$$|I_T(\omega)|^2 = e^2 |F(\omega)|^2 \left[\bar{N}T + \sum_{j \neq k}^{\bar{N}T} \sum_{k=1}^{\bar{N}T} e^{j\omega(t_k - t_j)} \right] \qquad (16.5.12b)$$

If we admit our ignorance and confess that we do *not* know the time delays, then the average result of many repeated applications of (16.5.12b) is the first term with the double summation averaging to zero:

$$|I_T(\omega)|^2 = e^2 |F(\omega)|^2 (\bar{N}eT) \qquad (16.5.13)$$

Now we can identify the product $\bar{N}e$ as the DC current [i.e., (16.5.10)] and substitute this expression into (16.5.6b) to find the spectral density function:

$$S_T(\nu) = \frac{2}{T} |I_T(\omega)|^2 = 2eI_{\text{DC}} |F(\omega)|^2 \qquad (16.5.14)$$

We are now in a position to handle many of the aspects of noise in a detection system.

16.6 SOURCES OF NOISE

Noise, by implication, is an undesirable feature of detection; it is a video current or voltage wave riding on top of the signal component. It is our purpose here to identify the sources of noise so as to estimate and to *accept* the *limiting performance* of our system. This limiting state is obtained when the noise as a result of the quantized nature of the electrical charge, $\pm 1.6 \times 10^{-19}$ C, generated by the absorption of quanta of light, overwhelms all other sources of noise. No

---
*The only way one could predict the arrival of the "next" photon is to have measured it—thereby ensuring that the "next" photon does not hit the detector.

Sec. 16.6 Sources of Noise

amount of circuitry tricks, cooling, design of optics, or other such skulduggery will help in this limit of "shot noise" or "quantum noise."

There are other noise sources that can be minimized but never completely eliminated. Any detector must observe a background that has a finite temperature, and thus there are always a few unwanted optical photons impinging on the detector, which, in turn, generates these quantized electrical carriers and their associated video noise. The video circuit itself consists of a resistor at a finite temperature, which is part of an amplifier that introduces its own brand of noise. By careful design, these sources can be minimized.

16.6.1 Shot Noise

There are two viewpoints as to the origin of this type of noise. One viewpoint is that the noise is the result of the quantized nature of electromagnetic energy; the other is that it is the result of the quantization of the electrical charge. It is impossible to distinguish between the two opinions, since both are facts of life and one (the charge) is the direct result of the absorption of the other (the photon). However, since there are other physical processes that generate a free carrier, such as thermionic emission from a photocathode or thermal generation of electron-hole pairs in a semiconductor, it is convenient to assign this noise to the discrete nature of the electrical charge.

In either case, shot noise is the result of the probabilistic nature of the generation of the electrical charge within the detector. One can specify the average generation rate, but one cannot specify when the next charge will be emitted given that the first started at $t = 0$. This is precisely the case considered in Sec. 16.5. If we neglect the time interval during which the charges move to the collector, then $\overline{F(\omega)^2} = 1$ in (16.5.14), and the equivalent circuit is as shown in Fig. 16.9.

It may seem strange at first that in Fig. 16.9, a direct current causes an AC noise, until one remembers that this average current is actually a sequence of many very short pulses. It is the spectral content of those pulses that is represented by the AC noise generator. Shown also in Fig. 16.9 is the realistic situation where there is some DC current even when the detector is not illuminated. As indicated above, this dark current may be due to thermionic emis-

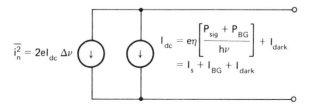

Figure 16.9 Equivalent circuit of the detector, showing the signal and noise current generators.

sion from the cathode of a photomultiplier, or it may be due to the thermal generation of electron-hole pairs in a semiconductor. Furthermore, there may be some current because of background photons riding along with our desired optical signal. But in either or any case, the current still comes in the form of impulses of charge and thus contributes to the noise power.

The signal power to a load R_L is $I_s^2 R_L$; hence, the signal-to-noise ratio is

$$\frac{S}{N} = \frac{I_s^2 R_L}{i_N^2 R_L} = \frac{[e\eta(P_{\text{sig.}}/h\nu)]^2}{2eI_{\text{DC}}\Delta\nu} \qquad (16.6.1)$$

Even if one hopes for the ideal situation where the dark current, I_D, and that due to the background, I_{BG}, is negligible, one is still faced with a finite S/N ratio due to the fact that signal itself produces a DC current.

$$\left(\frac{S}{N}\right)_{\text{max.}} = \frac{[e\eta(P_{\text{sig.}}/h\nu)]^2}{2e[e\eta(P_{\text{sig.}}/h\nu)]\Delta\nu} = \eta\left(\frac{P_{\text{sig.}}}{h\nu}\right)\frac{1}{2\,\Delta\nu} \qquad (16.6.2)$$

This is the best that can be done. Indeed, (16.6.2) is so obvious that we could have stated it based on common sense without all of the mathematical harangue of this chapter. If $\eta_{\text{qe}} = 1$ (the best), then $S/N = 1$ when there is one photon per sampling-time interval $T/2 = 1/(2\,\Delta\nu)$. Since we always require two samples to distinguish between a "1" or a "0" bit, (16.6.2) states that the maximum data rate with a video channel of bandwidth $\Delta\nu$ is $\Delta\nu/2$ bits per second (for $S/N = 1$).

There are other sources of noise, but (16.6.2) is the limiting value of S/N. Let us postpone further discussion until these sources are identified and combined with the above.

16.6.2 Thermal Noise

Thermal noise is due to the finite temperature of the elements of the detector system. As such, it can be partially alleviated by cooling these elements to dry-ice (195°K), liquid-nitrogen (77°K), or even liquid-helium (4°K) temperatures.

This radiation goes by many different names: (1) it is sometimes referred to as "white" noise, because its spectral distribution is uniform or "flat" at video frequencies at normal temperatures (the same is true of shot noise); (2) it is sometimes referred to as "Johnson" or "Nyquist" noise, after early pioneers in the field; and (3) it is also called "blackbody" noise, because it can be derived from Planck's blackbody formula discussed earlier.

Recall that the beginning of the quantum era was the successful explanation of the radiation emerging from a small "hole" in a heated cavity. The radiation emerging from this cavity is due to the energy in those cavity modes, which couple to the small aperture.

Although our initial discussion of this fact in Chapter 7 was in terms of a cavity that is huge compared to a wavelength, the same considerations apply at

Sec. 16.6 Sources of Noise

video frequencies for elements that are small compared to a wavelength. The maximum power that can be transferred from a blackbody at a temperature T in a bandwidth $\Delta \nu$ is the product of the following factors: (1) the number of modes in that bandwidth $\Delta \nu$, (2) the energy per photon, (3) the photons per mode, and (4) the bandwidth $\Delta \nu$.

At video frequencies, there is usually only one mode—the TEM mode extending down to zero frequency—and it surely has only one orientation (or polarization) of the field. Thus we obtain the low-frequency limit pertinent to our detector problem.

$$p_n = 1 \left[\frac{h\nu}{\exp(h\nu/kT) - 1} \right] \Delta \nu \qquad (16.6.3)$$

At normal temperatures (273°K) and reasonable frequencies (e.g., less than 1 GHz), the photon energy is much less than the characteristic thermal energy, $h\nu \ll kT$, and the approximate form of (16.6.3) is often used.

$$p_n = kT \, \Delta \nu \qquad (16.6.4)$$

It is this last form that is often referred to as "Johnson" noise, but it is an approximation to the more correct formula, (16.6.3).

Now, this is the maximum power that can be transferred from one blackbody circuit element to another blackbody element in the form of thermal noise. Fig. 16.10 illustrates the restrictions involved in applying (16.6.3). Resistor R_A at a temperature T_A emits noise power according to (16.6.3), to be absorbed by R_B, which, in turn, radiates power back to A. Only if $R_A = R_B$ will the two systems be "black" to each other's radiation, thereby absorbing all of the incident power.

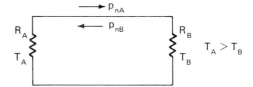

Figure 16.10 Interchange of power between two resistors.

Thus the characterization of an element being "black" is identical to the specification of a system being matched for maximum power transfer. To account for the fact that many systems are not matched, we construct an equivalent circuit with appropriate generators and *noiseless* resistors, which yields the correct answer for the power transfer when matched [i.e., (16.6.3)] and also properly allows for a mismatch. Such a model is shown in Fig. 16.11

We *define* a voltage (or current) generator of such a magnitude to yield the proper power transfer according to the laws of transmission-line theory, such that it agrees with the maximum power specified by Planck. Thus the mean of the squared rms voltage is

$$\overline{e_n^2} = 4R\left[\frac{h\nu}{\exp(h\nu/kT) - 1}\right]\Delta\nu \approx 4kTR\,\Delta\nu \qquad (16.6.5)$$

and the corresponding value for the current generator is

$$\overline{i_n^2} = \frac{4}{R}\left[\frac{h\nu}{\exp(h\nu/kT) - 1}\right]\Delta\nu \approx \frac{4kT}{R}\Delta\nu \qquad (16.6.6)$$

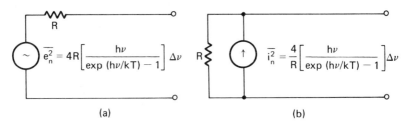

Figure 16.11 Equivalent circuits for thermal noise.

It is only the resistive part of the circuit that can accept power and, by the same token, can generate this noise power. A reactive element such as a capacitor affects the bandwidth $\Delta\nu$ under consideration but does not contribute to or accept the noise power.

16.6.3 Noise Figure of Video Amplifiers

Most optical systems require a video amplifier to amplify the output to a level where it can be used for communication and control purposes. In doing so these amplifiers contribute their own noise power, which must be *added* to the amplified value from the detector, thereby degrading the signal-to-noise ratio. It is imperative that one account for this reality.

The primary causes of noise in a video amplifier are shot noise from charge transport in the active devices (i.e., transistors) and amplified thermal noise from the resistive components. The net result is that there is a noise component to the voltage (or current) output of the amplifier even when there is no signal input or the input resistor is cooled to 0°K. This is the *excess* noise contributed by the amplifier.

The concept of a noise figure (of merit), F, was invented to describe this excess noise in a systematic manner. If one divides the output noise power by the gain, one obtains an equivalent noise source at the input terminals of the amplifier, which can now be considered noiseless. A noise temperature, T_A, is now defined to correctly specify the excess noise at the input terminals with its specified input resistance, which, in turn, contributes its own noise by virtue of its temperature. Since the excess noise and the resistor noise are independent quantities, one adds the two powers (or the mean of the squared voltages or currents). The noise figure, F, acknowledges the addition and, by convention, assumes that the input resistor is at 290°K (or 17°C).

Sec. 16.6 Sources of Noise

$$F = 1 + \frac{T_A}{290} \qquad (16.6.7)$$

An example should help to illustrate the procedure.

Example

Suppose that we had an amplifier of 40-dB gain, a noise figure of 13 dB, an input and output impedance of 50 Ω, and a bandwidth of 500 MHz. From the noise-figure specifications, $F = 20$ or the equivalent temperature of $19 \times 290°K = 5510°K = T_A$. The total *output* noise power in the 500-MHz bandwidth is 10^4 (i.e., the gain) times the total input noise power.

$$\langle P \rangle_{\text{out.}} = Gk(T_A + T_R) \Delta \nu = 0.4 \ \mu W$$

The rms value of the AC voltage across the 50-Ω output is 4.47 mV. If we cooled the input resistor to liquid-helium temperatures (4°K), we change T_R but do not affect T_A; thus we make a minimal improvement.

It should be clear from this example that the noise temperature, T_A, bears little, if any, relation to the ambient temperature of the amplifier. T_A is defined to correctly predict the noise contributed by the amplifier, and nothing more.

16.6.4 Background Radiation

Background radiation is probably the most obvious source of noise in an optical system and one that can be reduced with minimum effort. The background is the stray optical photons impinging on the detector. At the very minimum, any detector will be subjected to the blackbody radiation of the background at a finite temperature T_{BG}. We again invoke Planck's law* to estimate this unwanted optical radiation impinging on the detector.

$$\langle P_{\text{opt.}} \rangle_{BG} = \int_{\Delta \nu_{\text{opt.}}} \frac{1}{4} I(\nu) \, d\nu \, A_{\text{det.}} \frac{d\Omega}{4\pi} \qquad (16.6.8)$$

where $I(\nu)$ equals the intensity per frequency interval $d\nu$ and is given by (7.2.10) (repeated here for convenience):

$$I(\nu) = \frac{8\pi}{\lambda_0^2} \frac{h\nu}{\exp(h\nu/kT) - 1} \qquad (7.2.10)$$

$\Delta \nu_{\text{opt.}}$ represent the optical bandwidth of the detector, A is the area, and $d\Omega/4\pi$ is specified by the field of view (FOV).

If the signal has a specific sense of polarization, one can eliminate one half of the power represented by (16.6.8) by eliminating the orthogonal polarization. It is also desirable to minimize the optical bandpass $\Delta \nu_{\text{opt.}}$ and to restrict the field of view. But in doing so, it is important to recognize that the optical elements used to accomplish these purposes can also radiate blackbody radiation at their own temperature. For instance, if one used an absorptive polarizer at 300°K

---

*The factor $\frac{1}{4}$ is used when the *isotropic* blackbody flux given by (7.2.10) is used.

to eliminate the vertically polarized noise from a 200°K background source, the noise level would increase rather than decrease. This is because any component—the polarizer in this case—is equally good as a radiator as it is as an absorber (Kirchhoff's law).

16.7 LIMITS OF DETECTION SYSTEMS

Now that we have identified the major sources of noise in an optical detection system and have become familiar with the different types of detectors, it is time to combine the two to ascertain the limit and to establish realistic signal-to-noise ratios. As such, this section is more like a sequence of examples of previous concepts rather than a section in which new ground is broken. However, these examples point out the extreme importance of high quantum efficiency and low noise in the first stage of amplification in obtaining the highest signal-to-noise ratio.

The major new topic is that of optical heterodyning. With this technique, one can approach the quantum limit of detection given by (16.6.2), even without the benefit of extreme-low-noise amplifiers, such as the secondary emission amplifier in a photomultiplier.

16.7.1 Video Detection of Photons

The most straightforward system for detection is shown in Fig. 16.12, where a *p-i-n* photodiode is being irradiated by an optical signal P_s and its output is being fed into the video amplifier chain with finite noise temperatures T_{A1} and T_{A2}. The analysis of the signal-to-noise ratio at the output of this system follows from a straightforward application of the previous sections.

The desired signal power is the square of the signal current times the input resistance, and this power is then amplified by a factor $G_1 G_2$.

The *output* noise consists of a sum of three terms: (1) the second stage contributes $G_2 k T_{A2} \Delta \nu$; (2) the noise from the input resistor at T_R and the first amplifier contributes $G_1 G_2 k (T_R + T_{A1}) \Delta \nu$; and (3) the shot noise produced by

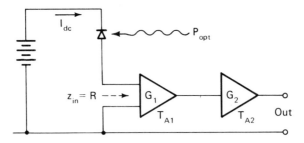

Figure 16.12 Photodiode video detection circuit.

Sec. 16.7 Limits of Detection Systems

the current in the detector and subsequently amplified by G_1G_2 contributes a power of $G_1G_2(2eI_{DC}\,\Delta\nu)R$. Thus S/N is given by

$$\frac{S}{N} = \frac{G_1G_2[e\eta(P_{opt.}/h\nu)]^2 R}{k(G_2 T_{A2} + G_1G_2 T_{A1} + G_1G_2 T_R)\,\Delta\nu + G_1G_2(2eI_{DC}\,\Delta\nu)R} \quad (16.7.1a)$$

or

$$\frac{S}{N} = \frac{[e\eta(P_{opt.}/h\nu)]^2 R}{k[(T_{A2}/G_1) + T_{A1} + T_R]\,\Delta\nu + 2eI_{DC}\,\Delta\nu\,R} \quad (16.7.1b)$$

This last form is equivalent to summing all noise powers at the input of the first stage and then considering all amplifiers as noiseless. This is the conventional procedure and will be followed hereafter.

Equation (16.7.1b) also emphasizes the importance of a low noise/high gain front end. If G_1 is large, the second amplifier can be quite noisy and affect the S/N ratio hardly at all. Obviously, we want to keep the excess noise contributed by the first stage as small as possible.

In any case, we obtain the *limiting performance* when the shot-noise term (due to the detection of the signal) overwhelms the thermal noise. The minimum value of the current through the detector is that caused by the optical power. When a detector system is dominated by shot noise, the system is said to be at the "quantum limit." (Obviously, we do not increase the DC current arbitrarily just to obtain a shot-noise–dominated system.) This fact explains why a photomultiplier and an avalanche photodiode are the premium detectors for video detection and why the heterodyne system reaches that limit without the benefit of an apparent amplifier.

Example: Noise in a Photomultiplier

Consider first the photomultiplier as described in Fig. 16.2. The incident photons create a cathode current of $e\eta(P_{opt.}/h\nu) = I_k$, which is subsequently amplified by the N stages of secondary emission or $G = \delta^N$. By the same token, the shot-noise current is amplified by the same value. Even though the secondary emission amplifier can be considered perfect insofar as thermal noise is concerned, the fact is that the secondary electron emission is a statistical phenomenon with considerable variance about the mean value of δ. Thus the emission from each dynode is also an independent statistical process and therefore creates its own brand of shot noise. The total squared noise current out of the photomultiplier is

$$\overline{i_n^2} = 2e\,\Delta\nu(\delta^{2N} I_k + \delta^{2N-2} I_1 + \cdots + I_N) \quad (16.7.2a)$$

where $I_1, I_2, \ldots, I_N$ are the currents emitted by the various dynodes. Now the ratio of the dynode emission currents is the mean value of the secondary emission ratio $\delta = I_N/I_{(N-1)} = I_1/I_k$. Hence, (16.7.2a) can be simplified and summed:

$$\overline{i_n^2} = 2eG^2 I_k \,\Delta\nu(1 + \delta^{-1} + \delta^{-2} + \cdots + \delta^{-N}) \quad (16.7.2b)$$

$$= 2eG^2 I_K \,\Delta\nu\,\frac{1 - \delta^{-(N+1)}}{1 - \delta^{-1}} \approx 2eG^2 I_k \,\Delta\nu\,\frac{1}{1 - \delta^{-1}} \quad (16.7.2c)$$

for δ a reasonable number ($2 \to 4$) and N large.

Now we add the thermal noise contributed by the load resistor or conventional amplifiers. The signal-to-noise ratio is

$$\frac{S}{N} = \frac{[Ge\eta(P_{\text{opt.}}/h\nu)]^2 R_L}{k(T_R + T_A)\, \Delta\nu + 2eG^2 I_k\, \Delta\nu[1/(1-\delta^{-1})]R_L}$$

$$I_k = \frac{e\eta P}{h\nu} + I_{\text{dark}}$$

(16.7.3)

For anything reasonable, the second term in the denominator overwhelms the thermal noise and the photomultiplier is entirely dominated by shot noise. If we cool the photocathode so as to eliminate thermionic emission and neglect other extraneous electron emission causes, $I_{\text{dark}} \doteq 0$ and we come close to the quantum limit rather easily.

$$\frac{S}{N} = \eta\left(\frac{P_{\text{opt.}}}{h\nu}\right)\frac{1}{2\,\Delta\nu}(1 - \delta^{-1}) \qquad (16.7.4)$$

We are able to do this by virtue of the secondary emission amplifier.

Example: Detection with an Avalanche Photodiode

A similar advantage is present in an avalanche photodiode, although one cannot, in practice, achieve as high a current gain as can be used in a photomultiplier. Typical avalanche gains of 30 to 100 are used, and this is sufficient for the system performance to be greatly improved over that with a simple photodiode.

In an avalanche photodiode, the carrier production rate $\eta(P/h\nu)$ is increased by a factor of M because of the ionization by the drifting electrons and holes; hence, the photocurrent and the dark current are enhanced by the same value:

$$i = Me\left[\eta\left(\frac{P}{h\nu}\right) + g_{\text{th}}\right] \qquad (16.7.5)$$

Each creation of a new carrier by the avalanche process is a statistical process, one whose average rate (M) can be measured and thus predicted. However, we cannot predict the occurrence of each individual event, and thus the avalanche multiplication contributes to the shot noise.

At first, one would hope for the same situation as found in a photomultiplier with its "noise-free" gain provided by the secondary emission amplifier (which is an avalanche of a sort). Unfortunately, noise power from an avalanche diode increases as M^n, where the exponent n is between 2 and 3.

$$p_n\,(shot\ noise) = M^n\left\{2e\left[\eta\left(\frac{P}{h\nu}\right) + g_{\text{th}}\right]\Delta\nu\right\}R \qquad (16.7.6)$$

where g_{th} is due to the thermal generation of carriers, and the term in brackets is the optical generation rate. Both processes create shot noise, and each must be multiplied by M_n.

The fact that the avalanche diode creates excess noise in the multiplication process (i.e., $n > 2$ in the equation above) means that one cannot use an arbitrarily large multplication factor. This follows directly from the expression for S/N for an avalanche photodiode and video amplifier combination.

Sec. 16.7 Limits of Detection Systems

$$\frac{S}{N} = \frac{M^2[e\eta(P/h\nu)]^2 R}{k(T_R + T_A)\Delta\nu + M^n\{2e^2[\eta(P/h\nu) + g_{th}]\Delta\nu\}R} \quad (16.7.7)$$

If the internal generation rate g_{th} and the optical power are very small, the thermal noise dominates and an increase in M from 1 helps considerably. However, if M is too large, the shot-noise term dominates and the signal-to-noise ratio degrades for increasing M.

$$\left.\frac{S}{N}\right|_{M\text{ large}} \sim \frac{1}{M^{n-2}}$$

Obviously, there is an optimum value of M for maximum S/N—an obvious problem for students.

Even though one cannot go to the extreme gain as with a photomultiplier, the multiplication does alleviate the necessity of conventional amplifiers with their own brand of noise. Furthermore, high-quantum-efficiency avalanche photodiodes can be constructed from various combinations of semiconducting materials for different regions.

For instance, Fig. 16.13 shows the construction details and measured quantum efficiency of a GaAlAsSb avalanche photodiode that has its peak response in the wavelength region 1.0 to 1.4 μm. If you will recall, this is precisely the wavelength region in which the absorption loss and material dispersion of fiber-optic cable is at a minimum. Hence, many of the future fiber communication systems will utilize wavelengths there and utilize detectors such as those illustrated in Fig. 16.13. Since the quantum efficiency of any photocathode in that region is extremely bad (if not zero), the avalanche photodiode is a clear choice for this application.

Although the discussion above acknowledges the existence of the excess noise, it does not address the question of why $n > 2$ in (16.7.7). To answer that requires us to delve more deeply into the multiplication itself. It would lead us far astray to give a complete theory of this excess noise, so we will have to be content with a gross physical picture.

The excess noise can be attributed to the fact that the multiplication process is the result of the ionization (or avalanche) by *both* carriers—electrons and holes. If only one carrier contributed to the multiplication, there would be noise caused by the statistical nature of the avalanche, as in a photomultiplier, but n would equal 2 and there would be no excess noise. But, alas, both do contribute and our hopes are dashed.

To appreciate the implications, consider the case where an electron-hole pair is generated by the photons in the high-field region of the *p-n* junction diode of Fig. 16.5 and assume that the electron can make two ionizations before being collected by the *n* contact. This would be true if the field were high enough, and the path length long enough, so that the electron could gain enough energy to create another electron-hole pair as it drifted. Because of the specification that two ionization events can occur, the second electron can also gain enough energy to create another electron-hole pair, as the first generation can create still another pair. Thus we have four electron-hole pairs in response to the absorption of one photon, and $M = 4$. If that was the end of it and the electrons were collected by

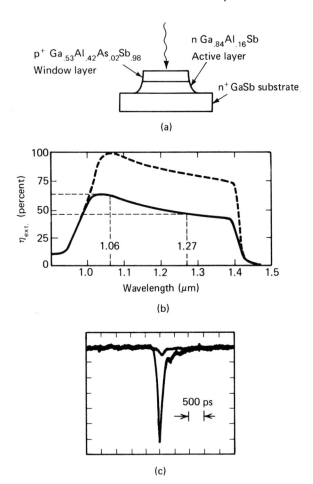

Figure 16.13 Details of a GaAlAsSb avalanche photodiode (APD). (a) Structure of a 1.0- to 1.4-μm GaAlAsSb avalanche photodetector. (b) Experimental photoresponse of the GaAlAsSb APD shown in (a) at low bias. The dashed curve shows the projected quantum efficiency for a detector with a suitable antireflection coating. (c) Pulse response to a mode-locked Nd-YAG laser. Lower trace $M = 17$, upper trace $M = 1$. The estimated photodiode response time is 120 ps (FWHM). (After "High Sensitivity Optical Receivers for 1.0–1.4 μm Fiber Optic Systems," Louis R. Tomasetta, H. David Law, Richard C. Eden, Ira Deyhimy, and Kenicht Nakano, IEEE J. Quantum Electronics, *QE-14*, No. 11, 800–804, 1978.)

ohmic contact (where they recombine) on the *n* terminal, and the hole collected by the *p* contact, we would have noise-free multiplication.

Unfortunately, that is not the end of it—we have ignored the holes. While the initial hole produced by the photon may be collected by the *p* terminal without causing any additional ionization, the other three holes are created much farther away from the collecting terminal. They can gain energy from the field and cause ionization in the same manner as the electrons.

Sec. 16.7 Limits of Detection Systems

If the probability of one hole making an ionization is greater than $\frac{1}{3}$, that of three holes is greater than 1. Thus there is greater than unity probability that a new electron-hole pair will be created in the area near the n region but still in the high-field. To follow that electron, one jumps back three paragraphs and starts all over. (Obviously, we are in an infinite DO loop!)

What is the result? The device is broken down with the external current infinite (or limited by the external circuit) and the device is useless as a detector. Obviously, we cannot use a value of $M = \infty$. But the point is that any small fluctuation in these probabilities causes a greatly enhanced fluctuation in M. That is the origin of the excess noise and $n > 2$.

It is interesting to note that the avalanche process (and breakdown) is identical to that occurring in a gas discharge. Indeed, the original avalanche photodiode was a gas-filled phototube. Although some of the details change for gas discharges, the end result is the same—a device passing current limited by the external circuit.

16.7.2 The Heterodyne System

The heterodyne detector system is the optical analog of the very common radio receiver, and its schematic is shown in Fig. 16.14. Its operation is trivial to understand, but the realization of the principle requires considerable effort and care. One combines a weak signal at an optical frequency ω_1 with a strong local oscillator at another optical frequency ω_2 to obtain a "beat" or IF frequency $(\omega_2 - \omega_1)$, which is then used for communication or control. Obviously, $\omega_2 - \omega_1$ must be in the passband of the video circuits for the beat note to perform this function; hence, this receiver has an extremely narrow optical bandpass. This has many advantages from a detection standpoint, but it places rather stringent requirements on the frequency stability of the source and local oscillator. The critical issues to ensure success are (1) to have a detector whose output current is proportional to the optical power (fortunately, all quantum detectors fall into this category) and (2) to align the signal and local oscillator waves to be collinear and coincident on the detector (a not so trivial, but not impossible task).

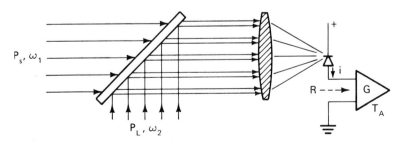

Figure 16.14 Optical heterodyne system.

To analyze the situation, we revert back to a classic description of the optical field and assume that the output current is proportional to the square of the *total* optical electric field of the waves absorbed by the detector.

$$i = e\eta_{qe}\frac{1}{h\nu}\left[\left(\frac{E_T^2}{\eta_0}\right)A\right] \tag{16.7.8}$$

One might tend to assume that this classic procedure ignores the "photon" characteristics of the optical signal, and therefore one could not hope for an accurate description of the process in the limit of a few photons. The fact is that this procedure will predict the correct answer, even in this limit,* because we *define* the classic field so that the quantity in the brackets is equal to the *photon power*.

Equation (16.7.8) also relates the peak optical field to the average or DC current in addition to relating it to the photon power.

$$I_{DC} = \left\langle e\eta_{qe}\frac{1}{h\nu}\frac{E^2}{\eta_0}A\right\rangle = e\eta_{qe}\frac{E^2}{2\eta_0}A = e\eta_{qe}\frac{\langle P_{opt.}\rangle}{h\nu} \tag{16.7.9a}$$

Therefore

$$E^2 = \frac{2\eta_0}{A}\langle P_{opt.}\rangle \tag{16.7.9b}$$

where A is the area of the optical beam and $\eta_0 = 377\ \Omega$. In the heterodyne system shown in Fig. 16.14, the total electric field has components associated with the signal and the local oscillator. We express the field amplitudes in terms of the powers of each wave.

$$E_T = \left(\frac{2\eta_0}{A}\right)^{1/2}(P_S^{1/2}\cos\omega_1 t + P_L^{1/2}\cos\omega_2 t) \tag{16.7.10}$$

Thus the current is given by (16.7.8) with the field given by (16.7.10).

$$i = \frac{2e\eta_{qe}}{h\nu}(P_S\cos^2\omega_1 t + 2P_S^{1/2}P_L^{1/2}\cos\omega_1 t\cos\omega_2 t + P_L\cos^2\omega_2 t) \tag{16.7.11a}$$

or

$$i = e\eta_{qe}\frac{1}{h\nu}[\underbrace{P_S + P_L}_{\text{DC terms}} + \underbrace{2P_S^{1/2}P_L^{1/2}\cos(\omega_2 - \omega_1)t}_{\text{intermediate frequency (IF)}} \tag{16.7.11b}$$

$$+ \underbrace{P_S\cos 2\omega_1 t + P_L\cos 2\omega_2 t + 2P_S^{1/2}P_L^{1/2}\cos(\omega_2 + \omega_1)t}_{\text{optical frequencies}}]$$

---

*An excellent, clear, readable, and enjoyable discussion of some of the issues involved in invoking the photon nature of the electromagnetic waves can be found in Scully and Sargent.[8]

Sec. 16.7   Limits of Detection Systems

The last line of (16.7.11b) represents current at optical frequencies, a result of our classic analysis. These would not be present if a full-blown quantum description had been used, but in any case, the low-frequency detector circuit would not respond anyway—so we drop them.

The first two terms represent a DC current and are overwhelmingly dominated by the much stronger local oscillator terms. This DC current causes "shot" noise, as before. The IF or "beat" frequency term contains the information about the presence or absence of the signal wave. Thus the video signal current can be expressed as

$$i_s = \frac{2e\eta_{qe}}{h\nu} P_S^{1/2} P_L^{1/2} \cos(\omega_2 - \omega_1)t \qquad (16.7.12)$$

where P_s is the peak value of the modulated wave used for communication.

One can make the signal current arbitrarily large by using a very large local oscillator power. This is how the heterodyne system obtains its gain. The penalty to be paid is that the DC current becomes arbitrarily large also:

$$I_{DC} = e\eta_{qe}\frac{P_L}{h\nu} \qquad (16.7.13)$$

and thus the shot noise also increases.

Now it is just a matter of elementary computation to compute the S/N ratio of the detector system shown in Fig. 16.13. For a 100% modulated wave, corresponding to the "on-off" code analyzed in Sec. 16.6.1, the average signal power is one half of the peak signal power, and thus

$$\frac{S}{N} = \frac{\frac{1}{2}i_s^2 R_L \cdot \frac{1}{2}}{k(T_A + T_R)\Delta\nu + 2e[e\eta_{qe}(P_L/h\nu)]\Delta\nu} \qquad (16.7.14)$$

By making the local oscillator power sufficiently strong, the shot noise dominates the thermal noise of the amplifier and we obtain the same quantum limit specified by (16.6.2).

$$\frac{S}{N} = \eta_{qe}\frac{\langle P_s \rangle}{h\nu}\frac{1}{2\Delta\nu} \qquad (16.6.2) \rightarrow (16.7.15)$$

Note that the heterodyne system provides (thermal) "noise-free" gain in much the same manner as the secondary emission amplifier does for the photomultiplier. Thus we should not be surprised at the same quantum limit. But note that this "noise-free" gain is not restricted to wavelengths where the vacuum photoelectric effect is applicable. Any square-law detector and any strong local oscillator can be used—indeed, the common household AM/FM radio operates on this principle. (It does not have the advantage of using quantum detectors. Nevertheless, it is incredibly sensitive.)

PROBLEMS

16.1. (a) What is the current density emitted by a photocathode with $\eta_{qe} = 0.2$ caused by an optical power of 1kW/cm² at $\lambda = 8000$ Å?

(b) Assume constant quantum efficiency *and* constant power. Plot the photocathode current as a function of wavelength between 2000 Å and 1.0 μm.

(c) What should the anode-to-cathode voltage be to ensure that space-charge effects are not important. (*Hint:* Evaluate the Child-Langmuir limit for the current; see Problem 12.2.)

(d) If the *A-K* spacing were 2 mm for a vacuum photodiode, what would be the transit time of a photoelectron? (Use $V_{A\text{-}K} = 2000$ V.)

16.2. One usually approximates the current associated with the transport of charge from the cathode to the anode of a photodiode as being a delta function, thus producing a "white" shot-noise spectra. Evaluate this approximation for a vacuum photocathode with $d = 2$ mm and $V_{A\text{-}K} = 2000$ V.

16.3. (a) Consider a surface charge $-\rho_s$ located at point x between two conducting planes at $x = 0$ and $x = d$ that are connected together by an external circuit. As this surface layer of charge moves with velocity $+v$ toward $x = d$, show that the current flow in the external circuit is given by $(\rho_s A)(v/d)$. (*Hint:* Use Gauss's law to find the time rate of change of the induced charge on the contacts at $x = 0, d$.)

(b) Plot the wave shape of the current from a vacuum diode and contrast it to that produced by a semiconductor diode. Assume equal charge transported in the same time for both cases.

(c) Now let an electron-hole pair be generated at some arbitrary point x. Let the electrons travel to $x = d$, the holes to $x = 0$. What is the current? [Find the induced charge by superposition of the solution found in (a).]

16.4. Suppose that N electron-hole pairs were created in the center of the intrinsic region of a *p-i-n* diode. Assume that bias voltage across this region of width w is V_0 and that the mobility of the electron is twice that of the hole (typical of silicon; see Streetman,[1] p. 87). Plot the output current as a function of time.

16.5. Consider the simple picture of a mode-locked laser at 6328 Å in which $N = 9$ equal-amplitude modes are centered about ν_0. The average power is 27 mW, and the mode spacing is 125 MHz. The output of this laser is detected with a vacuum photodiode with a quantum efficiency of $\eta = 0.15$. Neglect space-charge effects but do not assume ideal circuitry.

(a) What is the distance between the mirrors of this laser?

(b) If the load for the diode consists of a 1-kΩ resistor shunted by $C = 0.01$ μF, what is the output voltage?

(c) If $R_L = 50 \, \Omega$ and $C = 10$ pF, use (9.4.36), Fig. 9.17, and the material of this chapter to predict the display on an oscilloscope.

16.6. There is a considerable difference in the "$c/2d$ beats" between the various modes of a laser, depending on whether the system is mode-locked or not. Assume the system described in Problem 16.5 and $R_L = 50 \, \Omega$ and $C_x = 20$ pF for the radio receiver used to measure these beats. (Remember that mode 9 can beat with 8, which can beat with 7, and so on—each giving a $c/2d$ beat note.) Derive an expression for this beat-note amplitude, assuming:
   (a) The laser is mode-locked, and ideal circuitry (i.e., $C_x = 0$).
   (b) Nonideal circuitry.
   (c) The laser is *not* mode-locked (i.e., each mode has a phase ϕ_j, which has no correlation with that of any other mode).

16.7. The purpose of this problem is to analyze the degradation in signal-to-noise ratio in a heterodyne system when the incoming beams are *not* perfectly aligned.
   (a) Consider the case shown in the accompanying diagram, where the beam splitter is misaligned. We assume that the operator has some intelligence, and thus the beams will overlap at the detector. Assume that the detector is a vacuum photodiode whose photocathode emits a current density (amperes/area) in response to an optical power density (watts/area) and that the optical beams are limited-extent uniform plane waves. Plot the degradation in the signal-to-noise ratio as a function of misalignment angle θ.

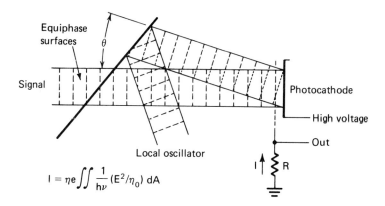

   (b) Suppose that the surface of the detector (say, a vacuum photocathode) is tilted with respect to the direction of the incoming beams. Show that misalignment does *not* affect the S/N ratio.
   (c) Suppose that the *centers* of the beams are collinear and coincident on the detector but that the *size* of the local oscillator beam is

different from that of the signal beam. Plot the degradation of the signal-to-noise ratio as a function of the ratio w_{LO}/w_S.

(d) Using the results found in (a) to (c), *discuss* other real-life effects that might degrade the signal-to-noise ratio from the ideal situation. For instance, we know that the beam from the local oscillator will be a Gaussian with a finite radius of curvature of the phase front, which will not be matched to the incoming signal. Is that a serious problem?

16.8. Suppose that one follows the avalanche of electron-hole pairs in a region of a semiconductor w units wide. Start with one pair at some point x between 0 and w and assume an equal probability, P, of an electron or a hole crossing the region w, creating a new pair. Show that the multiplication factor is $M = 1/(1 - P)$.

16.9. Consider a multimode laser with nine equal amplitude modes similar to that used in Sec. 9.4.1:

$$e(t) = \sum E_0 \cos\left[(\omega + n\omega_c)t + \phi_n\right] \quad -4 < n < 4$$

where

$$\omega_0 = \text{center frequency of the laser}$$
$$\omega_c = 2\pi \cdot (c/2d), \text{ the cavity mode spacing}$$
$$\phi_n = \text{phase of the } n^{th} \text{ cavity mode}$$

This field is incident onto a quantum detector, and the amplitude of the beat note at a frequency of $c/2d$ is measured in addition to the normal quantities. Assume the following numerical values: $\eta_{qe} = 0.25$; $\lambda = 6328$ Å; $[(E_0^2/2\eta_0) \cdot \text{area}] = 1$ mW; and $c/2d = 100$ MHz.

(a) If all of the phases are equal, the laser is mode-locked. Compute:
  (1) The time averaged current
  (2) The current as a function of time (identify the peak value and the FWHM)
  (3) The amplitude of the $c/2d$ beat note

(b) Now let the phases be distributed according to the following: $\phi_n/2\pi = 0.2; 0.7; 0.9; 0.0; 0.2; 0.6; 0.3; 0.5;$ and 0.9 for n running from -4 to $+4$ (a random number sequence chosen from the Social Security numbers of nine class members). Compute the beat note at a frequency of $c/2d$ and compare with (a, 3) above.

(c) Use the comparison indicated in (b) to indicate the amplitude of the beat note if the number of modes were large and the phases varied slowly with time (with respect to each other) in a random and uncorrelated fashion.

REFERENCES AND SUGGESTED READINGS

1. B. G. Streetman, *Solid State Electronic Devices,* 2nd ed., Ed. Nick Holonyak, Jr. (Englewood Cliffs, N.J.: Prentice-Hall, Inc., 1980).
2. A. van der Ziel, *Solid State Physical Electronics,* Ed. Nick Holonyak, Jr. (Englewood Cliffs, N.J.: Prentice-Hall, Inc., 1968).
3. Jacques I. Pankove, *Optical Processes in Semiconductors,* Ed. Nick Holonyak, Jr. (Englewood Cliffs, N.J.: Prentice-Hall, Inc., 1971).
4. A. Yariv, *Optical Electronics,* 2nd ed. (New York: Holt, Rinehart and Winston, 1971).
5. *RCA Electro-Optics Handbook,* Technical Series EOH-11, Copyright 1974 by the RCA Corporation.
6. James F. Gibbons, *Semiconductor Electronics* (New York: McGraw-Hill Book Company, 1966).
7. Willis W. Harman, *Electronic Motion* (New York: McGraw-Hill Book Company, 1953).
8. M. O. Scully and M. Sargent III, "The Concept of the Photon," Phys. Today, 38–47, Mar. 1972.
9. G. Margaritondo, "100 Years of Photoemission," Physics Today, *41*, 66–72, 1988.

I

"Detailed Balancing" or "Microscopic Reversibility"

The purpose of this appendix is to develop theory to enable us to compute the "reverse" reaction rate given that the "forward" rate has been measured. We applied this principle in Chapter 7 when the Einstein coefficient for absorption, B_{12}, was shown to be directly related to that for stimulated emission by

$$g_2 B_{21} = g_1 B_{12} \qquad (7.3.10a)$$

Consequently, a measurement of one quantity—say, the absorption cross section—also determines the stimulated emission cross section.

However, the principle of detailed balancing can be applied to inverse processes involving radiation, electron collisions, chemical and nuclear reactions, or combinations. Therefore it is an extremely important principle of physics with many ramifications and important applications.

Let us consider an inelastic collision of an electron with neutral atoms in its ground state creating an excited state—say, the upper laser level.

$$e(\epsilon_i) + N \longrightarrow N^*(E_2) + e(\epsilon_i - E_2) \qquad (I.1)$$

Obviously, this can happen only if the initial energy of the electron, ϵ_i, exceeds E_2. Although this process is most desirable in a laser, the first idea to be

App. I "Detailed Balancing" or "Microscopic Reversibility"

emphasized is that there is *always* the *reverse* process that is deleterious to the operation. Thus an electron of energy ϵ' could collide with the atom in the excited state, $N^*(E_2)$, and deactivate it back down to the ground level:

$$e(\epsilon') + N^*(E_2) \longrightarrow N + e(\epsilon' + E_2) \tag{I.2}$$

Note that any electron of any energy ϵ' is allowed to undergo collisions of the second kind (I.2), but only a select group with $\epsilon' > E_2$ have enough energy to participate in (I.1).

The foregoing equations represent reverse processes, and it must be emphasized that detailed balancing applies to that situation *only*. For instance, the same electron participating in (I.2) could also deactivate the atom from state 2 to another state 1; detailed balancing provides no information about this alternative route.

The key is to recognize that the cross section (for whatever process) is an integral part of the atom. Even though the environment can have a major, if not dominant, effect on the ultimate behavior of an atom, the cross section for a particular process is independent of that environment. Consequently, we can imagine a hypothetical thermodynamic equilibrium situation involving these same atoms N, electrons, a radiation field (and anything else).

The word "equilibrium" implies that all quantities are independent of time. Thus the electron density, the ground-state density, the excited-state population, and the radiation flux must all be constant. It follows, then, that the net rate of production of state 2 by all causes—namely, by the absorption of a photon of energy corresponding to the $2 \rightarrow 0$ transition, or by the process of (I.1), or by whatever—must exactly balance the decay rate of state 2, that is, by spontaneous emission, by stimulated emission from $2 \rightarrow 0$, or by the process in (I.2). An equilibrium situation demands this equality between the rates but does not dictate which is most important.

A thermodynamic equilibrium demands that each interaction, taken *by itself* and *in the microscopic limit,* must yield the appropriate thermodynamic population. It is in the balancing of the rates in detail and at the microscopic limit that gives this principle its name—detailed balancing or microscopic reversibility.

Once we say that the system is in thermodynamic equilibrium, we know everything about it. For instance, we know the radiant energy density per frequency interval around $h\nu = E_2 = \Delta E$:

$$\rho(\nu) = \frac{8\pi h\nu^3}{c^3} \frac{1}{\exp(h\nu/kT) - 1} \tag{I.3}$$

and the ratio of the populations of states 2 and 0:

$$\frac{N_2}{N_0} = \frac{g_2}{g_0} \exp\left(-\frac{E_2}{kT}\right) \tag{I.4}$$

and the electron energy distribution function:

$$f(\epsilon)\,d\epsilon = \left(\frac{4}{\pi}\right)^{1/2}\left(\frac{\epsilon}{kT}\right)^{1/2}\exp\left(-\frac{\epsilon}{kT}\right)d\left(\frac{\epsilon}{kT}\right) \tag{I.5}$$

all being specified by a common thermodynamic temperature, T.

Let us apply the principle to the electron collision sequence and show that we can generate the cross section for process 2 given the energy dependence of σ for 1. Fig. I.1 shows the *detail* of the balance required by thermodynamic equilibrium. Obviously, there are some electrons with enough energy to excite N_0 to N_2, and all electrons can deexcite N_2 to N_0. As shown in Fig. I.1(a), electrons at $\epsilon = \epsilon_i$ make an inelastic collision reappear at $\epsilon_i - E_2$, whereas the superelastic processes push the group upward.

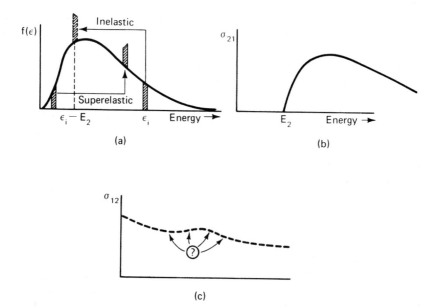

Figure I.1 (a) Electron energy distribution; (b) inelastic process (I.1); (c) superelastic process (I.2).

Equilibrium demands that the net rate of change of electrons in *any* energy interval be zero, and thermodynamic equilibrium demands that the electron distribution be given by (I.5). Thus we must balance, in detail, the rate *out* of a given energy interval by the rate *into* that same interval.

Consider the inelastic collisions by electrons in a small energy interval around ϵ_i. The total collision rate caused by that body of electrons is

App. I   "Detailed Balancing" or "Microscopic Reversibility"   603

$$dR_{02}(\epsilon_i) = \left[n\left(\frac{4}{\pi}\right)^{1/2}\left(\frac{\epsilon_i}{kT}\right)^{1/2}\exp\left(-\frac{\epsilon_i}{kT}\right)d\left(\frac{\epsilon}{kT}\right)\right] \cdot N_0 \cdot \sigma_{02}(\epsilon_i) \cdot \left(\frac{2\epsilon_i}{m}\right)^{1/2}$$

number of electrons at $\epsilon = \epsilon_i$
number of targets
cross section for inelastic process
speed of an electron with energy ϵ_i

(I.6)

This differential rate must be balanced by electrons leaving a correspondingly small interval around $\epsilon_i - E_2$ by the superelastic process:

$$dR_{20} = \left[n\left(\frac{4}{\pi}\right)^{1/2}\left(\frac{\epsilon_i - E_2}{kT}\right)^{1/2}\exp\left(-\frac{\epsilon_i - E_2}{kT}\right)d\left(\frac{\epsilon}{kT}\right)\right]$$
$$\cdot N_2\sigma_{20}(\epsilon_i - E_2)\left[\frac{2(\epsilon_i - E_2)}{m}\right]^{1/2}$$

(I.7)

As shown in Fig. I.1(c), σ_{20} is unknown, but we know it exists! Detailed balancing requires that the rates given by (I.6) and (I.7) be equal, which allows one to solve for σ_{20}. After canceling common factors and combining a few terms, we have

$$\epsilon_i\left[\exp\left(\frac{\epsilon_i}{kT}\right)\right]N_0\sigma_{02}(\epsilon_i) = (\epsilon_i - E_2)\left[\exp\left(-\frac{\epsilon_i - E_2}{kT}\right)\right]N_2\sigma_{20}(\epsilon_i - E_2) \quad (I.8)$$

Now (I.4) is used to express N_2 in terms of N_0.

$$\epsilon_i \exp\left(-\frac{\epsilon_i}{kT}\right) N_0\sigma_{02}(\epsilon_i)$$
$$= (\epsilon_i - E_2)\exp\left(-\frac{\epsilon_i - E_2}{kT}\right)\frac{g_2}{g_0}N_0\exp\left(-\frac{E_2}{kT}\right)\sigma_{20}(\epsilon_i - E_2)$$

Canceling common factors of N_0 and $\exp(-\epsilon_i/kT)$ leads to the desired relationship.

$$\sigma_{20}(\epsilon_i - E_2) = \frac{g_0}{g_2}\frac{\epsilon_i}{\epsilon_i - E_2}\sigma_{02}(\epsilon_i) \quad (I.9)$$

If we substitute $\epsilon = \epsilon_i - E_2$, the equation "looks" better.

$$\sigma_{20}(\epsilon) = \frac{g_0}{g_2}\frac{\epsilon + E_2}{\epsilon}\sigma_{02}(\epsilon + E_2) \quad (I.10)$$

Thus, knowing the cross section for the inelastic process, we can compute the superelastic cross section. Although we have derived this by considering a hypothetical situation of the thermodynamic equilibrium, this cross section is a part of the atom, and atoms have been "found" in nonequilibrium situations (most of the time). We can thus use this value for σ_{02} for these more interesting situations.

11

The Kramers-Kronig Relations

INTRODUCTION

The Kramers-Kronig relations relate the real and imaginary parts of a physical quantity of a generalized immittance function such as the resistance and reactance of a network, the gain (or loss) and the phase shift, or the real and imaginary parts of the dielectric constant. The significance of the relationships is that if you specify (or measure) *one* of these quantities as a function of frequency, then the other is also known (to within an additive constant term). Of course, some of the integrals may have to be evaluated numerically, but that is not a serious problem.

You are probably very familiar with the application of this theory in feedback amplifier design. We measure the gain (in dB) as a function of frequency and the result plotted as a function of the logarithm of frequency. From such "corner" or Bode plots, the phase shift can be determined. This is based on the theory to follow. The same theory is applicable to the optical domain, although its use requires considerably more sophistication.

Consider a function $F(s)$ regular in the right half of the s-plane. All

App. II The Kramers-Kronig Relations

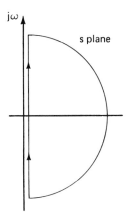

physical functions fall into this class in order to obey causality. Thus $F(s)$ can be an impedance function or the complex propagation constant of an electromagnetic wave.

$$\oint_C F(s)\, ds = 0 \qquad (\text{II}.1)$$

Along the real frequency axis,

$$F(j\omega) = u(\omega) + jv(\omega) \qquad (\text{II}.2)$$

where $u(\omega)$ and $v(\omega)$ are the real and imaginary parts.

We should recognize that any physical quantity $F(j\omega)$ must satisfy the following:

$$u(\omega) = u(-\omega) \qquad (\text{II}.3)$$

and

$$v(\omega) = -v(-\omega) \qquad (\text{II}.4)$$

This statement can be derived from the following: suppose that B, a real physical quantity, is related to A by the function $F(s)$. For points along the imaginary axis, $s = j\omega$, we have

$$B_{\text{complex}} = F(j\omega) A_{\text{complex}} e^{j\omega t} \qquad (\text{II}.5)$$

To construct the real quantity B, we add the complex conjugate:

$$B_{\text{real}} = F(j\omega) A e^{j\omega t} + F^*(j\omega) A^* e^{-j\omega t}$$

Thus B is real if $F^*(j\omega) = F(-j\omega)$, which in turn implies (II.3) and (II.4).

Now consider the integral along the path indicated below:

$$\oint \frac{F(s)\, ds}{s - j\omega_0} = I \qquad (\text{II}.6)$$

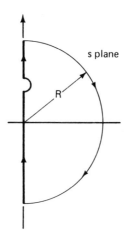

s plane

For any respectable function in the real physical world, $F(s) = 0$ as $|s| \to \infty$. Therefore there is no contribution along the closure line ($R \to \infty$). The function inside the integral is analytic along and within the curve shown; hence, $I = 0$. Since we are going halfway around the pole at $s = j\omega_0$, we obtain the following:

$$0 = \int_{-\infty}^{\omega_0-} \frac{F(j\omega)\,d(j\omega)}{j(\omega - \omega_0)} + j\pi F(\omega_0) + \int_{\omega_0+}^{\infty} \frac{F(j\omega)\,d(j\omega)}{j(\omega - \omega_0)}$$

Collecting real and imaginary parts, we obtain

$$\pi v(\omega_0) = \int_{-\infty}^{\omega_0-} \frac{u(\omega)}{\omega - \omega_0}\,d\omega + \int_{\omega_0+}^{\infty} \frac{u(\omega)}{\omega - \omega_0}\,d\omega$$

$$= \int_{-\infty}^{+\infty} \frac{u(\omega)}{\omega - \omega_0}\,d\omega \quad \text{(II.7)}$$

$$-\pi u(\omega_0) = \int_{-\infty}^{+\infty} \frac{v(\omega)}{\omega - \omega_0}\,d\omega \quad \text{(II.8)}$$

Equations (II.7) and (II.8) are one form of the Kramers-Kronig relations. Now let us use the symmetry properties of $u(\omega)$ and $v(\omega)$, that is, (II.3) and (II.4).

$$\pi v(\omega_0) = \int_{-\infty}^{0} \frac{u(\omega)}{\omega - \omega_0}\,d\omega + \int_{0}^{+\infty} \frac{u(\omega)\,d\omega}{\omega - \omega_0}$$

Since the variable of integration can be anything we choose, substitute $-x$ for ω in the first integral and $+x$ for ω in the second.

$$\pi v(\omega_0) = \int_{-\infty}^{0} \frac{u(-x)\,d(-x)}{-x - \omega_0} + \int_{0}^{\infty} \frac{u(+x)\,d(+x)}{x - \omega_0}$$

$$= \int_{0}^{\infty} u(x)\left(\frac{1}{x - \omega_0} - \frac{1}{x + \omega_0}\right)dx$$

App. II The Kramers-Kronig Relations

Collecting terms yields

$$v(\omega_0) = \frac{2\omega_0}{\pi} \int_0^\infty \frac{u(x)}{x^2 - \omega_0^2} \, d\omega \qquad (\text{II.9})$$

Manipulation of (II.8) proceeds in the same fashion:

$$\begin{aligned}
-\pi u(\omega_0) &= \int_{-\infty}^0 \frac{v(\omega) \, d\omega}{\omega - \omega_0} + \int_0^\infty \frac{v(\omega) \, d\omega}{\omega - \omega_0} \\
&= -\int_0^{+\infty} \frac{v(-x) \, d(-x)}{-x - \omega_0} + \int_0^{+\infty} \frac{v(+x) \, d(+x)}{x - \omega_0} \\
&= \int_0^\infty v(x) \left(\frac{1}{x + \omega_0} + \frac{1}{x - \omega_0} \right) dx
\end{aligned} \qquad (\text{II.10})$$

$$u(\omega_0) = -\frac{2}{\pi} \int_0^\infty \frac{x v(x) \, dx}{x^2 - \omega_0^2}$$

Equations (II.9) and (II.10) are the usual forms quoted for the Kramers-Kronig relations. In all but the simplest cases, one must resort to numerical integration to utilize these relations, but the fact that they exist is important in itself.

Example

One of the few cases that can be handled by analytic techniques is that of the parallel combination of a resistor and a capacitor. The driving-point impedance is given by

$$Z(j\omega) = \frac{R(1/j\omega C)}{R + (1/j\omega C)}$$

$$R(\omega) + jX(\omega) = \frac{R}{1 + \omega^2 \tau^2} - \frac{j\omega \tau R}{1 + \omega^2 \tau^2}$$

Thus a suitable candidate for $u(\omega)$ of (II.9) is $R(\omega)$, and $v(\omega)$ is identified with $X(\omega)$. Let us be "blind" to the equation above and assume that we know $R(\omega)$ only.

$$X(\omega_0) = \frac{2\omega_0}{\pi} \int_0^\infty \frac{R}{1 + x^2 \tau^2} \frac{1}{x^2 - \omega_0^2} \, dx$$

Use a partial-fraction expansion for the integrand:

$$\frac{R}{1 + x^2 \tau^2} \frac{1}{x^2 - \omega_0^2} = \frac{A}{x + \omega_0} + \frac{B}{x - \omega_0} + \frac{C}{x\tau + j} + \frac{D}{x\tau - j}$$

where

$$A = \frac{R}{1 + \omega_0^2 \tau^2} \frac{1}{(-2\omega_0)}, \qquad C = \frac{R}{2j} \frac{\tau^2}{1 + \omega_0^2 \tau^2}$$

$$B = \frac{R}{1 + \omega_0^2 \tau^2} \frac{1}{2\omega_0} = -A, \qquad D = -\frac{R}{2j} \frac{\tau^2}{1 + \omega_0^2 \tau^2} = -C$$

Now one can integrate this expansion term by term:

$$\int_0^\infty \frac{A\,dx}{x + \omega_0} + \int_0^\infty \frac{-A\,dx}{x - \omega_0} = -2\omega_0 \int_0^\infty \frac{dx}{x^2 - \omega_0^2}$$

$$= 2\,\ell n \left|\frac{\omega_0 - x}{\omega_0 + x}\right|\Bigg|_0^\infty = 0$$

$$\int_0^\infty \frac{dx}{x^2 \tau^2 + 1} = \frac{1}{\tau^2} \int \frac{dx}{x^2 + (1/\tau^2)} = \frac{\pi}{2\tau}$$

Therefore

$$X(\omega) = \frac{2\omega_0 R}{\pi(1 + \omega_0^2 \tau^2)} \left(\tau^2 \frac{\pi}{2\tau}\right) = \frac{\omega_0 \tau R}{1 + \omega_0^2 \tau^2}$$

Thus after considerable work, we obtain the expected result.

Author Index

Aiki, K., 510
Alferov, Zh. I., 412
Ana'ev, Yu. A., 509
Anderson, J. B., 229
Andreev, V. M., 412
Arecchi, F. T., 5, 412
Arrathoon, R., 504, 509
Asawa, C. K., 358

Balmain, K. G., 33
Barger, R. L., 204, 229
Barnes, Frank S., 547
Bass, M., 5, 358
Beck, R., 5
Bennett, W. R., Jr., 229, 304, 325, 358, 359
Bernard, M. G., 412
Bevacqua, S. T., 411
Bloembergen, N., 5
Bloom, A. L., 229
Borde, C. J., 229
Botez, D., 108, 412
Boyd, G. D., 61, 80, 509
Boyle, W. S., 108
Brau, C. A., 346, 359
Breene, R. G., Jr., 182
Brown, H., 68, 321, 359
Buczek, C. J., 461, 509
Burnham, R. D., 510
Burns, G., 411
Butler, J. K., 412, 510

Casey, H. C., Jr., 382, 391, 412, 510
Chang, W. S. C., 181, 229, 304, 546
Chebotayev, V. P., 230
Chenausky, P. P., 461, 509
Cheng, D. K., 33
Cheo, P. K., 358
Cherin, A. H., 108
Chinn, S., 510
Cobine, J. D., 451
Coleman, J. J., 412
Cook, D. D., 508, 510
Crompton, R. W., 438, 451
Cronemeyer, D. C., 358

Danielmey, H. G., 358
Dapkus, P. D., 412
Day, G. W., 108
DeBenedictis, L. C., 358
D'Haenens, I. J., 358

DeMaria, A. J., 358
Demtroder, W., 229
Denes, L. J., 434, 435, 451
Deyhimy, Ira, 392
Dienes, A., 61
Dills, F. H., 411
Dumke, W. P., 411, 412
Dunn, M. H., 148, 181, 229, 546
Dupris, R. D., 412
Duraffourg, G., 412

Eden, Richard C., 592
Eisberg, R. M., 33, 181, 546
Emanuel, M. A., 412
English, W., 5
Evtuhou, V., 358
Ewing, J. J., 360

Feynman, R. P., 181, 547
Fox, A. G., 61, 108, 113, 121, 148, 303, 467, 468, 469, 472, 473, 509
Franzen, D. L., 108
Frieberg, R. J., 461, 509

Gerber, R. A., 359
Gerry, E. T., 167, 168
Geusic, J. E., 358
Gibbons, James F., 599
Givens, M. E., 412
Gordan, J. P., 61, 80, 218, 509
Gordon, E. I., 229, 358
Grassman, M. W., 229
Gürs, K., 5

Hall, J. L., 205, 229
Hall, R. L., 229
Hänsch, T. A., 358
Harman, Willis W., 599
Haus, H. A., 80, 108, 128, 148, 509, 510
Hays, G. N., 359
Hayt, W. H. Jr., 33
Hellworth, R. W., 304
Herriott, D. R., 80, 325, 358, 359
Herskowitz, G. J., 108, 412
Herzberg, G., 181, 359, 546, 547, 553
Hoffman, J. M., 359
Holonyak, N. Jr., 411, 412, 599
Hoskins, R. H., 358
Hsia, J. C., 359
Huber, D. L., 182
Huestis, D. L., 304

Hummer, D. G., 229
Hunsperger, R. G., 108, 412, 510
Huxley, L. G. H., 438, 451

Ikegami, T., 480, 510
Ippen, E. P., 61
Irwin, B. W., 437, 438, 441, 444, 451

Kaminiski, A. A., 317, 359
Kapon, E., 510
Katz, J., 510
Keiser, G., 108
Keyes, R. J., 411
Kieffer, L. J., 447, 451
Koechner, W., 310, 313, 317, 318, 320, 358, 359
Kogelnik, H., 61, 80, 108, 121, 148, 303, 509, 510
Kolbas, R., 412
Korol'koy, V. I., 412
Kressel, H. K., 412, 510
Kuezenga, 304
Kunasz, C. V., 229
Kushida, T., 358

Labuda, E. F., 358
Lacy, E. A., 108
Lamb, W. E. Jr., 5, 204, 229, 547
Laqueskenko, R., 229
Lasher, G., 411, 412
Lau, K. Y., 405, 406, 412
Law, H. David, 592
Lax, B., 411
Leighton, R. B., 181, 547
Lengyel, B. A., 304
Leonard, D. A., 167, 168
Levine, A. K., 358
Li, T., 61, 80, 108, 113, 121, 148, 303, 467, 468, 469, 472, 473, 509
Lin, C., 409
Liou, K. Y., 409
Lowke, J. J., 434, 435, 436, 437, 438, 441, 444, 451

Madey, John M., 359, 360
Maiman, T. H., 358
Maitland, A., 148, 181, 229, 546
Marcateli, E. A. J., 108
Marcos, H. M., 358
Marcuse, D., 108
Margaritonso, G., 599
Marshall, Thomas C., 359
Matovich, E., 358
Matveyev, A., 547
Maust, L. J., 412
Maya, J., 229
Merzbacker, E., 546
Meyer, G. E., 358
Miller, S. E., 108

Mitchell, A. C. G., 181, 546
Moore, C. A., 550
Motz, H., 359
McCall, S. L., 510
McWhorter, A. L., 411

Nakamura, M., 510
Nakano, Kenicht, 592
Nash, F. R., 508, 510
Nathan, M. I., 411
Nighan, W. L., 451
Nussbaum, A., 33

Oster, L., 182

Panish, M. P., 391, 412, 510
Pankove, J. I., 412, 599
Pantell, R. H., 547
Paoli, T., 192, 193, 230, 408
Pasour, J. A., 359
Patel, C. K. N., 359
Patterson, E. L., 359
Phelps, A. V., 437, 438, 441, 444, 451
Phillips, R., 33
Pierce, J. R., 61
Piernier, M., 358
Platzman, P. M., 510
Portnic, E. L., 412
Presley, R. J., 5
Proffitt, W., 297, 304
Puthoff, H. E., 547

Quish, T. M., 411

Ramo, S., 33, 148
Rao, N. N., 33
Rigrod, W. W., 241, 297, 304
Rokni, M., 359

Sands, M., 181, 547
Sargent, M. III, 547, 599
Schalow, A. L., 5, 218, 229, 358
Schmeltzer, R. A., 108
Schultz, G. J., 340, 358
Schultz-DuBois, E. O., 266, 304, 412
Scifres, D. R., 510
Scully, M. O., 547, 599
Setser, D. W., 359
Shank, C. V., 61, 510
Shockley, W., 371, 412
Siegman, A., 61, 80, 122, 148, 181, 229, 303, 304, 359, 454, 504, 509
Smith, Peter W., 304, 546
Smith, W. V., 229
Snavely, B. B., 322, 323, 358
Snitzer, E., 358
Sorokin, P. P., 229
Steier, W. H., 510
Stern, F., 382, 412
Stitch, M. L., 5

Author Index

Streetman, B. G., 87, 359, 412, 596, 599
Streiffer, W., 510
Sze, S. M., 412

Thompson, G. H. B., 181, 412, 510
Tien, P. K., 108
Tilleman, M., 297
Tisone, G. C., 359
Tomasetta, Louis R., 592
Townes, C. H., 3, 5, 218, 229
Tret'yakov, D. N., 412

Ultee, C. J., 360
Umceda, J., 510

van der Ziel, A., 547, 599
VanDuzer, T., 33, 148
VanVleck, J. H., 182

Velazco, J. E., 359
vonEngel, A., 451

Wagner, W. G., 304
Waymouth, J. F., 229, 451
Whinnery, J. R., 33, 148
White, H. E., 33, 61
White, R. L., 547

Yariv, A., 61, 80, 122, 148, 181, 229, 304, 359, 405, 406, 412, 489, 510, 546, 599
Young, C. G., 358

Zeiger, H. J., 411
Zemansky, M. W., 181
Zmudzinski, C. A., 412

Subject Index

A_{21}, Einstein's coefficient for spontaneous emission, 156
  classical value, 513, 516
  derivation from Schrödinger's Equation, 526
    relation to B coefficient, 158
    relation to oscillator strength, 516
$ABCD$ law (for Gaussian Beams)
  definition, 91
  applied to a fiber, 81
  applied to free space, 91
  applied to stable cavities, 112
  applied to a thin lens, 93
$ABCD$ matrix
  of a continuous lens, 54
  of a dielectric discontinuity, 57
  of a dielectric slab, 57
  for a flat-curved mirror system, 45
  of a length of homogeneous dielectric, 35
  of a mirror, 38
  periodic sequence of lens, 40
  of a thin lens, 37
Absorption coefficient, 184, 515 (see also Attenuation)
Absorption cross-section, 175
Absorption oscillator strength
  relation to A Coefficient, 516
Acceptor: Hg in Ge, 571
Active cavity, 137
AlAs barrier in quantum well lasers
$Al_xGa_{1-x}As$ material, 390
  band gap and index, 391(see also Heterostructures)
  TE and TM mode in a symmetrical junction, 85
Algebraic Form of Maxwell's Equations, 9
Alkali-halogen reaction, 343
$Al_2O_3:Cr^{3+}$ (see Ruby Laser)
α, symbol for, 184
  distributed losses in a cavity, 184
  Townsend ionization coefficient, 441
α_e, correction to rotational constant due to vibration, 553
Amplified spontaneous emission (ASE), 210
  effect on Q-switching, 249
  limitations on gain, 212–215
Amplitude oscillations (in gain switched lasers), 262
Anistropic media, 15
Anomalous dispersion, 288, 519
  derivation from electron oscillator model, 517

mode pulling, effects in, 288
Anti-resonance, 130
ArCl, ArF, 346 (see also Excimer lasers)
(Ar^+F^-), excimer state, 344
Argon
  electron scattering cross section, 424
  ion laser, 333
Astigmatism, 49
  beams from gain guided lasers, 485
  Brewster angle windows, 50
  spherical mirror, 51
Atomic response to an applied field, 514
Atomic transitions
  in Argon-ion, 333
  in Cr^{3+}, 309
  in isotope shift, 301, 563
  in mercury, 560
  in neodymium (Nd^{3+}), 318, 321
  in neon, 328
  in sodium, 220
Attachment of electrons in a CO_2 laser mixture, 441
Attenuation
  evanescent decay of fields, 83
  of plane wave, 184, 516
Avalanche photodiode, 576
  avalanche gain, role in detection, 590
  excess noise, 592
Average electron, 418 (see Electron gas)

$B_{21}(B_{12})$, Einstein coefficient for stimulated emission (or absorption), 157–158
Ballast, of gas discharge, 433
Bands, in semiconductors, 363
B_e; rotational constant, 553
Beam slicer, 459
Beams, Hermite-Gaussian, 62
  spot, 75 (see also Spot-size)
  spreading, 14, 69
  transformation by a lens, 56, 92 (see also $ABCD$ law)
  waist of, 69
Blackbody radiation law
  Einstein's derivation, 159–160
  experiment, 150
  Planck's theory, 155
  Rayleigh-Jeans, 154
Bohr orbits; relation to spectroscopic notation, 544

Subject Index

Boltzmann
　distribution of allowed states, 155
　equation for the electron gas, 430
　factor, 159, 307
Bound electrons, 513
Boundary value problems, 18
Bragg, 488 (*see also* Distributed feedback)
　frequency, 496
　phase constant, 490
Branching ratio; $\phi_{21} = \tau_2/\tau_{21}$, 195
Brewster's angle, 20
　astigmatism caused by, 50, 146
　of a dye stream, 325
　on helium-neon lasers, 33
　in semiconductor lasers, 479
Built-in potential, 574

c axis of Ruby, 309
Carbon dioxide laser, 334–341
　Einstein A coefficients, 338
　electron beam sustained operations, 445–447
　energy level diagram, 336
　energy transfer from N_2, 336–337
　excitation by a discharge, 434–444
　gain switching, 338
　vibrational-rotational data, 338
Carbon monoxide (CO) molecule vibration
　rotational constants, 553
　spectra of 7-6 band, 556
　vibrational constants, 555
　VR spectra on 7-6 band, 376
Carrier drift
　electron hole pair in *pn* junction, 573
　electrons in argon, 425
　electrons in CO_2 laser mixture, 438
Cascade of optical components, 36–38
Cathode dark space (CDS), 414, 417
Causality, 289, 605
Characteristic energy (ϵ_k), 423
　relation to E/N, 423
　in CO_2 laser mixtures, 439
　relation to kT_e, 427
Chemical lasers (HF), 341
　hot/cold reaction, 342
Child-Langmuir Law, 448
　space charge limit, 596
Chirp
　due to phase modulation in modelocking, 286
　in semiconductor lasers, 409
Chromium (as a dopant in Al_2O_3), 308
Classical model of an atom, 512
Coherence
　measurement of, 21
　spectral representation, 27
　time, 22
　transverse, 29

Collisions
　atom-atom, 166
　broadening of a line, 166 (*see also* Pressure broadening)
　electrons in argon, 424
　electrons, elastic, 419
　inelastic, 421
　superelastic, 602
Complex beam parameter q, 65
　definition of $1/q$, 67, 92
　differential equation, 66
　relation to radius of curvature, 67
　relation to spot-size, 67
　transformation, 91 (*see also ABCD* law)
　variation in free space, 66
Complex dielectric constant
　relation to susceptibility, 515
Conductivity
　of electron gas, 420
Confocal parameter, 69
Confocal resonator (stable)
　by *ABCD* law, 43
　analysis by integral equation, 468
Confocal unstable resonator, 460
　burn pattern (experiment), 462
　losses in, 460
　radiation pattern (theory), 461
Contact potential, 574
Constitutive relations
　between D, E, and P, 8
Continuous lens, 51–53
　negative lens, 55
　positive lens, 55
Coupled modes
　for distributed feedback, 491
　laser arrays, 500
　self-consistency or characteristic equation, 493
Coupled oscillators, 499 (*see also* Laser arrays)
Coupling, 236
　optimum for a ring laser, 237
　Rigrod analysis, 241
Critical fluorescence power, 307
Current amplification
　in avalanche photodiodes, 576, 591
　in photomultipliers, 568, 589
Current spreading
　in semiconductor lasers, 387

Damping constant
　of a driven oscillator, 514
Dark current, 577
Debye (a measure of a dipole moment), (1 D = 3.33×10^{-30} C-m), 529
Degeneracy ratio $g2/g1$, 195
δ, electron secondary emission ratio, 568
　energy loss per elastic collision, 422

Density of electrons/holes
  relation to quasi-Fermi levels, 371
Density matrix
  definition, 531
  equation of motion, 533
  for a two-level system, 535
Density, optical density (of film blackening), 127
Density of states, 366
  area density, 394
  reduced density of states, 377
Detachment
  in excimers, 349
Detailed balancing, 60
Diatomic molecules
  rotational structures and transitions, 553
  typical energy level diagram, 552
  vibrational-rotational transitions, 556
  vibrational structure, 555
Dielectric matrix, 14
Difference equation
  for a ray, 40
Diffraction losses
  beam mode (hole coupling), 137
  finite mirror aperture, 469
  of Hermite-Gaussian beams, 136
  unstable resonators, 457, 474
Diffusion length, 388
Diffusion, Transverse, coefficient laser mixtures, 439
Dipole moment μ_{21}
  of an antenna, 512
  relation to gain coefficient, 541
  relation to wave functions, 523
Direct bandgap semiconductors, 363
Discharge pumping
  of excimer lasers, 348
Dispersion
  anomalous, 288, 517
  effect of gain/loss, 517
  in fibers, 101
    material dispersion, 102, 286
    modal dispersion, 102
    mode pulling, 288
  in semiconductors, 287
Displacement vector
  for a uniaxial crystal, 15
Dissociation energy
  of a molecule, 551
Distributed feedback, 487
Distribution function for electrons, 426
  computed for CO_2 laser mixture, 437
  definition, 427
  spherical symmetric part, 428
  time dependence, 430
Divergence equation
  applied to optical beams, 62
"Donut" mode, 78

Doppler effect, 170
  broadening, 171
  line shape, 172
  width, 172, 529
Double heterostructure laser, 392
Driven oscillator
  absorption by, 514
  index of refraction of, 518
Dye lasers, 321
  absorption and fluorescence of rhodamine 6G, 324
  configuration for CW oscillation, 325
  simplified energy level diagram, 324
Dynodes, 509, 568

E-Beam
  excitation of excimer lasers, 347
  in a FEL, 348
  sustained operation of CO_2, 445
Effective index
  for extraordinary wave, 17
  for ordinary wave, 17
EH (or HE) modes
  in a step index fiber, 86
Einstein A and B coefficients, 156
  derivation from Schrödinger's equation, 524
  rate equation use, 157, 176, 338, 380
Einstein relation, 439
Elastic collision frequency
  between neutrals, 167
  of electron with neutrals, 419
  power lost in, 421
Electric current
  in terms of the average electron, 420
  in terms of the distribution function, 431
Electrical length of a cavity, 127
Electron affinity
  of the halogens, 346
Electron beam
  in excimer lasers, 347
  ionization in gas, 445
Electron gas, 418
  average electron, 418–423
  distribution function, 426
Electron lifetime, 432
Electron mobility, 431
  in CO_2 laser mixtures, 438
  in terms of the average electron, 420
Electron oscillator model, 514
Electron scattering cross-section in argon, 424
Electron temperature
  of He:Ne laser, 328
  relation to characteristic energy ϵ_k, 423
  relationship to E/N, 327
Electron transport quantity, 431

Subject Index

Electronic states
  in molecules, 559
Elliptical beams, 51
E/N of CO_2 laser mixture, 435
  experiment, 435
  theory for, 441
E/N of a discharge
  of a $CO_2/N_2/He$ laser mixture, 435
  relation to the characteristic energy, 423
  of a self-sustained discharge, 433
Energy balance equation
  for a gas discharge, 421
Energy extraction
  by an optical pulse, 270
Energy level diagram, general (see Atomic transitions)
Energy storage capability
  of CO_2, 339
  of glass, 319
Equivalent circuit
  of noise generators, 580, 583
  of a p-i-n junction, 575
  of a vacuum photodiode, 567
Equivalent-lens waveguide, 39
Excimer lasers, 343
  ArF, XeF, XeCl, XeBr, KrF data, 346
  formation processes in ArF, 344
Expansion angle, 14, 70
Extra-ordinary wave
  in uniaxial crystal, 17
Extraction efficiency
  in pulse excited amplifier, 244 (see also Optimum coupling)
  in Q switching, 255
  from a ring laser, 238

f-number ($f\#$) of a lens, 94
Fabry-Perot Cavity, 217, 289, 409
  finesse, 132
  with gain, 137
  photon lifetime, 133
  Q, 133
  reflections, 129
  transmission characteristics, 129
  tuning by gas pressure, 147
  varying length, 145
Facet reflectivity of semiconductor lasers, 480
FEL (free electron laser), 349
  wavelength of, 352
Fermat principle, 56
Fermi function, 370
Fibers
  graded index, 95, 97
  step index, 85
    TE and TM modes, 85, 87
Finesse, 132
  relation to Q, photon lifetime, 133

Flashlamp
  efficiency, 315
  spectral emission of Xe or Kr, 313
Fluence, optimum, 247
Fluorescence power, critical, 307
Fluorescent lamp
  terminal characteristics, 415
  theoretical calculations, 425
Flux
  electron flux, 430
  photon flux Γ_p, 266
Four-level laser, 306
Fourier transforms, 8
  of a Gaussian beam, 32
  in mode locking, 281
  uncertainty relations, 12
Fox and Li approach to optical cavities, 113, 467
Franck-Condon principle
  in diatomic molecules, 560
  in dye lasers, 323
  in semiconductor lasers, 364
Free spectral range (FSR), 127
Frequency chirp
  in mode locking, 286
  of semiconductor lasers, 409
Frequency pulling, 290
Fresnel number
  definition, 470
  losses as a function of, 473
Fringe
  contrast, 26
  degradation of visibility, 26
Fundamental frequencies of the CO_2 laser, 338

g_0, single pass integrated gain $\gamma_0 l_g$ (in mode locking), 283
$g_{1,2} = 1 - d/R_{1,2}$, the g parameters of a simple cavity, 42
$g_{1,2}$ = statistical weights on energy states, $1,2 = 2J + 1$, 159
$\bar{g}(\nu)$, the line shape normalized to be unity at line center, 195
$g(\nu)$, the line shape obeying $\int g(\nu)\,d\nu = 1$, 163
Gain, power, 176
  relation to small signal gain coefficient, 183
  small signal G_0, 187
Gain/absorption
  for a normal population, 373
  self-reversed lines, 212
  in a semiconductor, 372
Gain coefficient
  relation to pumping rates and lifetimes, 195
  relation to susceptibility, 515

Gain coefficient (*Cont.*)
  saturated, 195
  small signal, 175
  in words, 379
Gain coefficient, complex
  from the mechanical model, 515
  in mode pulling, 289
  from quantum theory, 541
Gain coefficient in a semiconductor, 375
  relation to spontaneous emission, 381
Gain guiding, 481
Gain saturation
  by broad band radiation (ASE), 213
  homogeneous broadening, 188
    mathematical model, 194
    physical description, 185
  inhomogeneous broadening, 199
    mathematical model, 206
    physical description, 199
Gain switching, 260
GaAs; material properties, 365
$Ga_{1-x}Al_xAs$ laser, 391 (*see also* $Al_xGa_{1-x}As$)
Gas discharge lasers, 325
Gas heating
  by a discharge, 52, 444
  effect on gain of VR lasers, 554
  index of refraction, change due to, 52
Gas lens, 51
  ray matrix, 55
Gaussian beam
  in fibers, 95
  higher order modes, 72
    "dot" pattern, 75
    mathematical description, 73
    spreading of, 76
Gaussian beams
  fiber optic waveguide, 95
  in gain guided semiconductor lasers, 483
  in graded index fiber, 97
    exact solution, 106
  in stable resonators, 111
Gaussian beams $TEM_{0,0}$
  center of curvature, 72
  differential equation, 65
  phase velocity, 70
  physical description, 69
  radial phase, 71
  radius of curvature, 67
  spot-size of, 67
  spreading of, 69
  transverse variation, 67
"Gaussian-like" beam, 13
Giant pulse (see *Q* switching), 248
Green's function
  for free space, 465
  use in electromagnetics, 464
GRIN cavity, 147

GRIN lens, 120
Group index, 102, 154, 379
Group velocity, 102, 283
  dispersion, 103
Guided mode, 88

Hamiltonian operator, 521
Harmonic oscillator, 551
  classical, 514
  vibrational frequencies, 551
Harpooning reaction, 345
  with F_2, 349
  with NF_3, 347
Heavy (or normal) hole, 366
Heisenberg uncertainty relations, 12
Helium metastables, 328
Helium-Neon laser, 326–332
  common transitions, 326
  Einstein *A* coefficients, 328
  fill pressure, 327
  simple model, 330
Hemispherical cavity, 117
Hermite-Gaussian beam modes, 62–79
  differential equation, 99
  Hermite polynominals, 73
  orthogonality of, 78
  in a parabolic index fiber, 97
Heterodyne delection, 593
  quantum limit, 595
Heterojunction lasers, 389
  band diagram, 382
  confinement factor, 400
  electromagnetic modes (symmetrical junction), 85
  electromagnetic modes in a 3 layer diode, 475
  gain guiding, 481
HF laser, 341 (*see also* Chemical lasers)
Higher order modes, 72 (*see also* $TEM_{m,p}$ modes)
  dot patterns, 75
Hole burning
  hole width, 207
  phenomena, 200
Hole coupling, 74
Homogeneous broadening, 164
  collision broadening, 166
  line width, 167
  natural broadening, 166
Homojunction laser, 386
Homonuclear molecules, 556
Huygen's principle, 113, 486

Ideal laser, 198
  quantum efficiency for, 232
  rate equation definition, 194
Images in unstable resonators, 454, 455
Index ellipsoid, 18
Index of refraction

GaAs, $Al_xGa_{1-x}As$, 391
of CO_2, 61
of helium, 61
Indirect semiconductors, 364
Induced emission (see Stimulated emission)
Inelastic collision
  power lost in, 421
  relation to superelastic rates, 600
Infinite sequence of lens, 39
Inhomogeneous broadening, 169 (see also Doppler or isotope effects)
Initial condition for ray tracing
  stable cavities, 44, 47
  unstable cavities, 48
Injection lasers, 361
Integral equation
  application to cavities, 463
Integrated absorption, 516
Integrated gain, $g_0 = \gamma_0 lg$, 237
  net integrated gain, 238
Intensity, definition of, 25 (see also Poynting vector)
Interference, measurement of coherence length, time, 25
Internal reflection
  in a slab waveguide, 82
Inverted population, 183, 465
  lifetime of, 541
  from rate equations, 194
  in semiconductors, 375, 382
Iodine transition $\lambda_0 = 1.315\mu m$, 550
Ion lasers, 333
  argon, energy levels, 333
  cadmium, 354
Ionic bonding, 343
  of the rare gas-halide, 343
  of the salts, 343
Ionic channel
  for excimer formation, 344
Ionization potentials of rare gases, 346
  balance equation, 432
  rates in a gas, 432
Isotope effect
  in line broadening, 169
  in mercury, 563

J, angular momentum quantum number, 159, 553
Jello, as a laser, 305
"Johnson's" noise, 150, 584
Junction capacitance, 574
Junction detectors
  avalanche photodiode (APD), 576
  photo diode, 572

k conservation rule (see also Frank-Condon principle)
  for semiconductor, 364

$\mathbf{k}_0 = \omega/c$ phase constant in vacuum, 9
kcal/mol., energy unit, 342
Kirchhoff's radiation law, 152, 211
Kramer-Kronig
  in mode pulling, 289
  relations, 604
KrCl, KrF, excimer lasers, 346
k-space, 368
  of a Gaussian beam, 14
  of momentum states in a semiconductor, 367

Lamp dip, 203
  inverted in methane, 204
  normal, 203
Laser amplifier
  saturated power gain, 187
  small signal gain, 183
Laser arrays, 497
Laser Oscillation
  damped oscillation in gain switched lasers, 262
  homogeneous broadened transition, 185
  optimum coupling, 237
  inhomogeneous broadened transition
    doppler broadened transition, 206
    hole burning, 201
  isotope effect on Hg, 563
  in three- four-level lasers,
Lens
  f number, 94
  focal spot size, 94
  focal length, 36
  -like medium, 51
  ray matrix, 37
Lifetime
  of inversion, 537
  of states, 190
Lifetime broadening, 166
Lifetime ratio, favorable, 195
Light hole state, 368
Line shape, $g(\nu)$
  definition, 163
  Doppler, 171
    physical interpretation, 172
  Lorentzian, 165
  normalization, 163
  relation to spectral representation of field, 27
Local time, 266
Longitudential phase
  of a Gaussian beam, 68
Lorentzian line shape, 165
  from classical driven oscillator model, 517
  from density matrix, 544
  from a lifetime model, 27

Losses in cavities
　pro-rated over gain length, 184
　in strip mirrors, 469
　in unstable cavities, 474
L-S coupling, 549

Magnetic dipole transition, 543, 550
Matrix element, 523
Maxwell-Boltzmann distribution function, 170
　of atomic velocities, 170
　electron distribution function, 426
Maxwell's equations, 7
　algebraic form, 9
　in anisotropic media, 15
　for free space, 9
Mechanical periodicity, 491
Mercury
　energy levels, 550
　isotope effects, 563
Metastable
　A state in N_2, 559
　channel for excimer formation, 345
　level in helium, 328
　levels in neon, argon, xenon, 346
　in neon, 330
Michelson interferometer, 24, 25
Microscopic reversibility, 600 (*see also* Detailed balance)
Minimum spot size, 72
Minority carriers, 386
　in photodiodes, 573
Mobility
　of electron and holes in common semiconductors, 365
　of the electrons in a gas, 420 (*see also* Drift velocity)
Modal gain, 485
Mode density
　of a cubical cavity, 153, 154
Mode locking, 270
　equal amplitude modes, 273
　forced mode locking, 280
　Gaussian distribution, 275
Mode locked pulse
　peak power, 274, 277
　pulse width, 275, 277
Mode locking, active
　amplitude modulation, 279
　phase modulation, 285
Mode pulling, 290, 517
Mode volume of stable resonator, 116
Modes, characteristic, definition of, 112
　from *ABCD* law, 115
　.in a fiber
　　by *ABCD* law, 95
　　by an exact analysis, 97
　　in a slab waveguide, 82
　　classification, 86

Modulation
　frequency response, 403
　of semiconductor lasers, 399
　transfer characteristic, 405
Modulation frequency
　for active modelocking, 283
Molecular constants, definition of
　for rotation, 553
　for vibration, 555
Momentum of electrons in semiconductors, 364
　conservation of, 376
Momentum states in a semiconductor, 367
Multimode laser, 271 (*see also* Mode locking)
Multimode oscillation, 205
Multiple quantum well, 397

N_p number of photons in a cavity
　photon lifetime, 132
　in a Q-switched laser, 253
N_2 laser
　energy level diagram, 559
　molecular data, 561
　vibrational/rotational data, 336
Natural broadening, 166
Negative branch of unstable resonator, 459
Negative glow, 417
Nd^{3+} in glass, energy levels, 320
　absorption spectra, 320
　energy storage, 319
　line shape, 320
　physical properties, 321
Nd^{3+} in YAG
　energy levels, 317
　$^4F_{3/2}$-$^4I_{13/2}$, etc. transition data, 317, 318
　physical parameters, 318
Neodymium laser, 315
Neon
　energy level diagram, 326
　transition rates, 328
NF_3, role in excimer lasers, 348
Nitrogen gas (N_2)
　electron excitation, 340
　energy storage in vibrational states, 339
　molecular laser, 382–383
　role in CO_2 laser, 337
　vibrational levels, 338
Noise
　background, 587
　equivalent circuit, 580
　excess noise in APD, 591
　spectral distribution, 580
　thermal, 584
Noise figure, 587
Normal mode (*see* Characteristic mode)
Normalization condition
　for electron distribution function, 428
　for line shape, 163

Nuclear potential
  in KrF laser, 357
Numerical aperture (NA), 84, 85
Nyquist noise, 150, 584

Occupation probability, 528
Optic axis, 17
Optical confinement factor, 400, 401
Optical diode, 233
Optical fiber
  ray matrix, 55
  step index, 85
  variation of the index n with r, 52
Optical pumping, 308
Optical resonators, 109
  diffraction losses, 136
  losses, 127
  mode separation, 126
    Hermite-Gaussian beam modes, 134
    uniform plane wave, 126
  photon lifetime, 127
  stability of, 116
Optical transitions in semiconductors, 375
Optical transparency (in ruby), 314
Optically "thin" media, 211
Optimum pressure for He:Ne lasers, 327
Ordinary wave
  in uniaxial crystal, 16
Orthogonality
  of Hermite-Gaussian beam functions, 78
  of wave functions, 522
Oscillation
  bandwidth, 185
  frequency, 187
  in homogeneously broadened transition, 185
  in inhomogeneously broadened transition, 199
  threshold for, 3, 183
Oscillator strength, 516

P branch (see also VR transitions)
  in CO_2 laser, 339
  general, 555
Paraxial rays, 35
  ray equation, 54
Partial inversion, 557
Paschen notation
  for neon, 327
Periodic discontinuities
  in distributed feedback lasers, 489
Periodic lens waveguide, 39
Permittivity, ϵ_0, or permeability, μ_0, of free space, 6
Phase constants
  of Hermite-Gaussian beam modes, 72, 73
  for supermodes, 501
  uniform plane wave, 9

Phase interrupting collisions (see also T_2)
  effect on coherence, 23
  pressure broadening, 169
Phase modulation (modelocking), 285
Phase velocity in fibers, 102
  of Gaussian beam, 72, 73
Phasor diagram
  modelocked laser, 272
  of an optical cavity, 125
Photoconductor, 571
  photoconductive gain, 573
Photodiode
  avalanche (APD), 576
  junction, 572
  vacuum, 566
Photoelectric effect, 566
Photomultiplier, 568
  gain, 568
  limits of detection, 589
  noise in, 578
Photon lifetime
  of a simple passive cavity, 133
  relation to net losses, 251
  relation to Q, 133
Photons
  time dependence in cavity, 215
Photons per mode, 156
Photovoltaic mode, 574
p-i-n diode, 575
Planck's radiation law, 150, 211
Poisson brackets, 534
Polarization
  from a collection of atoms, 514
    classical expression, 514
  from density matrix, 532
    current (in Maxwell's equations), 537
  field in semiconductor lasers, 479
  in Maxwell's equations, 7
Positive branch of unstable resonator, 459
Positive column, 417
Power balance
  in a gas discharge, 421
Power broadening, 529
Power carried by a Gaussian beam, 72
Power cycle, 248 (see also Q switching)
Power gain (G), 183
Power output, optimum coupling, 237
Poynting vector, 10 (see also Intensity)
  for free space, 9
Pressure broadening
  of CO_2 laser line, 168
  de-phasing collisions, 537
Pulse propagation om, 265
  in amplifiers, 265
  in fibers, 184
Pulse width
  mode locked lasers, 275, 277
    loss modulation, 284
    phase modulation, 286

Pulse width (*Cont.*)
 from numerical integration, 258
 of a Q switched laser, 256
Pumping agents, 305
Pumping cycle, in regards to Q switching, 248
Pumping of a laser, general scheme, 189
 effective rate, 198
Pumping techniques, 305
Pythagorean relation, 86
 for TE and TM modes, 153

Q branch of VR transitions, 556
Q, quality factor for a cavity, 117–121
 in terms of energy, 133
 in terms of spectral response, 131
 relation to
  photon lifetime, 120, 133
  finesse, 120, 133
q, complex beam parameter
 basic equation, 65
 in a fiber, 95
 relation to R and w, 67
 transformation, 91 (*see also ABCD* law)
Q- switching (*see also* "Giant pulse")
 mathematical model, 251
 methods, 259
 physical description, 248
 of Ruby (example), 256
Quadratic gain profile, 481 (*see also* Gain guiding)
Quadratic index
 fibers, 53
 in GRIN lens, 58
 in heterostructure lasers, 481
 propagation in, 95–101
 ray matrix for, 55, 95
Quantum detectors, 566
Quantum efficiency
 of detectors, 566
 of a laser, 232
 of photocathode, 569
 of pumping in Ruby, 310, 315
Quantum hypothesis, 154
Quantum size effects, density of states, 395
Quantum well lasers, 395
Quartz fused, Brewster angle windows, 21
Quasi-Fermi levels, 371
Quenching, effect on rate equations, 161

R branch (in CO_2 lasers), 340
 of VR transitions, 555
$R\#$ of slab fiber $= (2\pi a/\lambda_0)NA$, 90
R_1, R_2 transitions in Ruby, 308
Rabi flopping frequency, 527, 538, 544, 546
Radial phase of a Gaussian beam, 71

Radiation damping, 542
Radiation pattens of laser arrays
 of supermodes, 503
Radiative lifetime, 162
Radius of curvature
 astigmatic semiconductor laser, 486
 of a Gaussian beam, 67
Raman scattering, 541
Rare gas-halide excimer combinations
 basic data, 345 (*see also* ArF, KrF, XeF, XeCl)
 excitation routes, 347
Rate equations, 160
Rates
 electron collisions, 429
Ray matrix (*see also ABCD* matrix)
 for continuous lens, 54
 for convention components, 35–38
 for GRIN lens, 58
 for index steps, 57
 stability of, 39
Ray paths repetitive, 47
Ray propagation or tracing
 in cavities, 38, 44
 in inhomogeneous media, 53
Rayleigh-Jeans, 154
Rays, definition, 34
Recombination
 of electron holes, 363, 373
 spectra, 382
 of electrons and ions, 440
Recombination rate
 in GaAs, 385
Reflection coefficient
 of a Fabry-Perot cavity, 129
Refractivity
 of a gas, 53
Relativistic factor $\gamma = (1-\beta^2)^{1/2}$, 352
Relaxation oscillation
 in a semiconductor laser, 262
Relaxation terms
 in density matrix, 536
Resonance
 general, 123
 Hermite-Gaussian beam modes, 134
 resonant wavelengths, 125
Resonant cavities, 121–139
Resonators, optical, 109–121
Rhodamine 6G, absorption/fluorescence, 323
Rigrod analysis, standing wave lasers, 241
Ring laser, 233
 cavity stability, 59
 optimum coupling, 236
Rotating wave approximation, 538
Rotational states in diatomic molecules
 distribution in CO, 554
 wave number of traansitions, 554

Ruby laser (Cr^{3+}:Al_2O_3)
  absorption bands, 311
  energy level diagram, 309
  line shape for R_1R_2, 311
  optical pumping scheme, 309
  physical data for, 310
$R(\nu)$, spontaneous recombination in a semiconductor, 381
  relation to absorption, 381
Russel-Saunders coupling (L-S), 549

Saturated gain coefficient
  differential equation for, 196
  relation to loss, 216
  relation to saturated gain, 197
Saturation
  of a Doppler line, 202
    mathematical treatment, 206
  of an inhomogeneous line, 200, 208
  of pumping in He-Ne lasers, 332
  via density matrix, 541
Saturation absorption
  in methane, 204
Saturation energy, 268
Saturation intensity, 191
  homogeneous broadening, 187
    effect on gain coefficient, 195
    exact expression, 198
    lifetime definition, 191
    rate equation, 190, 194
    simplified expression, 191
  inhomogeneous broadening, 207
Scale length for a Gaussian beam w_0, 66
  of $TEM_{m,p}$ modes, 76
Scaling laws, for a discharge
  for He-Ne laser, 327
Scattering matrix of a lossless mirror, 128
Schottky Junction, 391
Schrödinger's equation, 53, 413, 522, 533, 534
  perturbation by fields, 521
Secondary emission
  amplifier, 569
  noise in, 576
  coefficient, δ, 568
Selection rules
  for electric-dipole transitions, 550
Selective pumping
  of CO_2, 337
  of neon, 329
Self-reversed line, 212
Semiconductor cavities
  gain guided, 481
  modes in, 475
  symmetrical junction, 85
Semiconductor lasers, 361
  data for common semiconductors, 365
Semiconductor photodiodes, 572

Separation of variables
  in a parabolic index fiber, 98
Shot noise, 583
SI (or MKSA) system of units, 6
Simple harmonic oscillator
  model of an atom, 514
  vibrational model for a molecule, 552
Singlet state in dyes, 323
Small signal gain coefficient V_0
  in terms of pumping rates, 191
Snell's law, 19, 20, 53
S_0, S_1, T_1, T_2, states in a dye, 323
Sodium lamp, 211
Space charge
  limited current, 448
  neutrality of a plasma, 418
Spatial coherence of higher order modes, 76
Spectral density function, 579
  equivalent circuit for, 580
Spectral distribution of power
  in an active cavity, 138
  due to ASE, 213
  in a laser, 217
    Schalow-Townes formula, 218
Spectral narrowing
  of cavity modes, 138
  of spontaneous emission, 213
Spectral representation (of **E, H, P, D** . . . etc.), 8
Speed (of an electron)
  relation to characteristic energy, 426
Spherical wave
  relation to Gaussian beams, 71
  unstable resonators, 455
Spontaneous emission
  coefficient for (A_{21}), 157
  from a semiconductor, 382
Spontaneous lifetime, 162
Spontaneous power
  of excimer lasers, 347
  relation to critical fluorescence power, 315
Spot size (*see* Gaussian beam)
  definition, 67
  at focal plane of a lens, 93
Spreading of a beam
  of a Gaussian beam, 70
  general, 13
Stable cavities, 39
  stability diagram
    internal lens cavity, 59
    two curved mirrors, 43
Standing wave lasers, 239
  high Q approximation, 240
Stark effect, 171
Statistical weights ($g_{2(1)} = 2J_{2(1)} + 1$)
  atomic states, 159
  rotational states, 553

Stimulated emission
  characteristics of, 158
  cross-section, 178
  definition of rates, 157
Superelastic collisions
  in excimers, 349
  relation to inelastic collisions, 601
Supermodes, 499 (see also Laser arrays)
Susceptibility, complex
  of atoms, classical expression, 515
  relation to gain coefficient, 515
  relation to index of refraction, 8

T_2, mean time between phase interrupting collisions
  impact on line shape, 169
  inclusion in density matrix, 536
Tapered mirrors
  with $ABCD$ law, 107
  in confocal cavity, 506
TEA (Transverse Electric Atmospheric), 260
  discharge circuit, 434
TEM mode
  Gaussian beams, 63
TEM_{oo} modes, 65, 68, 72
TEM_{mp} modes, 73
  in a coaxial cable, 112
  waves in general, 62
TE, TM modes
  resonant frequency of a rectangular cavity, 152
  in semiconductor cavities, 477
  in slab fiber, 85
TE or "s" polarization, 21
TM or "p" polarization, 21
Thermal (Johnson) noise, 585
Thermal population
  in rotational manifold, 553
  of states in YAG, 317
Thermal velocity of electrons, 421
Thermionic emission
  effect on gas discharge, 415
Thermodynamic equilibrium, 151–156
  for blackbody radiation, 155
  for population densities, 159
Thin lens, 37
Three- four-level lasers, 306
Threshold current, for a homojunction laser, 388
Threshold inversion, 184, 252
Townsend ionization coefficient (α), 440
Transfer characteristic, 405
Transition rates
  definition in terms of Einstein coefficients, 157
  for monochromatic waves, 172
Transmission coefficient
  at antiresonance, 130
  of a Fabry-Perot cavity, 129
  at resonance, 130

Transverse divergence operator, 63
Tridiagonal matrix, for coupled laser arrays, 500
Triplet system
  in dye lasers, 324

Ultraviolet catastrophe, 154
Uncertainty
  definition, 12
  relations, 11
Uniphase front, 76
Unit cell
  application of $ABCD$ law, 116
  stability analysis, 39
Unstable cavities
  definition, 454
  general analysis, 454
    beam slicer, 459
    confocal geometry, 460
    losses, 457–459

V number
  of round fiber $(2\pi a/\lambda_0)NA$, 90
Velocity distribution, Maxwellian, 426
Velocity of propagation, mean
  in a Q switched cavity, 252
V-I characteristics
  experiment
    of CO_2 laser mixture, 435
    fluorescent lamp, 415
  theory of CO_2 laser mixture, 442
Vibration excitation
  of N_2 (data), 336
  role in CO_2 laser, 337
Vibration-rotational transitions, 555
Vibrational states, 555
Video detection, 588
Voight function, line shape, 171
VSWR (voltage standing wave ratio), 22

W value, 347
Wave equation
  in dielectrics, 10
  for free space, 8
  for optical beams, 65
Wave impedance, 10
Wave vector, **k**, 9
Waveguide, dielectric
  semiconductor lasers, 475
  slab guide, 82
"White" noise, 150
Wiggler, in FEL, 350
Work function, photoelectric, 566

XeF, XeBr, XeI lasers (see Rare gas-halide lasers)

YAG (Yttrium Aluminum Garnet) $Y_3Al_5O_{12}$, 317
  energy level diagram, 317
  pumping by semiconductors, 319